Metal Oxide-based High-K Dielectrics

This book provides chronological advancement of metal oxide high-K dielectrics up to contemporary scenarios, synthesis with suitability and challenges, and diverse properties with emerging technological applications. It helps readers select metal oxide–based high-K dielectrics with large band-gap, cost-effective, and highly efficient material properties for plausible applications. It provides up-to-date research findings on established synthesis techniques, easy processing, characterization, properties, and prospective practical applicability, including hybrid materials.

Features:

- Exhaustively covers synthesis, physical properties, and the applications of the high-K dielectrics.
- Focuses on synthetic routes of preparation, properties, and their various practical applications from bench to field.
- Discusses functionalization of novel metal oxides and flexible polymeric composite materials for superior dielectric and electrical performance.
- Explores facile synthesis techniques for high-K dielectrics and their hybrid composites, properties, and technological applications.
- Includes future perspectives and possible challenges for applying high-K dielectric materials.

This book is aimed at researchers and graduate students in materials science and engineering, physics, and electrical engineering.

Metal Oxide-based High-K Dielectrics

Fundamentals and Applications

Edited by
Srikanta Moharana
Santosh Kumar Satpathy
Tuan Anh Nguyen
Ram K. Gupta
Parvej Ahmad Alvi

CRC Press
Taylor & Francis Group
Boca Raton London New York

CRC Press is an imprint of the
Taylor & Francis Group, an **informa** business

Designed cover image: Shutterstock

First edition published 2025
by CRC Press
2385 NW Executive Center Drive, Suite 320, Boca Raton FL 33431

and by CRC Press
4 Park Square, Milton Park, Abingdon, Oxon, OX14 4RN

CRC Press is an imprint of Taylor & Francis Group, LLC

Library of Congress Cataloging-in-Publication Data
Names: Moharana, Srikanta (Chemist), editor. | Satpathy, Santosh Kumar (Chemist), editor. | Nguyen, Tuan Anh (Chemist), editor. | Gupta, Ram K., editor. | Alvi, Parvej Ahmad, editor.
Title: Metal oxide-based high-K dielectrics : fundamentals and applications / edited by Srikanta Moharana, Santosh Kumar Satpathy, Tuan Anh Nguyen, Ram K. Gupta, and Parvej Ahmad Alvi.
Description: First edition. | Boca Raton : CRC Press, 2025. | Includes bibliographical references and index.
Identifiers: LCCN 2024033769 (print) | LCCN 2024033770 (ebook) | ISBN 9781032402956 (hbk) | ISBN 9781032402963 (pbk) | ISBN 9781003352365 (ebk)
Subjects: LCSH: Dielectrics. | Metal oxide semiconductor field-effect transistors.
Classification: LCC QC585.7.C6 M47 2025 (print) | LCC QC585.7.C6 (ebook) | DDC 537/.24--dc23/eng/20241205
LC record available at https://lccn.loc.gov/2024033769
LC ebook record available at https://lccn.loc.gov/2024033770

ISBN: 978-1-032-40295-6 (hbk)
ISBN: 978-1-032-40296-3 (pbk)
ISBN: 978-1-003-35236-5 (ebk)

DOI: 10.1201/9781003352365

Typeset in Times
by SPi Technologies India Pvt Ltd (Straive)

Contents

Preface

Welcome to the amazing *metal oxide high-K dielectrics*. Metal oxide high-K dielectrics have attracted tremendous attention in the scientific and technology communities due to their broad range of potential applications in diverse fields such as capacitors, microelectronics, dynamic random access memory (DRAM), complementary metal oxide semiconductor (CMOS), resistive RAM memories, metal-insulator-metal (MIM) diodes, ferroelectric logic, memory devices, etc. They also have attracted great interest for construction of large-area electronics, especially thin film transistors (TFTs), for their high optical transparency and excellent chemical and thermal stability. Metal oxide high-K dielectrics are an important component for producing low-voltage and high-performance TFTs. The book comprehensively covers all the current and emerging fields of physics, chemistry, materials science, and technology of metal oxide–based high-K dielectrics.

Metal Oxide-based High-K Dielectrics: Fundamentals and Applications focuses on the most advanced, urgent, and fundamental phenomena in designing high-K dielectrics concerned with novel metal oxides, flexible polymeric composite materials for superior dielectric and electrical performance. It also highlights various facile processing techniques to prepare metal oxide–based high-K dielectric materials for practical applicability from bench to field. The emergence of such novel sophisticated high-K dielectric materials will be the gateway for tailoring and developing future technologies for flexible electronic devices, lightweight transportation systems, and smart cities. The fundamental understanding of the concepts with an eye on contemporary technological applications of metal oxide–based high-K dielectrics will be of great value in terms of providing a systematic and coherent picture for professionals and academic researchers, material scientists, engineers, graduate students, physicists, and chemists.

The purpose of this book is to provide a state-of-the-art overview of metal oxide high-K dielectrics. The book consists of several contributions from internationally recognized experts. Chapter 1 covers the fundamental theories and criteria for metal oxide–based high-K dielectrics. Chapter 2 describes the synthesis, characterization, and recent applications of perovskite oxide high-K dielectrics. Chapter 3 touches on the ferrite and perovskite nanocomposite-based materials synthesis techniques, characterization, and dielectric applications. Chapter 4 offers a brief outlook on various characterization techniques of metal oxide high-K dielectric materials. Chapter 5 describes the alternation of electrical and optical performance of different metal oxide high-K dielectric materials. Chapter 6 provides coverage of the basics, reviews, and overview of silicon-based high-K dielectrics. Chapter 7 introduces the current state of research into rare earth oxide–based high-K dielectrics. Chapter 8 addresses the reporting transition metals and nonmetals of high-K dielectric double perovskite-based materials with further opportunities and challenges. Chapter 9 deals with the applications of memory devices for metal oxide–related high-K dielectric materials. Chapter 10 addresses the sol-gel area printed metal oxide–based high-K dielectrics. Chapter 11 introduces and explains the functioning of high-K pocket in tunneling FET devices. Chapter 12 presents the overview and MIM capacitor applications of metal oxide high-K dielectric materials. Chapter 13 provides the properties and recent energy storage applications of polymer-based high-K dielectric materials. Chapter 14 describes the surface alternation of different metal oxide materials with superior high-K dielectrics. Chapter 15 provides the overview and current applications of high-K dielectrics. Chapter 16 covers polymer-based dielectric composites for advanced energy applications. Chapter 17 offers an introduction and current state of research on flexible and stretchable electronic applications of metal oxide–based high-K dielectrics. Overall, the book deals with the different aspects of metal oxide high-K dielectrics, providing a concise outline of the combination of physics, chemistry, materials science, and electrical engineering.

The challenges and future prospects of metal oxide high-K dielectric materials are among the key concerns in energy storage research. Based on the research experience, contributions collected in this volume address this topic in a timely and appropriate manner, providing a resourceful monograph on both the fundamentals and key integration of high-K dielectric materials in future science and technology. Extensive references sections accompanying each chapter point the readers to more cutting-edge research from all around the world, making it possible to continue their studies on the subject. This book is an informative and valuable reference for both experts in academia and industry as well as those just starting out in the sector.

About the Editors

Srikanta Moharana is an Associate Professor in the Department of Chemistry, School of Applied Sciences, Centurion University of Technology and Management, Odisha, India. He obtained his PhD and MPhil degrees in chemistry from the School of Chemistry, Sambalpur University, India. He received his MSc degree in chemistry from the National Institute of Technology, Rourkela. Dr Moharana has published more than 60 research papers in international peer-reviewed journals, contributed more than 65 book chapters, and coedited four contributed volumes. Under his guidance, two PhD students have successfully received their degrees, and another two are currently working toward theirs. He has actively participated in and presented his research work at several international conferences and has been a reviewer for reputed international journals. He has been awarded Prof. GB Behera's Best PhD thesis award under the auspices of Orissa Chemical Society, Odisha, India. He has more than five years of teaching and research experience. He has extensive experience in the field of perovskite metal oxide and nanocomposite synthesis based on polymer and fabrication of flexible thin-film capacitors for advanced energy storage. He is a life member of the Orissa Chemical Society.

Santosh Kumar Satpathy is an Associate Professor in the Department of Physics, School of Applied Sciences, Centurion University of Technology and Management, Odisha, India. He has completed his MSc (Physics) at Sambalpur University and his PhD at Sambalpur University in collaboration with Central Glass & Ceramic Research Institute, Kolkata. He has received his MPhil in Physics from the School of Physics, Sambalpur University, India. During his PhD studies, he worked on the double perovskite materials. He has published more than 50 research papers in various international and national journals, as well as 15 book chapters. Under his guidance, six MPhil and two PhD students have successfully received their degrees and three PhD candidates are continuing their research. He has more than five years of teaching and five years of research experience. His current area of research includes ferroelectrics, advanced ceramic materials, multiferroics, polymers, ceramic composites, and other ceramic materials for the application of perovskite materials. He has gained recent expertise in the field of nano-structured perovskite systems and has moved on to high-energy storage devices. He is a life member of the Orissa Physical Society.

Tuan Anh Nguyen, PhD is principal research scientist at the Institute for Tropical Technology, Vietnam Academy of Science and Technology, Vietnam. His research focuses on advanced nanomaterials and nanotechnology for corrosion and materials integrity in transportation systems. His research activities include smart coatings, conducting polymers, corrosion and protection of metals/concrete, antibacterial materials, and advanced nanomaterials. He has edited and coedited nine CRC Press books. He is editor-in-chief of *Kenkyu Journal of Nanotechnology & Nanoscience*.

Ram K. Gupta is an Associate Professor of Chemistry at Pittsburg State University, KS, United States. Dr. Gupta has been recently named by Stanford University among the top 2% of research scientists worldwide. Before joining Pittsburg State University, he worked as an Assistant Research Professor at Missouri State University, Springfield, MO, and later as a Senior Research Scientist at North Carolina A&T State University, Greensboro, NC. Dr. Gupta's research spans a range of subjects critical to current and future societal needs, including semiconducting materials and devices, biopolymers, flame-retardant polymers, green energy production and storage using nanostructured materials and conducting polymers, electrocatalysts, optoelectronics and photovoltaics devices, organic-inorganic heterojunctions for sensors, nanomagnetism, biocompatible nanofibers for tissue regeneration, scaffold and antibacterial applications, and biodegradable metallic implants. Dr. Gupta has published more 270 peer-reviewed journal articles, made more than 350 national/international/regional presentations, chaired/organized many sessions at national/international meetings, wrote numerous book chapters (90+), edited and coedited many books (30+), and received millions of dollars for research and educational activities from external agencies. He is also serving as editor, associate editor, guest editor, and editorial board member for various journals.

Parvej Ahmad Alvi has received his PhD in Applied Physics from the Department of Applied Physics, Faculty of Engineering & Technology, Aligarh Muslim University, Aligarh, India. His major area of research is MEMS/NEMS technology (nanotechnology). He is currently an Associate Professor in the Department of Physics at Banasthali University, Rajasthan, India. He has more than 14 years of teaching and research experience in the area of modern physics, semiconductor physics, and nanotechnology. He has served as an editorial board member and reviewer for several international/national journals and conferences. He has published more than 100 research papers in peer-reviewed international journals and conference proceedings. He has also published chapters in more than six textbooks in the area of physics. Approximately 20 PhD scholars have achieved their degrees under his guidance.

Contributors

Aashebir Taresa
Department of Electronics and Communication Engineering,
SOEEC, ASTU
Adama, Ethiopia

Ai-Dong Li
Department of Materials Science and Engineering
College of Engineering and Applied Sciences, National Lab of Solid-State Microstructures, Collaborative
Innovation Center of Advanced Microstructures, Nanjing University
Nanjing, Jiangsu Province, China

Arpita Patro
Department of Physics
School of Applied Sciences, Centurion University of Technology and Management
Odisha, India

Avtar Singh
Department of Electronics and Communication Engineering
SOEEC, ASTU
Adama, Ethiopia

B. N. Parida
Department of Physics
Central Institute of Technology Kokrajhar, BTR
Assam, India

Bibhuti B. Sahu
Department of Basic Science and Humanities
Nalanda Institute of Technology (NIT), Bhubaneswar
Odisha, India

Chandan Kumar Pandey
School of Electronics Engineering
VIT-AP University
Amravati, India

D. Vignesh
Birla Institute of Technology Mesra
Ranchi, India

Debajani Tripathy
School of Applied Sciences
Centurion University of Technology and Management
Odisha, India

Dereje Tekilu
Department of Electronics and Communication Engineering
SOEEC, ASTU
Adama, Ethiopia

Ela Rout
Birla Institute of Technology Mesra
Ranchi, India

Gyanendra Kumar Mishra
Department of Physics
School of Applied Sciences
Centurion University of Technology and Management
Odisha, India

Harshpreet Cheema
School of Engineering and Technology
Jaipur National University
Jaipur, Rajasthan, India

Jun-Gyu Choi
Department of Electrical and Computer Engineering
Ajou University
Suwon, Republic of Korea

Khalid Sultan
Solid State Research Lab
Department of Physics
Central University of Kashmir
Ganderbal, India

L. Biswal
Department of Physics
School of Applied Sciences
KIIT Deemed to be University
Bhubaneswar, India

Laxmi Mallik
Department of Chemistry
Larambha College, Larambha
Bargarh, India

Lipsa Priyadarshini
Department of Physics
School of Applied Sciences, KIIT Deemed to be University
Bhubaneswar, India

Myung-Han Yoon
School of Materials Science and Engineering
Gwangju Institute of Science and Technology (GIST)
Gwangju, Republic of Korea

Nibedita Nath
Department of Chemistry
D.S Degree College, Laida
Sambalpur, Odisha, India

Nilaya Kumar Mohanty
Department of Physics
School of Applied Sciences
Centurion University of Technology and Management
Odisha, India

Parvej Ahmad Alvi
Department of Physical Sciences
Banasthali Vidyapith
Banasthali, Rajasthan, India

Prafulla Kumar Pradhan
Department of Physics
School of Applied Sciences
Centurion University of Technology and Management
Odisha, India

Priyambada Mallick
School of Applied Sciences
Centurion University of Technology and Management
Odisha, India

Puspender Singh Beniwal
Department of Physics
J C Bose University of Science and Technology, YMCA
Faridabad, India

R. K. Parida
Faculty of Engineering and Technology
Department of Physics, ITER, S'O'A
Deemed to be University
Bhubaneswar, India

Ram Sundar Maurya
Department of Physical Sciences
Banasthali Vidyapith
Banasthali, Rajasthan, India

Rimpi Chakravarty
Department of Physics
Central Institute of Technology Kokrajhar, BTR
Assam, India

S. Mishra
Department of Physics
ITER, Siksha 'O' Anusandhan Deemed to be University
Bhubaneswar, India

S. K. Satpathy
School of Applied Sciences
Centurion University of Technology and Management
Odisha, India

Santanu Sen
Department of Physics
Central Institute of Technology Kokrajhar, BTR
Assam, India

Santosh Kumar Parida
Department of Physics
ITER, Siksha 'O' Anusandhan Deemed to be University
Bhubaneswar, India

Satyanarayan Dhal
Department of Physics
School of Applied Sciences, Centurion University of Technology and Management
Odisha, India

Shohaib Abass
Solid State Research Lab
Department of Physics
Central University of Kashmir
Ganderbal, India

Song Sun
Department of Materials Science and Engineering
College of Engineering and Applied Sciences
National Lab of Solid State Microstructures
Collaborative Innovation Center of Advanced Microstructures
Nanjing University, Nanjing, Jiangsu Province, China

Srikanta Moharana
School of Applied Sciences
Centurion University of Technology and Management
Odisha, India

Subhendu Chakroborty
Department of Basic Sciences
IITM, IES University
Bhopal, India

Sudhanshu Dwivedi
S.S. Jain Subodh P.G. (Autonomous) College
Jaipur, India

Sungjun Park
Department of Electrical and Computer Engineering
Ajou University
Suwon, Republic of Korea

Upendra Kumar
Advanced Functional Materials Laboratory
Department of Applied Sciences, IIIT Allahabad
Prayagraj, India

Vedika Yadav
Department of Physical Sciences
Banasthali Vidyapith
Banasthali, Rajasthan, India

Won-June Lee
Department of Chemistry
Purdue University
West Lafayette, IN, United States

Yan-Qiang Cao
Institute of Micro-nano Photonics and Quantum Manipulation
School of Science, Nanjing University of Science and Technology
Nanjing, China

Yogita Batra
Department of Physics
J C Bose University of Science and Technology, YMCA
Faridabad, India

1 Metal Oxide-based High-K Dielectric

An Introduction

Gyanendra Kumar Mishra, Prafulla Kumar Pradhan, and Nilaya Kumar Mohanty

1.1 INTRODUCTION

The modern era of research is based on the development of new smart materials for device applications. The material's properties change in response to external conditions. The smart material properties may be linked with human activities as the smart sensors resemble the human eye, electronics resemble the human brain, and actuators resemble human extremities. As most of the physical properties are structure and composition dependent, these criteria make them ready for application in various fields. The main focus of the current research is to save the existing energy and create a new source of energy. One such renowned material is called dielectric materials, which are used extensively in energy storage devices [1]. The meaning of the word "dielectric" comes from its prefix, "dia," which means "across" or "through." A dielectric is a material that allows the passage of an electric field or flux but not particles (electrons). It does not conduct electric current and is generally considered a non-conductor or an insulator. No ideal dielectric exists on this planet because the ideal dielectric requires the conditions of perfect vacuum, which is unachievable in any natural environment on Earth. It is found that a vacuum of 10^{-14} torr still contains about 300 particles per cubic centimeter [2].

Theoretical origins of dielectric materials can be traced as far back as early as 600 BC, when Greek philosopher Thales discovered that amber, when rubbed with cloth, attracted light objects such as bits of chaff (dry outer covering of some fruits or seeds). Similar results can be achieved today by rubbing a glass rod with a polyester sheet; the rod will attract a light piece of paper. This phenomena may be explained by the rubbing object creating an electric charge on the rod tip, polarizing the paper nearby. Electric polarization produces an opposite charge on the rod tip, resulting in this attraction. The electromagnetic wave (EM) also induces polarization in dielectric materials and magnetization in magnetic materials

1.2 BEHAVIOR OF DIELECTRIC MATERIALS IN ELECTRIC FIELD

When two parallel plates have different potential differences, an electric field is created between the two, as shown in Figure 1.1. The charges on the plates are called free charges $\left(q_0\right)$, and the electric field generated is called electric field due to free charges $\left(\vec{E}_0\right)$. When the dielectric materials are placed in an external electric field, the opposite charges develop on the two surfaces due to the effect of the electric field (E_0). The charges developed on the surfaces are called bound charges or induced charges $\left(q_b\right)$ and are capable of producing the electric field $\left(\vec{E}_d\right)$, which is called depolarizing field or electric field due to bound charges [3]. Hence, E_d is opposite to E_0. So, the net electric field within the dielectric is given by

$$E = E_0 - E_d \tag{1.1}$$

DOI: 10.1201/9781003352365-1

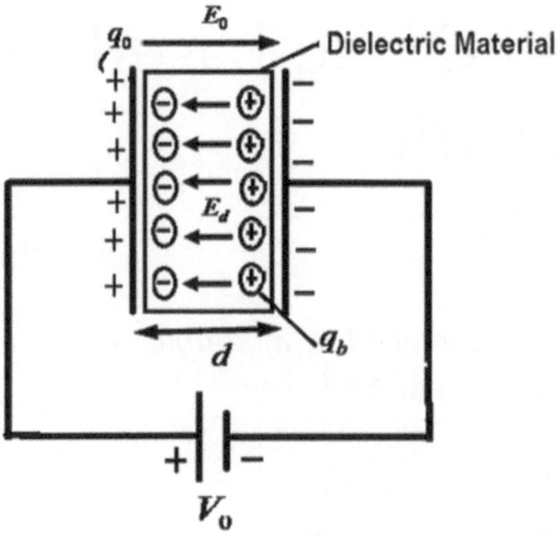

FIGURE 1.1 Schematic diagram for dielectric material placed in an external electric field.

Due to this opposite nature, the net field in the dielectric is reduced. The process of separation of charges is called polarization of dielectric. The dielectric materials also can be utilized as a temperature sensor [4], measuring instruments for the flow of liquid, liquid level regulator, and current and voltage limiter. Some of the high-K materials show high conductivity [5].

1.3 THEORIES OF DIELECTRIC POLARIZATION: REASONS BEHIND THE HIGH-K VALUE

Materials that do not have certain electrons in the conduction band are called dielectrics. The insulating behavior of dielectrics is reduced when an outer electric field is applied to them. This particular electric field is known as the breakdown strength of the material [2, 6]. In general, dielectric materials can broadly be divided in non-ferroelectric and ferroelectric. The non-ferroelectric dielectrics are further subdivided into nonpolar and polar. The schematic diagram is shown in Figure 1.2. In practice, at the atomic level, all materials consist of both positive and negative charges having equal magnitudes. Both positive and negative charges have their own respective centers of gravity. In a nonpolar dielectric, centers of gravity of positive and negative charges coincide. Hence the net dipole moment is zero. In the presence of the electric field $\left(\vec{E}_0\right)$, the molecules are realigned and the centers of gravity of both charges are pulled toward one another so that each molecule behaves as a dipole. This produces an induced dipole moment (P_{induced}). These types of dielectrics have

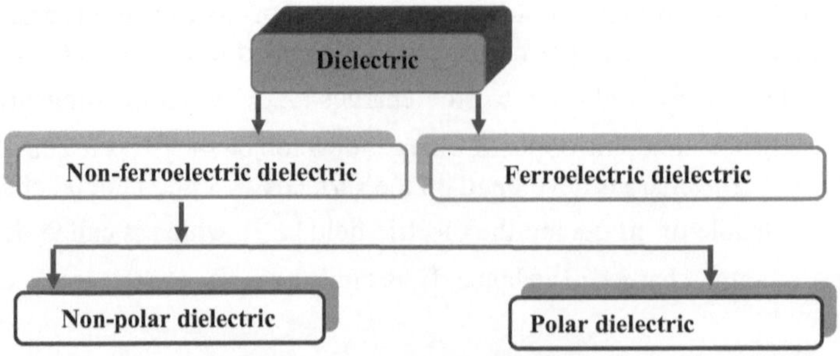

FIGURE 1.2 Schematic diagram for types of dielectric materials.

symmetrical shapes. Examples are CO_2, H_2, N_2, O_2, etc. In polar dielectrics, centers of gravity of positive and negative charges do not coincide. The molecules of the polar dielectrics also behave as dipole when no field is applied where the dipoles are ambiguously aligned with the existence of some dipole moment, called permanent dipole moment ($P_{\text{permanent}}$). When polar dielectrics are kept in an electric field, the centers of the positive and negative dipoles are pulled toward the negative and positive electrodes, respectively, and the charges are oriented with an increase in a dipole moment. This change in a dipole moment is called induced dipole moment (P_{induced}). The net dipole moment is given by $P = P_{\text{permanent}} + P_{\text{induced}}$. These types of dielectrics have asymmetrical shapes. Examples are CO, HCl, H_2O, etc.

1.4 DIELECTRIC POLARIZATION

The main reason for high value of dielectric constants is due to the polarization. At the outside of the electric field the atoms remain in symmetry and the centers of negative and positive charges meet one another in case of nonpolar (and may not meet in case of polar) dielectric. When atoms are placed between two parallel plates connected through a battery such that the plates are charged and both plates create an electric field, electrons of the atoms are displaced toward positive plate and the nucleus is shifted toward the negative plate. As the positive and negative charge centers are separated, they cause an electric dipole and separation of two opposite charges toward the plates, which is called dielectric polarization and is measured as the dipole moment per unit volume.

In dielectric materials, while an external electric field is applied, different kinds of polarizations are seen, specifically:

1. Induced or electronic polarization
2. Atomic or ionic polarization
3. Dipolar or orientational polarization
4. Interfacial or space-charge polarization

With the application of an external electric field, polarization takes place due to the shifting of the negative charge electrons and positive charge nucleus in opposite direction. The phenomenon by virtue of which such phenomenon takes place is called induced polarization. The induced dipole moment in presence of electric field E is given by

$$P_{induced} = \alpha_{induced} E \tag{1.2}$$

where $\alpha_{induced}$ = electronic polarizability or induced polarizability

1.4.1 Induced or Electric Polarization

The following calculation may help evaluate electronic polarizability.

Considering an atom of atomic number Z consisting of a nucleus having charge $+Ze$ where the electrons $-Ze$ are revolving around the nucleus in a circular orbit of radius r in the absence of an electric field, the charge density ρ, defined as the total number of negative charges of the atom to the volume of the atom, is given by

$$\rho = \frac{\text{Number of charges of the atom}}{\text{Volume of the atom}}$$

$$\rho = \frac{-Ze}{\frac{4}{3}\pi r^3} = \frac{-3Ze}{4\pi r^3} \tag{1.3}$$

When an external electric field is applied to the dielectric materials, two forces are introduced: Coulomb's force and Lorentz force. The Coulomb's attractive force develops because of the presence of the positive charge $+Ze$ of the nucleus and negative charge $-Ze$ of the electrons. Mainly this force maintains the equilibrium position of the nucleus and the electron clouds. The Lorentz force works opposite to the Coulomb's force by distracting the electron clouds from their mean position where the electrons move toward the opposite direction of the electric field while the nucleus is directed along the the applied electric field.

Let d = displacement of the electron cloud from the nucleus as shown in the Figure 1.3.

As the distance d is proportional to the electric field E, the induced dipole moment P_{induced} is the product of the magnitude of the charge and the displacement d. It is given by

$$P_{induced} = Zed \tag{1.4}$$

Here the mass of the nucleus is very large as compared to the mass of the electron cloud, so the displacement of the electron will be negligible with the radius of atom, and hence $d << r$.

We know that Coulomb's force and Lorentz force are of same magnitude but in opposite direction, and the magnitudes of both forces are equal at equilibrium.

$$\therefore \text{Lorentz force} = \text{Coulomb's force}$$

Coulomb's force,

$$f_c = \frac{qZe}{4\pi\varepsilon_0 d^2} \tag{1.5}$$

Here within the sphere of radius d the total number of negative charges q is equal to the product of the volume of the sphere $v = \frac{4}{3}\pi d^3$ and charge density of electrons ρ.

Hence, $q = \rho \times v$

Using Equation (1.3) in above equation,

$$q = \frac{-3Ze}{4\pi r^3} \times \frac{4}{3}\pi d^3$$

$$q = \frac{-Zed^3}{r^3} \tag{1.6}$$

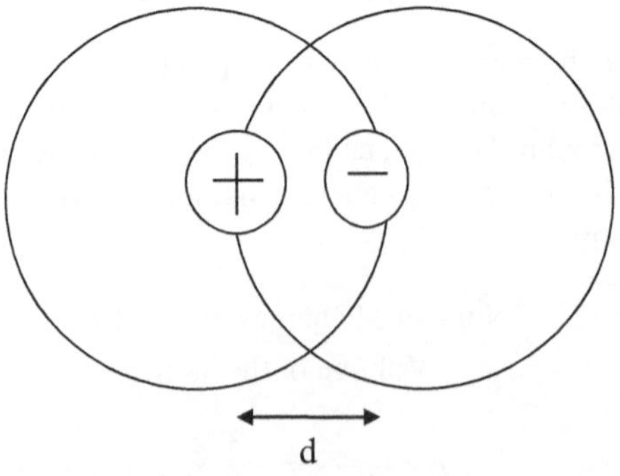

FIGURE 1.3 Separation of electron clouds from the nucleus.

Putting Equation (1.6) in Equation (1.5) we get

$$f_c = \frac{Ze}{4\pi\varepsilon_0 d^2} \times \frac{-Zed^3}{r^3}$$

$$f_c = \frac{-Z^2e^2d}{4\pi\varepsilon_0 r^3} \tag{1.7}$$

Lorentz force f_l is given by the product of the charge and the applied electric field.

$$f_l = -ZeE \tag{1.8}$$

At equilibrium condition, $f_c = f_l$.

Using Equations (1.7) and (1.8) we can write,

$$-ZeE = \frac{-Z^2e^2d}{4\pi\varepsilon_0 r^3}$$

Or

$$d = \frac{4\pi\varepsilon_0 r^3 E}{Ze} \tag{1.9}$$

Using Equation (1.9), Equation (1.4) can be written as

$$P_{induced} = \frac{Ze4\pi\varepsilon_0 r^3 E}{Ze} = 4\pi\varepsilon_0 r^3 E \tag{1.10}$$

From Equations (1.2) and (1.10) we get

$$\alpha_{electronic} E = 4\pi\varepsilon_0 r^3 E$$

$$\alpha_{electronic} = 4\pi\varepsilon_0 r^3 \tag{1.11}$$

where $\alpha_{electronic}$ is electronic or induced polarization.

Some important facts can be identified from Equation (1.11), specifically:

1. $\alpha_{electronic}$ does not depend on temperature.
2. Induced or electronic polarization is directly proportional to the volume of the atoms of dielectric materials.
3. $\alpha_{electronic}$ is seen in all kinds of dielectric materials.

1.4.2 Atomic or Ionic Polarization

Similar to electronic polarization, when an external electric field is applied, instead of electrons, positive and negative ions are displaced in opposite directions to one another. Polarization that occurs because of the reverse displacement of ions relative to each other is called atomic or ionic polarization. Atomic or ionic polarization occurs for such ionic molecules as KCl, KBr, NaCl, NaBr, etc.

Consider KBr crystal, in which each K+ and Br $^-$ ion has their own dipoles as shown in Figure 1.4. When an external electric field is applied, ions are forced in opposite directions. The ion *K+* moves left for a short distance d_1 while Br$^-$ moves right for a short distance d_2.

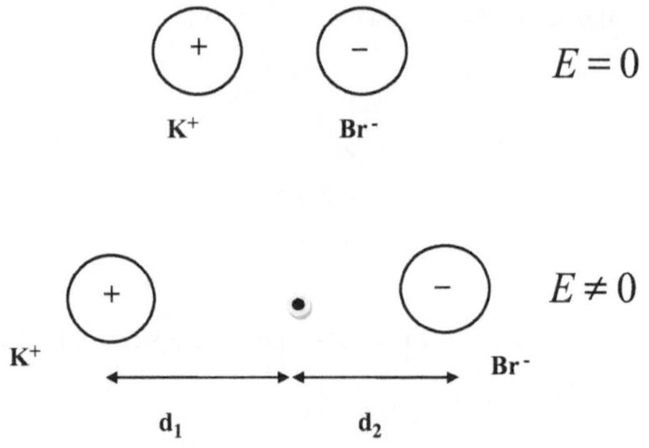

FIGURE 1.4 Separation of two types of ions in the absence and presence of electric field E.

Therefore, the induced dipole moment is given by

$$P_{ionic} = \left(d_1 + d_2\right) \times e \tag{1.12}$$

where P_{ionic} = ionic dipole moment due to ions.

In the presence of external electric field, the restoring force and the separation of positive charge ions are directly proportional to one another.

The restoring forces for respective ions are given by

$$f_1 = k_1 d_1 \tag{1.13}$$

$$f_2 = k_2 d_2 \tag{1.14}$$

where $k_1 = m_1\omega^2$ and $k_2 = m_2\omega^2$ are the restoring force constants. Here m_1, m_2 are masses of positive and negative ions and ω is angular frequency.

Using the above expression for k_1 and k_2 in Equation (1.13), one can write,

$$f_1 = m_1\omega^2 d_1 \tag{1.15}$$

We know that electric field,

$$f_1 = Ee \tag{1.16}$$

From Equations (1.15) and (1.16),

$$Ee = m_1\omega^2 d_1$$

$$d_1 = \frac{Ee}{m_1\omega^2} \tag{1.17}$$

Similarly for negative ions,

$$d_2 = \frac{Ee}{m_2\omega^2} \tag{1.18}$$

Now,

$$d_1 + d_2 = \frac{Ee}{\omega^2}\left(\frac{1}{m_1} + \frac{1}{m_2}\right) \tag{1.19}$$

Using Equation (1.19), we can write Equation (1.12) as

$$P_{ionic} = (d_1 + d_2) \times e$$

$$p_{ionic} = \frac{Ee^2}{\omega^2}\left(\frac{1}{m_1} + \frac{1}{m_2}\right) \tag{1.20}$$

Also,

$$P_{ionic} = \alpha_{ionic} \times E \tag{1.21}$$

From Equations (1.20) and (1.21),

$$\alpha_{ionic} \times E = \frac{Ee^2}{\omega^2}\left(\frac{1}{m_1} + \frac{1}{m_2}\right)$$

$$\alpha_{ionic} = \frac{e^2}{\omega^2}\left(\frac{1}{m_1} + \frac{1}{m_2}\right) \tag{1.22}$$

From Equation (1.22) it is observed that ionic polarization is inversely proportional to the mass of the nucleus in negative ions. Also it is inversely proportional to that of angular frequency and is independent of temperature.

1.4.3 Dipolar or Orientational Polarization

Dipolar or polar molecules have a permanent dipole. Examples of such molecules are methyl chloride (CH_3OH), water, etc. The net dipole moment of the substance becomes zero in the absence of an applied electric force field, and the dipoles are arranged randomly. In case of an applied force field, the dipoles of the substance are oriented along the direction of the electric force field as shown in Figure 1.5. Polarization that occurs due to orientations of the dipoles toward the applied field is called dipolar or orientational polarization.

Due to thermal agitation, the molecules do not approach to orient the dipoles, but in a state of equilibrium, various dipoles are aligned at different angles.

If the material contains n number of dipoles, then the mean sharing to polarization is given by

$$\bar{p} = \frac{Total\ polarization}{All\ the\ dipoles} \tag{1.23}$$

$$\bar{p} = \frac{\int_0^{\pi} 2\pi p \cos\theta \sin\theta e^{(pE\cos\theta/KT)} d\theta}{\int_0^{\pi} 2\pi \sin\theta e^{(pE\cos\theta/KT)} d\theta}$$

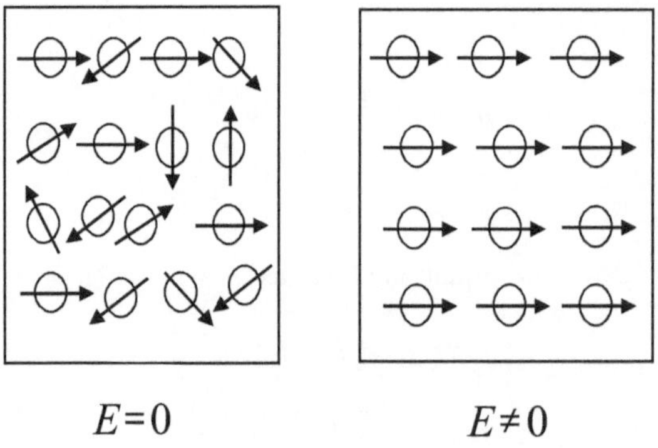

FIGURE 1.5 Situation of dipoles in the absence and presence of the electric field E.

Where p is the dipole moment, θ is the angle between the field and dipole, T is the temperature, and K is the Boltzmann constant.
Using $pE/kT = x$ and $\cos\theta = y$,

$$\bar{p} = \frac{p\int_{+1}^{-1} y \; e^{xy} dy}{\int_{+1}^{-1} e^{xy} dy}$$

$$\frac{\bar{p}}{p} = \frac{e^{x} + e^{-x}}{e^{x} - e^{-x}} - \frac{1}{x}$$

$$\frac{\bar{p}}{p} = \coth x - \frac{1}{x} \tag{1.24}$$

$$L(x) = \frac{\bar{p}}{p} \tag{1.25}$$

where $L(x) = \coth x - \frac{1}{x}$ is a Langevian function.

At high temperature, $pE/kT = x << 1$,

$$L(x) = \frac{x}{3} = \frac{pE}{3kT} \tag{1.26}$$

Using the above equation, Equation (1.25) becomes

$$\frac{\bar{p}}{p} = \frac{pE}{3kT} \Rightarrow \bar{p} = \frac{p^{2}E}{3kT} \tag{1.27}$$

Therefore, the dipolar polarization is given by

$$P_{Dipolar} = n\bar{p}$$

$$P_{Dipolar} = \frac{np^{2}E}{3kT} \tag{1.28}$$

From the above it is seen that the orientational polarization is inversely proportional to the absolute temperature of the dielectric substance, which is given by

$$\alpha_{Dipolar} = \frac{\overline{p}}{E} = \frac{p^2}{3kT} \tag{1.29}$$

The dipolar polarizability is temperature dependent.

1.4.4 Interfacial Polarization or Space Charge Polarization

In this type of polarization, in the presence of an applied electric field, the ions are diffused along the applied field and the charges are redistributed in the dielectrics as shown in Figure 1.6.

There is an induced dipole moment because of the displacement of the ions. The induced dipole moment per unit volume develops, and this type of polarization is known as interfacial polarization.

Net or total electric polarization is the sum of all polarization. Thus, total polarization is given by

$$\alpha_{Total} = \alpha_{electronic} + \alpha_{ionic} + \alpha_{orientational} + \alpha_{spacecharge} \tag{1.30}$$

1.5 NEED AND SELECTION RULE OF HIGH-K MATERIALS

The dielectric properties play a crucial role in the miniaturization of dielectric components of the devices such as capacitors, transistor, memories, sensors, resonators, etc. [7–10]. Taking a capacitor as an example, the dimensional capacitance of a parallel plate capacitor is given by

$$C_0 = \frac{\varepsilon_0 A}{d} \tag{1.31}$$

where ε_0 is the permittivity of the free space, A is the area of the plates, and d is the distance between two parallel plates. Here the capacitance can be increased by increasing A and decreasing d. There are some limitations in variation of A and d. When a dielectric material having a dielectric constant or relative permittivity ε_r is placed between the two parallel plates of the capacitors, the capacitance increases per the relation given by

$$C = \frac{\varepsilon_0 \varepsilon_r A}{d} \tag{1.32}$$

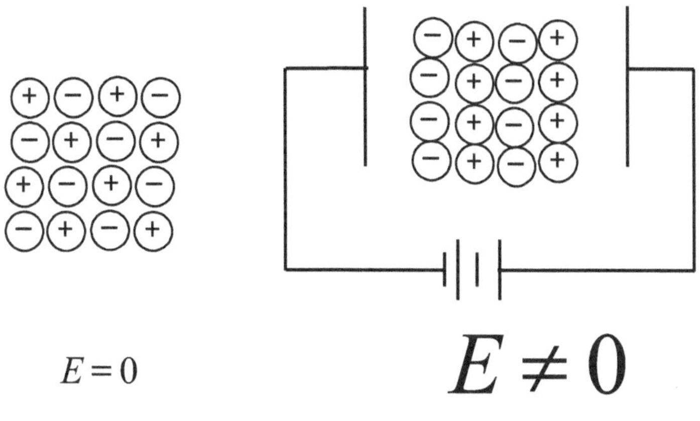

FIGURE 1.6 Distribution of ions in the absence and presence of the electric field.

Silicon dioxide (SiO_2) is one of the standard materials in MOSFET (metal–oxide–semiconductor field-effect transistor) devices, which are one of the key components in microelectronic circuit. In order to reduce the size of such a device, the thickness of the SiO_2 needs to be reduced. When the thickness is reduced, the leakage current due to the tunneling effect in a MOSFET increases, because the tunneling probability is inversely proportional to the thickness of the silicon dioxide gate. Hence, there is some limit on the decrease in the size of the layer [11]. Increasing A also creates issued with the miniaturization.

When SiO_2 is used as a dielectric medium between two conducting plates of a capacitor, per the above relations (1.31) and (1.32), capacitance is increased by decreasing the thickness of SiO_2. Once the thickness of SiO_2 reaches the threshold value, any further reduction in thickness leads to quantum tunneling effects and an increase in leakage current. Hence, it is not possible to reduce the thickness of SiO_2 indefinitely. To overcome the above issues, the materials with high dielectric constant value have to be used [12]. Generally, with reference to the K values of SiO_2 whose dielectric constant value is 3.9, the dielectric materials are divided into two types high-K and low-K materials. The K values of high-K and low-K materials are greater than and less than 3.9, respectively. High-K dielectrics materials do not allow a steady current can flow, and as a result, the static electric field in not zero [13]. Dielectric polarization occurs in the materials due to the displacement of the electron cloud with respect to positive ions when the external electric field is applied. Due to these polarization properties, dielectric materials help store the electromagnetic energy, which also explains the mechanism associated with electronics, optics, etc. [14–15]. Since the dielectrics are natural insulators, they are used widely in various electronic devices because of the energy stored in them [16]. For example, high-K materials are generally used in capacitors.

The electromagnetic energy stored in a dielectric is given by

$$U = \frac{1}{2}CV^2 \tag{1.33}$$

where C is capacitance and V is the potential developing within the dielectric. This potential V is assumed to be the electric field, as the electric field is negative of potential gradient, i.e.,

$$E = -\frac{dV}{dl} \tag{1.34}$$

where dl is the small change in length of the materials.

From Equation (1.33) it may be noted that, in order to have greater energy stored in materials, one has to increase the capacitance or potential/electric field. First, to increase the capacitance of parallel plate capacitor, d can be decreased and A can be increased as stated in Equation (1.32). But when d is decreased to around 1 nm, it causes leakage current due to quantum mechanical tunneling, and increasing A limits the miniaturization of devices. Therefore, one of the options to increase capacitance is to use dielectric materials with a high dielectric constant with limited A and d, which helps suppress leakage current. The high-K material must possess a dielectric constant between 10 to 30 [17]. Second, to increase the applied electric field without causing a dielectric breakdown, the following fitness model may help. The universal relationship [18–19] between intrinsic breakdown field E_{BI} and the band gap E_g is given by

$$E_{BI} = 1.36 \times \left(\frac{E_g}{E_{gc}}\right)^{\alpha} \tag{1.35}$$

where $E_{gc} = 4.0\,\text{eV}$, which is the critical band gap value separating materials into semiconductors and insulators. If $E_{gc} < 4.0$, then it is semiconductor, and if $E_{gc} > 4.0$, then it is an insulator. Here,

$\alpha = 3$ for semiconductors, and $\alpha = 1$ for insulators. Hence, the applied electric field can be increased by placing the semiconducting materials with a large band gap between the plates of capacitors. The high-K materials should have a large band gap generally greater than 4.0 eV, because the breakdown field can be very high.

The maximum energy density can be expressed by using Equations (1.32) and (1.34) in Equation (1.33), and one can have the relation,

$$U_{ED} = \frac{1}{2}\varepsilon_0 k E_{BI}^{\ 2} \tag{1.36}$$

Using Equation (1.35) in Equation (1.36),

$$U_{ED} = 0.9248\varepsilon_0 k\left(\frac{E_g}{E_{gc}}\right)^{2\alpha} \tag{1.37}$$

This is the fitness model that connects the dielectric constant, the band gap, and the breakdown field during optimization. Hence, it helps search the optimal dielectric materials for the different application purpose.

1.6 EXAMPLES OF HIGH-K MATERIALS

Examples of high-K dielectric materials with different types of a broad crystal structure include perovskite, tungsten-bronze, aurivillious, pyrochlore, and ZrCuSiAs, among others. Perovskites possess two cations and one anion (mostly oxide). Their general formula is ABX_3, where A and B sites are occupied by cations and the X site is occupied by anions [20–23]. Structural, dielectric, electric, magnetic, and other properties these materials can be greatly affected by choosing and altering the cations on A and B sites [24]. Barium titanate ($BaTiO_3$) is one of the perovskite materials and is a candidate for a high dielectric insulator; the dielectric constant of cubic BaTiO3 is 100 times larger than that of SiO_2 or V_2O_5 [25]. Some of the double perovskite high-K materials are $0.5BiFeO_3$–$0.5PbTiO_3$, $0.5BiDy_yFe_{1-y}O_3$–$0.5PbTiO_3$: (y = 0.05, 0.10, 0.15, 0.20), $0.5BiGd_yFe_{1-y}O_3$–$0.5PbTiO_3$:(y = 0.05, 0.10, 0.15, 0.20), and $0.5BiSm_yFe_{1-y}O_3$–$0.5PbTiO_3$:(y = 0.05, 0.10, 0.15, 0.20) at 10, 50, and 100 kHz. The dielectric constant and tangent loss at room temperature are listed in Table 1.1. These types of materials are prepared using solid-state reaction methods. These materials are the examples of untilted and tilted perovskite structures with a tetragonal system. The dielectric constant is found to be above 120 at room temperature and at 10 kHz. The dielectric constant is found to be higher for a rare-earth doped $BiFeO_3$-$PbTiO_3$ sample as compared to its parent. Further, the dielectric constants for Sm and Dy doped samples are greater compared to Gd doped sample. These high dielectric constants in the samples can be explained by the polarization mechanism. In the case of Gd^{3+} ion, an electronics configuration $4f^7\ 5s^2\ 5p^6$ gives $L = 0$, $S = 7/2$. The orbital angular momentum of Gd^{3+} is zero, which helps with the spherical distribution of the electric charge. In the case of Dy^{3+} and Sm^{3+}, a non-zero value for the orbital angular momentum causes an asymmetric charge distribution, which in turn is responsible for increasing the polarization in the sample as compared to Gd^{3+} and, hence, a higher value of ε_r.

Some other rare-earth doped perovskite materials also show a high dielectric constant [26].

The tungsten-bronze (TB) materials have the general formula $[(A_1)_2(A_2)_4C_4][(B_1)_2(B_2)_8]O_{30}$, which generally comes out to $A_6B_{10}O_{30}$, where A sites, i.e., A_1, A_2, can accommodate monovalent, divalent, or trivalent ions; B sites, i.e., B_1, B_2, can accommodate highly charged cations usually Nb+, Ta^{+5}, and V^{+5}; and a small C site can be vacant or occupied by Li [27–29]. Some of the TB materials show a high dielectric constant, such $LiSr_2Nb_5O_{15}$, $Na_2Pb_2La_2W_2Ti_4Ta_4O_{30}$, and $Na_2Pb_2Pr_2W_2Ti_4Ta_4O_{30}$ [30–32].

TABLE 1.1
Select Perovskite Materials with Their Dielectric Constant and Tangent Values at Room Temperature for 10 and 100 kHz

Sample	Frequency (kHz)	ε_r at Room 25°C	tan δ at 25°C
0.5BF–0.5 PT	10	122.06	0.056
	100	117.44	0.021
$0.5BDy_{0.05}Fe_{0.95}$–0.5PT	10	138.25	0.026
	100	135.87	0.013
$0.5BDy_{0.10}Fe_{0.90}$–0.5PT	10	187.25	0.065
	100	179.95	0.024
$0.5BDy_{0.15}Fe_{0.85}$–0.5PT	10	195.70	0.023
	100	192.06	0.012
$0.5BDy_{0.20}Fe_{0.80}$–0.5PT	10	195.87	0.034
	100	192.26	0.015
$0.5BSm_{0.05}Fe_{0.95}$–0.5PT	10	438.21	0.047
	100	412.44	0.041
$0.5BSm_{0.10}Fe_{0.90}$–0.5PT	10	489.27	0.056
	100	456.57	0.043
$0.5BSm_{0.15}Fe_{0.85}$–0.5PT	10	503.15	0.044
	100	475.98	0.039
$0.5BSm_{0.20}Fe_{0.80}$–0.5PT	10	517.33	0.049
	100	4885.33	0.041
$0.5BGd_{0.05}Fe_{0.95}$–0.5PT	10	135.10	0.062
	100	130.50	0.022
$0.5BGd_{0.10}Fe_{0.90}$–0.5PT	10	157.35	0.065
	100	151.44	0.023
$0.5BGd_{0.15}Fe_{0.85}$–0.5PT	10	208.54	0.098
	100	196.01	0.035
$0.5BGd_{0.20}Fe_{0.80}$–0.5PT	10	248.92	0.125
	100	236.42	0.033

Pyrochlore materials have the general formula $A_2B_2O_7$, where A and B site ions are cations. The A and B site ions are generally rare-earth or transition-metal species. The ferroelectric and dielectric nature of this compound is roughly related to the average interaction of the A site cations. Pyrochlore ferroelectricity was first discovered in $Cd_2Nb_2O_7$ in 1952 [33]. Some other examples of these materials are $Y_2Ti_2O_7$, $La_2Zr_2O_7$, $Gd_{1.9}Ca_{0.1}Ti_2O_{6.9}$,$Bi_2Ru_2O_{7-y}$, $Dy_2Ti_2O_7$, $Y_2Mo_2O_7$, $Tl_2Ru_2O_7$, $Cd_2Re_2O_7$, and $Gd_2Zr_2O_7$ [34–35].

Aurivillius materials contain a bismuth oxide layer $(Bi_2O_2)^{2+}$ and have the general formula $Bi_2A_{m-1}B_mO_{3m+3}$ where $m = 1, 2, 3, 4$, or $Bi_2O_2(A_{m-1}B_mO_{3m+3})$ [36]. These materials are formed by alternating the layers of $(Bi_2O_2)^{2+}$ and $(A_{n-1}B_nO_{3n+3})^2$. The second layer is a pseudo-perovskite material and has the capacity to accommodate the ions of different sizes in it to explore different properties including structural, dielectric, ferroelectric, etc. The bismuth oxide layer helps stabilize the structure and allows for high operating temperature and frequency. Some of these materials that show higher dielectric constant values are bismuth titanate ($Bi_4Ti_3O_{12}$), lead-bismuth niobate ($PbBi_2Nb_2O_9$), and $Ba_{0.5}Sr_{0.5}Bi_2V_2O_9$ [37]. The ZrCuSiAs materials are tetragonal structures in nature, which are quaternary silicide arsenides. These materials are of great importance due to their superconducting properties at low temperatures [38]. The iron-based ceramic material RFeAsO combines with the rare-earth ions *R*, which includes lanthanum (La), cerium (Ce), praseodymium

(Pr), neodymium (Nd), samarium (Sm), gadolinium (Gd), and so on. Among these, the materials GdFeAsO, DyFeAsO, NdFeAsO are the most promising for achieving excellent magnetic, semiconducting, optical, and superconducting properties, among others. The material GdFeAsO is believed to exhibit ferromagnetic and superconducting properties after slight modification at low temperatures. This material has a tetragonal structure with lattice constants $a = b = 3.9213$ Å and $c = 8.4676$ Å. The dielectric constant of this material at room temperature is found to be around 180 for 10 kHz. The sample of NdFeAsO exhibits superconducting properties with transition temperature 50 K. The tetragonal crystal structure has been confirmed for this sample, with lattice parameter values $a = 3.9763$ Å, $c = 8.6469$ Å, $c/a = 2.174$. The dielectric constant at room temperature is found to be 38 for 10 kHz [39].

1.7 SUMMARY AND CONCLUSION

The metal oxide–based high-K materials have proved to be of great importance in the field of electronic and energy storage device applications. The basic mechanism behind the high value of the dielectric constant is the dielectric polarization. The dielectric polarization value depends on the nature of the dielectric, i.e., whether it is polar or nonpolar. The polarization in the materials occurs due to the displacement of electrons with respect to mean position, displacement of positive and negative ions, and orientation of the dipole, among other factors. The polarizability obtained from different types of polarization explores many physical parameters related to the increasing value of polarization. High-K materials are necessary to store maximum energy in a small space, which helps miniaturize the device. One can select the optimal dielectric materials for different application purposes based on the fitness model that connects the dielectric constant, band gap, and breakdown field during optimization in a single equation. The various structure types of materials including perovskite, tungsten-bronze, aurivillious, pyrochlore, and ZrCuSiAs, which show high dielectric constant values and are widely used in many device applications.

Dielectric constant values in these materials can be altered by accommodating various ions at different sites of the parent structure. The dielectric constant can also be optimized by carefully considering the factors such as crystalline structure, imperfection, and impurities, nature of the external surface, etc.

REFERENCES

1. Satpathy S. K., Sen S., Behera B., Dielectric, electrical and magnetic properties of La doped $BiFeO_3$–$PbZrO_3$ composites, *Journal of Materials Science: Materials in Electronics*, 2017, 28, 9102–9113.
2. Kao K.C., *Dielectric phenomena in solids:with emphasis on physical concepts of electronic processes*, Elsevier Academic Press, 2004, ISBN 0-12-396561-6.
3. Fischer-Cripps A. C., *The materials physics companion*, CRC Press Taylor and Francis, 2008, ISBN 978-1-58488-680-8.
4. Sahu A. K., Mallick P., Satpathy S. K., Behera B., Effect on structural, electrical and temperature sensing behavior of neodymium doped bismuth ferrite, *Advanced Materials Letters*, 2021, 12, 1–7.
5. Sahu A. K., Satpathy S. K., Behera B., Dielectric and frequency-dependent transport properties of lanthanum-doped bismuth ferrite, *Journal of Advanced Dielectrics*, 2019, 9 (04), 1950031.
6. Tarrev B., *Physics of dielectric materials*, Mir Publisher, 1979.
7. Tummala R., *Electronic packaging for high reliability, low cost electronics*, Springer Sci. Business Media, 1999, 57.
8. Lau J. H., *Chip on board: Technology for multichip modules*, Springer Science and Business Media, 1994.
9. Bai Y., Cheng Z. Y., Bharti V., Xu H., Zhang Q., High dielectric constant ceramic – Powder polymer composites, *Applied Physics Letters*, 2000, 76 (25), 3804–3806.
10. Facchetti A., Yoon M. H., Marks T. J., Gate dielectrics for organic field-effect transistors: New opportunities for organic electronics, *Advanced Materials*, 2005, 17 (14), 17051725
11. Frank D. J., Dennard R. H., Nowak E., Solomon P. M., Taur Y., Wong H., Device scaling limits of simosfets and their application dependencies. *Proceedings of the IEEE*, 2001, 89 (3), 259288.

12. Houssa M., *High-K gate dielectrics*, CRC Press, 2003.
13. Sri Harsha P., Venkata Saravanan K., Madhurima V., *High-k dielectric materials: Structural properties and selection, high-k gate dielectric materials applications with advanced metal oxide semiconductor field effect transistors (MOSFETs)*, Taylor and Francis, 2021.
14. Aspnes D. E., Studna A. A., Dielectric functions and optical parameters of Si, Ge, GaP, GaAs, GaSb, InP, InAs, and InSb from 1.5 to 6.0 eV, *Physical Review B*, 1983, 27, 985–1009.
15. Zhao X., Vanderbilt D., First-principles study of structural, vibrational, and lattice dielectric properties of hafnium oxide, *Physical Review B*, 2002, 65, 233106.
16. Hybertsen M. S., Louie S. G., *Ab initio* static dielectric matrices from the density-functional approach. I. Formulation and application to semiconductors and insulators *Physical Review B*, 1987, 35, 5585–5601.
17. Choi J. H., Mao Y., Chang J. P., Development of hafnium based high-*k* materials – A review. *Materials Science & Engineering R: Reports*, 2011, 72, 97–136.
18. Wang L. M., *25th* International Conference on microelectronics, Belgrade, 2006, 576–579.
19. Ieong M., Doris B., Kedzierski J., Rim K., Yang M., Silicon device scaling to the sub-10-nm regime, *Science*, 2004, 306, 2057–2060.
20. Szuromi P., Grocholski B., Natural and engineered perovskites, *Science*, 2017, 358(6364), 732–733.
21. George G., Rao S., Luo Z., *Fundamental of perovskite oxides, synthesis structure properties and applications*. CRC Press, Taylor and Francis Group, 2021.
22. Satpathy S. K., Mohanty N. K., Moharana S., *Progress of perovskite materials with surface additives, chemical modification of solid surfaces by the use of additive*, Bentham Science Publishers, 2021, 208–226.
23. Barik R., Satpathy S. K., Behera B., Biswal S. K., Mohapatra R. K., Synthesis and spectral characterizations of nano-sized lithium niobate ($LiNbO_3$) ceramic, *Micro and Nanosystems*, 2020, 12(2), 81–86.
24. Sahu A. K., Satpathy S. K., Rout S. K., Behera B., Dielectric and frequency dependent transport properties of gadolinium doped bismuth ferrite, *Transactions on Electrical and Electronic Materials*, 2020, 21, 217–226.
25. Naito M., Hosokawa M., Yokoyama T., Nogi K., *Nanoparticle Technology Handbook*, Elsevier, 2008, 473–477.
26. Ray A., Behera B., Basu T., Vajandar S., Satpathy S. K., Nayak P., Modification of structural and dielectric properties of polycrystalline Gd-doped BFO–PZO, *Journal of Advanced Dielectrics*, 2018, 8 (05), 1850031.
27. Van Uitert L. G., Levinstein H. J., Rubin J. J., Capio C. D., Dearborn E. F., Bonner W. A., Some characteristics of niobates having “filled” tetragonal tungsten bronze-like structures, *Materials Research Bulletin*, 1968, 3(1), 47–57.
28. Abrahams S. C., Jamieson P. B., Bernstein J. L., Ferroelectric tungsten bronze-type crystal structures. III. potassium lithium niobate $K_{(6-x-y)}Li_{(4+x)}Nb_{(10+y)}O_{30}$, *The Journal of Chemical Physics*, 1971, 54(6), 2355.
29. Gross, U., Zone purification of alkali iodides, *Materials Research Bulletin*, 1970, 5 (47), 117–128.
30. Mohanty N. K., Satpathy S. K., Behera B., Nayak P., & Choudhary R. N. P., Structural and dielectric properties of $LiSr_2Nb_5O_{15}$ ceramic, *AIP Conf. Proc.*, 2011, 1372, 11–14.
31. Das P. R., Behera S., Padhee R., Nayak P., Choudhary R. N. P., Dielectric and electrical properties of $Na_2Pb_2La_2W_2Ti_4Ta_4O_{30}$ electroceramics, *Journal of Advanced Ceramics*, 2012, 1(3), 232–240.
32. Behera S., Parida B. N., Nayak P., Das P. R., Synthesis and characterization of $Na_2Pb_2Pr_2W_2Ti_4Ta_4O_{30}$, *Journal of Materials Science: Materials in Electronics*, 2013, 24, 1132–1140.
33. Yang Z., Gu R., Wei L., Ren H., Phase formation, microstructure and dielectric properties of $Sr_{0.53}Ba_{0.47}Nb_{2-x}Ta_xO_6$ ceramics, *Journal of Alloys and Compounds*, 2011, 504(1), 211–217.
34. Subramanian M. A., Aravamudan G., SubbaRao G. V., Oxidepyrochlores – A review, *Progress in Solid State Chemistry*, 1983, 15(2), 55–143.
35. Sharma S. K., Mohanty H. S., Pradhan D. K., Ajay K., Shukla V. K., Singh F., Kulriya P. K., Structural, dielectric and electrical properties of pyrochlore-type Gd2Zr2O7 ceramic, *Journal of Materials Science: Materials in Electronics*, 2020, 31, 21959–21970.
36. Aurivillius B., Mixed bismuth oxides with layer lattices. 1. The structure type of $CaNb_2Bi_2O_9$, *Ark. Kemi.*, 1949, 1, 463–480.

37. Mohanty N. K., Pradhan R. N., Satpathy S. K., Behera A. K., Behera B., Nayak P., Electrical transport properties of layered structure bismuth oxide: $Ba_{0.5}Sr_{0.5}Bi_2V_2O_9$, *Journal of Materials Science: Materials in Electronics*, 2014, 25, 117–123.
38. Johnson V., Jeitschko W., ZrCuSiAs: A "filled" PbFCltyp, *Journal of Solid State Chemistry*, 1974, 11, 161–166.
39. Mishra G. K., Mohanty N. K., Behera B., Analysis of dielectric and electrical transport properties of NdFeAsO ceramic, *Cerâmica*, 2022, 68(386), 181–187.

2 Perovskite Oxides as High-K Material

Synthesis, Characterization, and Application

Vedika Yadav, Harshpreet Cheema, Ram Sundar Maurya, Parvej Ahmad Alvi, and Upendra Kumar

2.1 INTRODUCTION

As the world's population is rising, the living standard of a new generation has risen, industrialization and deforestation are increasing due to all these, and demand for energy and energy-storing devices is at the peak. Storage captures energy produced at one time to be used later. In the rapidly developing electronic industry, the demand for materials with high dielectric constant and permittivity is increasing rapidly. A substance with a higher dielectric constant than that of SiO_2 is generally referred to as a high-K dielectric. The need for a high-K nano-dielectric material is becoming more urgent and significant as electronic devices get smaller. Many methods have been used to improve the dielectric constant and permittivity and to lower the dissipation factor of materials. The ceramics of $BaTiO_3$ doped show high permittivity, but they are highly dependent on temperature [1]. Interestingly, CCTO ($CaCu_3TiO_{12}$) has high permittivity (~10^5), but its high dielectric loss has limited its applications [2]. Researchers extensively study perovskites for energy storage applications.

Gustav Rose, a German chemist and mineralogist, discovered natural perovskite in 1839. Calcium titanate ($CaTiO_3$) is the first known perovskite structure, though the term "perovskite" was applied to $CaTiO_3$ in honor of Russian mineralogist Count Lev Alekseyevich Perovski. The term was later applied to synthetic substances with the exact stoichiometry and crystal structure of $CaTiO_3$ rocks [3]. The chemical formula ABO_3 is commonly accepted for perovskite oxides, where A and B are cations and O is an anion. In perovskite oxides, the atomic size of cations A and B varies, and their bonds with anion produce distinct perovskite structures.

The valencies of cation A and cation B vary from +1 to +3 and +3 to +5, respectively, to compensate for the total charge obtained by oxygen. In a Perovskite structure, atom A is at the corner of the cells, atom B is in the middle with sixfold coordination that encircles the anion and forms octahedrons, and oxygen atoms are in the center of the faces. Cation A is shared by 12 oxygen atoms in this structure, forming a dodecahedral site. Perovskites are classified into inorganic and alkali halide. A pictorial representation of perovskites is shown in Figure 2.1. Inorganic perovskite-type oxides are potential materials for various applications due to their excellent properties [4]. These materials have been used as catalysts in numerous reactions because of their sensitivity, exceptional long-term stability, and interference resistance. Some inorganic perovskite materials are up-and-coming candidates for enhancing the effectiveness of anodic catalysts. Due to their unique diversity of characteristics, perovskite-phase metal oxides have found new uses in the electrochemical sensing of alcohols and glucose.

Halide perovskites have emerged as one of the most promising materials for photovoltaic and light-emitting devices during the past two decades [5]. Alkali metals are frequently employed to passivate the surface and grain boundaries, increasing the crystal's stability. Moreover, organo-metal

 DOI: 10.1201/9781003352365-2

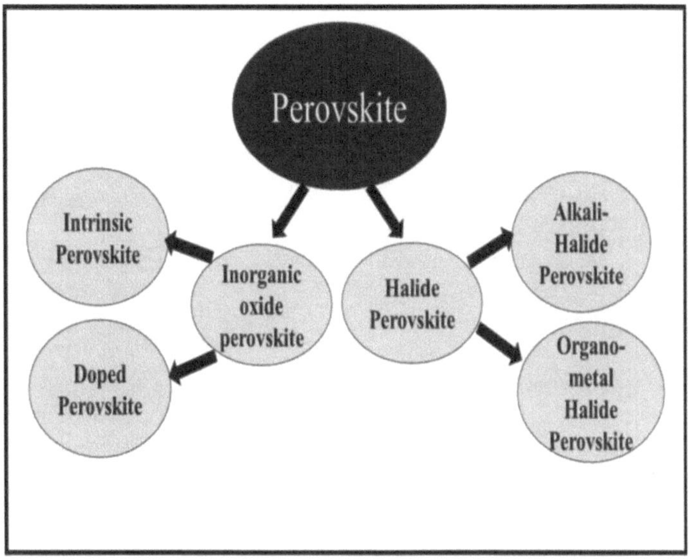

FIGURE 2.1 Classification of different types of perovskites.

halide perovskites with direct band gaps have demonstrated exceptional capabilities. Based on the outstanding results, the perovskites have received significant attention for various applications in all optoelectronic devices, particularly emphasizing the photovoltaic and luminous aspects. But among all types of perovskites, oxide perovskites have received the most significant attention due to their stable phase, high crystallinity, and chemical and physical stability. For further improvement in the properties of perovskites, which can be achieved with the formula (ABO_3), doping can be done on either site A or B. A Hume-Rothery criterion must be satisfied to achieve doping. One of the criterion's parameters is that the difference in ionic radii must not exceed 15%. The host and dopant must be of the same group with similar electronegativity [6]. Also, the ionic radii of A/B and their electronic configurations can be used to understand the different forms of crystal structure caused by distortion in the perovskite structure. Two types of distortion are frequently observed in perovskite structures. The first distortion in perovskite is B ion off-centering in the BO_6 octahedral, leading to phase transition, notably structural transformation. Because of the importance of octahedral BO_6 in ferromagnetism and ferroelectricity, this can be predicted by calculating Goldschmidt's tolerance factor t. The following equation can be used to compute the Goldschmidt tolerance factor t for the perovskite structure:

$$t = \frac{r_a + r_o}{\sqrt{2(r_b + r_o)}}$$

where r_a and r_b are the ionic radius of cations A and B, respectively, and r_o is the anion O's ionic radius. The numerical value of t indicates the possibility of different crystal structure; for example, tolerance factor t equal to 1 indicates perfect cubic structure, while $t > 1$ or < 1 represents tilting in the octahedron that represents the possibility of other crystal structures such as tetragonal, orthorhombic, and so on. Table 2.1 summarizes the crystal structure possibilities based on the numerical value of t.

The fascination with ferroelectric-based materials on Ba, Ca, and Sr can be used in many scientific and technological fields, such as microelectronic devices, capacitors, catalysts, and storage devices. $SrTiO_3$ is a perovskite material used in various applications such as microelectronic and microwave devices. So, in the present case, $SrTiO_3$ has been investigated. The general chemical

TABLE 2.1
Values of Tolerance Factor *t* of the Different Compound, Crystal Structure, and Their Description

S.N.	Compound	Crystal Structure	Tolerance Factor t	Interpretation
1.	$SrTiO_3$	Cubic	0.90–1.00	Both A and B ions are of a perfect size.
2.	$CaTiO_3$	Orthorhombic/Rhombohedra	0.71–0.90	Small size of A ions, which is adjusted into B ions slots.
3.	$SrTiO_3/BaTiO_3$	Cubic	0.90–1.00	Both A and B ions are of a perfect size.
4.	$BaNiO_3$	Hexagonal or Tetragonal	>1.00	Size of the A ion is very large, and the size of the B ion is too small.
5.	$GdFeO_3$	Orthorhombic/Rhombohedral	0.71–0.90	The tiny size of A ions compared to B ions.
6.	$FeTiO_3$	Different structures	<0.71	Ionic radii of both A and B ions are very close.
7.	$SrSnO_3$	Orthorhombic	0.71–0.90	A atom has larger ionic radii than B atom
8.	$CaSnO_3$			
9.	$SrHfO_3$			

formula of strontium titanate ($SrTiO_3$), which has a perovskite structure, is ABO_3, where A and B are cations and O is an anion [7]. It has crystallized into a simple cubic perovskite structure under space group (pm3m) at room temperature and undergoes a phase transition from high to low symmetry [8].

$SrTiO_3$ has quantum paraelectric material and exhibits low dielectric loss, excellent tunability, and high dielectric constant at room temperature. Many attempts have been made to improve the electrical properties and microstructure of $SrTiO_3$ ceramics by doping simultaneously at the Sr site, Ti-site, or both sides. There are two types of doping: homovalent and heterovalent. In homovalent substitution, the dopant and host atom have the same valency, and no additional charge is generated to compensate for all-over charge neutrality. The second type of doping is heterovalent substitution. In this type of substitution, the dopant has valency other than the host atom. This is called donor-type substitution. If the dopant has lower valency than the host, it is called acceptor-type substitution. For example, the following charge compensation mechanism occurs for both types of doping [9].

$$M_2O_3 \xrightarrow{2SrTiO_3} 2M'_{Ti} + 2Sr^{x}_{Sr} + 3O^{x}_{O} \tag{2.1}$$

$2M'_{Ti}$ is the excess charge on a lattice is compensated by either creation of a hole or oxygen vacancies,

$$2M'_{Ti} = \left[2h^{\cdot}\right] = \left[V_{O}^{\cdot\cdot}\right] \tag{2.2}$$

$$M_2O_3 \xrightarrow{2SrTiO_3} 2Nd^{\cdot}_{Sr} + Ti^{x}_{Ti} + 3O^{x}_{O} \tag{2.3}$$

where $2Nd''_{Sr}$ is the excess charge on the lattice, which has been compensated by either creation of electron or cationic vacancies, $\left[2Nd^{\cdot}_{Sr}\right] = \left[2e'\right] = \left[V''_{Sr}\right]$

In the present case, V^{5+} is a donor-type substitution because the dopant valency is higher than the host (Ti^{4+}). $SrTiO_3$ has also been found as a high-K dielectric material for the possible application in electronic devices. A substance with a higher dielectric constant than that of SiO_2 is generally referred to as a high-K dielectric. Baek and colleagues have reported $SrTiO_3$ as a high-K material, while with doping of Al, the capacitance and dielectric constant decreased and the equivalent oxide thickness increased [10]. C. C. Wang and colleagues examined the dielectric properties of ceramic and single-crystal $SrTiO_3$. They also discovered benefits from oxygen vacancy clusters developing in a single crystal [10].

2.2 APPLICATIONS OF SRTIO$_3$

2.2.1 Energy Storage Devices

Systems that store energy in various ways, such as an electrochemical, electromagnetic, kinetic, potential, pressure, thermal, and chemical, include fuel cells, capacitors, batteries, flywheels, pumped hydro, compressed air, hydrogen, and super magnets [11]. The ability to store energy generated at one time for use at another time may help smooth out fluctuations in the supply and demand for electricity. Simply said, energy storage is the process of transforming energy from its original form into one that is easier or more convenient to store. The act of transforming electrical energy from a power grid into a form that can be stored and then converted back into electricity when it is required is known as energy storage (EES). By using this strategy, electricity can be generated when demand is low, when the cost of generation is low, or when there are no other options for generating electricity, and then used when demand is high, when the cost of generation is high, or when there are no other options for generating electricity. EES may be used for both portable and stationary power plants. An accumulator, often known as a battery, is a device used to store energy. Various forms of energy storage are shown in Figure 2.2.

2.3 ELECTROCHEMICAL PROPERTIES

Reinforcing phase nanoparticles were strontium titanate ceramic ($SrTiO_3$). $SrTiO_3$ is a complex oxide with a perovskite structure. Its large charge storage capacity, superior insulating characteristics, and outstanding optical transparency in the visible range have piqued the curiosity of researchers because of its potential application in electrical devices. $SrTiO_3$ was a popular substrate for epitaxial development of high-temperature superconducting films in the past. In addition to its usefulness in electrical devices as a high dielectric constant material, $SrTiO_3$ also has a band gap of 3.2 eV, making it a wide gap semiconductor. Quantum fluctuation dampen the ferroelectric transition in $SrTiO_3$, making it a ferroelectric material. Similar to the other perovskite oxides with the general formula ABO_3, strontium titanate crystallizes into a simple cubic lattice Pm3m with a lattice parameter of 0.3905 nm, with the strontium atoms occupying the corners, the oxygen atoms the faces, and the titanium atoms the bodies. The ease with which the oxygen octahedra may rotate around the central titanium atom opens the door to a wide variety of potential flaws in the ideal crystal. There are four TiO_6 octahedra around each Sr^{2+} ion, and the O_2 ions coordinate the Ti^{4+} ions in a sixfold fashion. Twelve oxygen disulfide (O^{2-}) ions act as coordinates for the strontium $(Sr)^{2+}$ ions. In addition to the ionic bonding characteristic between Sr^{2+} and O^{2-}, $SrTiO_3$ also has strong covalent bonding among the TiO_6 octahedra. This structure is due to chemical bonding. At a certain

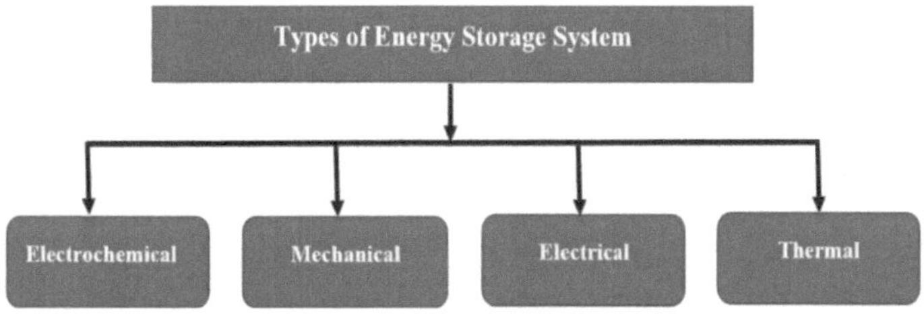

FIGURE 2.2 Various types of energy storage systems.

temperature, $T_c = 104$ K, $SrTiO_3$ undergoes an anti-ferro distortive phase change from a cubic to a tetragonal form. Electrocatalytic and conductivity findings for composites with doped strontium titanates that are suitable for SOFC applications have recently been published. Doped strontium titanate composite anodes showed excellent durability in oxygen, carbon, and sulfur-rich conditions. Several research teams have demonstrated that titanate-based anodes can oxidize CH_4 at high temperatures without the use of steam. Some *n*-type doped titanates have been shown to be dimensionally phase stable during redox cycling, and they are also exceptionally resistant to very high-sulfur (up to 1% H_2S) fuel atmospheres. The thermodynamic stability of $SrTiO_3$ in a solution of 10% H_2S, 3% H_2O, and 87% H_2 was validated by both theoretical and experimental studies. In spite of being heated to 950°F with sulfur for five days, the titanate did not react in any way. $SrTiO_3$-based anode performance is influenced by the temperature profile and calcination (sintering) conditions under which they were manufactured. However, the Sr-vacancy compensated Nb-doped $SrTiO_3$, with a nominal composition of $Sr_{0.94}Ti_{0.9}Nb_{0.1}O_3$, was evaluated for usage as an anode material in solid oxide fuel cells based on its redox stability and electrochemical properties (SOFCs). For the production of $Sr_{0.94}Ti_{0.9}Nb_{0.1}O_3$, a glycine-nitrate method with certain tweaks was developed. $Sr_{0.94}Ti_{0.9}Nb_{0.1}O_3$ and a composite electrode of $Sr_{0.94}Ti_{0.9}Nb_{0.1}O_3$/YSZ were analyzed using electrochemical impedance spectroscopy (EIS) on cone-shaped electrodes and on electrodes in a symmetrical cell configuration. Trial findings revealed that Nb-doped $SrTiO_3$ electrodes were redox stable and might be useful as a SOFC anode. It seems that the electrochemical activity around the triple phase boundaries was governed by the concentration of defect species (especially Ti^{3+} and $VO^{••}$). However, Nb-doped $SrTiO_3$-based materials lacked enough electrocatalytic activity; this property must be improved before they can be used as SOFC anodes. The perovskite-structured metal oxide $SrTiO_3$ has been proven to have several applications in microelectronics due to its high charge storage capacity, excellent insulating property, transparency, and chemical stability. A number of advantages may be gained by using the sol-gel process rather than more conventional synthetic strategies. It allows for consistent molecular mixing of reactants and may be used to control things like particle size, shape, and morphology. Applications as photo-catalysts and electro-active materials for $SrTiO_3$ nanoparticles (STNPs) manufactured using a sol-gel technique and calcined at 550°C have been studied by Deshmukh and coworkers. Pseudo-capacitance in nanoparticles of $SrTiO_3$ with a perovskite structure was reported in scientific literature. The chemical make-up, structure, shape, size, and crystallinity of $SrTiO_3$-the prototype perovskite-type oxide have a major impact on the material's physical properties. Electronics, photoanode for quantum dot-sensitized solar cells, and electro-ceramics are only a few of the many applications for strontium titanate, one of the most significant multifunctional perovskites. Dielectrics, thermistors, multilayer capacitors, electro-optical appliances, electromechanical devices, and field effect transistors are only some of the many applications for $SrTiO_3$ NPs. $SrTiO_3$ has a cubic perovskite structure and is an n-type semiconductor with an indirect band gap of 3.2–3.4 eV; however, recent studies have shown that this gap may be increased to 3.77 eV, leading to photon emission in the blue–green region of the spectrum.

One of the most exciting developments in electrode materials for batteries is the emergence of perovskite-based metal oxides. A large metal cation in a 12-coordinated habitat with oxygen ions and a smaller metal ion in a 6-coordinated environment with oxygen ions make up the usual formula for perovskite-based metal oxides, ABO_3. Titanate compounds include the often-encountered perovskite titanate strontium titanate ($SrTiO_3$). $SrTiO_3$ is a mechanically robust, chemically inert, high-melting-point material with a considerable nonlinear optical coefficient and dielectric constant (300). Although the $SrTiO_3$ has only been looked at once as an anode material for LIBs, that study boosted the reversible specific capacity to 24.8 mAh g^{-1} while retaining excellent charge-discharge stability. Even though $SrTiO_3$ as a cathode for RAZBs is a promising area of study, further investigation is required. $SrTiO_3$ may be synthesized in a number of different ways, including in a solid state, via a polymeric complex, by sol-gel combustion, or via hydrothermal synthesis. Because of its mild temperature (250°C), improved nucleation control, simplicity, and ability to produce highly crystalline material, the hydrothermal technique stands out as the most promising of these approaches.

Pellegrino and colleagues, for instance, produced $SrTiO_3$ cubes and polyhedral in the presence of polyethylene glycol (PEG) as a capping agent, which affects the shape of $SrTiO_3$. As a result of the increased control over nucleation that hydrothermal techniques provide, $SrTiO_3$ of precisely controlled shape and size may be synthesized for improved photochemical behavior in photocurrent applications. Single-crystal $SrTiO_3$ nanocubes of uniform dimensions were successfully synthesized by hydrothermal technique, demonstrating that high crystalline materials may be produced. There have been claims that $SrTiO_3$ can be synthesized using the hydrothermal route at temperatures between 40°C and 200°C from a variety of different precursors. Physical and chemical properties of $SrTiO_3$ are determined by its hydrothermal temperature. Thus, the RAZBs' efficiency was significantly affected by the hydrothermal temperatures. Hydrothermal treatment has also been linked to the formation of oxygen vacancies ($Vo^{\cdot\cdot}$). A boost in electrochemical performance can be attributed in part to the presence of $Vo^{\cdot\cdot}$, which plays a crucial role in ionic transport. $SrTiO_3$ has received much attention as an anode material for intermediate and low-temperature SOFCs. C.D. Savaniu and colleagues synthesized La-doped $SrTiO_3$ and reported it as anode materials with particle densities of 0.54 W/cm^2 at 750°C and humidified H_2 as fuel [12]. Jia and colleagues have reported La, Co-substituted $SrTiO_3$ samples as cathode materials for SOFCs [13]. They also demonstrated that by inserting La at the Sr-site and Co at the Ti-site, conductivity increased from 0.21 S/cm to 1.06 S/cm at 800°C [13]. Yousaf Shah and colleagues have reported Fe-doped $SrTiO_3$ as an electrolyte for low-temperature SOFCs; they discovered that Fe-doping in $SrTiO_3$ increases the concentration of oxygen vacancies, which improves ionic conductivity and oxygen ion transport [10]. Pseudocapacitive electrode behavior was observed in $SrTiO_3$ produced by Karthick and colleagues, who used a perovskite structure. Because of its chemical composition, structure, shape, size, and crystallinity, $SrTiO_3$ displays characteristic perovskite-type oxide physical properties. One of the most significant multifunctional perovskites, strontium titanate ($SrTiO_3$) is employed in the production of electronics, as a photo anode in quantum dot-sensitized solar cells, and as an electrode in electro-ceramics. Dielectrics, photocatalysis, thermistors, multilayer capacitors, electro-optical appliances, electro-mechanical devices, and field effect transistors are just a few of the many uses for $SrTiO_3$ perovskite materials. $SrTiO_3$ is an n-type semiconductor with a cubic perovskite structure with an indirect band gap ranging from 3.2 to 3.4 eV; however, newer research indicates an increase in the band gap to 3.77 eV, which results in photon emission in the blue-green spectral range. Recently, metal elements like iron doped into $SrTiO_3$ have led to a variety of changes, all of which are intimately tied to the valence state of a dopant and its connection to closest oxygen vacancies and might have applications in memristive memory.

Improvements in piezoelectric and dielectric responses may also be achieved by using an artificial super lattice made from two types of perovskites. There has been a lot of interest in the development of a $BaTiO_3/SrTiO_3$ (BT/ST) super lattice featuring a heteroepitaxial interface between the ferroelectric BT phase and the paraelectric ST phase. Observations of a horizontal strain inside the 2D BT/ST heteroepitaxial interface have been linked to the occurrence of the BT/ST super lattices, which results in large increases in dielectric constants and remnant polarization. The lattice mismatch between the BT and ST phases accounts for the 2.2% horizontal strain. Increasing the dielectric and piezoelectric responses is only one application of strain engineering, which is the process of purposefully inducing strain in the interfacial lattice. Ferroelectric materials may have their Curie temperature and residual polarization increased by interfacial lattice strain as well. When compared to MPB-derived strain, lattice strain resulting from a mismatch in the crystal lattice would be less sensitive to temperature. The BT-ST super lattice was created using molecular beam epitaxy (MBE). In spite of this, such a process is intricate and requires a very delicate substrate and a costly stacking structure.

2.4 MULTILAYER CERAMIC CAPACITORS

The capacitor of choice for applications requiring small-value capacitances is often a multilayer ceramic capacitor (MLCC). They are utilized as filters, op-amp circuitry, bypass capacitors, and more. One way to think of an MLCC is as a collection of several single-layer capacitors that have been packaged together. All MLCC chips begin with a blend of finely ground paraelectric or ferroelectric raw material granules that precisely chosen additions have altered. The mixture's composition and the size of the powder particles, which can be as fine as 10 nm, demonstrate the manufacturer's skill. The capacitance value will rise with thinner dielectric, more significant electrode areas, and higher-permittivity dielectric materials. The size, voltage requirement, and dielectric of MLCC chips affect capacitance (rated voltage).

Luckily, $SrTiO_3$ possesses all the properties required for an electrode of a capacitor [14].

The following are the advantages of MLCC:

- Comparing small parasitic inductance to aluminum electrolytic capacitors, the performance at high frequencies is improved.
- Improved temperature stability, depending on the temperature coefficient.

The following are the disadvantages:

- Small capacitance per volume, particularly for materials with class 1 dielectrics (NO/COG).
- Bias instability in DC.

2.5 PIEZOELECTRIC DEVICES

The piezoelectric effect is a phenomenon that mainly affects dielectric materials, such as quartz and tourmaline, where pressure-induced voltage results in the production of an electric charge. On the other hand, these dielectrics change size when voltage is applied, either expanding or contracting. Piezoelectric devices are passive gadgets that use the piezoelectric effect. Aside from gas stoves and lighters, a range of goods utilize piezoelectric components. For instance, facilities have been set up in several locations where the vibration and weight of a person walking may be transformed into energy via piezoelectric materials and utilized for door opening and shutting, as well as for intuitive nighttime illumination.

2.6 SENSORS

Any input from the physical world may be detected by a sensor, which then responds to it. Inputs include light, heat, motion, moisture, pressure, and other environmental events. Sensors are employed in commonplace items like touch-sensitive appliances. Sensors are employed in applications that most people are unaware of, like touch-sensitive elevator buttons (tactile sensors) and lights that may be dimmed or brightened by touching the base. The usage of sensors has moved beyond the conventional disciplines of temperature, pressure, and flow measurement. Types of sensors include, among others, chemical sensors, biosensors, neuromorphic sensors, MOS (metal oxide semiconductor) sensors, image sensors, and monitoring sensors. $SrTiO_3$ has been studied as an oxygen sensor by many researchers. Modified as well as pure $SrTiO_3$ are also used as gas sensors [15].

2.7 TRANSDUCERS

A transducer is an electrical device that converts energy from one form to another [16]. Energy is transformed from one form to another through the process of transduction. The precise amount of physical forces, such as temperature and pressure, is difficult to quantify. However, it is possible to turn these physical forces into electrical impulses and quantify them with a meter. Transducers' main job is to change a physical force into an electrical signal so that it can be measured and controlled more conveniently.

Advantages of using transducers are as follows:

1. Transmission and processing of electrical signals for measurement are simple.
2. Electrical impulses process fewer friction mistakes.
3. The electrical systems can be controlled with very little electricity.
4. Electrical signals are easily amplified and attenuated.

The device utilized to measure the electrical signal is exceptionally accurate and small. Recently many studies have been reported to improve the dielectric properties, specifically impendance and charge transport studies on strontium titanate–based ceramics using various doping elements of different concentrations. V_2O_5 and $SrTiO_3$ solid solution has great potential for electronic applications such as thermoelectric materials and anode material for solid oxide fuel cells, and they exhibit good dielectric permittivity and quality factors [17].

This chapter reviews vanadium-doped $SrTiO_3$ ceramics that were synthesized by conventional ceramic route followed by heat treatment at 1,300°C.

2.8 SAMPLE PREPARATION

Vanadium-doped $SrTi_{1-x}V_xO_3$ (x = 0.00, 0.02, 0.04) samples were synthesized using the solid-state method. The raw materials used for this synthesis were $SrCO_3$ (Aldrich sigma, 99.9%), TiO_2 (Alfa-Aesar, 99.99%), and V_2O_5 (Alfa Aesar, 99.9%). The stoichiometric quantity of raw materials was mixed for four hours in an agate mortar and pestle with acetone as the mixing medium. Then the samples were transferred to an alumina crucible for calcination for 12 hours at 800°C. The furnace was cooled to room temperature, and the calcinated powders were transferred into agate mortar and ground again for 30 minutes to remove the microstrain generated by heat treatment. Polyvinyl alcohol (PVA) of 2wt % was used as a binder. The dried calcinated powder was then used for making cylindrical pellets in a stainless-steel die by applying pressure of 7kN with the hydraulic press. The pellets formed with a diameter of ~12mm and thickness of ~1.3mm were again shifted to the furnace and sintered at 1,300°C for 6 hours. The sintered pellets were coated with high-quality conducting silver paste for dielectric measurements. The silver-coated pellets were fired at 500°C for 30 minutes. Figure 2.3 depicts the followed synthesis route.

2.9 CHARACTERIZATION TECHNIQUES

2.9.1 X-ray Diffraction

In addition to providing information on unit cell dimensions, X-ray diffraction is an analytical characterization technique that has been around for a while and is mostly utilized for phase identification of crystalline materials. The XRD pattern is produced by microscopic poly-crystals in powder samples that are randomly orientated in all directions. Coherently diffracting domains are those that exhibit diffraction in a single direction. The basis of X-ray diffraction is constructive interference between X-rays and crystalline material. This approach involves using a cathode ray tube to generate X-rays, filtering them to create monochromatic radiation, collimating it to concentrate it, and then directing it at the sample while collecting diffracted rays. Only when Bragg's law is met when incident rays interact with the sample do we see constructive interference. The lattice spacing and diffraction angle of the crystalline sample are linked to the wavelength of electromagnetic radiation by Bragg's equation [18]. The geometric interpretation of Bragg's law is shown in Figure 2.4. Currently, all possible diffraction directions of the lattice may be attained by simply scanning the sample through a range of angles (a range of 20 angles), thanks to the powdered material's random orientation.

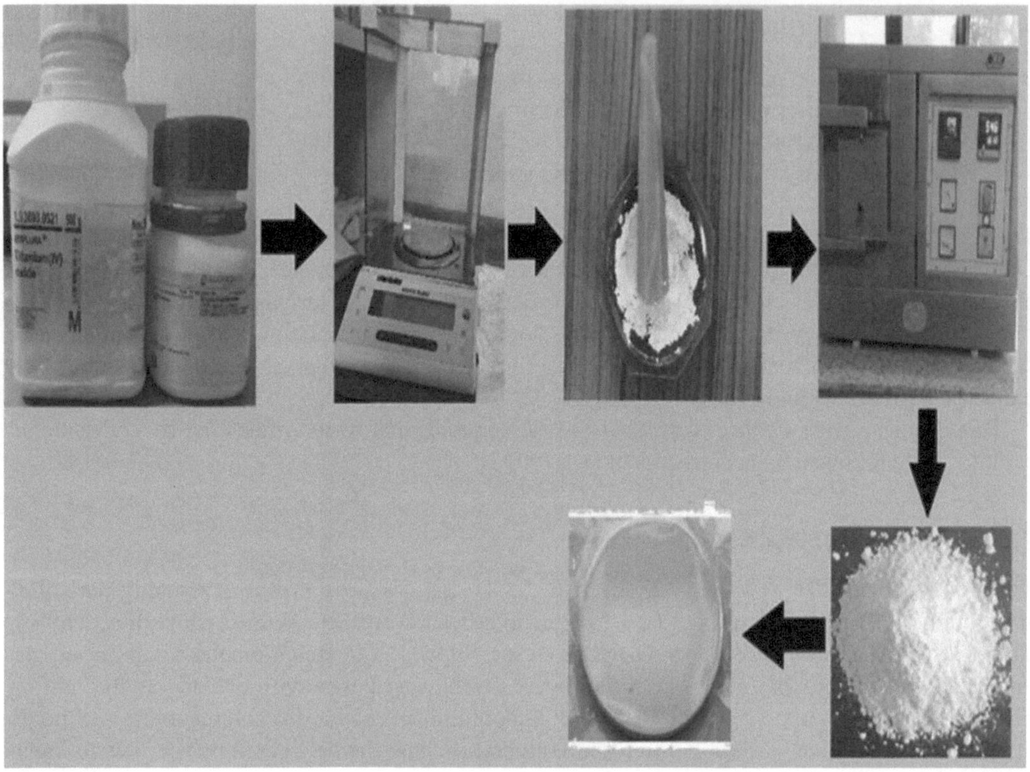

FIGURE 2.3 Steps involve in the synthesis of all STV samples using the solid-state reaction route.

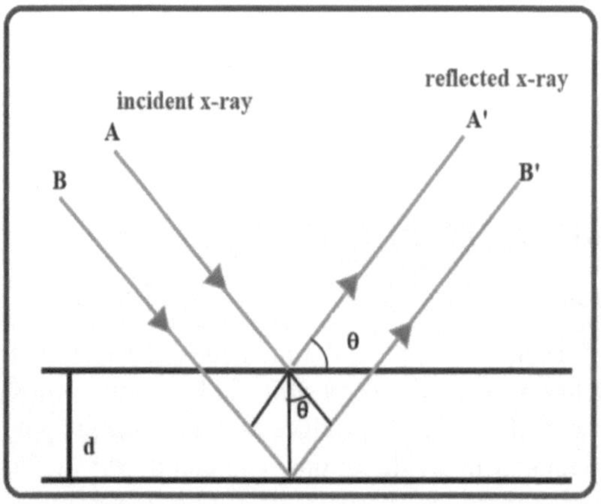

FIGURE 2.4 Description of Bragg's law.

2.9.2 Zeta Potential

Zeta potential refers to the electrostatic force that develops when a solid surface meets a liquid medium. To put it simply, the zeta potential is the surface charge of nanoparticles in solution. Understanding the nanoparticle surface's state is essential for figuring out a colloidal dispersion's long-term stability. Zeta potential is the name given to the electro-kinetic potential in colloidal dispersions in science. It is often represented in the literature on colloidal chemistry by the Greek

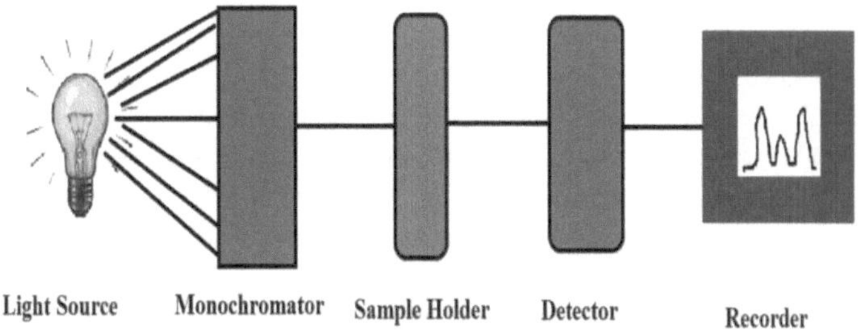

FIGURE 2.5 Components of UV visible spectroscope.

symbol zeta (ς). Volts (V) or millivolts (mV) are the accepted units. The zeta potential is the hypothesized electric potential between the site of the sliding plane in the interfacial double layer and a point in the bulk fluid distant from the interface. The stability of colloidal dispersions is promptly and effectively measured by the zeta potential [19]. The size of the zeta potential represents the strength of electrostatic attraction between charged particles in a dispersion. The surface charge of nanoparticles in a colloidal solution may be calculated by measuring their zeta potential. Insight into the factors that lead to colloidal particles, aggregation, or dispersion gained by its measurement may be utilized to enhance the manufacturing of dispersions, emulsions, and suspensions. A zeta potential value is practically useless without knowing the details of the solution. Measuring procedures are only appropriate when the measuring conditions are known. These factors include pH, buffer volume, temperature, and ionic strength.

2.9.3 UV Visible Spectroscopy

Ultraviolet (UV) visible spectroscopy is used to study the absorption and reflectance of a sample in the UV region and adjacent visible region. It is based on the phenomenon of electronic excitation [20]. When UV radiation is passed through the sample, atoms and molecules absorb energy and move from the ground to an excited state. A spectrum with a gap is formed when the residual radiation is passed through a prism. It is called the absorption spectrum [21]. Components of the UV visible spectrophotometer are shown in Figure 2.5. There are two types of light sources: (1) deuterium lamp, used when radiation is required in the UV spectrum (200–400 nm) and (2) tungsten halogen lamp, used when radiation is required in the visible spectrum (400–700 nm). Monochromators resolve a broad band of polychromatic light radiation to a narrow band of monochromatic radiation. There are three monochromators: (1) filter, (2) prism, and (3) grating. In addition, a cuvette made of quartz is used as a sample holder. The detector will detect the light energy and convert it to an electrical signal displayed on read-out devices. Generally, the photo-multiplier tube is used as a detector. Then amplifier is used to amplify the incoming signal from the detector and record them as displayed on the read-out device.

2.9.4 Electrical Properties

For studying the dielectric constant and dissipation factor's behavior with respect to temperature (40–300°C) at 1 kHz, the pellets were sintered at 1300°C for 6 hours at a rate of 5°C/ minute. The electrical behavior was investigated by utilizing two probe methods after electroding with a highly conductive silver paste on both sides of the pellets.

2.10 RESULTS AND DISCUSSION

In the present document, $SrTi_{1-x}V_xO_3$ ($x = 0, 0.02, 0.04$) would be denoted by STV-0, STV-2, and STV-4, respectively

2.10.1 X-ray Diffraction Analysis

$SrTi_{1-x}V_xO_3$ (x = 0, 0.02, 0.04) samples were recorded using X-ray diffraction at room temperature, as shown in Figure 2.6(a). The $SrTiO_3$ XRD patterns show diffraction peaks that agree with the JCPDS file no. 35-0734 from the joint committee on powder diffraction standards. As expected, spectra for $SrCO_3$, TiO_2, and V_2O_5 showed no peaks. This revealed the perfect incorporation of vanadium into the lattice. Moreover, the incorporation of dopant at the site of Ti in $SrTiO_3$ has been analyzed by enlarging the XRD peak (110) observed around 31.716°C in Figure 2.6(b). It has been observed that all peak positions on doping of vanadium shift toward a lower angle than in the pristine sample. This could be understood by the relationship of Bragg's angle with respect to interplanar spacing *d*. Bragg's diffraction law states that the interplanar spacing *d* is inversely proportional to Bragg's angle. Based on this observation, the unit cell must expand. Further, the ionic radii of V and Ti are 0.54 Å and 0.605Å, respectively, in the same coordination state. Then, according to this, if the dopant's ionic radii are lower than the host's, it can be concluded that the unit cell must contract. This contradicts the situation explained by Bragg's law. To determine crystallite size, the W-H formula was used. According to it, crystallite size is calculated using the following equation:

$$\left(\frac{\beta\cos\theta}{\lambda}\right)^2 = \frac{1}{D^2} + \left(\frac{4\,\varepsilon\sin\theta}{\lambda}\right)^2 \tag{2.4}$$

where β = width; θ = Bragg's diffraction angle; D = crystallite size; ε = micro strain; λ= 1.54Å, and k is the Scherer constant.

Using equation (2.4), the plots have been generated for all the samples depicted in Figure 2.7. The micro-strain was calculated using the slope of the linear equation, and the intercept determined the crystallite size. Numerical values of the results concluded from W-H plots are shown in Table 2.2. It has been noticed that the value of crystallite size and micro-strain is increased for STV-2, and after that it decreases for STV-4 compared to STV-2. The variation in crystallite size can be correlated with the ionic radii of dopant and then the host, surface defects, and processing parameters. Initially, the crystallite size was found to be increased, which might be due to higher ionic radii and surface defect, while further increase in the doping concentration reduces the crystallite sizes. Similar trend was observed for the variation in micro-strain.

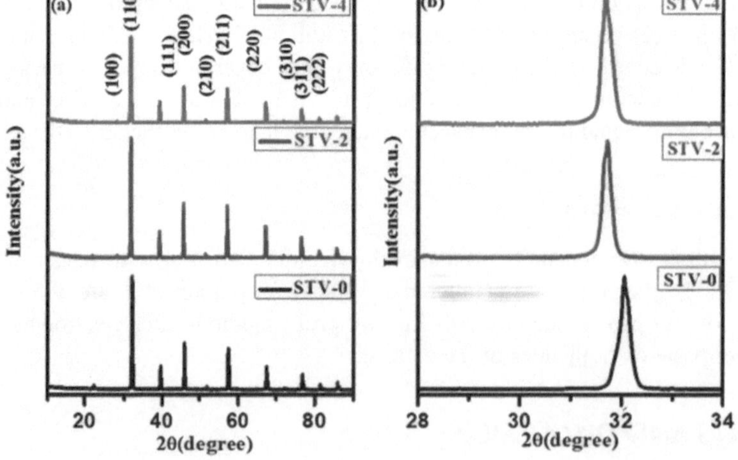

FIGURE 2.6 (a) Room-temperature powder X-ray diffraction pattern of all samples; (b) enlarged view of (110) peak.

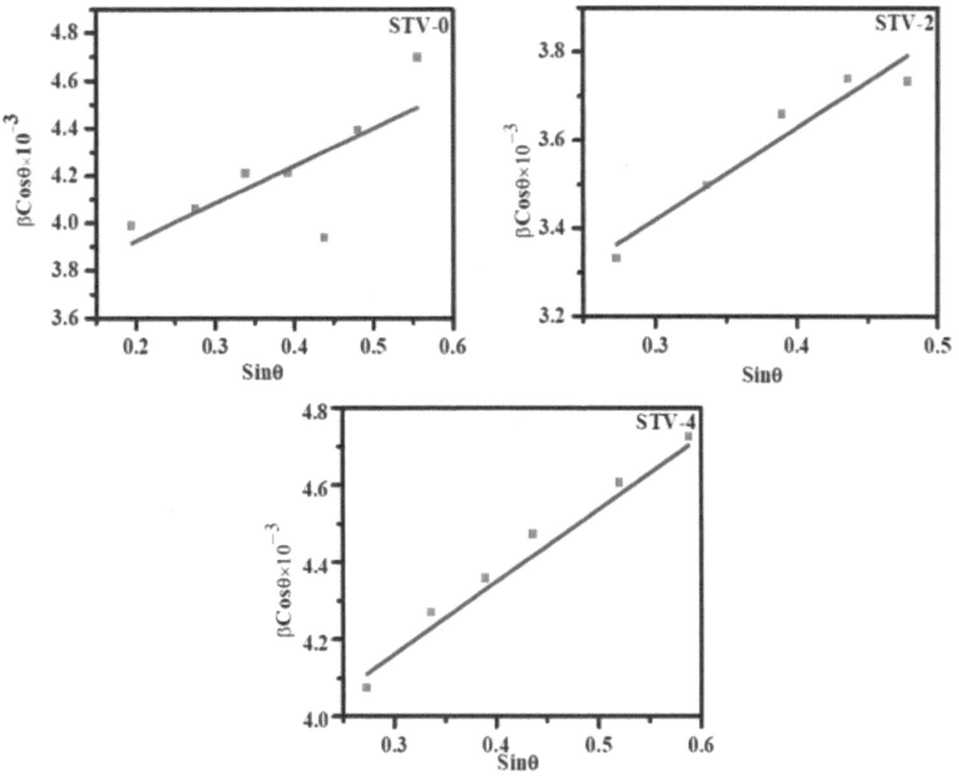

FIGURE 2.7 W-H plots of all the samples.

2.10.2 Zeta Potential Analysis

The zeta potential of a solid particle in a conducting liquid is the potential difference between the particle's surface and the bulk of the liquid. Nanoparticles may have an apparent surface charge, which will attract ions with opposite charges. The electro-kinetic potential of all the samples was measured by dispersing the powder pinch into 5 ml of distilled water solutions.

All samples' zeta potentials, hydrodynamic sizes, and dispersity indices (PdI) were calculated using zeta potential measurements. Figure 2.8 shows zeta potential, while Figure 2.9 shows the hydrodynamic size. An isoelectric point is a pH value at which the electro-kinetic potential is zero. Further, electrostatic repulsion is insufficient to counteract colloidal coagulation. The value of zeta potential ±30 mV is thought to be the limit value that regulates the stability of the dispersion in the water system. It was noticed that hydrodynamic size initially increases from un-doped to STV-2 as crystallite size increases in W-H and then is reduced for STV-4 compared to STV-2. Table 2.2 illustrates the numerical values obtained from the zeta analysis.

2.10.3 UV-Vis Spectroscopy Analysis

The ultraviolet-visible (UV-Vis) spectroscopy has been used to study the absorption and reflectance of sample in the ultraviolet to visible spectral regions. Optical band gaps of the sample have been calculated from the UV-Vis data. The absorption or reflectance in the visible range directly affects the perceived color of the chemicals involved. In this region of electromagnetic spectrum, the molecules undergo an electronic transition. It measures the intensity of light before it passes through the sample (I) and compares it to the intensity of light after it passes through the sample (I_0). The ratio (I/I_0) is called transmittance expressed in percentage (%T). However, the absorbance (A) is defined

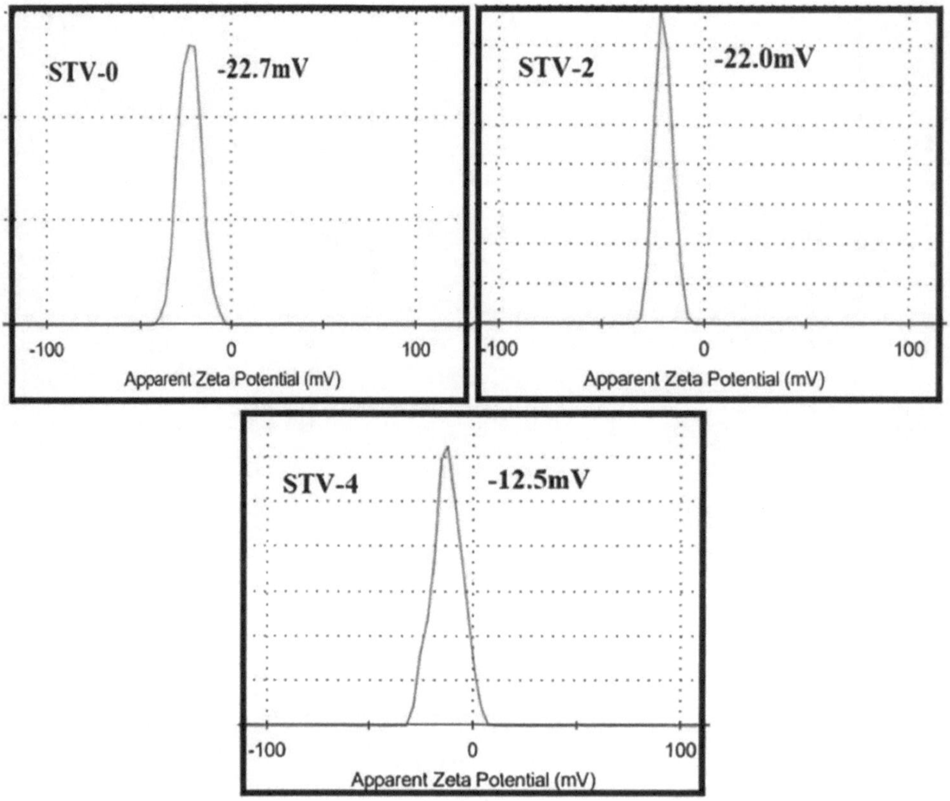

FIGURE 2.8 Zeta potential distribution of all samples.

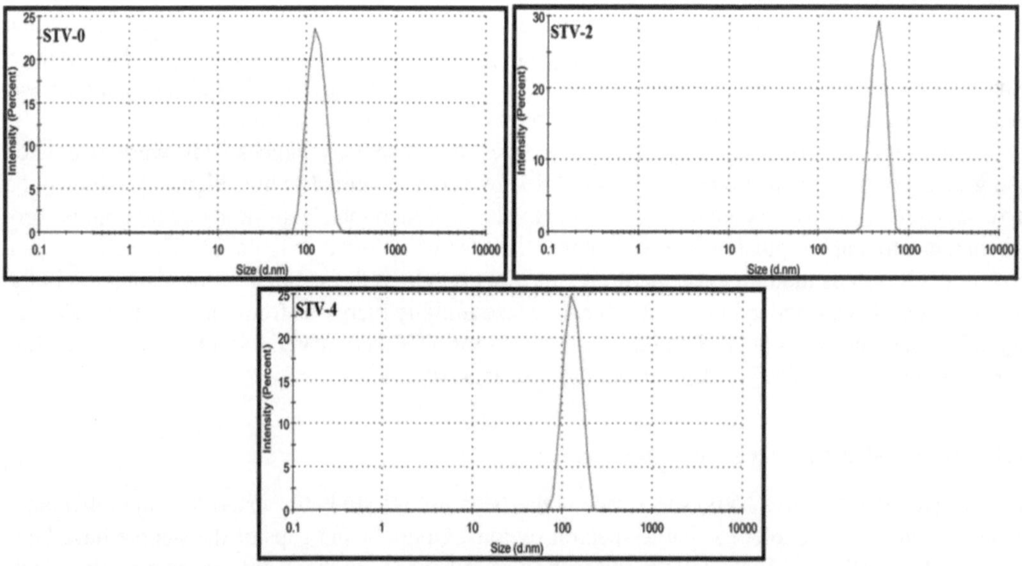

FIGURE 2.9 Hydrodynamic size distribution of all samples under dynamic light scattering.

TABLE 2.2
For All Samples, Values of Crystallite Size, Micro-Strain, Hydrodynamic size, Zeta Potential, PdI, and Direct and Indirect Band Gap

Materials	Crystallite Size (nm)	Micro Strain	Hydrodynamic Size (nm)	Zeta Potential (mV)	PdI	Direct Bandgap (eV)	Indirect Band Gap (eV)
STV-0	38.81	0.00158	130.9	−22.7	0.689	3.16	3.002
STV-2	50.22	0.00209	462.3	−20.0	0.480	3.202	3.07
STV-4	38.92	0.00188	130.3	−12.5	0.118	3.36	2.73

in terms of transmittance: $A = -\log\left(\frac{T}{100}\right)$. The UV-Vis spectrometer can also be used to measure reflectance. In case of reflectance mode, the spectrometer measures the intensity of light reflected from sample (I) to intensity of light reflected from a reference material (I_0). The ratio (I/I_0) is called reflectance expressed in percentage (%R).

In this chapter, to determine the optical band gap of the sample, the absorption data obtained from UV-Vis is correlated with the band gap (E_g) by the following equation:

$$(\alpha h\nu)^m = B(E - E_g) \tag{2.5}$$

where B is the independent temperature constant, $E = h\nu$ is the energy of the incident photon, and m is an index parameter, which gives information about the nature of the transition in the energy band diagram [22]. The value of both direct and indirect band gaps for all samples is displayed in Table 2.2 by condensing the linear component of the curve to the energy axis at $(\alpha h\upsilon)^{1/2}$, $(\alpha h\upsilon)^2=0$. The samples' onset absorption is determined by connecting the absorption to the wavelength axis. Figure 2.10(a) shows that the absorbance of vanadium-doped materials changes noticeably in the visible region with a slight shift in the UV area, while Figure 2.10(a), (b) and (c) are Tauc plots. The electronic transition between the 2p orbit of O and the 3d orbit of Ti may be responsible for the absorption peak [23]. The direct band gap ranges from 3.16 eV to 3.36 eV. Indirect band gaps are smaller than direct band gaps, which vary from 2.73 eV to 3.07 eV. The direct and indirect band gap values of produced samples decrease with dopant concentration. Both pure and doped samples exhibit semiconducting properties that are exploited in semiconductor device applications based on their optical values.

2.10.4 Dielectric Analysis

Different polarizations control the dielectric polarization of a material. This polarization includes interfacial, orientation, ionic, and electronic polarization [24]. The separation of mobile positively and negatively charged particles under an applied field forms positive and negative space charges in the bulk of the material or at the interfaces between different materials, known as interfacial polarization. In turn, these space charges alter the field distribution. Orientation polarization includes the molecules with permanent dipole moments orienting themselves in the direction of the applied electric field when an external electric field is applied. This is due to the external electric field's torque on each molecule's permanent dipole moment. Orientational polarization is the process of permanent dipole moments being oriented along the applied electric field axis [25]. Only those dielectric materials with ionic-bonded atoms in ionic polarization experienced ionic polarization. When such a material is exposed to an external electric field, positive and negative charges are separated across a greater distance than initially. At the same

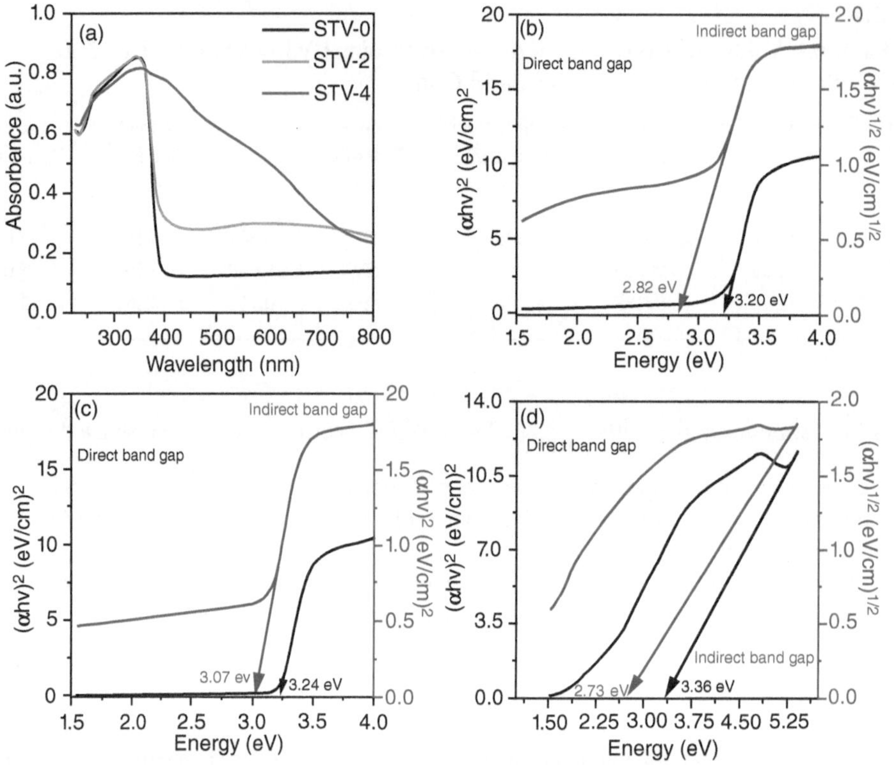

FIGURE 2.10 (a) UV-Vis. Absorption spectrum of all samples, Tauc plot for all samples generated for both direct and indirect band gap: (b) STV-0, (c) STV-2, (d) STV-4.

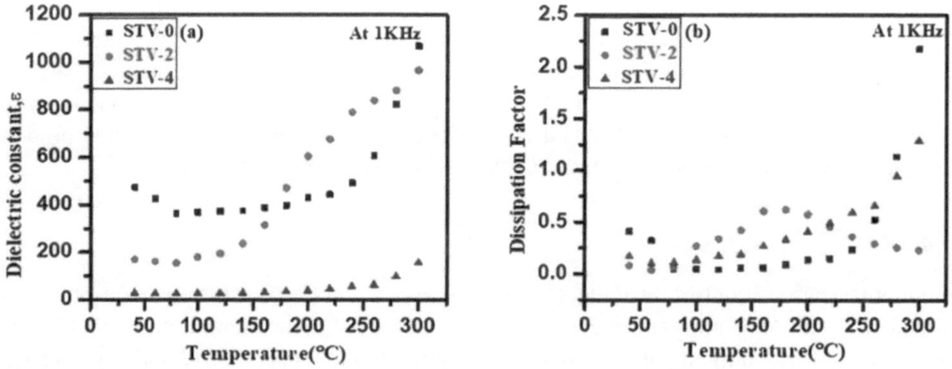

FIGURE 2.11 Variation of (a) dielectric constant and (b) dissipation factor with the temperature at 1 KHz frequency.

time, electronic polarization results from an external electric field that is applied to the bulk material, which causes an electric strain to be created in an atom [26]. It happens when an atom's negatively charged electrons and positively charged nucleus move in opposite directions. These types of polarization are active throughout a spectrum of frequency bands, such that the interfacial polarization active between 1 MHz and 1 kHz, orientational polarization is found to be active between 1 KHz and 1 MHz, ionic polarization is found to be active between 1 MHz and 1 GHz, and electronic polarization is found to be active above 1 GHz [27]. Within the temperature range of 40–300°C, the dielectric characteristics of each sample have been examined as a function of temperature at a fixed frequency (1 KHz). Figure 2.11(a) shows the variation of the dielectric constant with respect to temperature at 1 KHz. The dielectric constant is found to be increased with temperature for each sample. This increase is due to an increase in the number of dipoles. The presence of Ti^{2+} at Ti^{4+} performed as a negative defect and was denoted by ${Ti^{2+}}_{Ti^{4+}}''$, and oxygen vacancy $V_o^{\cdot\cdot}$ acts as a positive defect; the two defects are situated far away from each other and form an electric dipole $\left({Ti^{2+}}_{Ti^{4+}}'' - V_o^{\cdot\cdot}\right)$, responding in the presence of small AC field, which has significant a role on the variation of dielectric properties [28]. The interface, which separates the conducting regions, results in a high value of dielectric constant [29]. Figure 2.11(b) displays the fluctuation of the dissipation factor for all STV samples. In the case of STV-0 and STV-4, the dissipation factor continuously increases with temperature, but in the case of STV-2 sample, the dissipation factor increases up to 175°C and then decreases. The Koop's hypothesis can be used to explain the variance shown in the sample STV-2. According to Koop's theory, grain boundaries contribute significantly when resistivity is high, whereas grain contributes when resistivity is low.

The dissipation factor first increased to a specific temperature before decreasing to the maximum temperature; however, a transition peak separated these two variations. Since the present sample has been synthesized at 1,350°C, at this temperature the possibility of oxygen loss can't be ruled out. The presence of oxygen vacancy can be understood as follows:

$$O_o \rightarrow \frac{1}{2}O_2(g) + V_o^{\cdot\cdot} + 2e' \tag{2.6}$$

These electrons may be captured by Ti in $SrTiO_3$ and reduced to a lower valency state of multivalent ion as follows:

$$Ti^{4+} + 2e' \rightarrow Ti^{2+} \tag{2.7}$$

In a perovskite structure, the Sr occupied the dodecahedral site, oxygen vacancy occurred at the face-centered site, and Ti^{2+}/Ti^{4+} occurred at the body-centered site. The occupation of Sr^{2+}, Ti^{2+}/Ti^{4+}, and oxygen vacancy is an entirely random process, so the micro-regions having different atoms at their respective sites create micro-heterogeneity with varying conductivity, resulting in interfacial polarization. Further, the decrement in a dissipation factor can be attributed to the hopping of electrons through Ti sites in the grains.

2.11 FUTURE PROSPECTIVE AND CHALLENGES

2.11.1 Dielectric Capacitor

High-power energy storage applications have a high demand for dielectric capacitors with desirable characteristics such as a high charge/discharge rate, low cost, and extended life cycle. Dielectric capacitors' energy densities may be calculated from their polarization against electric field (P vs E) curves. Therefore, the most significant factors for assessing the dielectric energy storage capacitors are the materials' large P_{max}, low P_r, and high breakdown strength (BDS). Due to the energy density being proportional to the square of the applied electric field (on the premise that the permittivity is electric field independent), increasing BDS is an effective method of increasing energy density. However, at high electric fields, energy efficiency often declines owing to the higher leakage current, which creates more heat, hence speeding up the thermal breakdown. The increased BDS and high energy efficiency at high electric field are attributed to the huge band gap and low dielectric loss. Furthermore, owing to its low dielectric loss at high temperatures, the ceramic displays excellent thermal stability of energy efficiency (variation is less than 5%) across a temperature range from room temperature up to 200°C. These benefits show that ceramics based on SBT show promise as high-temperature energy storage materials.

2.11.2 Battery

Herein, we rationally constructed a heterostructure of hollow $SrTiO_3$ microsphere surrounded by reduced graphene oxide ($SrTiO_3$ @rGO) as the sulfur host to attach LiPSs and efficiently speed its conversion reaction kinetics, combining the benefits of graphene and polar $SrTiO_3$. The benefits of this kind of content are as follows. (1) The $SrTiO_3$ can adsorb LiPSs and effectively impede the LiPS shuttle through the ferroelectric effect, resulting in remarkable extended cycle performance. Good initial capacity and rate capabilities are achieved because (2) the electronic conductivity of the $SrTiO_3$ @rGO/S cathode is enhanced by the wrapped reduced graphene network. As a result, (3) the kinetics of the LiPSs conversion reaction are increased because the hollow structure of $SrTiO_3$ and wrapped rGO may increase the loading content of sulfur and efficiently accommodate the volume variations during the charge/discharge process.

2.11.3 Supercapacitor

Because of their huge surface area, and double-layer charge storage mechanism, supercapacitors manufactured from high-surface-area carbons are highly appropriate for the perovskite structure of strontium titanate ($SrTiO_3$). The dielectric constant is 300 in its cubic form at ambient temperature. Point defect chemistry is often used to characterize the conductivity mechanism of single-crystal $SrTiO_3$. We've settled on this metal oxide as the dielectric material in the composites because of its high dielectric constant and strong thermal stability (melting point 2,080°C). Both composites' thermal resilience is improved by the addition of $SrTiO_3$, allowing them to be used in high-temperature situations where graphene-polypyrrole composites would otherwise fail. The electrochemical analysis demonstrates that the composites have great potential as a supercapacitor electrode. Certain properties of the electrode materials used in supercapacitors are necessary for the generation of high capacitance and high-quality results. Research problems for the future include how to make supercapacitor electrodes with desirable properties including high capacitance, high power, easy procedures, cheap materials, and high energy density. The electrodes of supercapacitors are typically made of metal oxide or conductive polymer. The exceptional qualities of strontium titanate ($SrTiO_3$) made it an ideal candidate for use as an electrode material in supercapacitors. Paraelectric $SrTiO_3$ has a cubic structure at room temperature and may produce perovskite. Great dielectric constant, high thermal stability, high permittivity, minimal dielectric loss, and ultra-high charge density are some of the properties of $SrTiO_3$. As a result, $SrTiO_3$ ceramics may be put to good use in a number of prospective supercapacitor electrode applications. $SrTiO_3$ ceramics' dielectric characteristics will change if doping is introduced.

2.11.4 Fuel Cell

Redox instability, low sulfur tolerance, and carbon deposition from burning hydrocarbon fuels without enough steam are often identified as the primary issues with cermet materials. This has sparked research into potential replacement materials that might sidestep these issues without sacrificing performance. However, the mechanism of donor doping in $SrTiO_3$ remains unclear. Due to the close-packed structure of the perovskite anion sublattice, it is widely assumed that in oxidizing circumstances donors are compensated by cation vacancies rather than by absorption of extra oxygen. At low oxygen partial pressure, the compensating process changes to one involving electrons (often linked with the reduction of Ti^{4+} to Ti^{3+}), which makes for very good electrical conductivity. As the oxygen chemical potential and charge compensation mechanism shift, a material created at a certain PO_2 will need reformulation. To do this with $SrTiO_3$, a trace quantity of the second phase may need to be precipitated or dissolved, depending on the initial stoichiometry. However, when the temperature is lowered, these modifications may be kinetically constrained, especially if they entail cation diffusion.

2.11.5 Sensor

$SrTiO_3$ is appealing because of its high thermal and chemical stability, minimal dielectric loss, and large dielectric constant (300). Because of these benefits, the material may be used in a variety of settings, including sensors, actuators, electro-optical devices, random-access memory devices, and multilayer capacitors. Its most typical use is in oxygen sensors, but it also finds widespread usage in temperature sensors and as the cantilever foundation for a wide range of other sensors. Thanks to advancements in deposition technology, nanoscale strontium titanate is now a practical possibility. Researchers were able to get the operating temperature down to 40°C by shrinking the device. The authors hypothesize that cation vacancies are important not just at high temperatures but also in the creation of grain resistance boundaries. Because of mobility, the shift in the concentration of cationic vacancies across the interface may be confined to a very small number of monolayers on each side. For these temperatures, strontium vacancies are thought to be essentially immobile across long distances, hence the concentration of strontium vacancies only varied by one cell on each side of the border. Therefore, in this research, we solely considered electrons and strontium vacancies, two types of point defects. When taken along with the dense interface between the defect state and the surrounding material, this explains the dramatic shift in sensor resistivity. The authors suggested a point defect model in which a space charge region forms close to the phase border, greatly affecting the equilibrium of local defects. This means that the total value has an outsized impact on the prevalence of each defect type. The quick response of the sensor may be due to the space charge inducing a higher concentration of the cationic vacuum close to the contact. High electron depletion is due to the presence of cationic voids.

2.12 CONCLUSIONS

All the samples have been successfully synthesized via the solid-state ceramic route. The X-ray diffraction pattern shows that all samples were in a single phase and crystallized into a cubic structure. W-H plots were employed to evaluate crystallite size and micro-strain. Zeta potential values show that the surface of the material is negatively charged. Indirect and direct band gaps were calculated using the Tauc relation and found to exhibit semiconducting properties. The electrical properties were studied in the temperature range of 40–300°C at a fixed frequency (1 KHz). The dielectric constant is found to be increased with temperature for each sample. This increase can be attributed to an increase in the number of dipoles. In samples STV-0 and STV-4, the dissipation factor increased with temperature, but in the STV-2 sample, the dissipation factor increased to 175°C and then a decrement was observed due to hopping of electrons through Ti sites in the grains. The results lead to a conclusion that this material could potentially be used as an electrode material in SOFC, memory devices, and supercapacitors.

ACKNOWLEDGMENT

The authors thank the Head Department of Physical Sciences and Head Department of Pharmacy, Banasthali Vidyapith, for providing the experimental facilities required to characterize the synthesized samples. Upendra Kumar enthusiastically acknowledges the DST-SERB through a project (File No: DST-SERB/EEQ/2021/000132).

REFERENCES

[1] S. Nayak, B. Sahoo, T.K. Chaki, D. Khastgir, Facile preparation of uniform barium titanate ($BaTiO_3$) multipods with high permittivity: Impedance and temperature dependent dielectric behavior, *RSC Adv.* 4 (2014) 1212–1224. https://doi.org/10.1039/c3ra44815k

[2] R. Tararam, I. Bdikin, J. Varela, P.R. Bueno, A.L. Kholkin, Local electromechanical properties of $CaCu_3Ti_4O_{12}$ ceramics, *MRS Proc.* 1255 (2010). https://doi.org/10.1557/proc-1255-m03-19

[3] A. Manikandan, Y. Slimani, A. Dinesh, A. Khan, K. Thanrasu, A. Baykal, S.K. Jaganathan, H. Dzudzevic-Cancari, A.M. Asirid, *Perovskite's potential functionality in a composite structure*, Elsevier Ltd., 2020. https://doi.org/10.1016/B978-0-12-819977-0.00008-1

[4] S.A. Kandjani, S. Mirershadi, A. Nikniaz, Inorganic–Organic perovskite solar cells, *Sol. Cells - New Approaches Rev.* (2015). https://doi.org/10.5772/58970

[5] J.N. Wilson, J.M. Frost, S.K. Wallace, A. Walsh, Dielectric and ferroic properties of metal halide perovskites, *APL Mater.* 7 (2019). https://doi.org/10.1063/1.5079633

[6] H. Cheema, S. Kumar, P.A. Alvi, B.L. Choudhary, U. Kumar, Synthesis and physical properties of nanopowder and electrical properties of bulk samples of $ZnFe_{2-x}NixO_4$ (x: 0, 0.05, 0.10), *Adv. Powder Technol.* 31 (2020) 4241–4252. https://doi.org/10.1016/j.apt.2020.09.001

[7] F. Alarab, J. Minár, P. Šutta, L. Prušáková, R. Medlín, O. Heckmann, C. Richter, K. Hricovini, Study and characterization of $SrTiO_3$ surface, *AIP Conf. Proc.* 1996 (2018). https://doi.org/10.1063/1.5048853

[8] V. Yadav, H. Cheema, R.S. Maurya, S. Kumar, P.A. Alvi, M. Sharma, U. Kumar, Study of structural, optical, dielectric, and electric properties of homovalently substituted Ce in $SrTiO_3$ perovskite oxide, *Ionics (Kiel).* 28 (2022) 5513–5524. https://doi.org/10.1007/s11581-022-04782-2

[9] U. Kumar, S. Upadhyay, Structural, microstructure, optical, and dielectric properties of $Sr_{1.99}M_{0.01}SnO_4$ (M: La, Nd, Eu) Ruddlesden–Popper oxide, *J. Mater. Sci. Mater. Electron.* 31 (2020) 5721–5730. https://doi.org/10.1007/s10854-020-03140-0

[10] M.A.K.Y. Shah, S. Rauf, N. Mushtaq, Z. Tayyab, N. Ali, M. Yousaf, Y. Xing, M. Akbar, P.D. Lund, C.P. Yang, B. Zhu, M.I. Asghar, Semiconductor Fe-doped $SrTiO_{3-\delta}$ perovskite electrolyte for low-temperature solid oxide fuel cell (LT-SOFC) operating below 520 °C, *Int. J. Hydrogen Energy.* 45 (2020) 14470–14479. https://doi.org/10.1016/j.ijhydene.2020.03.147

[11] S.S. Reddy, C.M. Jung, Overview of energy storage technologies: A techno-economic comparison, *Int. J. Appl. Eng. Res.* 12 (2017) 12872–12879.

[12] C.D. Savaniu, J.T.S. Irvine, La-doped $SrTiO_3$ as anode material for IT-SOFC, *Solid State Ionics.* 192 (2011) 491–493. https://doi.org/10.1016/j.ssi.2010.02.010

[13] R. Li, C. Zhang, J. Liu, J. Zhou, L. Xu, A review on the electrical properties of doped $SrTiO_3$ as anode materials for solid oxide fuel cells, *Mater. Res. Express.* 6 (2019). https://doi.org/10.1088/2053-1591/ab4303

[14] Y. Fukuda, K. Aoki, K. Numata, A. Nishimura, Temperature dependence of dielectric absorption current of srtio3thin-film capacitor, *Jpn. J. Appl. Phys.* 34 (1995) L1291–L1292. https://doi.org/10.1143/JJAP.34.L1291

[15] S.K. Hodak, T. Supasai, A. Wisitsoraat, J.H. Hodak, Design of low cost gas sensor based on $SrTiO_3$ and $BaTiO_3$ films, *J. Nanosci. Nanotechnol.* 10 (2010) 7236–7238. https://doi.org/10.1166/jnn.2010.2800

[16] J. Christenson, Sensors and transducers, *Handb. Biomechatronics.* 132 (2018) 61–93. https://doi.org/10.1016/B978-0-12-812539-7.00003-9

[17] S. Singh, R. Pandey, S. Presto, M.P. Carpanese, A. Barbucci, M. Viviani, P. Singh, Suitability of Sm3+-Substituted $SrTiO_3$ as anode materials for solid oxide fuel cells: A correlation between structural and electrical properties, *Energies.* 12 (2019) 1–16. https://doi.org/10.3390/en12214042

[18] A.A. Bunaciu, E. Gabriela Udriştioiu, H.Y. Aboul-Enein, X-ray diffraction: Instrumentation and applications, *Crit.Rev.Anal.Chem.* 45 (2015) 289–299. https://doi.org/10.1080/10408347.2014.949616

[19] H. Cheema, V. Yadav, R.S. Maurya, V. Yadav, A. Kumar, N. Sharma, P.A. Alvi, U. Kumar, Structural, optical and electrical properties of Mn-doped $ZnFe_2O_4$ synthesized using sol–gel method, *J. Mater. Sci. Mater. Electron.* 32 (2021) 23578–23600. https://doi.org/10.1007/s10854-021-06847-w

[20] A.F. Hussain, *UV-Visible Spectrometry*, 4 (2019).

[21] S. Mkata, UV vis spectroscopy practical, *Phys. Prakt.* I. (2017) 1–11. https://www.coursehero.com/file/76918963/UVVis-HS17pdf/

[22] A. Kumar, B. Khan, V. Yadav, A. Dixit, U. Kumar, M.K. Singh, Rietveld refinement, optical, dielectric and ac conductivity studies of Ba-doped $SrSnO_3$, *J. Mater. Sci. Mater. Electron.* 31 (2020) 16838–16848. https://doi.org/10.1007/s10854-020-04240-7

[23] C.H. Lee, N.J. Podraza, Y. Zhu, R.F. Berger, S. Shen, M. Sestak, R.W. Collins, L.F. Kourkoutis, J.A. Mundy, H. Wang, Q. Mao, X. Xi, L.J. Brillson, J.B. Neaton, D.A. Muller, D.G. Schlom, Effect of reduced dimensionality on the optical band gap of $SrTiO_3$, *Appl. Phys. Lett.* 102 (2013) 1–6. https://doi.org/10.1063/1.4798241

[24] U. Kumar, K. Ankur, D. Yadav, S. Upadhyay, Synthesis and characterization of Ruddlesden-Popper system $(Ba_{1-x}Srx)2SnO_4$, *Mater. Charact.* 162 (2020) 110198. https://doi.org/10.1016/j.matchar.2020.110198

[25] F. Kremer, W.K. Kipnusu, M. Fränzl, Orientation polarization spectroscopy—Toward an atomistic understanding of dielectric relaxation processes, *Int. J. Mol. Sci.* 23 (2022). https://doi.org/10.3390/ijms23158254

[26] W.B. Fowler, Influence of electronic polarization on the optical properties of insulators, *Phys. Rev.* 151 (1966) 657–667. https://doi.org/10.1103/PhysRev.151.657

[27] H. Cheema, V. Yadav, R. Sunder, S. Kumar, P.A. Alvi, U. Kumar, Materials Today: Proceedings (0 .16, 0 .20) synthesized using solid state ceramic route, *Mater. Today Proc.* 3 (2022). https://doi.org/10.1016/j.matpr.2022.06.349

[28] R. Moos, K.H. Härdtl, Defect chemistry of donor-doped and undoped strontium titanate ceramics between 1000° and 1400°C, *J. Am. Ceram. Soc.* 80 (1997) 2549–2562. https://doi.org/10.1111/j.1151-2916.1997.tb03157.x

[29] S. Singh, P. Singh, O. Parkash, D. Kumar, Synthesis, microstructure and electrical properties of Ti doped $SrSnO_3$, *Adv. Appl. Ceram.* (2007). https://doi.org/10.1179/174367607X202573

3 Ferrite and Perovskite-based High-K Dielectric Materials

Synthesis, Characterization, and Application

Harshpreet Cheema, Vedika Yadav, Parvej Ahmad Alvi, and Upendra Kumar

3.1 INTRODUCTION

Energy is considered to be a pillar of strength for national and economic growth and development. However, due to rapid urbanization and industrialization, energy consumption has increased at such a pace that energy production often struggles to meet demand [1]. The increased gap between the rate of energy production and consumption has put tremendous pressure on conventional energy sources [2]. With the ever-growing population and the improvement in living standards throughout the world, global energy demand is likely to rise even more significantly in the near future [3]. In India, maximum power generation is made via conventional energy sources, such as mineral oil- and coal-based power plants, that contribute massively to greenhouse gas emissions and pollution affecting the environment [4]. Figure 3.1 represents various available energy sources. Conventional energy sources have been used for a long time and have limited availability. Based on finite, nonrenewable resources, they are projected to be severely depleted in a relatively short time [5, 6]. In the meantime, the renewable energy sources such as wind, hydro, and solar are not currently at the level of technology what would allow to produce energy to meet consumption rates. Modern civilization's increased energy needs exert tremendous pressure on our natural resources and contribute to pollution. The need of the hour is to check the degradation and depletion of our precious natural resources without halting the development process [7]. The future of our planet and that of its people depend on our ability to maintain and sustain the life support system nature is providing for us. It is, therefore, our first and foremost duty to make efforts to utilize and sustain renewable resources and minimize the damage to the natural environment by changes in consumption patterns. We should also try to improve the quality of human life through cooperation among all nations. Poverty should ideally be eradicated to reduce the gap between developed and developing nations. Communities need to be educated to protect and care for their own environment. We should adopt an environmentally friendly strategy of economic growth that conserves our earth's diversity and vitality [8]. Hence, there is an urgent need to motivate people to switch from conventional energy sources to renewable energy sources that include solar, biomass, geothermal, wind, and ocean to minimize pollution and global warming and move the world toward a cleaner and healthier energy future [9].

The energy crisis in the future can be tackled through the judicious utilization of abundant renewable energy resources. There are great innovative ways available to store energy. Energy storage can be classified into the following categories:

(a) **Thermal**: During summers, thermal energy is available abundantly. Underground thermal energy storage is the best technology by which thermal energy can be stored on an enormous scale [10]. Underground thermal energy storage can be classified as closed-loop and

DOI: 10.1201/9781003352365-3

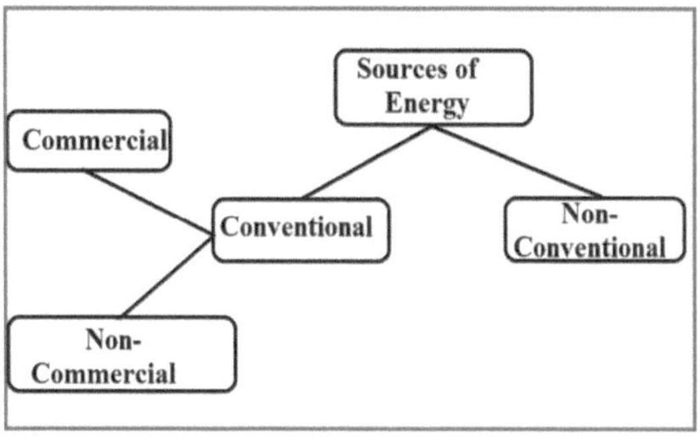

FIGURE 3.1 Different types of energy sources and their classification.

open-loop systems. In closed-loop systems, the working fluid is circulated through heat exchangers in the ground. In open-loop systems, groundwater is pumped out and forced back into the ground using wells after the extraction of thermal energy from it. Borehole thermal energy storage, tank thermal energy storage, and pit thermal energy storage are the most common systems for storing thermal energy [11].

(b) **Mechanical**: This simple method includes energy storage by applying force to the medium that results in displacement. The most commonly used mechanical energy storage systems include flywheels, compressed air energy storage, and pumped hydro storage systems [12]. In flywheel, ceramic superconducting materials are used. Compressed air storage systems involve the compression of air and its storage in an underground cavern [13]. A large amount of energy over a long time can be stored through pumped hydro storage systems. Its efficiency is also high (around 70–85%).

(c) **Electrical**: An electrical energy storage system is the conventional approach to building sustainable energy technology. Various daily usage gadgets and appliances depend on electrical energy, so the means to store electrical energy is essential. Primary examples of storing energy electrostatically are capacitors, supercapacitors, and superconducting magnets. In capacitors, energy is stored between two plates separated by a dielectric. A supercapacitor is a more advanced capacitor design and stores more energy than a capacitor [14]. It stores, on average, almost 100 times more energy than a standard capacitor. Energy in a supercapacitor is stored by polarizing an electrolytic solution. When voltage is applied to a positive electrode, it attracts negative ions from an electrolyte; when applied to a negative electrode, it attracts positive ions from the electrolyte. These ions are stored near the electrode surface, and because of this, the distance between the electrodes decreases. This, in turn, results in a large capacitance. Superconducting magnets store energy within a magnet in the magnetic field generated by the direct current when the superconducting coil is cooled below the critical temperature. Examples of superconducting materials are vanadium and mercury [15].

(d) **Electrochemical**: This category includes batteries that store chemical energy and release it in electrical form. Rechargeable batteries consist of one or more electrochemical cells. Rechargeable batteries have a comparatively lower cost of use than disposable batteries [16]. Some of the common examples of rechargeable batteries are:
- Lead-acid batteries
- Ni-Cd batteries
- Lithium-ion battery
- Lithium-ion polymer battery

(e) **Chemical**: Energy can also be stored in chemical form. Chemical energy storage is defined as storage of potential energy in the arranged way of atoms [17]. The best example of energy storage in this category are fossil fuels. They are naturally occurring hydrocarbons found in the earth's crust, formed through a biochemical process of decomposition of organic matter over the millennia of the planet's existence. In fossil fuels, the energy is stored as a form of the hydrogen and carbon bond. Various mining techniques extract these fossil fuels from the earth's crust, and the energy stored in them is utilized for transportation, electricity generation, etc. [18]. Selecting the best material for energy storage system applications has been the prevailing problem for decades. Scientists have made countless efforts and successfully studied various materials' physical properties to solve this mystery [19].

Advanced dielectric materials with high K are considered a hot topic of today's research into energy storage because the material with high K can store more charge. Dielectric materials are divided into two subcategories: (1) linear dielectric systems and (2) nonlinear dielectric systems. Nonlinear dielectrics are further subcategorized into (1) paraelectric, (2) ferroelectric, (3) anti-ferroelectric, and (4) relaxor ferroelectric. It has been discovered that linear dielectric materials have less energy loss, good thermal stability, and high breakdown strength, so they are considered a good option for energy storage applications. $CaTiO_3$ (with dielectric strength of 4.2 $mVcm^{-1}$ [20], dielectric constant of 175, and breakdown strength of 300 $Kv\ cm^{-1}$), $CaZrTiO_3$, $SrTiO_3$, and $CaSrTiO_3$ are some linear dielectric materials that are studied for their applications in energy storage systems. In comparison to linear dielectrics, nonlinear dielectrics have more ability to get polarized. Thus, the focus of energy storage research has shifted to nonlinear dielectrics.

Ferroelectrics have spontaneous polarization, and the dipoles are aligned in the direction of the applied electric field and have high polarization (P_{max}), high dielectric constant, and acceptable breakdown strength. So, ferroelectrics can be used in energy storage devices. Among the ferroelectrics, barium titanate is investigated more often because its high P_{max} value makes it a suitable candidate for energy storage applications [21].

In the case of anti-ferroelectrics, initially there is zero spontaneous polarization. Still, when a powerful external field is applied, a phase transition occurs from anti-ferroelectric to ferroelectric, so it can be rightly said that anti-ferroelectrics also possess good energy storage properties because of the existence of a double hysteresis loop. Suppose the comparison is made between the energy storage capacities of ferroelectrics and anti-ferroelectrics. In that case, surprisingly, anti-ferroelectrics have more energy storage ability because of lower residual polarization, less dielectric loss, double hysteresis loop, lower coercive speed, relatively efficient discharge speed, moderate breakdown strength, and high energy efficiency. $PbZrO_3$, $PbHfO_3$, and $AgNbO_3$ are some materials that exhibit anti-ferroelectric properties at room temperature. Relaxor ferroelectrics also have spontaneous polarization, but this spontaneous polarization decreases slowly above a certain temperature. The narrow hysteresis loop, lower remanent polarization, and high P_{max} value also contribute to make relaxor ferroelectrics a suitable choice for energy storage. The present work, discussed in this chapter, focuses on $BaTiO_3$ and $ZnFe_2O_4$ because of their exceptional energy storage capabilities.

3.2 ROLE OF $ZNFE_2O_4$ IN ENERGY STORAGE APPLICATIONS

The ferrites were generally defined by the chemical formula AFe_2O_4, where A can be a divalent ion such as Zn^{2+}, Co^{2+}, Ba^{2+}, Sr^{2+}, Ca^{2+}, and so on [22]. The ferrites were crystallized into a cubic spinel group with the space group $Fd\bar{3}m$. Among the studied ferrites, $ZnFe_2O_4$ has shown exciting properties due to the zinc's diamagnetic nature and iron's ferrous behavior. The presence of both Fe^{2+} and Fe^{3+} ions allows the material to exhibit diamagnetic and ferrous properties, respectively, which results in a highly chemically stable structure [23–25]. In the $ZnFe_2O_4$ structure, the octahedral sites have higher crystal field splitting energy (CFSE) than the tetrahedral sites while both the cations Zn^{2+} and Fe^{3+} possessed zero, and Fe^{2+} ions have a higher value of CFSE [26]. Therefore, it is essential to research $ZnFe_2O_4$ because the cationic occupancy of all cations Zn^{2+}, Fe^{3+}, and Fe^{2+} in the material, for both tetrahedral and octahedral sites, are mostly random.

Electrochemical properties of pure and modified $ZnFe_2O_4$ are considered excellent for various processes such as oxidation, degradation, and reduction of hydrogen peroxide and carbon-related gases [27, 28]. The fuel cell has enormous potential in energy storage systems to convert chemical fuel energy into electrical energy. Intermediate temperature solid oxide fuel cell includes three main components: cathode, anode, and electrolyte, working from 500°C to 1,000°C. The cathode and anode materials of IT-SOFCs must be porous and possess electronic and hole conductivity, while electrolytes possess good ionic conductivity [19]. To improve the ionic conductivity of electrolytes below 600°C, IT-SOFCs have been a challenge for researchers over the last few decades. Various types of doping can enhance the ionic conductivity of spinel ferrites. The modification in the valence states of Fe^{3+} by a different type of substitution modified the catalytic properties of spinel $ZnFe_2O_4$ [19, 29]. However, the modification of divalent ions at the Fe^{3+} site of $ZnFe_2O_4$ has been reported to be an excellent ionic and electronic conductor for mixed ionic and electronic conductor applications in IT-SOFCs [30]. Also, in the area of storage devices, $ZnFe_2O_4$ has been studied as a promising electrode material for supercapacitor applications because of its electronic conductivity, high porosity, high thermal stability, and higher surface reactivity. Araveeti Eswar Reddy and colleagues reported that $ZnMn_2O_4$@$ZnFe_2O_4$ microspheres fabricated on Ni foam possess high specific capacitance of about 1024.66 F/gm at current density 10 mA/cm^2 [31]. M. Vadiyar and colleagues reported that the single phase of $ZnFe_2O_4$ nanoflake possesses a specific capacitance of 781 F/gm, while the nanoflake of $ZnFe_2O_4$/C demonstrates an excellent specific capacitance of 1,884 F/gm at current density 5 mA/cm^2 [32, 33]. Therefore, the literature survey suggests that the modified $ZnFe_2O_4$ possesses excellent electrochemical properties favorable for supercapacitor application.

3.3 ROLE OF $BaTiO_3$ PEROVSKITES IN ENERGY STORAGE APPLICATIONS

The perovskite oxides were also studied for many technological applications like energy storage devices, energy harvesters, renewable energy sources, etc. The perovskite oxides were generally defined by a chemical formula ABO_3, where the valency of A and B cations are taken in such a way that it compensates the overall charge of oxygen, for example, ($A^{1+}B^{5+}$, $A^{2+}B^{4+}$, and $A^{3+}B^{3+}$) [34]. Among the perovskites, the $BaTiO_3$ was found to be a multifunctional material and used in many electronic devices, gas sensors, piezoelectric ceramics, ferroelectric, relaxers, etc. [35]. It mainly crystallized into cubic and tetragonal crystal structures, among which the tetrahedral phase was reported as multifunctional. The crystal structure of $BaTiO_3$ consists of dodecahedral sharing of Ba^{2+} at the corners of the cube, Ti^{4+} as the body-centered position, and shared by six oxygen atoms available at the faces of a cube. The bond length of Ti-O1 (along axial) was lower than Ti-O2 (along apical), making it ferroelectric due to the presence of non-centrosymmetric.

3.3.1 Multilayer Ceramic Capacitors (MCLs)

Generally, a capacitor consists of two parallel metal plates separated by a free space. When a voltage is applied across these metal plates, a charge is produced on them directly related to the applied voltage. If the insulating material is placed in between the plates, the charge produced on it increases by relative dielectric constant. The high value of barium titanate's dielectric constant makes it a suitable choice for its use in a capacitor. Zhang and colleagues prepared the nanocomposites of cellulose and $BaTiO_3$ and found they exhibit a high energy density of 13.14 J/cm^3 [36]. Wen-Bo Li and colleagues prepared the composite of $BaTiO_3$ and $Bi(Li_{0.5}Nb_{0.5})O_3$ and found it to have outstanding energy storage performance for high-temperature capacitor applications [37].

3.3.2 Data Storage Devices

The ferroelectric behavior of $BaTiO_3$ makes it useful for data storage devices because of its spontaneous polarization behavior, which means that on the application of an electric field, maximum polarization is achieved. At the same time, it gives the lowest value of polarization as the electric field is removed from the system. These ferroelectric properties can be used for data storage in computer hard drives. Shen and coworkers prepared the solid solution of $BaTiO_3$ and $BiYbO_3$ and found it to be an excellent data storage material [38]. Yang and colleagues reported that the novel $BaTiO_3$-based Ag/Pd-compatible lead-free relaxor exhibits superior energy storage performance [39].

3.4 APPLICATIONS OF FERRITES AND $BaTiO_3$ NANOCOMPOSITES

Before the discussion of ferrites and $BaTiO_3$ nanocomposites, let's shine some light on nanocomposites. Nanocomposites are synthesized with the idea of improved physical properties because nanocomposites are a mixture of different materials with each material good at a certain property, so their combination would yield the best properties and a wider range of applicability. On the basis of the matrix (the material with more amount), nanocomposites have been categorized into three classes: (1) metal matrix nanocomposites, (2) polymer matrix nanocomposites, (3) ceramic matrix nanocomposites. As the name suggests, in metal matrix nanocomposites, the matrix is metal. It might include Al, Ti, and Mg as the matrix. In comparison to other metals mentioned here, aluminum is preferred over others due to its properties such as effective strength, toughness, and resistance to corrosion. Polymer matrix nanocomposites include polymer as matrix material. Polymer matrix nanocomposites have been actively researched for the past few years. In ceramic matrix nanocomposites, the matrix is ceramic, and these types of nanocomposites have many applications. These are used in heat exchangers, turbine blades, heating elements, bulletproof armor, gas-fired burner parts, etc. The nanopowders of $ZnFe_2O_4$ exhibit ferrimagnetic properties, while $BaTiO_3$ exhibits ferroelectric properties with 120 K as curie temperature [22, 40]. Therefore, the composite among the ferrimagnetic and ferroelectric may exhibit a new interface that would be more useful than their individual one. The nanocomposites are known to mix the two phases without forming any intermediate phase, resulting in individual and interfacial properties called *multiferroic*. Multiferroic involves simultaneous ferroelectric and ferri/ferromagnetic characteristics that are widely studied for spintronics applications [19]. To initiate the magneto-electric (ME) coupling, one needs a nanostructural multiferroic composite with significant magnetism and polarization. The ferroelectric/ferromagnetic composite forms a perfect interface that overwhelms the two charges and makes multiferroic properties, such as mixing between stages, sticking of ferromagnetic/ferroelectric spaces and transmission of strain between segment stages, proximity field impacts, etc., subordinate [41]. Therefore, studying the nanocomposites of modified ferrimagnetic material $ZnFe_2O_4$ and ferroelectric $BaTiO_3$ is necessary not only for developing new materials for solid-state electronic devices but also for the basic understanding of solid-state physics.

The ferrimagnetic and ferroelectric properties can be combined, producing multiferroic properties. In multiferroic materials, the magnetic properties can be tuned by applying an electric field, and the electrical properties can be tuned by applying a magnetic field. Selvi and colleagues prepared the nanocomposite of $CoFe_2O_4$-$BaTiO_3$ and found the value of ME coupling susceptibility of 0.6 mV/cm-Oe [42]. Jian-Ping Zhou and coworkers prepared the nanocomposite of $NiFe_2O_4$-$BaTiO_3$ and found it exhibited enhanced magnetic permeability due to magneto-electric coupling [43]. Jawed Ansaree and colleagues developed the varied nanocomposites of the relaxor phase of $BaTi_{0.85}Sn_{0.15}O_3$ and ferrite $NiFe_2O_4$ in the composition (1-x) $BaTi_{0.85}Sn_{0.15}O_3$ – x $NiFe_2O_4$ (with x = 5, 10, 15, and 20 wt%). The relaxor phase of the composite was made via the solid-state ceramic pathway, while the ferrites were made using the sol-gel combustion technique. For the RT-200°C temperature range and the RF range, they looked at the AC conductivity, dielectric constant, and dissipation factor for the all-composites sample (20 Hz–1 MHz). Strong frequency dispersion in the dielectric constant is seen between 20 Hz and 1 kHz, which is ascribed to Maxwell-Wagner interfacial polarization at the grain–grain boundary interface of grains in BTS-NF. All the composites' M-H curves exhibited the classic hysteresis loop seen in ferromagnetic materials. $NiFe_2O_4$'s presence as a component causes the composites to be ferromagnetic, and the bound magnetic polarons (BMP) model is used to explain this phenomenon. The BMP model has been used to fit the experimental magnetic data. The composites' Ms values are lower than $NiFe_2O_4$'s, but Hc and Mr are greater. For ZFC and FC, the composites have a higher temperature at which the M vs. T curve diverges than for $NiFe_2O_4$. Miniaturization in solid-state electronics is possible through downscaling and multifunctionality. In the past decade, researchers have poured a great deal of time and energy into studying multifunctional materials. Spintronics research has focused extensively on multiferroic materials, which exhibit both ferroelectric and ferri/ferromagnetic properties at the same time.

3.5 SYNTHESIS ROUTES OF NANOMATERIALS

There are many methods to prepare perovskite, ferrites, and their composites, and each synthesis route has its advantages and disadvantages. One can achieve different properties with a mere change in the material preparation method. So, choosing and adopting a synthesis method with benefits is crucial. Some of the preparatory techniques are described in the following subsections.

3.5.1 Solid-State Ceramic Route

The chemical breakdown processes that occur when a solid combination of reactants is heated to high temperatures result in a new solid composition and gases. Complex oxides may be synthesized from simpler oxides, carbonates, nitrates, hydroxides, oxalates, alkoxides, and other metal salts by this approach. This process typically involves numerous annealing operations with various intermediate milling steps to improve mixture homogeneity and reduce powder particle size. When the powder is finely ground, it becomes more reactive (i.e., more sinter-active) during the following heat treatment. The equipment needed for the solid-state method is minimal and the process is inexpensive. Additionally, it is easy to make massive quantities of powder. The resultant powder, however, displays a rather strong agglomeration, leading to a relatively large particle size and somewhat restricted homogeneity as compared to the wet powder manufacturing procedures discussed later in the chapter. The conventional solid-state ceramic route is one of the simplest and most commonly used methods to synthesize materials. It is also known as conventional high-temperature synthesis, shake-and-bake, or heat-and-beat [44]. This is because the solids are generally nonreactive at room temperature, so they must be heated to high temperatures.

In this method, raw materials were weighed and ground in agate mortar-and-pestle to achieve a fine powder. The reaction starts while grinding, and acetone is added as a mixing agent to achieve homogeneous powder. The well-ground sample is then transferred to crucibles and calcined at high temperatures for a specific period of time. All the steps involved in synthesizing nanomaterials are shown in Figure 3.2.

Advantages of this method are as follows:

1. Nontoxic and inexpensive organic solvents are used in this method.
2. This method includes synthesizing the material at room temperature and under atmospheric pressure.
3. It is the simplest method to synthesize nanomaterials.

The disadvantages are as follows:

1. This is a time-consuming technique.
2. The samples prepared through this method must be sintered at relatively higher temperatures.

3.5.2 Ball Milling

This technique of synthesizing nanomaterials is just an advanced version of the solid-state method. This method is an environmentally friendly, cost-effective technique, so it sees a wide application in

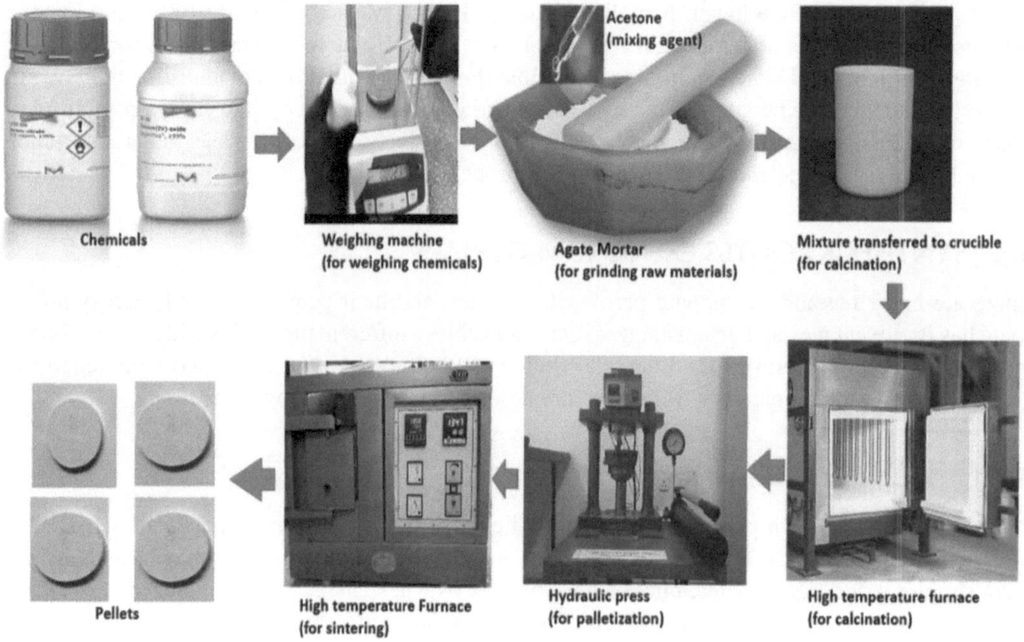

FIGURE 3.2 The steps involved in synthesizing nanomaterials by the solid-state ceramic route.

the industry worldwide [45]. In this method, powder precursors are kept inside a tightly closed vial with balls made up of the same material as the vial. The vials are rotated with a predefined rotation speed for some particular duration. The formation of new phases and compounds occurs by mixing the precursor powder. The precursor powder undergoes several processes, such as fracturing, welding, and activation, and the finally formed nanopowders are collected from the vials. Selectivity of the material of vials and balls depends on the nature of the chemical used in sample preparation so that the precursors do not react chemically with the vial and balls. Conventional materials for vials and balls are stainless steel, zirconia dioxide, and tungsten carbide, among others.

Advantages of this methos are as follows:

1. Low-cost installation and grinding medium.
2. One pot, one step mechanosynthesis.
3. Fabrication of materials at room temperature.
4. Revolutions per minute (rpm) and milling time can be programmed by the user.
5. Suitability of choosing a milling atmosphere.
6. Applicable to materials having all degrees of hardness.
7. Synthesis of a new phase is possible with localized deformation.
8. Formation of particles in the nano dimension or even in the amorphous phase; control over particle size of the final product.
9. Formation of a complete solid solution.
10. Complex powder formation in a predetermined amount of time.
11. Various milling parameters can control shock energy.

Disadvantages are as follows:

1. Contamination from the milling medium.
2. The stickiness of the sample with the container results in the loss of materials.
3. Excessive heating during milling.

Optimization of the milling parameters is achieved by minimizing the above-mentioned disadvantages. The duration of milling must be optimized from time to time and sample to sample to avoid contamination from the milling medium. Excess heat generated from the milling can be controlled by controlling the rpm of the vials. The balls-to-powder mass ratio and the milling atmosphere are chosen to ensure the purity of the final product.

3.5.3 Hydrothermal Method

This technique is employed for large-scale production of nano- to micro-size particles. This method involves the formation of nanomaterials for a wide range of temperatures (from room to very high temperatures) [43]. In this method, raw materials are dissolved in water and placed in an autoclave (pressurized vessel) at a high temperature. It is also popularly used to synthesize single crystals under high temperature and pressure conditions. The method is advantageous, as it is simple and yields high-quality products with minimal loss. The drawbacks include long processing time, utilization of basic solution of very high concentration during the process, and high cost of autoclaves.

3.5.4 Sol-Gel Auto-combustion

The evolution of sol-gel science and technology is quite impactful and fruitful, as nowadays it plays a crucial role in the development of nanotechnology in various applications. In 1845, Ebelman carried out a slow hydrolysis of an ester of silicic acid and reported the creation of a transparent material. In 1930, Geffcen and Berger introduced the proper steps to be followed for coating oxide layers on industrial glasses using the sol-gel technique. After this outstanding achievement, the invention of more methods of preparation were developed for more homogeneous melts and glasses by D. Roy and R. Roy using the sol-gel method [46]. Later on, thin film physics for sol-gel was introduced by Schroeder with the help of single-oxide and mixed-oxide layers. Thus, sol-gel became a popular method for synthesizing materials. The sol-gel method includes two components, the "sol" and the "gel," as suggested by its name. "Sol" is generally a solid particle in liquid while "gel" is a continuous network of particles filled with pores. In this method, first, a "sol" is formed in liquid, and then sol particles get connected to form a "gel" network. Then, by evaporating the powder, thin films or monolithic solids can be obtained.

Sol-gel is considered a flexible way to synthesize materials under mild conditions. Some of the advantages are:

1. Sol-gel is a low-temperature process, meaning it has low energy consumption.
2. This method can generate materials of high purity.
3. Sol-gel is a reliable technique, as a material with good homogeneity is produced via this method.
4. In this method, particles do not get agglomerated in the network matrix; as a result, uniform particle size distribution is achieved.
5. Nanomaterials with good optical properties can be achieved using this technique.
6. It is cost effective, as no delicate vacuum system is required in this process.
7. It can produce an excellent powder.

Some disadvantages of the Sol-gel method are:

1. It is a time-consuming process.
2. As this process is a pH-sensitive technique, pH must be checked repeatedly during processing.

In this chapter, the sol-gel auto combustion technique is adopted for synthesizing $ZnFe_2O4$, and the solid-state ceramic route is employed for $BaTiO_3$ and the composites 95ZF-5BT and 90ZF-10BT.

3.5.5 Synthesis of $ZnFe_2O_4$ Using Sol-gel Auto-combustion

Dissolve zinc nitrate hexahydrate, iron (III) nitrate nanohydrate, and citric acid in 25 ml of distilled water in separate beakers. Mix them well and stir until a transparent solution is formed. Now put the beaker on a hot plate using a fume hood and put a magnetic bead in it. Heat the solution at 80ºC until polymerization takes place. Check for pH after some intervals. It was noticed that the solution was acidic initially. To accelerate the reaction, add some ammonia solution drops and maintain a pH of 8 to 10. The flowchart and pictorial representation of the process followed are shown in Figure 3.3.

3.5.6 Solid-State Method for Synthesis of Barium Titanate and Composites of ZF and BT

For synthesizing $BaTiO_3$, appropriate amounts of $BaCO_3$ and TiO_2 were mixed for 3.5 hours using acetone as a mixing agent. After mixing, samples were transferred to crucibles, and calcination was done at 1,200°C for 6 hours with a 5°C /minute heating rate. From this calcined powder of BT and ZF, the composites were also made by grinding them in appropriate ratios for 4 hours. The resulting pristine as well as composite powders were then characterized. For dielectric measurements, the addition of 2% polyvinyl alcohol (PVA) (that acts as a binder) was made to the samples and pelletized using a 12-mm die set. The pressure applied for making the pellets was 80 kg/cm^2. These composed pellets were heated to 500°C with a heating rate of 5°C/minute for an hour to burn out the PVA completely. After that, the pellets were heated to 1,350°C and kept at this temperature for 6 hours for sintering. The sintered pellets were then coated with silver paste on both surfaces to study the dielectric properties. The silver-coated pellets were fired at 500°C for an hour to make perfect electrical contact. Now the prepared pellets were used for dielectric properties measurement.

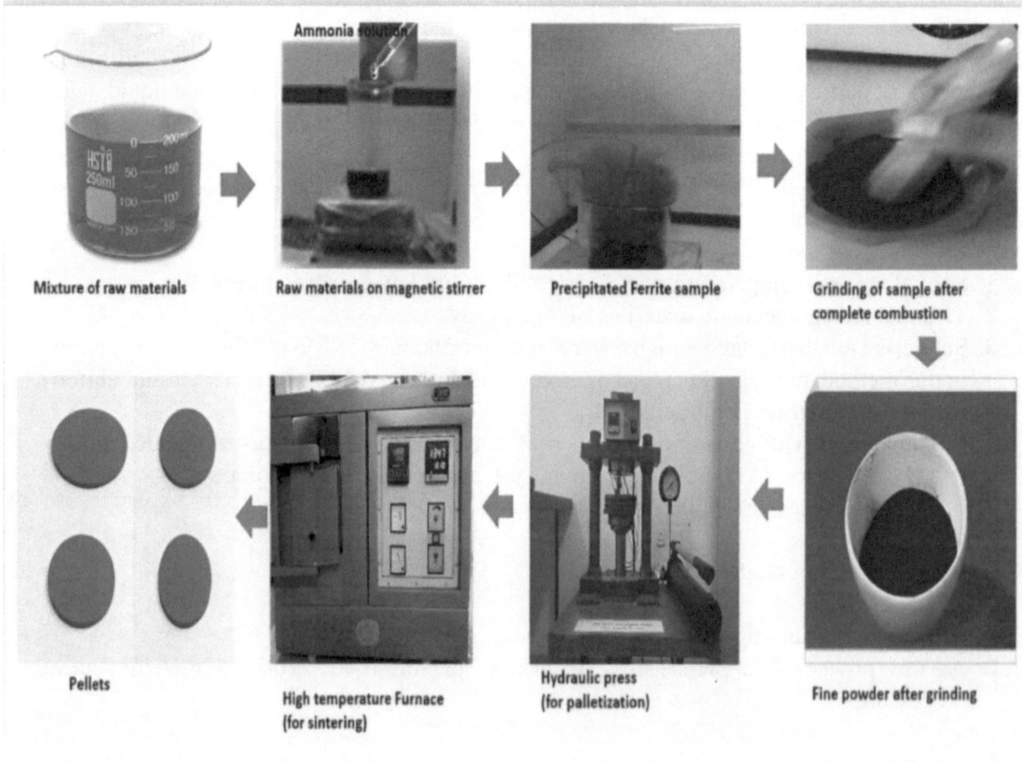

FIGURE 3.3 Pictorial representation of the synthesis of ZF via the sol-gel auto-combustion method.

3.5.7 Characterization Techniques

3.5.7.1 X-Ray Diffraction (XRD)

X-ray diffraction (XRD), is an *a priori* technique used to study a material's structural properties. The most information we can get from XRD is about the unit cell dimensions of the crystal as well as about the microstrain in the crystal. The analyzed material is ground and homogenized, and then the average bulk composition is determined. The basic principle underlying the XRD is constructive interference of the monochromatic X-rays with the crystalline sample. A cathode ray tube is responsible for producing the X-rays when the electrons having some sufficient energy displace the electrons in the inner shell of the target material bombarded with them. The most commonly used target material is copper with a wavelength $\lambda = 1.54056$ Å, and the rays are then filtered to produce monochromatic radiation, collimated to concentrate, and directed toward the sample. Constructive interference is produced when the sample interacts with the incident rays, and Bragg's condition must be satisfied. All possible diffraction directions of the lattice are obtained by the random orientations of the powdered material by scanning the material through a range of 2θ, and all the different crystallographic phases present in the sample are obtained from scanning at room temperature. The peaks obtained in the XRD pattern from the powdered material should be indexed and matched by the standard phase identification method that contains Miller indices, relative intensity, with similar interplanar spacing or diffraction angles, and other crystallographic information. The standard database is maintained by the International Center for Diffraction Data (ICDD), formerly known as the Joint Committee for Powder Diffraction Standards (JCPDS). There is another database available freely for users, the Crystallographic Open Database (COD). In the work associated with this chapter, the experimental XRD pattern is matched with the COD database. Another alternative method besides this search match method is the simulation method in which the theoretical XRD pattern is generated by a computer program through structural information available from the ICDD database. This method is applicable only when the crystal phase or similar structure file is available in the ICDD database. It is important to note that only X-rays can be used to study the structural properties of a material. This is because X-rays have a wavelength that is comparable to interatomic distances, so for diffraction to occur, X-rays are used. While visible light has a lower wavelength and so cannot penetrate inside a crystal, X-rays can do so. This is why they are utilized in order to get crystallographic information.

3.5.7.2 Zeta Potential

The samples' powder form was first dispersed in distilled water for 30 minutes in an ultra-sonicator. Ultra-sonicator is a device that uses high-frequency sound waves for dispersion, breaking apart particle agglomerates. Zeta potential values measure the total charge on the nanoparticle surface in a liquid state. When nanoparticles are dispersed in a liquid, a thin layer of charge is formed on the surface of the dispersed nanoparticle. This layer induces the formation of another layer. When the nanoparticles are in a liquid phase, and both of these layers have already been formed, an electrostatic potential develops between them. This potential is called Zeta potential. While measuring potential, an external electric field is applied across the nanoparticle sample. Due to the presence of the field, the charge particles start to move. It is known that the electric field is directly proportional to the velocity of the charged particle. While measuring Zeta potential, a laser source is used. The Zeta-sizer nano from Malvern uses the phase analysis light scattering technique. At first, the electric field is quickly reversed, making it possible to measure the mobility of nanoparticles. Then, the field is reversed slowly, which reduces electrode polarization. In this way, Zeta potential is calculated.

3.5.7.3 Electrical Properties

For the electrical properties measurement, the necessary electrode was created on both surfaces of the pellets using a high-temperature silver paste. The silver-coated pellets were then fired at 500°C for 30 min to form a proper electrode. The electrical response of the different samples was carried out in the presence of a signal by mounting the pellet in a sample holder inside a tube furnace in an air atmosphere using two probe methods. The frequency and temperature dependence of different electrical responses, such as dielectric constant and dissipation factor, were measured with the help of an LCR meter (Agilent E-4980, USA) interface as a function of temperature (in the temperature range of 40–300°C) at a constant frequency of 1 kHz. In this method of measurement, the dissipation factor was evaluated by the apparatus itself, but the dielectric constant was evaluated by the formula, $\varepsilon_r = \frac{Cd}{\varepsilon_0 A}$, where C is the capacitance, d is the thickness of the pellet, ε_0 is the capacitance of free space, and A is the cross-sectional area of the pellet.

3.6 RESULTS AND DISCUSSION

In the present work, $ZnFe_2O_4$, $BaTiO_3$, $0.95ZnFe_2O_4$–$0.05BaTiO_3$, and $0.90ZnFe_2O_4$–$0.10BaTiO_3$ would be abbreviated as ZF, BT, 95ZF-5BT, and 90ZF-10BT, respectively.

3.6.1 X-RAY DIFFRACTION

The structural properties of BT, ZF, and their composites were studied using X-ray diffraction. Figure 3.4 illustrates the XRD pattern of all the samples. The diffraction peaks at positions 21.81°, 31.24°, 38.51°, 44.92°, 50.70°, 55.91°, and 65.60° relevant to reflections (100), (110), (111), (200), (102), (211), and (220) of Barium titanate according to the COD file no. 1507756, while the diffraction peaks at positions 29.23°, 34.80°, 36.21°, 42.37°, 52.63°, 56.34°, and 61.89° associated with reflections (022), (113), (222), (004), (224), (115), and (144) were identified as the peaks of ZF according to the COD file no. 1010130. The composites 95ZF-5BT and 90ZF-10 BT mark the presence of all peaks of BT and ZF. The broadening in XRD peaks appeared due to (1) instrumental broadening, (2) small crystallite size, (3) accumulation of lattice strain, and (4) the presence of different kinds of stacking faults in the lattice. The diffraction pattern is analyzed by the integral breadth method from which the composition's finite particle size and microstructure can be predicted. To do so, Scherer introduced a method and later modified it using the work of Stokes and Wilson [47].

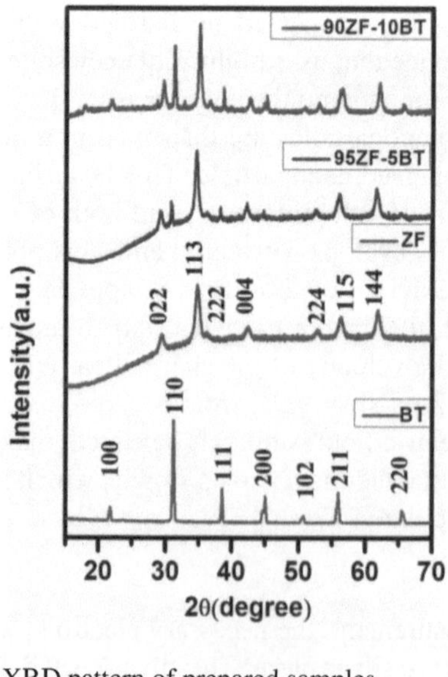

FIGURE 3.4 Room-temperature XRD pattern of prepared samples.

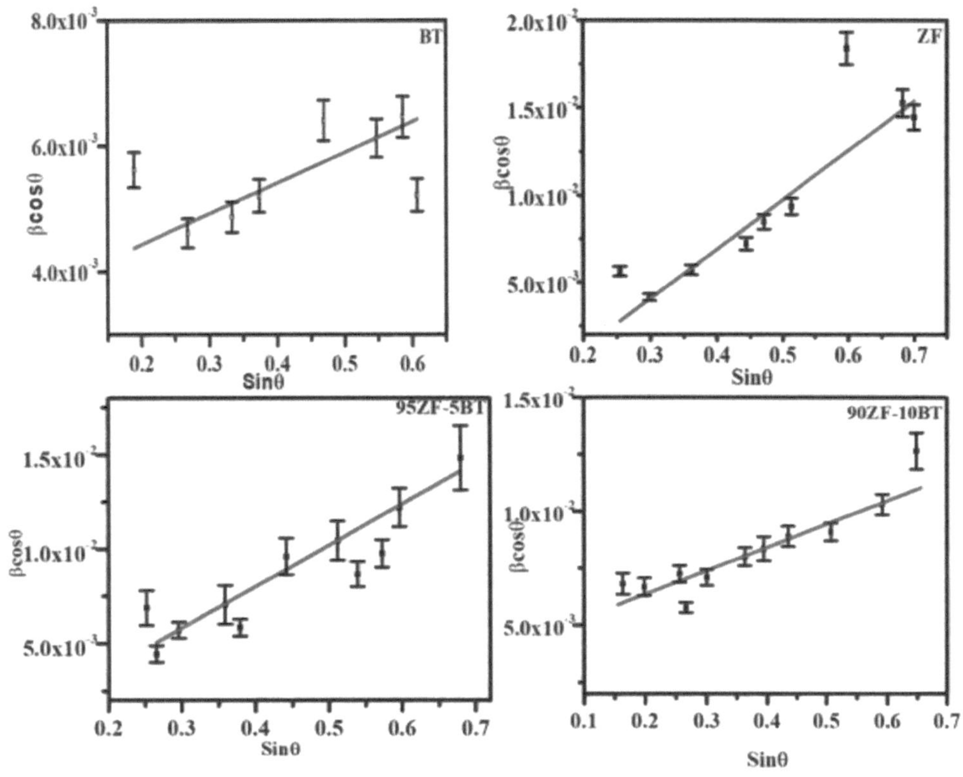

FIGURE 3.5 W-H plots generated for all samples (symbols represent experimental data, and the solid line shows the linear fitting of Eq. [3.3]).

In 1949, Hall defined integral breadth as

$$\beta = \beta_D + \beta_S \tag{3.1}$$

where β_D and β_S are the broadenings of the peak due to small crystallite size and the microstrain in the lattice, respectively. According to Scherrer, $\beta_D = \frac{k\lambda}{D\cos\theta}$, and $\beta_S = 4\varepsilon\tan\theta$.

Where λ is the X-ray wavelength, k is the crystallite shape constant (0.89 for spherical shape), D is the crystallite size, θ is the Bragg angle, β_S is the broadening arising from the microstrain in the lattice, and ε is the microstrain in the lattice. Substituting the values of β_D and β_S in Eq. (3.1) results in:

$$\beta = \beta_D + \beta_S = \frac{k\lambda}{D\cos\theta} + 4\varepsilon\tan\theta \tag{3.2}$$

According to Williamson-Hall,

$$\beta\frac{\cos\theta}{\lambda} = \frac{1}{D} + 4\varepsilon\frac{\sin\theta}{\lambda} \tag{3.3}$$

Therefore, measuring the FWHM (β) and the corresponding Bragg angle θ in the XRD pattern is linearly fitted using Eq. (3.3), which is the Williamson-Hall (W-H) plot. W-H plots for all the samples are shown in Figure 3.5, and the conclusions drawn from them are summarized in Table 3.1, which also includes lattice parameters and volume calculated using Unit Cell Win software.

TABLE 3.1
Numerical Values of Results Drawn from XRD Data

Material	Crystallite Size (nm)	Micro-strain	Lattice Parameter (Å)		c/a	Volume (Å³)
			BT	ZF		
BT	40.50	0.004	a = b = 4.028 c = 4.025	—	0.999	65.34
ZF	31.70	0.021	—	a = b = c = 8.541	1	623.05
95ZF-5BT	33.52	0.019	a = b = 4.046 c = 4.072	a = b = c = 8.595	1.006 (for BT) 1 (for ZF)	66.68 (for BT) 635.07 (for ZF)
90ZF-10BT	32.59	0.010	a = b = 4.021 c = 4.061	a = b = c = 8.520	1.009 (for BT) 1 (for ZF)	65.68 (for BT) 618.54 (for ZF)

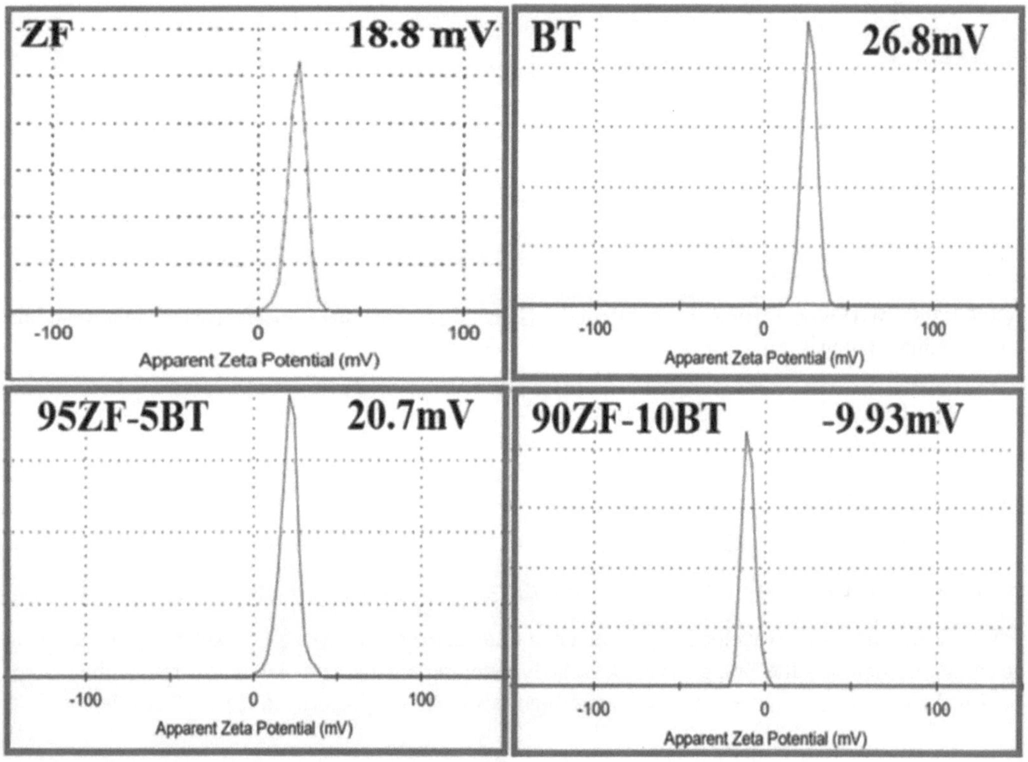

FIGURE 3.6 Zeta potential distribution for all samples.

3.6.2 Zeta Potential and Hydrodynamic Size Measurement

The difference in potential across the phase boundaries of solid (sample powder) and liquid in which the sample is dispersed is called Zeta potential [48]. Simply it is a measure of the stability of the colloidal dispersion. In the present investigation, a pinch of undoped and doped samples each was dispersed in distilled water to obtain Zeta potential, hydrodynamic size, and polydispersity index (PdI). According to the literature, the limit value of Zeta potential should be +/–30 mV for the stability of dispersion in the water system [49]. Figure 3.6 shows the curves for Zeta potential, while Figure 3.7 depicts the curves for hydrodynamic size

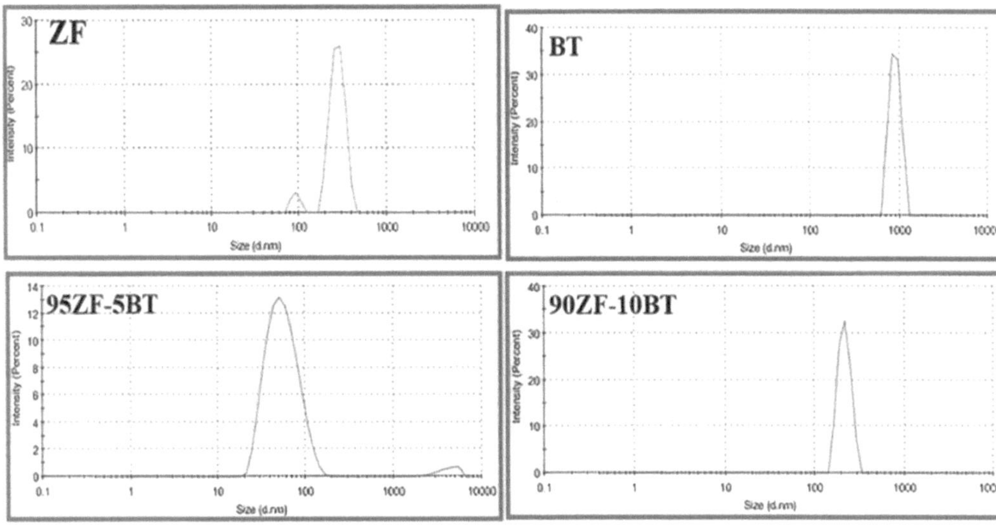

FIGURE 3.7 Hydrodynamic size distribution for all samples.

distribution. It was observed that the numerical value of Zeta potential was found to be 18.8 for ZF [19], and it was 26.8, 20.7, and –9.93 mV for BT, 95 ZF-5BT, and 90ZF-10BT, respectively. As shown, all the samples except 90ZF-10BT have positive values for Zeta potential. This in turn shows that on the dispersion of these samples, the stability of electrostatic attraction was certain. For 95ZF-10BT, the surface of the material becomes negatively charged, which means the defect has a positive charge; therefore, it attracts a negative charge on the surface, and the resultant net charge is negative. This negative charge might be due to excess oxygen vacancy in the present case. With Zeta potential, hydrodynamic size, and polydispersity index have also been determined. The values of hydrodynamic size are 504.6, 1396, 52.03, and 211.4 nm, while the values for PdI are 0.709, 0.325, 0.192, and 0.400 for ZF, BT, 95ZF-5BT, and 90ZF-10BT, respectively [19]. It is noticed that hydrodynamic size follows the same trend as crystallite size except for 90ZF-10BT.

3.6.3 Dielectric Analysis

To understand the dielectric properties, it should be clear that four types of polarizations affect the dielectric properties of glass, glass-ceramics, and polycrystalline ceramic materials. These polarizations are under effect at various ranges of frequency. For example, the dielectric properties are governed by interfacial polarization up to 1 kHz, and from 1 kHz to 1 MHz, the properties are governed by orientational polarization; from 1 MHz to 1 GHz, ionic polarization comes into effect, and above 1 GHz, electronic polarization rules. To ensure a clear understanding, it is necessary to understand these types of polarization individually.

There will often be defects in natural insulating materials, such as lattice vacancies, impurity ions, and so on, as well as some free electrons. Any or all of these can migrate through the materials under the influence of an applied field to the electrode with the opposite polarity to their charge. If they enter the electrode and discharge there, the ions can acquire electrons from the electrode and escape through the electrode, causing a current loss through the dielectric. However, if not all of them discharge, a thin layer of charge of opposite types forms close to the surface of each electrode, resulting in a pileup of charge. This gives the dielectric dipole moment and constitutes a separate mechanism of polarization, known as interfacial polarization and shown in Figure 3.8(a). Only polar molecules with a permanent dipole moment experience dipolar

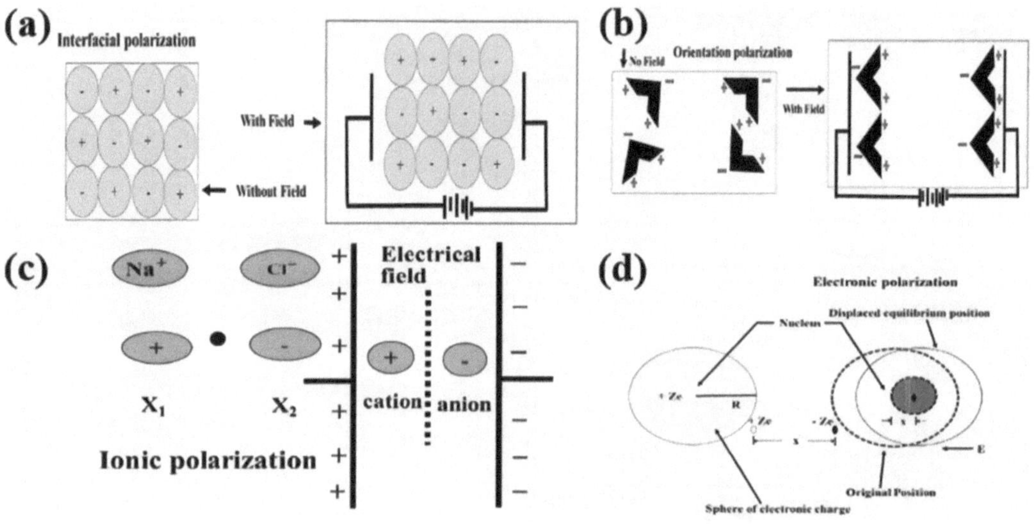

FIGURE 3.8 Illustration and schematic presentation of all types of polarization. (a) Interfacial polarization. (b) Orientational polarization. (c) Ionic polarization. (d) Electronic polarization.

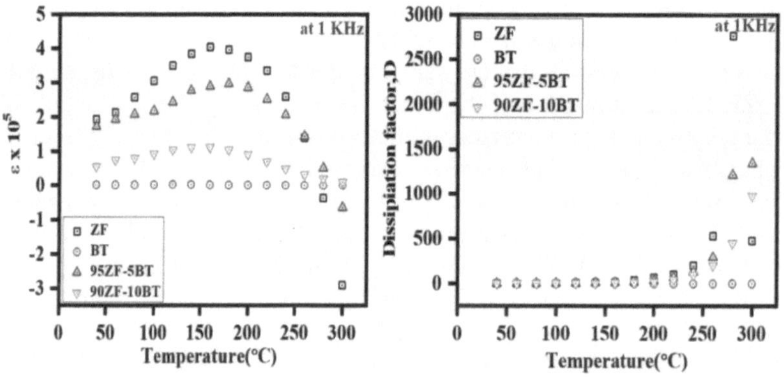

FIGURE 3.9 Variation of (a) dielectric constant and (b) dissipation factor for all samples as a function of temperature at constant frequency 1 kHz.

polarization. When a polar molecule is exposed to an electric field, the randomly oriented dipoles align (or orient) themselves in the direction of the applied field. Figure 3.8(b) depicts the mechanism of orientation polarization. The movement of a charged ion relative to other ions in the solid induces ionic polarization. Figure 3.8(c) illustrates the ionic polarization process. The displacement of an atom's electron cloud relative to its nucleus in an applied electric field causes electronic polarization, as shown in Figure 3.8(d).

To all pellets sintered at 1,350°C for 6 hours, highly conducting silver paste was applied on both sides and fired at 500°C to study the dielectric behavior of all samples. The samples' dielectric constant and dissipation factor have been measured from 40°C to 300°C at 1 kHz and shown in Figure 3.9(a–b). Considering the frequency range, only interfacial and orientational polarizations affect the dielectric properties in the present case. The dielectric constant initially increases up to 150°C, then decreases with increasing temperature for pure ZF. However, the dielectric constant doesn't show any variation with temperature for the BT sample. The dielectric properties of the composite phase were found to lie between the BT and ZF phases. The following relation can understand the overall dielectric properties of the phase of the composite:

$$\varepsilon = (x) \times BT + (1 - x) \times ZF \tag{3.4}$$

Since the present sample has been synthesized at 1,350°C, the possibility of oxygen loss cannot be ruled out. The presence of oxygen vacancy can be understood as follows:

$$O_o \rightarrow \frac{1}{2} O_2(g) + V_O^{\cdot\cdot} + 2e' \tag{3.5}$$

The presence of these electrons may be captured by the Fe of ZF and Ti of BT and reduced to a lower valency state of multivalent ion by following equations:

$$Ti^{4+} + 2e' \rightarrow Ti^{2+} \tag{3.6}$$

$$2Fe^{3+} + 2e' \rightarrow 2Fe^{2+} \tag{3.7}$$

In a perovskite structure, the Ba is occupied at the dodecahedral site, oxygen vacancy at the face-centered (fcc) site, and Ti^{2+}/Ti^{4+} at the body-centered site. The occupation of Ba^{2+}, Ti^{2+}/Ti^{4+}, and oxygen vacancy is an entirely random process, so the micro-regions having different atoms at their respective sites create microheterogeneity with varying conductivity, resulting in interfacial polarization. Similar to perovskite, the spinel also has a random distribution of Fe^{2+}, Fe^{3+}, and $V_O^{\cdot\cdot}$, which results in grains with varying conductivity. The interface, which separates the conducting regions, results in a high value of dielectric constant [50]. Interfacial polarization mainly dominates over others at lower frequencies, resulting in a large dielectric constant. However, the presence of Ti^{2+} at Ti^{4+} performed as a negative defect and was denoted by ${Ti^{2+}}_{Ti^{4+}}''$, Fe^{2+} at Fe^{3+} by $2{Fe^{2+}}_{Fe^{3+}}''$, and oxygen vacancy $V_O^{\cdot\cdot}$ acts as a positive defect, instances of which are situated far away from each other, and forms an electric dipole $\left({Ti^{2+}}_{Ti^{4+}}'' - V_O^{\cdot\cdot}\right), \left(2{Fe^{2+}}_{Fe^{3+}}'' - V_O^{\cdot\cdot}\right)$ that responds in the presence of small AC field, which has a significant role on the variation of dielectric properties.

Furthermore, the dissipation factor has been showing to be almost constant up to 200°C, increasing after that with increasing temperature. The variation can be understood in terms of DC conduction, which occurs either due to the conduction of electrons through the multivalent site or their direct conduction. Since both materials are wide bandgap semiconductors, this rules out any direct conduction in the sample. Here, the dipoles present in samples work as a bridge for the conduction of charge carriers through the transferring of electrons between the dipoles energy required to transfer electrons as provided by thermal energy, as seen above at 300°C. The dielectric constant and dissipation factor values at room temperature for ZF, BT, 95ZF-5BT, 90ZF-10BT samples are 1.81, 0.037, 2.9, 1.25 and 192913.20, 1781.13, 172443.88, 56129.39, respectively which makes them suitable as high-K dielectrics in the barrier layer, multilayer ceramic capacitors, and sensor applications.

3.6.4 Challenges and Future Perspectives

In today's world, experimental and theoretical studies of physical properties of nanocomposites as high-K dielectric materials are focused on extreme levels, and the reason for this can be found in the vast range of application areas for these nanomaterials and nanocomposites. As already stated nanomaterials, as individuals and in nanocomposite form with other nanomaterials have enormous potential in several major fields, but with this potential come some challenges in the proper and effective utilization of these nanocomposites. Some of the major issues are:

1. The very foremost challenge has its roots in the cost of the raw materials required for synthesizing the nanomaterials. Sometimes, with high costs, only a small number of samples can be synthesized. Fulfilling the proper environmental requirements for effective synthesis is also a major issue because high-quality instrumentation required to achieve good yield is expensive. So the issue of cost cannot be ignored.
2. Selection of the best synthesizing method is another serious issue. Each synthesizing route has its own pros and cons, so choosing a synthesis method with maximum advantages is also essential. While selecting the method of preparation, it should be kept in mind that resultant samples should be of highest possible purity and the process is less time consuming. Factors that can affect the yield of samples should also be considered. Thus, selecting an appropriate method of synthesis requires focused efforts.
3. The cleanliness at every step, from weighing of raw material to the last process, is also a challenge. If this factor is ignored at any stage, then all the hard work done before and after the misstep will be in vain, as impure products will lead to poor output results.
4. The demand for nanocomposites with good energy storage capacities (as demonstrated in the present work) is growing at a rapid rate. In a race to satisfy this demand, some current practical research is ignoring the issue of toxicity, which, if continuously dismissed, could lead to serious issues. Bridging the gap of knowledge between toxicity and the rapid development of nanomaterials and nanocomposites is crucial.
5. In the sol-gel auto-combustion method employed in this work for the synthesis of ZF, it is mandatory to keep the solution basic in order to enhance the rate of reaction, which requires constant vigilance when it comes to checing the PH level of the mixture throughout the synthesizing process.
6. Real-world application of the created nanocomposites requires expert minds and advanced facilities and technology, neither of which is available in abundance even in well-developed industrialized nations. Problems of staffing and means of production will be overcome only with a close and vigorous cooperation between the public and private sectors of any country truly interested in advancing these technologies.

3.7 CONCLUSIONS

ZF was successfully synthesized via sol-gel auto-combustion and BT, and the composites were also successfully synthesized via the solid-state ceramic route. Phase identification for all samples was confirmed by X-ray diffraction. It was observed that ZF crystallized into a cubic spinel group, while a tetragonal phase was observed for pristine BT. Composites 95ZF-5BT and 90ZF-10BT express all significant diffraction peaks of ZF and BT. Further W-H plots were generated for the estimation of crystallite size and microstrain. Lattice parameters and volume were found using Unit Cell Win software. Hydrodynamic size followed the same trend as crystallite size except for 90ZF-10BT. The dielectric behavior of pellets sintered at 1,350°C was studied as a function of temperature at 1 KHz. The dielectric constant followed the increasing trend for ZF until 150°C, while there was no significant increment in the case of BT. The dissipation factor was found to be increased after 200°C but was constant before that. Based on these studies, it can be reasonably claimed that these materials have the potential to be used in capacitors, memory devices, power systems, and energy storage devices like electrode material for supercapacitors and SOFCs.

ACKNOWLEDGMENTS

The authors thank the Head Department of Physical Sciences and Head Department of Pharmacy, Banasthali Vidyapith, for providing the experimental facilities required to characterize the synthesized samples. Upendra Kumar enthusiastically acknowledges the DST-SERB through a project (File No: DST-SERB/EEQ/2021/000132).

REFERENCES

[1] A. Sharma, Potential of Non-conventional Source of Energy and Associated Environmental Issue: An Indian Scenario, *Int. J. Comput. Appl.* (0975–8887) Recent Trends Electron. Commun. (RTEC 2013). (2017) 43–47.

[2] S. Bansal, B.N. Agrawal, Conventional and renewable energy scenario of India: present and future, *Invertis J. Renew. Energy*. 7 (2017) 29. https://doi.org/10.5958/2454-7611.2017.00004.2

[3] A.R. Osmani, Conventional Energy to Renewable Energy: Perspectives for India, *NEHU J.* XII (2014) 41–60.

[4] H. Gupta, S.K. Gupta, V. Dwivedi, Conventional and Non-Conventional Energy Resources of India: Present and Future Drilling Burrs View Project Interfacial Tribology in Metal Working Processes View Project, (2011). https://www.researchgate.net/publication/331276188

[5] A. Kumar, K. Kumar, N. Kaushik, S. Sharma, S. Mishra, Renewable energy in India: Current Status and Future Potentials, *Renew. Sustain. Energy Rev.* 14 (2010) 2434–2442. https://doi.org/10.1016/j.rser.2010.04.003

[6] M. Hoel, 96/06588 Depletion of Fossil Fuels and the Impacts of Global Warming, *Fuel Energy Abstr.* 37 (1996) 460. https://doi.org/10.1016/s0140-6701(97)83963-x

[7] I. Mittal, K.R. Gupta, Natural Resources Depletion and Economic Growth in Present Era, *Soch.* 10 (2015) 24–28.

[8] J. Walker, Caring for the Earth, *Min. Surv.* 2 (1991) 42–52. https://doi.org/10.1142/9789813147720_0019

[9] J. Charles Rajesh Kumar, M.A. Majid, Renewable Energy for Sustainable Development in India: Current Status, Future Prospects, Challenges, Employment, and Investment Opportunities, *Energy. Sustain. Soc.* 10 (2020) 1–36. https://doi.org/10.1186/s13705-019-0232-1

[10] Y. Zhang, Research of Thermal Energy Storage Technology in the Solar Thermodynamic Power, *J. Power Energy Eng.* 04 (2016) 42–49. https://doi.org/10.4236/jpee.2016.47006

[11] E.M. Languri, G. Cunningham, Thermal Energy Storage Systems, *Lect. Notes Energy*. 70 (2019) 169–176. https://doi.org/10.1007/978-3-030-05636-0_9

[12] E.S. Systems, H. Development, S. Flywheels, S. Reserve, D. Trends, A. Accumulator, M SC PL O E – C EO M SC PL O E – C EO, I (n.d.).

[13] S.A. Rackley, Other geological storage options, *Carbon Capture and Storage*, (2017) 471–488. https://doi.org/10.1016/b978-0-12-812041-5.00018-0

[14] J.C. Ellenbogen, *Supercapacitors: A Brief Overview*, (2006).

[15] E. Elshazly, N. Emam, N. Eltayeb, Attributes affecting energy of superconducting magnetic energy storage systems, *Int. J. Sci. Eng. Res.* 6 (2015) 356–361. http://www.ijser.org

[16] L. Yao, B. Yang, H. Cui, J. Zhuang, J. Ye, J. Xue, Challenges and Progresses of Energy Storage Technology and Its Application in Power Systems, *J. Mod. Power Syst. Clean Energy*. 4 (2016) 519–528. https://doi.org/10.1007/s40565-016-0248-x

[17] R. Schlögl, Chemical Energy Storage: Part of a Systemic Solution, *EPJ Web Conf.* 148 (2017) 1–14. https://doi.org/10.1051/epjconf/201714800015

[18] M. Sati, M. Verna, Fundamentals of energy: its potential and achievements. In *Renewable Energy and Green Technology*. CRC Press (2021), 1–12.

[19] H. Cheema, S. Kumar, P.A. Alvi, B.L. Choudhary, U. Kumar, Synthesis and Physical Properties of Nanopowder and Electrical Properties of Bulk Samples of ZnFe2-xNixO4 (x: 0, 0.05, 0.10), *Adv. Powder Technol.* 31 (2020) 4241–4252. https://doi.org/10.1016/j.apt.2020.09.001

[20] G. Palareti, C. Legnani, B. Cosmi, E. Antonucci, N. Erba, D. Poli, S. Testa, A. Tosetto, Comparison between different D-Dimer cutoff values to assess the individual risk of recurrent venous thromboembolism: Analysis of results obtained in the DULCIS study, *Int. J. Lab. Hematol.* 38 (2016) 42–49. https://doi.org/10.1111/ijlh.12426

[21] Z. Sun, L. Li, S. Yu, X. Kang, S. Chen, Energy storage properties and relaxor behavior of lead-free Ba1-:XSm2 x/3Zr0.15Ti0.85O3 ceramics, *Dalt. Trans.* 46 (2017) 14341–14347. https://doi.org/10.1039/c7dt03140h

[22] L.I. Granone, R. Dillert, P. Heitjans, D.W. Bahnemann, Effect of the Degree of Inversion on the Electrical Conductivity of Spinel ZnFe2O4, *ChemistrySelect.* 4 (2019) 1232–1239. https://doi.org/10.1002/slct.201804062

[23] C. Yao, Q. Zeng, G.F. Goya, T. Torres, J. Liu, H. Wu, M. Ge, Y. Zeng, Y. Wang, J.Z. Jiang, ZnFe2O4 Nanocrystals: Synthesis and Magnetic Properties, *J. Phys. Chem. C.* 111 (2007) 12274–12278. https://doi.org/10.1021/jp0732763

[24] S.M. Hoque, M.S. Hossain, S. Choudhury, S. Akhter, F. Hyder, Synthesis and Characterization of $ZnFe_2O_4$ Nanoparticles and its Biomedical Applications, *Mater. Lett.* 162 (2016) 60–63. https://doi.org/10.1016/j.matlet.2015.09.066

[25] A. Ashok, T. Ratnaji, L. John Kennedy, J. Judith Vijaya, Magnetically Separable Zn1-xCuxFe2O4 (0 ≤ x ≤ 0.5) Nanocatalysts for the Transesterification of Waste Cooking Oil, *Adv. Powder Technol.* 4 (2020) 1–13. https://doi.org/10.1016/j.apt.2020.04.022

[26] A. Manohar, C. Krishnamoorthi, Site Selective Cu2+ Substitution in Single Crystal Fe3O4 Biocompatible Nanospheres by Solvothermal Reflux Method, *J. Cryst. Growth.* 473 (2017) 66–74. https://doi.org/10.1016/j.jcrysgro.2017.05.013

[27] E.A. Chavarriaga, A.A. Lopera, V. Franco, C.P. Bergmann, J. Alarcón, Gel Combustion Synthesis and Magnetic Properties of $CoFe_2O_4$, $ZnFe_2O_4$, and $MgFe_2O_4$ Using 6-Aminohexanoic Acid as a New Fuel, *J. Magn. Magn. Mater.* 497 (2020) 166054. https://doi.org/10.1016/j.jmmm.2019.166054

[28] S.P. Jiang, A Review of Wet Impregnation – An Alternative Method for the Fabrication of High Performance and Nano-Structured Electrodes of Solid Oxide Fuel Cells, *Mater. Sci. Eng. A.* 418 (2006) 199–210. https://doi.org/10.1016/j.msea.2005.11.052

[29] M. Yousaf, N. Mushtaq, B. Zhu, B. Wang, M.N. Akhtar, A. Noor, M. Afzal, Electrochemical Properties of $Ni_{0.4}Zn_{0.6}$ Fe_2O_4 and the Heterostructure Composites (Ni–Zn ferrite-SDC) for Low Temperature Solid Oxide Fuel Cell (LT-SOFC), *Electrochim. Acta.* 331 (2020) 135349. https://doi.org/10.1016/j.electacta.2019.135349

[30] A. Hakimyfard, S. Mohammadi, $ZnFe_2$ O_4 and ZnO-Zn 1−x M x Fe_2 O_4+δ (M = Sm 3+, Eu $^{3+}$ and Ho $^{3+}$): Synthesis, Physical Properties and High Performance Visible Light Induced Photocatalytic Degradation of Malachite Green, *Adv. Powder Technol.* 30 (2019) 1257–1268. https://doi.org/10.1016/j.apt.2019.04.005

[31] A.E. Reddy, T. Anitha, C.V. V. Muralee Gopi, I.K. Durga, H.-J. Kim, Facile Synthesis of Hierarchical $ZnMn_2O_4$@$ZnFe_2O_4$ Microspheres on Nickel Foam for High-Performance Supercapacitor Applications, *New J. Chem.* 42 (2018) 2964–2969. https://doi.org/10.1039/C7NJ04269H

[32] M.M. Vadiyar, S.K. Patil, S.C. Bhise, A. V. Ghule, S.H. Han, S.S. Kolekar, Improved Electrochemical Performance of a $ZnFe_2O_4$ Nanoflake-Based Supercapacitor Electrode by Using Thiocyanate-Functionalized Ionic Liquid Electrolytes, *Eur. J. Inorg. Chem.* 2015 (2015) 5832–5838. https://doi.org/10.1002/ejic.201500870

[33] M.M. Vadiyar, S.S. Kolekar, J.-Y. Chang, Z. Ye, A. V. Ghule, Anchoring Ultrafine ZnFe $_2$ O $_4$ /C Nanoparticles on 3D $ZnFe_2O_4$ Nanoflakes for Boosting Cycle Stability and Energy Density of Flexible Asymmetric Supercapacitor, *ACS Appl. Mater. Interfaces.* 9 (2017) 26016–26028. https://doi.org/10.1021/acsami.7b06847

[34] U. Kumar, M.J. Ansaree, A.K. Verma, S. Upadhyay, G. Gupta, Oxygen Vacancy Induced Electrical Conduction and Room Temperature Ferromagnetism in System BaSn1-xNixO_3 (0x 0.20), *Mater. Res. Express.* 4 (2017) 116304. https://doi.org/10.1088/2053-1591/aa9416

[35] M.J. Ansaree, U. Kumar, S. Upadhyay, Solid-State Synthesis of Nano-Sized Ba(Ti1−xSnx)O_3 Powders and Dielectric Properties of Corresponding Ceramics, *Appl. Phys. A Mater. Sci. Process.* 123 (2017) 1–12. https://doi.org/10.1007/s00339-017-1047-6

[36] C. Zhang, Y. Yin, Q. Yang, Z. Shi, G.-H. Hu, C. Xiong, Flexible Cellulose/BaTiO 3 Nanocomposites with High Energy Density for Film Dielectric Capacitor, *ACS Sustain. Chem. Eng.* 7 (2019) 10641–10648. https://doi.org/10.1021/acssuschemeng.9b01302

[37] W.B. Li, D. Zhou, R. Xu, D.W. Wang, J.Z. Su, L.X. Pang, W.F. Liu, G.H. Chen, BaTiO3-Based Multilayers with Outstanding Energy Storage Performance for High Temperature Capacitor Applications, *ACS Appl. Energy Mater.* (2019). https://doi.org/10.1021/acsaem.9b00664

[38] Z. Shen, X. Wang, B. Luo, L. Li, $BaTiO_3$ –$BiYbO_3$ Perovskite Materials for Energy Storage Applications, *J. Mater. Chem. A*. 3 (2015) 18146–18153. https://doi.org/10.1039/C5TA03614C
[39] H. Yang, Z. Lu, L. Li, W. Bao, H. Ji, J. Li, A. Feteira, F. Xu, Y. Zhang, H. Sun, Z. Huang, W. Lou, K. Song, S. Sun, G. Wang, D. Wang, I.M. Reaney, Novel BaTiO 3 -Based, Ag/Pd-Compatible Lead-Free Relaxors with Superior Energy Storage Performance, *ACS Appl. Mater. Interfaces*. 12 (2020) 43942–43949. https://doi.org/10.1021/acsami.0c13057
[40] M.J. Ansaree, S. Upadhyay, Study of Phase Evolution and Dielectric Properties of Sn-Doped Barium Titanate, *Emerg. Mater. Res*. 6 (2017) 21–28. https://doi.org/10.1680/jemmr.16.00013
[41] M.J. Ansaree, U. Kumar, S. Upadhyay, Structural, Dielectric and Magnetic Properties of Particulate Composites of Relaxor ($BaTi_{0.85}Sn0.15O_3$) and Ferrite ($NiFe_2O_4$) Synthesized by Gel-Combustion Method, *J. Electroceramics*. 40 (2018) 257–269. https://doi.org/10.1007/s10832-018-0127-0
[42] M.M. Selvi, P. Manimuthu, K.S. Kumar, C. Venkateswaran, Magnetodielectric Properties of $CoFe_2O_4$–$BaTiO_3$ Core–Shell Nanocomposite, *J. Magn. Magn. Mater*. 369 (2014) 155–161. https://doi.org/10.1016/j.jmmm.2014.06.039
[43] J.-P. Zhou, L. Lv, Q. Liu, Y.-X. Zhang, P. Liu, Hydrothermal Synthesis and Properties of $NiFe_2O_4$ @ $BaTiO_3$ Composites With Well-Matched Interface, *Sci. Technol. Adv. Mater*. 13 (2012) 045001. https://doi.org/10.1088/1468-6996/13/40450012
[44] Synthesis Methods.pdf, (n.d.).
[45] C.C. Piras, S. Fernández-Prieto, W.M. De Borggraeve, Ball Milling: A Green Technology for the Preparation and Functionalisation of Nanocellulose Derivatives, *Nanoscale Adv*. 1 (2019) 937–947. https://doi.org/10.1039/c8na00238j
[46] N. Khan, History of Science and Technology, *Bull. Sci. Technol. Soc*. 21 (2001) 510. https://doi.org/10.1177/027046760102100609
[47] H. Cheema, V. Yadav, R. Sunder, S. Kumar, P.A. Alvi, U. Kumar, Materials Today: Proceedings (0 .16, 0 .20) Synthesized Using Solid State Ceramic Route, *Mater. Today Proc*. 3 (2022). https://doi.org/10.1016/j.matpr.2022.06.349
[48] K. Rajpoot, R.K. Tekade, *Microemulsion as Drug and Gene Delivery Vehicle: An Inside Story*, Elsevier Inc., 2019. https://doi.org/10.1016/B978-0-12-814487-9.00010-7
[49] H. Cheema, Structural, Optical and Electrical Properties of Mn-Doped $ZnFe_2O_4$ Synthesized Using Sol-Gel Method, (n.d.).
[50] S. Singh, P. Singh, O. Parkash, D. Kumar, Synthesis, Microstructure and Electrical Properties of Ti Doped $SrSnO_3$, *Adv. Appl. Ceram*. (2007). https://doi.org/10.1179/174367607X202573

4 Methods for Synthesis of Metal Oxide-based High-K Dielectrics

Sudhanshu Dwivedi

4.1 INTRODUCTION

In the pursuit of downscaling of nanodevices to almost atomic dimensions, novel high-K dielectric materials with exceptional electronic properties are required for the integration of 1-nm ultrathin gate oxide layers in complementary metal oxide semiconductor (CMOS) technology. For advanced nanoscale devices, metal oxides such as hafnium oxide (HfO_2), zirconium oxide (ZrO_2), lanthanum oxide (LaO_2), and many others, in addition to silicon dioxide (SiO_2), are of prime importance. Important parameters in the formation of these oxidic layers are the pre-deposition treatment methodology of substrate surfaces, associated chemistry and kinetic of growth mechanism, effects of temperature, and, last but not least, gate metallization to form electrical contacts. Advanced metrological tools are required with due consideration given to specific characteristics generated because of ultrathin nanolayers along with the downsizing of features [1–14].

SiO_2 is a very well-known gate dielectric material viable with high-temperature integrated circuit (IC) manufacturing technology. Ever-increasing miniaturization requires ultrathin nanolayers of gate oxides with exceptional electrostatic gating control of the metal oxide field effect transistor (MOSFET) to effectively overcome barriers presented by drain voltages for minimization of short channel effects. For the realization of bulk-similar properties, three separate monolayers of SiO_2 generate an EOT with a downsized limit of ~0.7 nm. Here, EOT is the equivalent thickness of SiO_2 layer that has the same capacitance value on replacing with a high-K dielectric layer that is expressed as

$$t_{\text{oxide}} = \text{EOT} = \left(\frac{3.9}{k} t_k \right). \tag{4.1}$$

Here, t_{oxide} is the EOT, k is the dielectric constant of a high-K dielectric metal oxide, and 3.9 is the dielectric constant of SiO_2. It means that high-K metal oxides with the same EOT value will be able to reduce tunneling-based leakage current in gate dielectrics by a large order of magnitude. High-K dielectrics with large band gaps alleviate parasitic charge conduction by inhibiting charge carriers from both source and drain electrodes along with blocking of thermally induced inherent free charge carriers [6].

In ultrathin nanolayers (0.2–1 nm), quantum mechanical tunneling of electrons through the gate oxide is highly probabilistic and enhances stand-by power dissipation in ultra-large-scale integrated circuitry. In this surge, materials with high K are required to compensate gate capacitances of an equivalently thicker oxide to undermine leakage current without actually affecting the *ON*-current of the transistor. High-K materials must consist of a range of specific features that (1) have sufficiently high K along with a cost-effective and scalable commercially viable process, (2) are effectively integrable and thermodynamically stable with positive device reliability and testing procedures in consonance with existing Si-based CMOS technology, and (3) fulfill

DOI: 10.1201/9781003352365-4

relationship $E_g \propto \frac{1}{k}$ with band-offset considerations for electrons and holes to be greater than 1eV transforming into equivalent E_g greater than 5 eV. With these features, metallic oxide HfO_2 can be considered the best suited for the fabrication process. Hafnium silicates and aluminates possess crystallization characteristics only at temperatures beyond 450°C with minor reduction in K values [3, 4, 6, 10].

4.2 MECHANISM OF HIGH-K DIELECTRIC IN METAL OXIDES

Metal oxides are dielectrics or insulators that undergo polarization under the effect of an electric field. Dielectric polarization happens in the form of displacement of atomic oscillators with respect to their mean position. Types of polarizations include electronic, ionic, and orientational [6]. Electron-charged clouds-nucleus couples that experience rearrangements under the effect of a force exposed due to an externally applied electric field E create electronic polarization. This is caused by the generation of distortions in symmetrical orbital motions of negatively charged electron clouds around the positively charged nucleus to induce dipole formations and hence electronic polarization. Ionic polarization happens because of the pairing of oppositely charged species under an external E

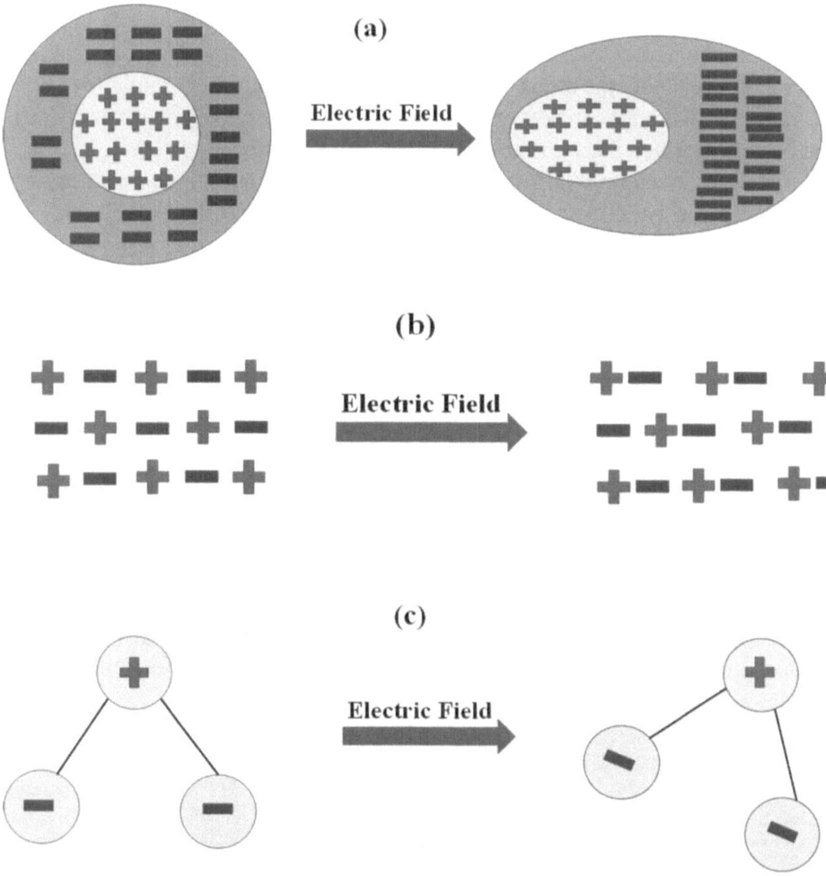

FIGURE 4.1 Schematic representation of (a) electronic polarization to show dipole formation in charged species in an asymmetrical manner under E from a symmetrical arrangement under $E = 0$, (b) ionic polarization that happens because of development of a net dipole moment owed to ionic displacements under E, and (c) orientational polarization that develops because of alignment of randomly oriented permanent dipole moments in molecules under the effect of E [6].

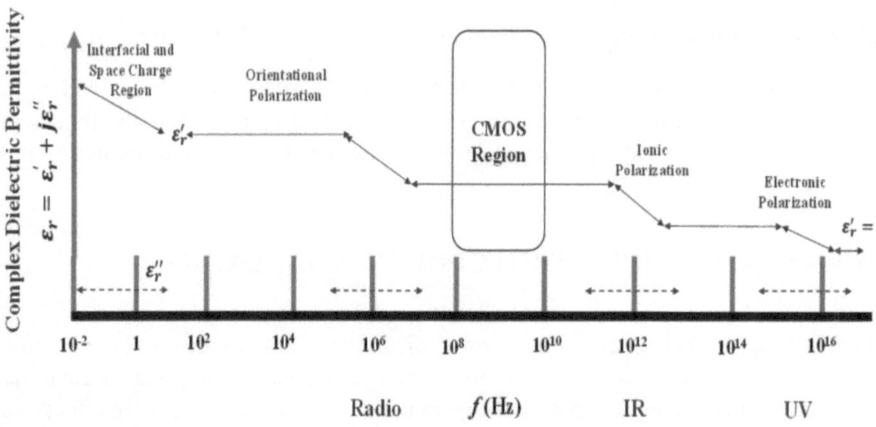

FIGURE 4.2 Schematic to show complex dielectric permittivity as a function of a range of frequencies depicting different regions of dielectric polarizations from low frequency to one as high as 10^{16} Hz. Complex dielectric permittivity ($\varepsilon_r = \varepsilon'_r + j\varepsilon''_r$) is plotted on the y-axis while frequency is displayed on the x-axis. Here, ε'_r is the real part and ε''_r is the imaginary part in the expression for frequency dependence of dielectric permittivity [4] (Open access image).

to induce a net dipole moment. Orientational polarization happens due to the rearrangement of permanent molecular dipoles under the effect of an externally applied E. These randomly aligned dipole moments in a range of materials already exist under zero E that are rearranged in the direction of E. This can be best represented through schematics shown in Figure 4.1 for (a) electronic, (b) ionic, and (c) orientational polarizations. Different forms of dielectric polarizations do not reflect in a uniform manner under a range of frequencies of the time-varying electric field, i.e., under *ac*-frequency. The complex dielectric permittivity is given as $\varepsilon_r = \varepsilon'_r + j\varepsilon''_r$, where ε_r' is the real part and ε_r'' is the imaginary part in the expression for the frequency dependence of dielectric permittivity. Stephen Hall has given an account of different operational frequency windows generating different types of dielectric polarizations as shown in Figure 4.2 [4].

At higher frequencies, electronic (around 10^{16} Hz) and ionic polarizations (around 10^{12} Hz) happen. Net permittivity ε_{ox} consists of contributions from electronic ε_∞ and lattice permittivities $\varepsilon_{lattice}$ where ε_{ox} is equivalent to K. Electronic contribution arises due to the electric field polarization of the atoms and is the major one in the case of SiO_2. Optical transparency of dielectric materials is expressed by the correlation of electronic permittivity ε_∞ refractive index n by the relationship, $n \sim (\varepsilon_\infty)^{1/2}$. This can be determined by spectroscopic ellipsometry under the relation $\varepsilon_{ox} \sim \varepsilon_\infty$. In that sense, a large lattice permittivity contribution is ensured by enhancing the value of K. Hall and colleagues have given ε_{ox} or K values as high as 32.81 for different forms of crystalline HfO_2 [4]. The dependency of K on structural variations in metallic oxides is given by Clausius-Mossotti (C-M) theoretical considerations. The C-M theory expresses variations in K as a function of polarizability α and volume of unit cell V_0 in the manner α/V_0,

$$\varepsilon_r = \frac{\left(1 + \frac{2}{3} 4\pi \frac{\alpha}{V_0}\right)}{1 - \left(\frac{1}{3} 4\pi \frac{\alpha}{V_0}\right)} \tag{4.2}$$

This variational dependency shows that K is heavily dependent on structural and crystalline parameters of the material. Another point is that large-sized atoms possess higher degree of polarizability and hence larger K values [4].

4.3 HIGH-K METAL OXIDES IN MOSFETS

Metal oxide field effect transistor has become a core device structure at the heart of modern smart devices that include smartphones, laptops, bioelectronic devices, defense applications, safety devices, smart watches, energy devices, and in many other fields [7–9]. MOSFET structure consists of a source, a drain, and a gate between them just above the channel region to provide for switching action in a solid-state device configuration. The gate is capacitively coupled to the channel region, and, under biasing conditions, carriers are pulled toward the channel region along with the flow of current from source to the drain. Biasing of capacitively coupled gate structure generates an electric field that electrostatically induces dielectric polarization in gate oxides so that charges are accumulated at the interface of gate oxides and semiconductor surfaces. Output or drain current (I_D) is expressed in case of transistors as [6]

$$I_D = \frac{W}{L}\mu C\left(\left(V_G - V_T\right) - \frac{V_D}{2}\right)V_D \tag{4.3}$$

Here, W is the channel width enclosed between source and drain, L is channel length in a direction perpendicular to W, μ is mobility of charge carriers, C is capacitance, V_G is gate voltage, V_T is threshold voltage, V_D is drain voltage, and I_D is output or drain current. Equation (4.3) is true under the condition $V_D < V_G - V_T$ so that a linear current conduction happens in the uniform I_D density configuration. On the other hand, under the condition, $V_D \geq V_G - V_T$, pinch-off happens at the drain so that saturation happens. In this condition, I_D under the saturation output current regime is expressed as [6]

$$I_D = \frac{1}{2}\mu C_{ox}\frac{W}{L}\left(V_G - V_T\right)^2 \tag{4.4}$$

Here, I_D is the drain current, μ is electron mobility, C_{ox} is oxide capacitance, W and L are channel width and lengths respectively, V_G is gate voltage under bias, and V_T is the threshold voltage. This situation is equivalent to the formation of a parallel-plate capacitor model. Hence,

$$\text{Capacitance}\left(C\right) = \text{Cox} = k\,\epsilon_0\,\frac{A}{d} \tag{4.5}$$

Here, A is the area of the capacitor that is enclosed in the channel region between source and drain so that $A = W \times L$, d is oxide thickness, $k = \frac{\varepsilon_d}{\varepsilon_0}$ is the relative dielectric constant with ε_0 as permittivity of free space and ε_d as permittivity of dielectric medium. From the above relationships it is clear that with reduction in gate oxide thicknesses in accordance with highly reduced dimension devices, C increases so that more of charge carriers can be attracted at the same gate bias or V_G and output current at the drain is enhanced for optimized circuit operational speeds. Thus, by biasing the gate oxide through a metal contact, gate oxide is polarized so that a net dipole moment is developed and hence surface charge density at two sides of capacitor. This polarization effect along with simultaneous modulation of channel resistivity lies at the core of operation of MOSFETs. Switching from linear regime to saturation regime happens is indicated in electrical curves of I_D versus V_D under different fixed gate voltages V_G.

Diffusion current I_{Diff} happens because of concentration gradient induced by application of drain voltage until formation of conduction channel takes place. Conduction mechanism under the subthreshold regime is ascribed to this diffusion current. Electrical curve of I_D versus V_G

provides information about MOSFET characteristics in form of subthreshold swing (SS), which entails V_G range essential to make enhancement in I_D by one order of magnitude in the plot of $\log(I_D)$ versus V_G. The SS is directly linked to the presence of deep traps with energies falling in the band gap energy range of semiconductors. Such deep traps are detrimental because they retard the movement of charge carriers introduced by capacitively coupled gating mechanism [4, 7],

$$SS = \frac{kTln10}{q}\left(\frac{q^2N_t}{C}+1\right) \tag{4.6}$$

Here, N_t denotes density of traps, k is Boltzmann's constant, q is elementary charge, and T is absolute temperature. Slope of SS determined from plot of $\log(I_D)$ versus V_G is actually an inverse function of capacitance as expressed by Equation (4.6). This entails the necessity of a highly capacitive gate structure in MOSFET configurations to experience very steep gate modulation of output currents I_D. Hence, dielectric materials with high K values can effectively tailor the ON/OFF mechanism in MOSFETs by enhancing the overall current switching slope. Defect-free metal oxide can possess substantially higher dielectric constant values at extremely reduced thicknesses of nanolayers without showing much tunnelling leakage current.

High-K metal oxidic nanolayers must be of high dielectric constant to have high capacitive features aerially, and be devoid of dangling bonds, intrinsic defects, dislocations, and imperfections at the level of unit cell volume along with higher degree of surface smoothness by ensuring a uniform deposition over large area substrates. High capacitance ensures larger amounts of charges that drive the current, effectively resulting in higher electron mobilities, while high film density alleviates the problem of leakage currents, and defect-free surfaces ensure that the mechanism of polarization is not retarded. Moreover, interface between the dielectric nanolayer and semiconductor surfaces must have surface energies that are in tandem with each other to ensure charges accumulate positively and are not trapped at the unwanted sites. Additionally, dielectric materials should possess highly reduced defect densities and reduced fringing effects in terms of capacitance, along with ease of band gap engineering and minimal traps at the interface [1–9].

4.4 PROPERTIES OF METAL OXIDES

Devices in highly advanced compaction are already being developed with much faster logic and memory operations. Integration of nanostructures is highly sought after for fabrication of next-generation devices. In the current CMOS technology, device scaling is possible by minimization of oxide-equivalent thickness (EOT) of metal oxides to form gate oxides. A major concern is the exponential increment in tunneling leakage currents with scaling of EOT of gate oxides. As outlined before, EOT can be best expressed with the expression,

$$EOT = \left(t_{MO}\frac{k_{SiO_2}}{k_{MO}}\right) = \left(\frac{k_{SiO_2\varepsilon_0}A}{C_{ox}}\right) \tag{4.7}$$

where t_{MO} is the thickness of high-K metal oxide nanolayer and k_{MO} is the dielectric constant of high-K metal oxide, while k_{SiO2} is the dielectric constant of SiO_2. Similarly, capacitance-equivalent thickness (CET) is another term that determines total capacitance of the metal oxide–semiconductor stack as opposed to C_{ox} appearing in Equation (4.7). CET provides a measure of capacitance in case it becomes difficult to determine capacitance arising from substrate via quantum mechanical mechanism [15].

Metal oxides as gate dielectrics must have high K values, suitable band gap values, compatibility with gate metal, and be thermodynamically stable with semiconductor substrates, preferably with silicon (Si), since Si fabrication technology is the prevalent commercial technology. Robertson has pointed out that band gap E_g has a reciprocal relationship with dielectric constant k in the manner, $E_g \propto \frac{1}{k^{0.65}}$ [15–17]. Based on this relationship, it can be easily deciphered that a very high K value will result in electrons crossing the energy barrier producing significant quantum-mechanical tunneling currents. HfO_2 with a typical ~20 of high K value exhibits E_g in the range of 5.5 eV to 6.0 eV, making it a potentially attractive material for application as a gate oxide with high dielectric properties [15]. Current fabrication technology nodes have come down to as low as 3 nm, reaching almost atomic scales. From this perspective, EOT ≤ 0.9 nm for continual gate length scaling after the fabrication limitation beyond 22 nm nodes has been crossed successfully. Threshold voltage V_T control is another important parameter tailored in highly performing logic circuits with the necessity of effective metal work function to be restricted under 0.3 eV in reference to conduction band edge E_C and valence band edge E_V in Si-based technology. Gate-first (GF) and gate-last (GL) are the techniques employed for successful integration into conventional CMOS processes. GF fabrication process is in high consonance with conventional CMOS process. In GF process, a high-K metal oxide is primarily required to bring down the costs involved due to thermal treatments [16]. GF fabrication process was first launched at the 32-nm process nodes by IBM [15, 18]. GL fabrication technology relies on dual metal gates with vacuum work functions in correlation to E_C and E_V of Si substrate [15]. This process also restricts the operating temperature of thermal treatment to under 500°C. In this technique, junctions are deposited before the fabrication of high-K metal gate stacks termed *GL integration*.

Different types of high-K metal oxides can be developed, but these must be compatible with the prevailing Si fabrication technology. Wilk and colleagues framed ternary phase diagrams pertaining to M-Si-O systems to discover the stability features of different metal oxide nanolayers deposited over Si substrates. Here, M in M-Si-O is the high-K metal oxide gate dielectric for implementation in MOSFETs. Both HfO_2 and ZrO_2 form high-K nanolayers over Si substrates that are thermally stable. Hf and Zr nanolayers may form silicates as well and have become the materials of choice for gate oxides. As a matter of fact, high-K metal oxides that do not form stable ternary compounds are actually separated into M_xO_y and M_xSi_y phases due to temperature-based processing.

4.5 SYNTHESIS OF HIGH-K METAL OXIDES

A range of physical vapor deposition techniques can be employed for the deposition of high-quality defect-free uniform high-K metal oxide layers confined in nanodimensions nearing atomic scales. Molecular beam epitaxy or MBE is a highly sophisticated technique relying on evaporation of precursor material to be deposited over semiconductor substrates with very high degree of control of composition, structure, and ultra-high purity epitaxial films. Similarly, atomic layer deposition (ALD) is another technique for the preparation of high-K oxide nanolayers with high degree of conformalities at the commercial level as well. ALD can be applied for the formation of high-K gate dielectrics compatible with Si technology in III–V gate stacks configuration.

4.5.1 Atomic Layer Deposition

ALD has played a crucial role in the development of high-K metal oxides integrable with conventional CMOS technology. ALD is pertinent for the dawn of highly advanced devices based on Fin-field effect transistors or Fin-FETs relying on multiple gate-stack formation, wide band gap semiconductor materials, and even in fabrication of high-quality interconnects, contacts, and metallization [19–21].

ALD is focused on repetition of a series of self-limiting surface reactions in a distinct manner in a rather chemistry-based approach for the deposition of nanolayers [19]. Two or more reactants are fed into the ALD chamber in a sequential manner. Each of these steps contribute to form t_1, t_2, t_3, t_4 ... etc. so that the pulse durations for two reactants constitute the "deposition cycle" that is total of each pulse durations i.e., t_1, t_2, t_3, t_4 ... In each of the serial steps, feed is pulsed inside the deposition chamber or the reactor for each of separate pulse durations. An associated term is growth-per-cycle (GPC) that remains constant for steady-growth regime of ALD. Under this GPC state, increment in film thickness is fixed in a unit deposition cycle. Co-reactants are allowed to involve in reactions that partake only with surface-active sites. Number of surface-active sites decreases with progression in reaction as a function of time, and the reaction is self-limited by a decrease in the number of surface-active sites. Reactor purging needs to be performed after introduction of each of the co-reactant to free reactor of by-products as well as unreacted species left behind. This is done to prevent incorporation of by-products and unreacted species, including cessation of gas-phase reactions, into the films under development. Sufficient enough longevity of a repeating chain of introduction of co-reactants and reactor purging ensures saturation GPC. Any increment in exposure of co-reactants or enhanced reactor purging has no bearing on saturation GPC.

Self-limiting surface reactions lie at the heart of ALD process technology that has numerous benefits including uniform layer deposition over large surfaces, control of film thickness at the sub-nanometer level, very low defects density and outstanding step-coverage of high-aspect ratio anisotropic features in non-planar geometry. Chemical vapor deposition is also an excellent technique for deposition of thin films over large area substrates. However, the major difference is that of self-restricting surface reactions in ALD so that saturation GPC is attained. ALD is a highly promising technique for the fabrication of high-K dielectric materials integrated logic and memory devices, FinFET-based devices, and in the development of next-generation nanoelectronic devices. The operation is performed with a substrate placed at a specific temperature with introduction of co-reactants in a timely manner. A precursor is a material that consists of metal-based species (organic or inorganic) along with a co-reactant (hydride, molecular gases, or even a plasma-enhanced species). Self-restricting thermodynamically stable surface reactions in case of precursor reactant and the co-reactants is maintained by appropriate choices of precursor pulse t_1, post-precursor purge t_2, reactant pulse t_3, and post-reactant purge t_4.

4.5.2 ALD: Precursor and Reactant Surface Reactions

Muneshwar and others have analytically discussed surface reactions involved in an ALD process [19]. Let the growth of a compound XY_2 be modeled using the precursor reactant of XL_4 and reactant YM_2. Here, XL_4 and YM_2 are the source of atomic species X and Y. This reaction is stated as follows:

$$XL_4 + 2.YM_2 \rightarrow XY_2 + 4.LM(\uparrow) \tag{4.8}$$

L in the precursor material and M in the co-reactant are the species that are removed in the form of LM in the ALD process reaction as pointed in the preceding sections. The above reaction happens on introduction of precursor and co-reactant one by one into the reactor. This process is schematically presented in Figure 4.3. Before the start of the reaction, the substrate surface is terminated with Y-M* species and is subsequently subjected to precursor pulses. The role of Y-M* species is that they function as core reacting sites to provide space for surface reaction of a molecule of XL_4 species with multiple Y-M* surface species to form $Y_\delta XL_{4-\delta}$ complex at the surface level. Here, it is considered that a δ number of species of L are liberated in the form of side products under the condition, $1 < \delta \leq 3$. The equation to represent release of surface complex product is

$$\delta.(Y - M^*) + XL_4 \rightarrow Y_\delta XL_{4-\delta} + \delta.LM \tag{4.9}$$

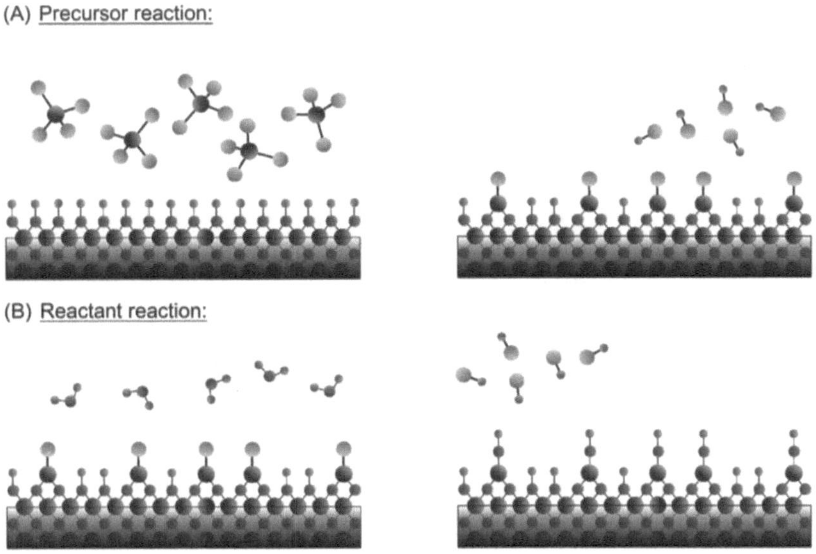

FIGURE 4.3 Schematic representation of reactions in an ALD process to fabricate XY_2 involving precursor material XL_4 and co-reactant YM_4. Here, X, Y, L, and M are atomic species represented in blue, green, yellow, and red colors, respectively. Here, (A) represents a schematic of a precursor material–based reaction and (B) is a schematic representation of the co-reactant reaction [19]. (Reprinted with permission from Elsevier Inc.)

Muneshwar and others have stated that the amount of XL_4 molecules can be increased by enhancing the precursor pulse time (t_1), which in turn increases the surface density of $Y_\delta XL_{4\text{-}\delta}$ complex molecules [19]. The precursor reaction attains a self-limiting state when the surface-active reaction sites of Y-M* species are reduced to a significant degree. However, leftover ligands after this self-terminating precursor reaction offer steric hindrance in surface exposure of $Y_\delta XL_{4\text{-}\delta}$ complex species. Hence, it is essential to effectively purge the remaining unreacted XL_4 molecular species along with the side product LM from the ALD reactor chamber. This ensures prevention of inclusion of side product LM into the film undergrowth as well as avoids possible gas-phase reactions on sequential $Y.M_2$ pulse treatments.

Now, after the precursor reaction, substrate surface is enhanced with $Y_\delta XL_{4\text{-}\delta}$ complex units along with the remaining unreacted Y-M* species. YM_2 molecular species partake in reactions only with $Y_\delta XL_{4\text{-}\delta}$ complex surface species in a single unit to form a unitary XY_2 molecular compound. Co-reactant YM_2 molecular species quantity can be enhanced by increasing the pulse time t_3; a self-limiting reactant reaction happens after each accessible unit of $Y_\delta XL_{4\text{-}\delta}$ is exhausted to compensate for the growth of monolayered XY_2. The Y-M* species are regenerated in reactant reaction that are exhausted during precursor reaction. It is this specific feature that is typical of a linear ALD growth regime characteristic of exhaustion and regeneration of an equal Y-M* species during a deposition cycle responsible for the production of a fixed GPC. Figure 4.3 illustrates surface chemistry involved in precursor and reactant surface reactions consisting of physisorption and desorption at the gas–solid interface, physisorption followed by surface diffusion, chemisorption, and elimination of produced by-products. Kinetics of surface reactions is best represented in Figure 4.4 illustrating gas–solid interfacial reactions between pulsed XL_4 molecular species and Y-M* enhanced substrate surfaces at a particular temperature involving precursor reaction. Gas–solid reactions are depicted in Figure 4.4(A) and expressed in form of chemical equations as follows:

$$\delta.(\mathrm{Y}-\mathrm{M}*)+\mathrm{XL4}\underset{k^{Desorp}\ \leftarrow}{\overset{k^{Physisorp}\ \rightarrow}{\longleftrightarrow}}\mathrm{XL4}(2)\underset{k^{Backward}\ \leftarrow}{\overset{k^{Transient\ State}\ \rightarrow}{\longleftrightarrow}}\mathrm{XL4}(3)\xrightarrow[\rightarrow]{k^{Forward}}\mathrm{Y}\delta\mathrm{XL4}-\delta* \\ +\ \delta.\mathrm{LM}(\uparrow) \tag{4.10}$$

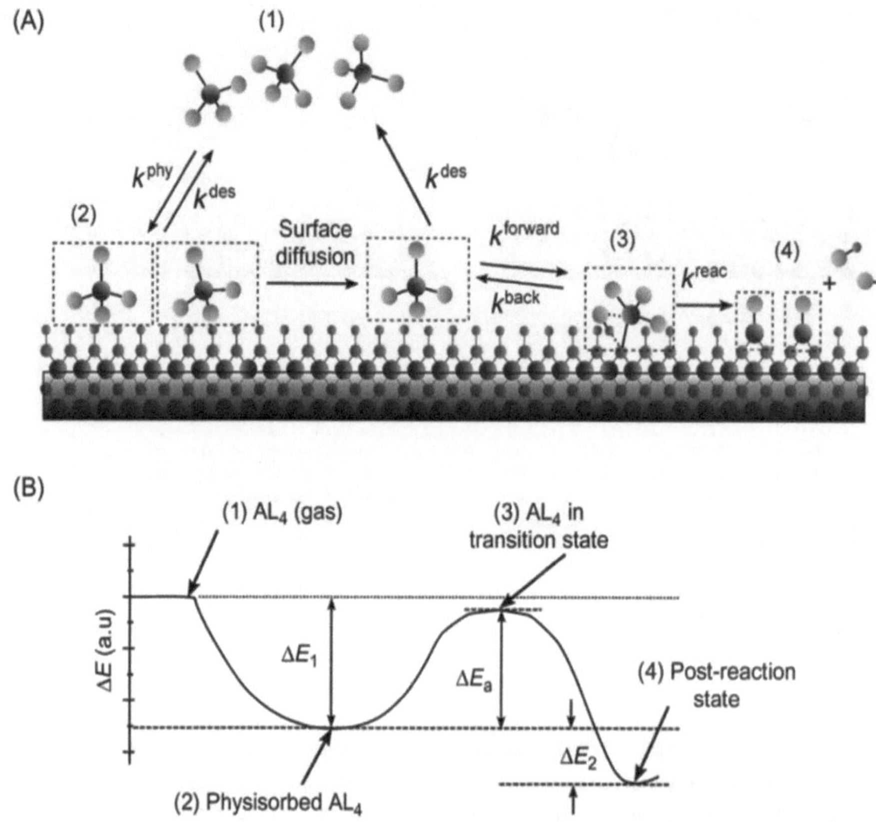

FIGURE 4.4 Schematic representation of (A) gas–solid interactions in precursor reaction in an ALD process chamber at a particular substrate temperature. Schematic in (B) shows associated reaction energy under the same conditions representing different chemical intermediate states in case of precursor XL_4 molecular species in different configurations of gas phase, physisorbed state, temporary chemical state, and post-reaction state of chemically sorbed molecular $Y_\delta XL_{4-\delta}$ species [19]. (Reprinted with permission from Elsevier Inc.)

Step-by-step explanations can be formulated for each of the process steps in the above equation. Initially, XL_4 molecular species are in the gaseous state, which is marked as step (1); they are physisorbed over the substrate, which is represented as step (2). It can also be observed that this physisorption process is associated with an energy change of $-\Delta E_1$. This is followed by formation of a temporary chemical state as visualized from step (3) due to either advanced reactions or desorption associated with a positive energy change of $(+\Delta E_a)$. After attainment of this temporary chemical state in step (3), precursor molecules may transform back to the physisorbed state. As another option, it may also move in a forward direction to form chemisorbed $Y_\delta XL_{4-\delta}$ complex species accompanied with a net energy change of $-(\Delta E_a + \Delta E_2)$.

Partial pressure of XL_4 gaseous molecules is directly proportional to the physisorption rate, and $k_{\text{desorption}}$ is proportional to $exp\left[-\frac{\Delta E_1}{k_B T_{Substrate}}\right]$. On the other hand, k_{TS}, or rate constant of formation of transient state, is proportional to $exp\left[-\frac{\Delta E_a}{k_B T_{Substrate}}\right]$. Lesser degree of speed of reaction or with the slowest kinetics out of these possible reaction steps form the governing equation or the rate-determining step for net surface kinetics [19].

Zhang and colleagues deposited HfO_2 thin films on a 4-inch *p*-type Si wafers by plasma-enhanced atomic layer deposition (PEALD) technique at temperatures of 100°C, 200°C, 300°C, 400°C, and 450°C. Plasma of oxygen/argon (O_2/Ar) was generated inside the PEALD system, and deposition was performed using tetrakis(ethylemethylamino)hafnium (TEMAH). Different parameters included bubbler temperature of 120°C, TEMAH pulse time of 1.6 sec, TEMAH purge time of 10 sec, O_2 pulse time of 10 sec, O_2 purge time of 12 sec, substrate temperature in a range of 100°C to 450°C, along with an applied plasma (O_2)Ar power of 2,500 Watts [22]. Different material property characterization techniques, such as grazing-incidence X-ray diffraction, were used for phase analysis of deposited thin layer, field emission scanning electron microscope for imaging of surface morphological features, and atomic force microscopy for surface topology along with roughness. X-ray photoelectron spectroscopy was performed to determine the chemical composition along with bonding orders in deposited HfO_2 thin films. Capacitance-voltage and current-voltage characterizations of Al/HfO_2/Si devices (formed by thermally evaporating Al contacts on top of HfO_2 films) were performed to obtain complete electrical characteristics. HfO_2 thin films were largely amorphous for deposition temperatures until 300°C at which point the onset of crystallization happens in the form of very faint X-ray diffraction (XRD) peaks. Fully crystallized thin films were derived for the deposition temperatures of 400°C and 450°C. The resulting change in density because of enhanced crystallization was measured by X-ray reflectivity (XRR) using a fitting model. The reported research work included an effective comparison of the obtained high K value of 18.21 at the deposition temperature of 300°C. It was better than other reported K values but slightly lower than the samples

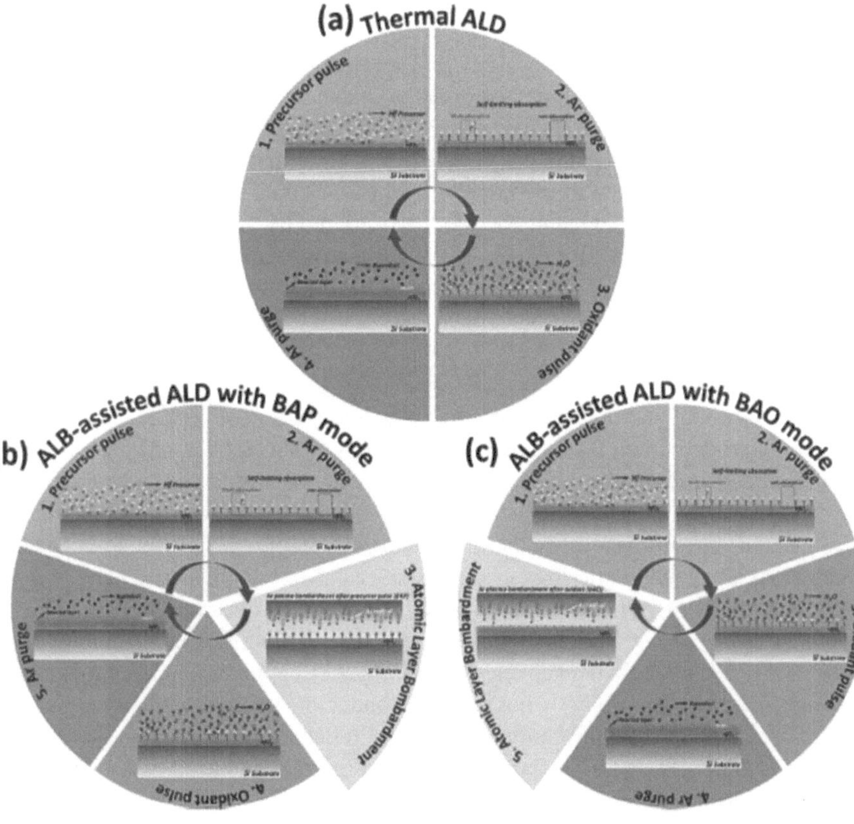

FIGURE 4.5 Schematic representations of (a) thermal ALD, i.e., routine ALD-based deposition, (b) ALB-assisted ALD-based deposition in BAP configuration, and (c) ALB-assisted ALD-based deposition in BAO configuration [23]. (Reprinted with permission from ACS Publications.)

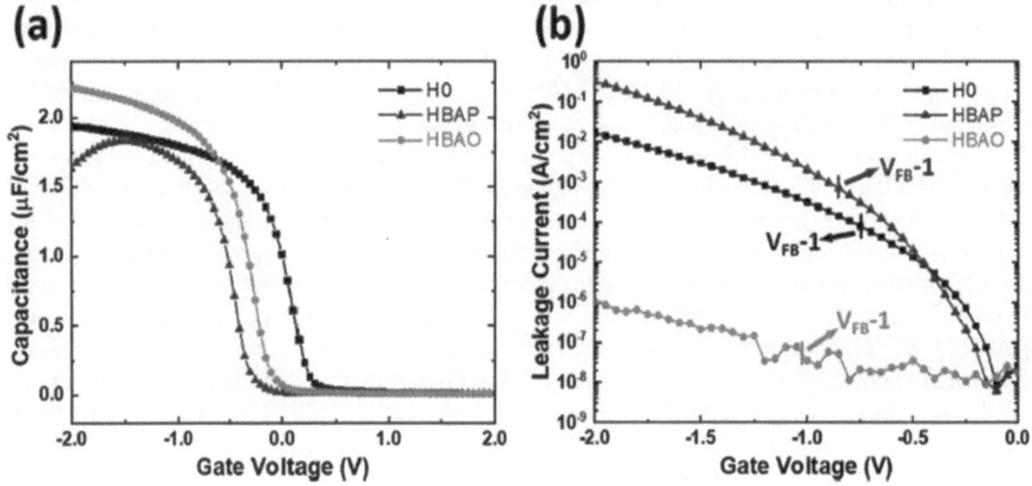

FIGURE 4.6 Capacitance-voltage (CV) measurements of H0, HBAP, and HBAO metal-oxide-semiconductor (MOS) capacitors. It is clearly visible from (a) that capacitance of HBAO is higher than that of H0 and HBAP thin film samples. (b) Leakage current density is highly reduced to three orders of magnitude in HBAO thin film sample in comparison to H0 and HBAP thin film samples [23]. (Reprinted with permission from ACS Publications.)

processed with post-deposition annealing [22]. Yin and colleagues applied in situ layer-by-layer nonreactive gas plasma treatment (atomic layer bombardment or ALB) in each the cycle of ALD (as shown in Figure 4.5) to obtain high-quality nanoscale HfO_2 thin films. ALB has special characteristics in that a reduction by three orders of magnitude in leakage current along with 26% enhancement of dielectric constant happens due to the bombardment-after-oxidation (BAO) process. Thus, this BAO ALD technique is an efficient method to produce ultrathin (typically 3 nm) high-K gate dielectric films with a low EOT of typically 1.1 nm along with reduced 6.77×10^{-4} A.cm^2 leakage current densities (as shown in Figure 4.6) [23].

4.5.3 Molecular Beam Epitaxy for Ultra-High-Quality High-K Epitaxial Films

Epitaxial films of ultra-high quality and defect-free characteristics are the penultimate requirements of the increasingly modern and ever-advancing fabrication of nanodevices. Oxidic dielectric films grown over the semiconductors must have isostructural features with the semiconductor, as well as a high degree of chemical adherence over the semiconductor along with good thermal stability. Moreover, oxidic dielectric films must be thermodynamically stable over the semiconductor to avoid interfacial reactions along with lattice parameters having a close resemblance to that of semiconductor substrates. Avoidance of formation of interfacial layer over Si substrate, silicate or silicide, is a major concern, be it the deposition of binary oxides or ternary oxides or even multicomponent oxide dielectrics. This poses a great challenge, since the interface offers higher complexities due to interactive transitions from purely covalently bonded semiconductors to predominantly ionically bonded oxide dielectric materials in typically shortened scales of less than 1 nm. As a result, semiconductor oxidation needs to be prevented effectively while providing sufficient oxygen to let flourish the growth of an oxide layer in a highly controlled manner induced by suitable thermodynamics reinforced by a highly suitable substrate temperature. On these terms, molecular beam epitaxy (MBE) is a technique that offers highly precise control for the growth of oxide dielectrics, growth rate, and in situ monitoring along with production of very high-quality, defect-free, electronic-grade, and even epitaxial layers [24–26]. MBE is an ideal technique for the development of crystalline oxide/semiconductor heterostructures for advanced nanoelectronics. MBE-based development of $SrTiO_3$/Si(001) is widely studied, while $SrHfO_3$, $BaTiO_3$, and $PbTiO_3$ are other good examples [27].

For next-generation highly advanced nanodevice processing modules, there is an urgent requirement to enhance the ON- or drive-current, highly efficient devices operating in high-frequency regime, and to achieve inherent MOSFET gate delay times in picoseconds or terahertz operating frequencies. High ON-current values, highly improved EOT thickness, channel mobility, threshold voltage and gate-length scaling are the needs of the hour. Nanometer-sized node technology aims at high-speed ultra-compact microprocessor units for high-speed electronics by implementing high-K dielectric/metallized electrodes–based stacks, or multigated stacks, with gate leakage currents at or below the acceptable maximum limits. Such advanced high-K dielectric multigated stacks experience mobility degradation. Low mobility problem can be eliminated by developing strained Si channels or even application of Ge or substrates of III–V compounds possessing high mobility characteristics in place of Si. For ultrathin EOT values, MBE offers a technique that can deposit ultraclean defect-free epitaxial oxide dielectric layers for nanometer-sized node processing technology [26]. Dimoulas has summarized the EOT of 1.2 nm, 0.8 nm, 0.7 nm, 0.6 nm, and 0.5 nm for physical gate lengths of 37 nm, 25 nm, 18 nm, 13 nm and 9 nm, respectively, for the fabrication nodes of 90 nm, 65 nm, 45 nm, 32 nm, and 22 nm arranged sequentially [26]. $La_2Hf_2O_7$, or LHO, can be grown epitaxially over Si(100) that has a close resemblance with crystal structure of Si, since it crystallizes in cubic pyrochlore geometry in space group symmetry of *Fd3m* [27]. A mismatch in lattice constant is only –0.74% at room temperature, with lattice constant of LHO being double to that of Si. Deposition is performed by co-evaporation of Hf and La emanating from e-beam and effusion cell sources in the presence of atomic O beam formed by application of an RF plasma source in O_2 flow inside the chamber. Atomic O becomes necessary to halt the growth of interfacial silicide formation during deposition and promote healthier oxide growth over Si. Temperature plays a decisive role in deposition so that nucleation of LHO crystallites happens in a preferred orientation for an ultimate growth in epitaxial configuration at high temperature of 770°C. Polycrystalline LHO films are grown at 700°C, and only amorphous phases can be obtained at temperatures less than 600°C. HfO_2 thin films can be deposited over reconstructed Ge(100) surfaces by heating the substrate at 360°C under vacuum. Reflection high-energy electron diffraction shows a 2 × 1 reconstruction indicative of an ultraclean Ge(100) surface. This is followed by growth of ultrathin GeO_xN_y enhanced substrate surfaces by a combined exposure to N_2 and O_2 atomic beams. This ultrathin starting layer becomes unstable above the temperature of 360°C, and hence growth is accomplished by evaporation of Hf atomic beams under atomic O beams in the usual temperature range of 60°C to 360°C [26]. Fissel and coworkers deposited high-quality crystalline films of rare-earth high-K dielectrics of Gd_2O_3 by MBE on 6H-SiC(0001) silicon carbide (SiC) substrates. Thin film formation took place in the form of islands in hexagonal configuration that eventually developed into flat layers consisting of a mixture of monoclinic and cubic bixbyite in (111) crystal orientation. MOS capacitors based on these films had a high dielectric constant of 22, leakage current density of 10^{-8} A.cm^2 at operational voltage of 1 V, and a breakdown electric field greater than 4.3 MV.cm^{-1} for Gd_2O_3 thin films with thickness higher than 14 nm [28]. Progressive downsizing of CMOS technology has compelled research on rare-earth metal oxides as potential dielectrics. Gd_2O_3 is an interesting rare-earth metal oxide that crystallizes in cubic α-Mn_2O_3 structure of a bixbyite type with a band gap of 5.2 eV under equilibrium conditions at highly reduced temperatures. Valence band offset of $\Delta E_V = 2.0$ eV was determined between Gd_2O_3 and Si that accounts for a conduction band offset of $\Delta E_C = 2.1$ eV in reference to Si. The band offset of Gd_2O_3 is represented schematically in Figure 4.7 in reference to Si, 3C-SiC, and 6C-SiC substrates. In this schematic, experimental offsets between Gd_2O_3 and Si, Si and 3C-SiC, and that 3C-SiC and 6H-SiC have been added. Comparative band lineups of Al_2O_3/6H-SiC and Al_2O_3/Si or HfO_2/SiC and HfO_2/Si follow the same configurational lineup as displayed in Figure 4.7 [28].

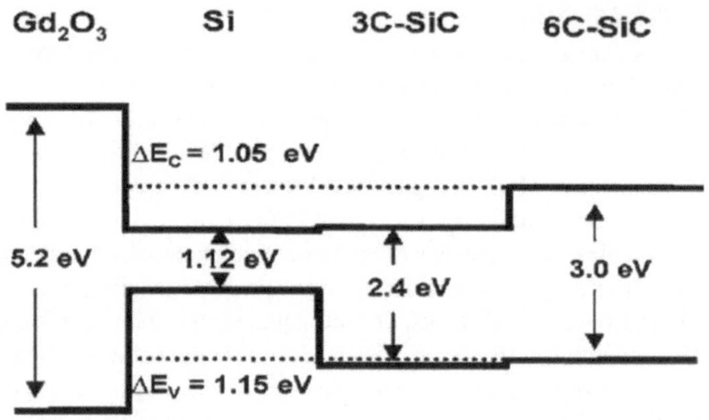

FIGURE 4.7 Band-offset of Gd_2O_3 films with Si, 3C-SiC, and 6C-SiC substrates [28]. (Reprinted with permission from AIP Publishing.)

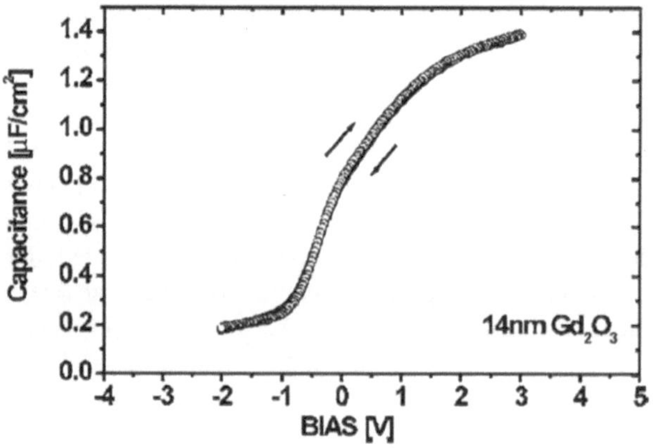

FIGURE 4.8 CV curve of a 14-nm thin Gd_2O_3 film–based MOS capacitor in structural configuration of Pt/Gd_2O_3/6H-SiC(0001) acquired at an operational frequency of 1 kHz. Data acquisition was not performed beyond 3 V for prevention of not-so-essential stress generation in film leading to breakdown [28]. (Reprinted with permission from AIP Publishing.)

In a real MOS capacitor, dipole formation in ionic oxides should be given due consideration at the interface of insulator and Si substrate based on specific interfacial structures in addition to the fixed and mobile charges. Dipolar field can induce shifts or higher misalignments in band offsets. Figure 4.8 shows a typical CV curve acquired at room temperature for a 14 nm Gd_2O_3 film at an operational frequency of 1 kHz using Pt-based electrodes, producing a high dielectric constant of 22. Both forward and reverse sweeps of CV curves showed only a very small hysteresis well below 0.1 V. This is owed to the negligible presence of defects density that induces low hysteretic characteristic.

Figure 4.9 shows an IV curve acquired at room temperature indicating leakage current density versus gate voltage plots for a 14-nm Gd_2O_3/6H-SiC capacitors reflecting leakage current density only in the range of 10^{-8} A.cm^{-2}. Wang and colleagues worked on deposition of epitaxial cubic-structured Gd_2O_3 thin films in (001) and (110) orientations grown over $SrTiO_3$ (100) and (001) oriented substrates possessing a uniform thickness of ~40 nm for dielectric applications [29].

Figure 4.10(a) shows the transmission electron microscope (TEM) image of a uniform Gd_2O_3 layer 45 nm in thickness deposited and annealed over highly flat STO substrate surfaces. Polycrystalline phases are marked in white, formed because of a lattice mismatch between Gd_2O_3 and STO or because of the low temperature of annealing. Figure 4.10(c) shows a clear lattice at the

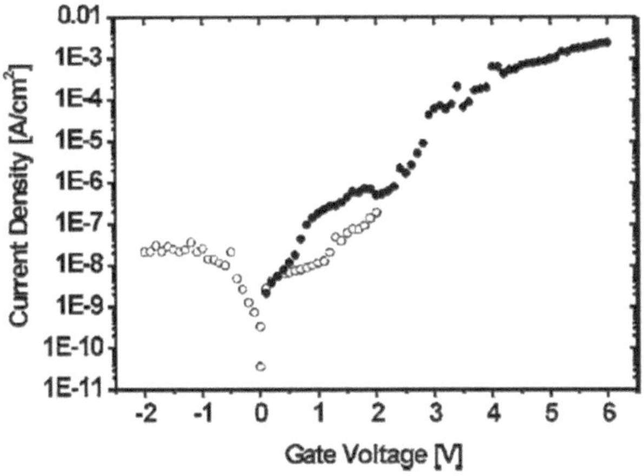

FIGURE 4.9 IV curves show an ever-increasing current density without any breakdown beyond 6 V. This value corresponds to an electric field strength of 10.2 $MV.cm^{-1}$ at surfaces of 6H-SiC with dielectric constant of 9.3, and that of 4.3 $MV.cm^{-1}$ over Gd_2O_3 surfaces. Leakage current density was determined to be ~10^{-8} $A.cm^{-2}$ in close resemblance with that of Gd_2O_3 over Si(001) surfaces in an approximate thickness. Gd_2O_3 successfully blocks electrons injected from SiC despite relatively lower band misalignment between Gd_2O_3 and 6H-SiC as displayed in Figure 4.8 [28]. (Reprinted with permission from AIP Publishing.)

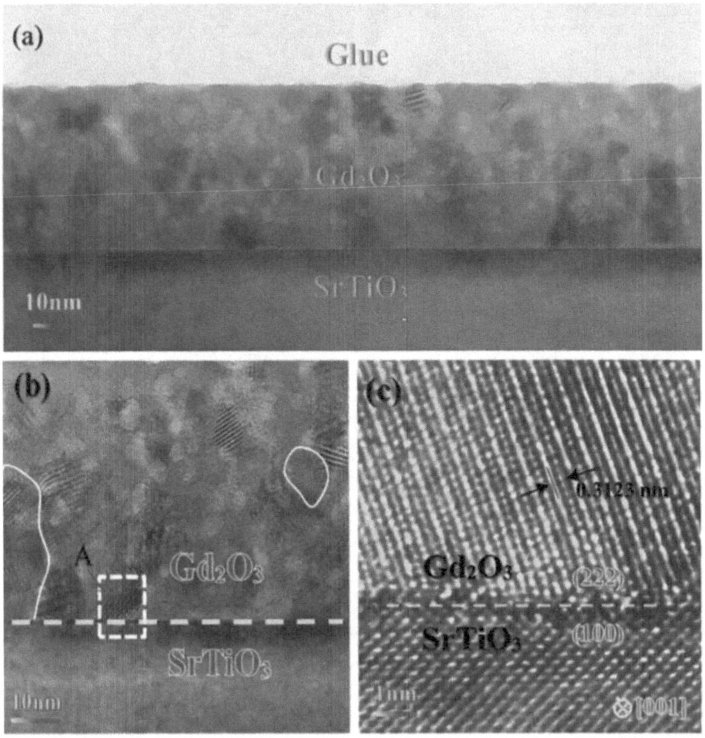

FIGURE 4.10 (a) TEM image acquired under low magnification of Gd_2O_3 thin films grown over STO followed by annealing at a temperature of 500°C. (b) HRTEM image of a cross-section showing the interface between Gd_2O_3 and STO. (c) A highly magnified area marked as A in (b), which is situated at the interface of Gd_2O_3 and STO. This shows a crystal orientation of (222) of Gd_2O_3 on (100) oriented STO with an interplanar spacing of 0.3123 nm [29]. (Reprinted with permission from Elsevier Inc.)

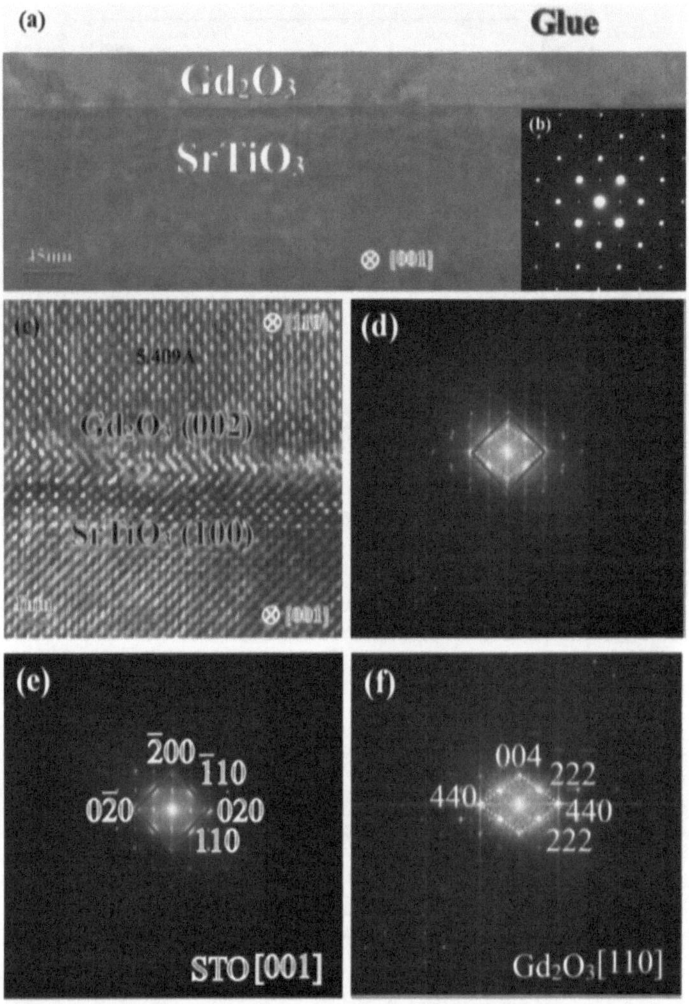

FIGURE 4.11 (a) TEM image of Gd_2O_3 thin films grown over STO followed by annealing. (b) The selected area diffraction pattern (SADP) of STO substrates. (c) HRTEM image of Gd_2O_3 over STO (100) substrates that have been annealed at 1,000°C. (d) Fast Fourier transform (FFT) image of the interface formed between Gd_2O_3 and STO. (e) FFT image of STO substrate in which arrays of diffraction spots from STO are marked in red. (f) FFT image of Gd_2O_3 films in which arrays of diffraction spots from Gd_2O_3 phase are marked in red [29]. (Reprinted with permission from Elsevier Inc.)

interface in a highly magnified HRTEM image of region marked as "A" shown in Figure 4.10(b), suggestive of a high degree of crystallinity of Gd_2O_3 thin films. Interplanar spacing was determined to be 0.3123 nm, which was in tandem with (222) crystal orientation of cubic-phased Gd_2O_3 in XRD.

Figure 4.11(a) shows a TEM image of Gd_2O_3 phase films grown over STO subjected to post-deposition annealing at 1,000°C. Figure 4.11(b) shows selected area diffraction pattern (SADP) of the STO (100) substrate acquired in a direction along [001] zone-axis. A clear-cut, well-demarcated, sharp, and coherent interface can be observed. A clear-cut, well-demarcated, sharp, and coherent interface can be observed. Figure 4.11(c) shows an HRTEM image as observed in a direction along [110] of the annealed film. Strong lattice film contrast is clearly visible along with sharp and distinct lattices at the interface. Interplanar spacing determined by digital micrograph (DM) technique is 0.5409 nm, which corresponds to (002) crystal orientation of cubic phase of Gd_2O_3, reflecting an epitaxial crystalline structure. Fast Fourier transform (FFT) micrographs are presented in

Figure 4.11(d), (e), and (f) to show electron diffraction patterns of interface, STO, and Gd_2O_3 phase. Figure 4.11(e) shows good alignment of Gd_2O_3 thin films with that of an STO substrate. Figures 4.11(e) and 11(f) show the epitaxial nature of grown films for (001)Gd_2O_3/(100)STO and (110) Gd_2O_3/(001)STO heterostructures. Su and coworkers grew amorphous MgO films with thickness in range of 3 nm to 30 nm over Si(001) substrates by MBE technique, applying high O_2 pressures of 5×10^{-3} Pa at a low substrate temperature of 200°C [30]. Dielectric constant of amorphous MgO films falls in the range of 8.74 to 9.8, with films grown using atomic O_2 displaying lower capacitances than the films grown using molecular O_2. Formation of a thin SiO_2 interfacial layer takes place in the presence of film growth using atomic oxygen that induces high defect densities. On annealing in forming gas at a temperature of 450°C for 15 minutes, formation of $MgSiO_3$ silicate happens, which decreases the capacitance while enhancing the leakage current. The silicate, $MgSiO_3$, possesses low dielectric constant of 6.6, band gap of 6.6 eV, along with a flat band voltage shift of ($V_{FB}= -0.6$ eV) assigned to diffusion of Mg into SiO_2 interfacial layer leading to appearance of O_2 vacancies.

Figure 4.12 shows the inverse capacitance versus MgO film thickness plot in which all points fall on a straight line with dielectric constant determined to be 8.74. This was determined to be lower than 9.8 of bulk-phase MgO owing to a 10% reduction in film density as determined by XRR measurements. A reduction in MgO film density leads to an apparent decrement in dielectric constant because of a reduction in the number of dipoles in the thin film. Figure 4.13(a) shows IV curves of 3-nm and 5-nm thin MgO film grown by MBE using atomic and molecular O_2, respectively. Lower leakage current densities are estimated for films grown under atomic O_2.

Figure 4.13(b) shows CV curves of 3-nm and 5-nm thin MgO film grown by MBE using atomic and molecular O_2, respectively. It can be seen that lower capacitances are observed for films produced by use of atomic O_2. The reduction in leakage current density is attributed to the formation of a thicker SiO_2 layer at the interface for film growth using atomic O_2, since SiO_2 has a higher band gap in comparison to MgO. A reduction in capacitance for film grown using atomic O_2 is due to the SiO_2 interfacial layer having a low dielectric constant.

Figure 4.14 shows the (a) electrical and (b) capacitance plots for MgO films that underwent post-deposition annealing at 450°C in the environment of forming gas. Annealing-induced changes in the form of increment in leakage current densities and decrement in capacitance values are clearly visible in case of post-deposition annealed thin films that are grown in the environment

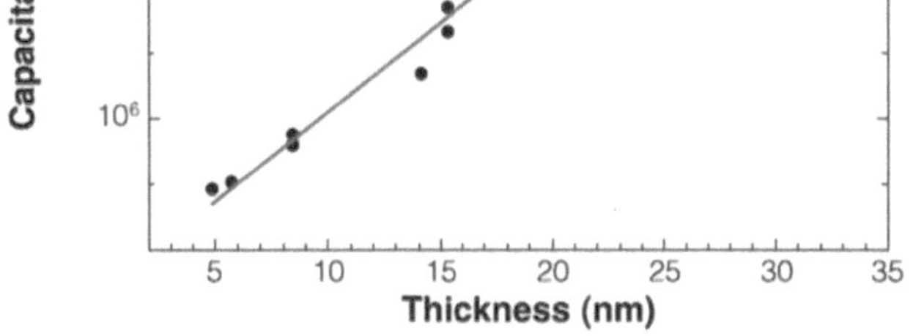

FIGURE 4.12 Inverse of capacitance (C^{-1}) ($cm^2.F^{-1}$) versus thickness of MgO films (nm) to determine a dielectric constant of 8.74. All points lie on a straight line as visible from the above curve [30]. (Reprinted with permission from Elsevier B. V.)

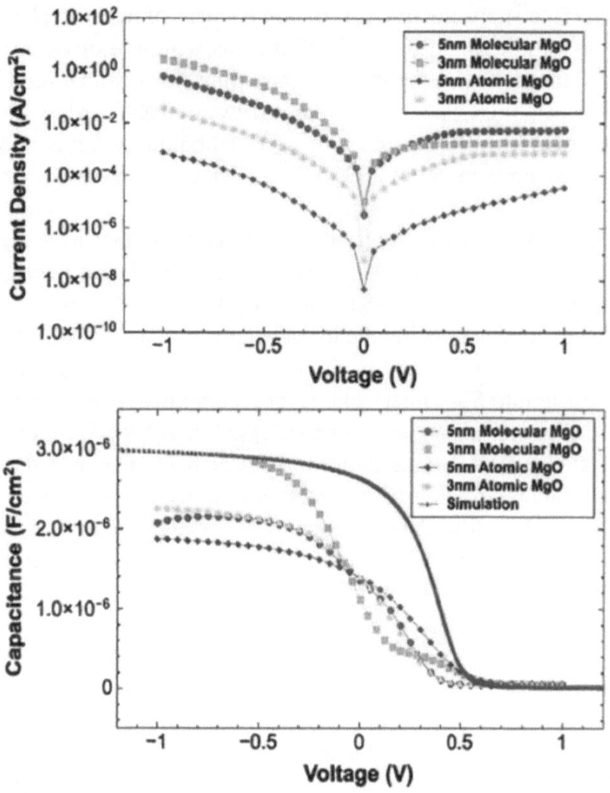

FIGURE 4.13 (a) IV plots for 3-nm and 5-nm thin MgO films grown over Si(001) substrates by MBE in the presence of atomic and molecular O_2. (b) CV plots for 3-nm and 5-nm thin MgO films along with a simulated curve line marked in dark yellow are shown [30]. (Reprinted with permission from Elsevier B. V.)

of molecular O_2. This was owed to the formation of $MgSiO_3$ silicate phase having dielectric constant of 6.6 and band gap of 6.6 eV [30].

4.6 PHYSICS OF HOMOGENEOUS NUCLEATION IN LATTICE-MATCHED CONFIGURATIONS

Homogeneity in epitaxial systems requires very low lattice mismatches between the depositing layer and the substrate along with tuning of strong interfacial bonding and low interfacial energy between them. Semiconductor epitaxial systems of IV–IV and III–V can be explained under the lattice mismatch epitaxy theoretical approach. The growing oxide layer acquires the in-plane lattice parameter identical to the substrate. In this process, the growing layer begins to experience an elastic deformation with an associated elastic energy. As the thickness of the growing layer continues to increase, this accumulated elastic energy may be compensated, initially partially and eventually completely, in the growth of dislocations in the depositing layer at a critical thickness. Hence, a mismatch can happen or even the higher degree of dislocations in the growing layer owed to plastic relaxation mechanism. This mechanism can severely degrade the quality of depositing layer for highly sophisticated electronic applications. Roughening of growing layers due to formation of Stranski-Krastanov growth model-based islands can happen in case of a high degree of lattice mismatch along with low surface energy of the growing layer. In this way, elastic strain is relaxed in the formation of three-dimensional (3D) islands accumulated because of the cessation of enhancement in surface energies [27].

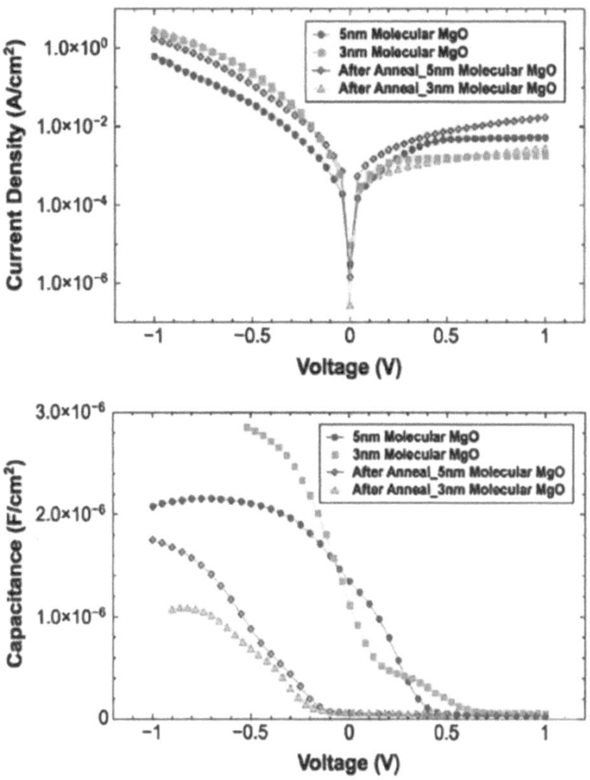

FIGURE 4.14 (a) IV plots for 3-nm and 5-nm forming gas annealed thin MgO films in the presence of molecular O_2. (b) CV plots for 3-nm and 5-nm forming gas annealed thin MgO films in the presence of molecular O_2 [30]. (Reprinted with permission from Elsevier B. V.)

4.7 PHYSICAL MODEL OF HETEROGENEOUS NUCLEATION SUBSTRATE ENHANCED EPITAXIES

In case of deposition of a material layer completely different than the substrate, 3D and two-dimensional (2D) structures-based growth of film can be defined in a spherical and cylindrical geometry, respectively. Two-dimensional structures in cylindrical geometry are classified to possess a height equal to a monolayer. These geometries are shown in Figure 4.5.

In 1901, Müller and coworkers produced nucleation-free enthalpies for a unitary volume mathematically, which was elaborated by Müller and Kern in 2000 and is stated as follows for 3D structures [27, 31]:

$$\Delta G^{3D} = \left(E_V - \Delta\mu\right)\times n + \left(\gamma_{AB} - \gamma_B\right)\times\frac{f(q)}{a}\times n^{\frac{2}{3}} + \gamma_i\times\frac{g(q)}{a}\times n^{2/3}$$

Here, n is the number of atoms in a gas phase condensing over substrate, $\Delta\mu$ is chemical potential between solid of substrate material and unitary volume of gas condensates, E_V is any associated energy (e.g., elastic energy), a is the height in a monolayer configuration, γ_{AB} and γ_B are, respectively, interface and condensate surface energies in J.m^{-2}, and $q = d/h$.

Similar formulation for 2D structures in cylindrical geometry is expressed as

$$\Delta G^{2D} = -n\times\left(\Delta\mu + \Delta\mu_C - E_V\right) + \frac{2\times\delta\times\sqrt{\pi}}{a^2}\times\sqrt{n}$$

Here, δ denotes the stepwise energy for each of atomic step ladder in the formation of 2D geometries in a growing layer and is of the order of 10^{-12} J.m^{-1} in case of III–V and IV–IV semiconductors.

Here, $f(q) = \frac{\pi}{4} \times \left(\frac{2}{\pi}\right)^{2/3} \times \frac{q^2}{\left(\frac{1}{3} + \frac{q^2}{4}\right)^{2/3}}$

$$g(q) = \frac{\pi}{4} \times \left(\frac{2}{\pi}\right)^{2/3} \times \frac{4 + q^2}{\left(\frac{1}{3} + \frac{q^2}{4}\right)^{2/3}}$$

and $\Delta\mu_C = -\frac{\gamma_{AB} - \gamma_{B+\gamma_i}}{a}$

ΔG becomes maximum for positive value assignments in case of n, which is denoted as n^* under the condition $\Delta\mu - E_V$ for 3D structures, and is written as

$$n_{3D}^* = \left(\frac{2}{3} \times \frac{\gamma_{AB} \times f(q) - \gamma_i \times g(q)}{a \times (\Delta\vartheta)}\right)^3$$

and $n_{2D}^* = \left(\frac{\delta \times \sqrt{\pi}}{a^2 \times (\Delta\mu + \Delta\mu_C - E_V)}\right)^2$

Here, n^* denotes the critical nucleation volume reflecting the formation of the smallest stable cluster over substrate surfaces during the initial growth of epitaxial layers. For positive values of ΔG happening for other values of n, nucleation does not take place. The nucleation energy barrier is represented by minimizing the highest value of $\Delta G(\Delta G^*)$ and is the most favorable condition. Formation of interfacial dislocations is compensated by $E_V = 0$, which enhances the interfacial energy. Mathews in 1989 stated that the formation of perfect Lomer-Cottrell dislocations could be represented as [27, 32]:

$$\gamma_{AB}^* = \gamma_{AB} + 2 \times m \times \frac{b^2}{a} \times \frac{E_{AB}}{2\pi} \times (1 + \ln(\chi))$$

Here,

(i) b denotes dislocation Burger vector,

(ii) m = mismatch strain = $\frac{(a_{Deposit} - a_{Substrate})}{(a_{Deposit})}$

(iii) $\frac{1}{E_{AB}} = \frac{1}{E_A} + \frac{1}{E_B}$ where E_A and E_B are Young moduli of substrate and growing layer, and E_{AB} is the combination of contribution of both of them simultaneously or net Young modulus of the substrate–layer system.

χ = Cutoff radius

d = Distance between two consecutive dislocations, where $\chi = \frac{h}{2a}$ holds true if $h < d$ and for other relationships of h and d, $\chi = \frac{h}{2d}$.

In case elastic deformation accommodates the lattice mismatch, this condition is written as

$$E_V = \varepsilon \times m^2 \times \left(1 - e^{-2kq}\right)$$

Here, $\varepsilon = \frac{E_V}{1 - \upsilon_A}$, where υ_A denotes Poisson's ratio of the layer under development while $\left(1 - e^{-2kq}\right)$ is the quantum of elastic energy given by nuclei surfaces. Value of *k*-is 0.082 for spherical caps, while it becomes reduces slightly to 0.073 to form cylindrical platelets.

4.8 SOL-GEL–DERIVED HIGH-K DIELECTRIC OXIDES

In this technique, a molecular precursor is dissolved in solution followed by polymerization at a temperature much below the melting temperature. Both glassy and non-glassy materials, especially thin oxide layers, have been synthesized using this method. This method is very simple, does not require cumbersome preparations, is cost effective, offers high stoichiometric control, and produces highly pure, homogeneous layers including nanoparticles in composite configuration. Sol is a stable solution consisting of molecular species of metal salts or alkoxides, which is eventually converted into a soft and fluffy solid-state interwoven network semisolid called a gel. This is due to hydrolysis-condensation reactions happening in sol that drive polymerization to form the gel. The obtained gel can be crystallized or even amorphous that can be recrystallized by heat treatment to form nanoparticles or composites in nanodimensions. A controlled heat treatment can drive enhancement in various lattice parameters of *a*, *b*, *c*, volume of unit cell V_0, and density ρ that generates an alteration in materials properties. It is also possible to make thin films of sol-gel–derived nanoparticles or nanocomposite particles by spin-coating and dip-coating. Spin-coating or centrifugation involves rotating or spinning the substrate consisting of molecular sol at an optimal rotations-per-minute (rpm) speed inside the chamber. The substrate is fixed on an axis by creating vacuum, and extra material over substrates is pushed outside by centripetal forces. Variations of rpm may result in variations of thickness of the growing layer; it is also affected by sol parameters such as viscosity, evaporation rate, and concentration.

Dip-coating is a process in which a substrate mounted on a stand is made to dip in a sol containing a molecular precursor solution. A fixed rate of dipping in a sol solution and taking it out results in the formation of a uniform layer by maintaining a suitable dipping rate. More uniform sol-gel–derived films over various types of substrates can be obtained by applying a scalpel followed by heat treatment. Heat treatment may be performed at room temperature under airflow or even in vacuum or may be done at a favorable condition to let the solution dry and acquire a solid deposit form in 2D geometry. Organic residues contained in the deposit are removed by calcinations or pyrolysis usually at a temperature between 300°C to 400°C, because their remaining in the material may increase its porosity. Selection of proper temperature is a critical matter, since higher temperatures may lead to crystallization of amorphous deposited phase or enhancement of the crystal size of the lattice structure. Rapid thermal annealing is another technique based on transmission of thermal power by radiation rather than use of conventional method of heating in a furnace based on convection. Each step can be repeated in a systematic manner to obtain the required film thicknesses. An extensive number of research reports on a variety of materials are available on synthesis of sol-gel–derived high-K dielectric thin films. It is practically impossible to mention all of them here, but a few are provided solely for illustration. Sarkar and colleagues used the sol-gel method to derive high-K dielectric titanium silicate ($Ti_xSi_{1-x}O_2$) films [33]. These films were fired at an optimum temperature of 600°C in the structural configuration of $Ti_{0.5}Si_{0.5}O_2$ that showed a high dielectric constant of ~23 along with minimized leakage current densities and dielectric losses. An important focus of this study is tuning of dielectric constant between that of SiO_2 (3.8) and TiO_2 (60) by ordinary control parameter of contributing TiO_2 content. Figure 4.15 shows the variation of dielectric constant,

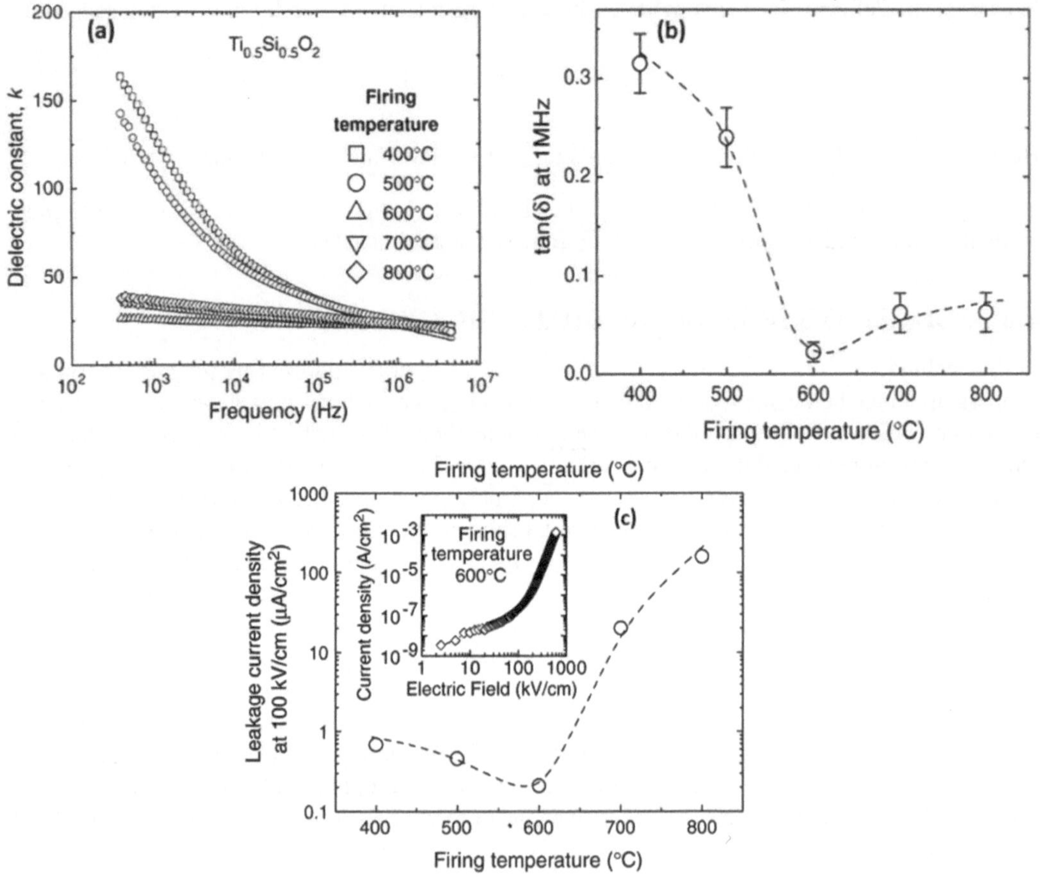

FIGURE 4.15 (a) The plot of dielectric constant (k) as a function of different calcination temperatures for $Ti_{0.5}Si_{0.5}O_2$ sol-gel–derived materials. It can be observed that the highest k value happens at a calcination temperature of 400°C followed by 500°C. Dielectric constant for films calcined at 600°C is well below 25 and does not show high dependence on operational frequency range. (b) Tangential loss (tan δ) plotted as a function of calcination temperature for the films. It is obvious that tangential loss (tan δ) is minimized at the calcination temperature of 600°C while it is maximized for calcination temperatures of 400°C and 500°C. (c) Leakage current density plotted as a function of calcination temperature for the films. Leakage current densities at 100 $kV.cm^{-1}$ are low in values for all calcination temperatures in microamperes of current for unit area ($\mu A.cm^{-2}$). Leakage current density is also minimized for films calcined at 600°C. Inset of (c) shows the variation of leakage current density as a function of applied electric field in $kV.cm^{-1}$. Based on low tangential loss and leakage current density, films calcined at 600°C are the optimized samples for whole study in this work [33]. (Reprinted with permission from Elsevier Inc.)

tangential loss (tan δ), and leakage current densities as a function of calcination temperature of sol-gel–derived films. Complete frequency-dependent behavior of dielectric constant is observed for calcined films in the temperature range of 400°C to 600°C. Typically, value of K changed from 130 to 25 for films calcined in this temperature range at a low operational frequency of 1 kHz. Dielectric constant did not show frequency dependence for operational frequencies greater than 1 MHz. For films calcined at 600°C, dielectric constant remains the same at ~23 for operational frequency range of 0.5 kHz to 13 MHz showing disappearance of relaxation. The tangential loss acquires the lowest value of 0.022 at an operational frequency of 1 MHz for this particular sample of a composite metal oxide film. This sample also exhibited the minimized leakage current density of 2.1×10^{-7} $A.cm^{-2}$ at an applied electric field of 0.1 $MV.cm^{-1}$. Therefore, films calcined at 600°C with high dielectric

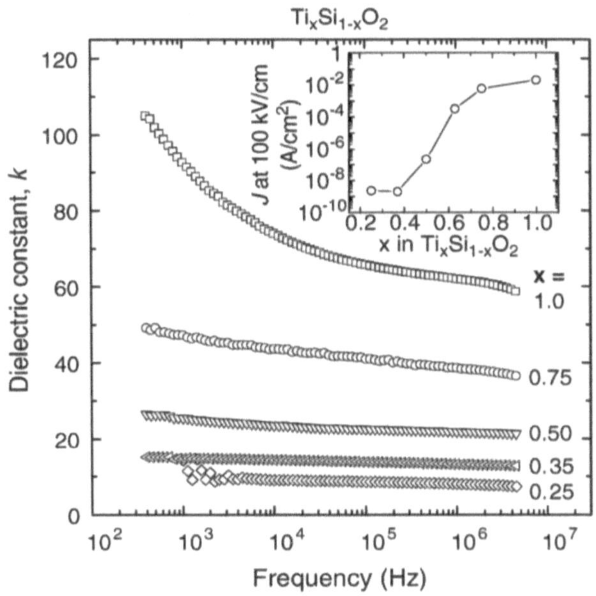

FIGURE 4.16 Dielectric constant is plotted as a function of applied frequencies (Hz) for $Ti_xSi_{1-x}O_2$ films calcined at 600° with x having values in the range of $0.25 \leq x \leq 1$. Inset shows the variation of leakage current density in $A.cm^{-2}$ at 100 $kV.cm^{-1}$ as a function of variation of x in $Ti_xSi_{1-x}O_2$ [33]. (Reprinted with permission from Elsevier Inc.)

constant and low tan δ along with low leakage current possess the best combination to form an optimized dielectric film sample.

Figure 4.16 shows the variation of dielectric constant as a function of applied frequency for different compositions of silicate by varying x in $Ti_xSi_{1-x}O_2$ so that k can be changed in the range of 8.5 to 38 by changing x in the range of 0.25–0.75. For $x = 1$, i.e., TiO_2, k value is determined to be ~60, and for $x = 0.5$, i.e., $Ti_{0.5}Si_{0.5}O_2$, optimized k value is ~23. With a decrease in TiO_2 content, dielectric constant also decreases, while TiO_2 induces percolation pathways in the composite film to give rise to higher leakage current densities. It can be observed from Figure 4.16 that TiO_2 shows the highest dependency of dielectric constant on applied frequencies, while this is very less so in case of $Ti_{0.5}Si_{0.5}O_2$ films with the lowest leakage current densities.

Figure 4.17 shows the variation of dielectric constant as a function of volume fraction of TiO_2 in which experimental values are analyzed in the framework of Maxwell-Garnet and Bruggeman theoretical models [27, 34]. The Maxwell-Garnet theoretical model states that the effective dielectric constant (k) of a material system in which spherical particles of a dielectric material (k_1) are randomly distributed in matrix of another dielectric material (k_2) is given as

$$k = k_1 * \frac{k_1 + 2k_2 + 2(k_1 - k_2)f}{k_1 + 2k_2 - (k_1 - k_2)f}$$

Here, f is the volume fraction of the dielectric material that is randomly distributed in the matrix of dielectric material. In this reported work, $k_1 = 60$ for TiO_2 and $k_2 = 3.8$ for SiO_2 matrix, and f belongs to volume fraction of TiO_2 in SiO_2 matrix.

The Bruggeman model gives the following formula for calculation of effective dielectric constant k:

$$(1-f)\frac{k_2 - k}{k_2 + 2k} + f\frac{k_1 - k}{k_1 + 2k} = 0$$

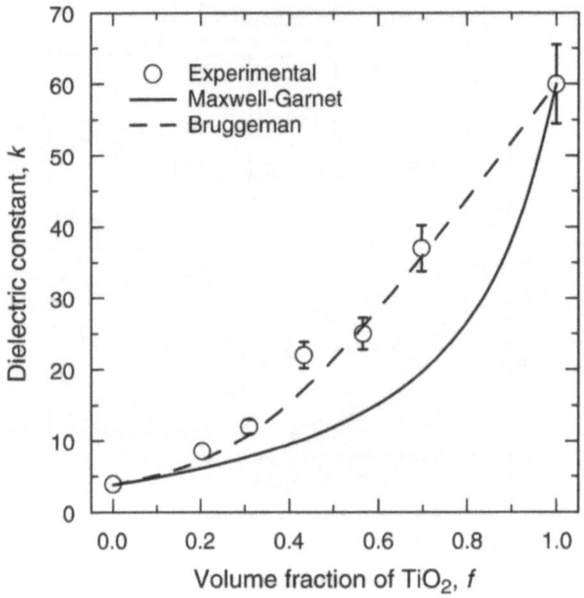

FIGURE 4.17 Dielectric constant (k) plotted as a function of volume fraction of TiO_2 (f) at an operational frequency of 1 MHz. Experimental data are presented in open circles while theoretical predictions are in solid line and dashed line for Maxwell-Garnet and Bruggeman theoretical fitting models [33]. (Reprinted with permission from Elsevier Inc.)

The difference between Maxwell-Garnet and Bruggeman theoretical models is that the Maxwell-Garnet model considers spherical inclusions in a host dielectric matrix, while the Bruggeman model does not take into account any shape of dielectric inclusions into a dielectric matrix but treats them in a completely symmetrical manner. In this study, experimental data fit well according to considerations of the Bruggeman model as visible from Figure 4.17.

Tetzner and colleagues synthesized HfO_2 dielectrics by the sol-gel method that could be annealed using the photonic curing technique to solve the problem of high temperature annealing that potentially damages organic substrates [35]. Reduction in leakage current density by more than three orders of magnitude was estimated due to photonic curing along with a dielectric constant of 26. Organic transistors based on these photonically cured dielectrics showed low operating voltages of less than 2 V, low subthreshold swing of 110 mV per decade along with mobilities of 1 cm^2.Vs by application of a semiconducting liquid-crystal polymer.

4.9 CONCLUSION

Development of high-K dielectrics is a core requirement for high-tech applications in the modern world. Semiconductor industry is essential not only for highly advanced smart electronic equipment but also for applications in other industries that include automobiles, defense sector, agriculture, environment protection, safety equipment, home automation tools, and many others. State-of-the-art nanofabrication technology has reached close to atomic dimensions as per the scaling law of Moore's law. As a result, ultrathin gate oxides are the need of the hour with exceptionally low leakage current densities to avoid the flow of electrons by quantum tunneling. A number of oxides are under study around the world to cater to these requirements and advance the miniaturization process. The candidates of particularly high potential for dielectric oxides are SiO_2, Si_3N_4, Al_2O_3, La_2O_3, ZrO_2, Ta_2O_5, HfO_2, $HfSiO_4$, TiO_2, and $SrTiO_3$. HfO_2 has dielectric constant of 25, TiO_2 of 80, with $SrTiO_3$ having dielectric constant as high as 2,000. Other complex materials are zirconium silicate ($ZrSiO_4$), hafnium silicate ($HfSi_xO_y$), gadolinium oxide (Gd_2O_3), yttrium oxide (Y_2O_3), lanthanum aluminate ($LaAlO_3$), and scandates of La, Gd, Dy, and many others.

Both chemical and deposition techniques can be applied for the synthesis of high-K oxides. For nanoelectronics, standard deposition techniques are used that can deposit oxides even on large area substrates. ALD and MBE are the preferred deposition techniques for clean device fabrication, while sol-gel–derived oxides and blends with polymers are also used for other applications. High-K dielectric oxides should have a band gap larger than 5 eV to act as a perfect insulator with high potential to resist quantum tunneling of electrons to prevent leakage current. Schottky emission to inject charge carriers into oxide bands must also be minimized to prevent leakage currents. Gate dielectric oxides must possess almost negligible electrically active defects. Charge carriers can get trapped in these electrically active defects. These electrically active defects are vacancies or impurities or consist of surplus or deficiency of oxygen on a general basis, and can even happen at interfaces. Such defects cause electrical failure, electrical breakdown of dielectric oxides, high leakage current densities, decreased electrical mobility due to scattering of carriers in channel by trapped charges, shift of gate threshold voltage leading to shift in turn-on voltages, and even instability in operational features. These are the parameters that must be given due consideration during the synthesis of high-K dielectric oxides.

REFERENCES

1. He, G., Zhu, L., Sun, Z., Wan, Q., Zhang, L. 2011. Integration and challenges of novel high-*k* gate stacks in advanced CMOS technology. *Prog. Mater. Sci.* 56, 475–572.
2. Clark, D. R. 2014. Emerging applications for high *k* materials in VLSI technology. *Materials* 7, 2913–2944.
3. Robertson, J. 2002. Band structures and band offsets of high *k* dielectrics on Si. *Appl. Surf. Sci.* 190, 2–10.
4. Hall, S., Buiu, O., Mitrovic, Z. I., Lu, Y., Davey, M. W. 2007. Review and perspective of high-*k* dielectrics on silicon. *J. Telecommun. Inf. Technol.* 2, 33–43.
5. Mohsenifar, S., Shahrokhabadi, H. M. 2015. Gate stack high-*k* materials for Si-based MOSFETs past, present and futures. *Microelectron. Solid State Electron.* 4, 12–24.
6. Hong, X., Liao, L. 2020. Metal oxide dielectrics, In: *Solution Processed Metal Oxide Thin Films for Electronic Applications*, Eds. Cui, Z., Korotcenkov, G. 31–39.
7. Dwivedi, S. 2022. Nanoelectronics, In: *Nanotechnology; Device Design and Applications*, Eds. Birla, S., Singh, N., Shukla, K. N. CRC Press, Taylor & Francis, 1–25.
8. Dwivedi, S. 2021. Graphene for flexible electronic devices, In : *Electrical and Electronic Devices, Circuits and Materials*, Eds. Tripathi, S. L., Alvi, P. A., Subramaniam, U. CRC Press, Taylor & Francis, 361–376.
9. Dwivedi, S. 2020. Fabrication and manufacturing process of solar cell : Part I, In : *Green Energy : Solar Energy, Photovoltaics, and Smart Cities*, Eds. Tripathi, S. L., Padmanaban, S. Scrivener Publishing LLC, Wiley, 1–37.
10. Biswas, D., Singh, M. N., Sinha, A. K. Bhattacharya, S., Chakraborty, S. 2016. Effect of excess hafnium on HfO_2 crystallization temperature and leakage current behaviour of HfO_2/Si metal-oxide-semiconductor devices. *J. Vac. Sci. Technol.* 34, 022201 (1–6).
11. Dwivedi, S., Biswas, S. 2013. Effects of annealing on pulsed laser deposited TiO_2 thin films. *Appl. Mech. Mater.* 446–447, 306–311.
12. Dwivedi, S., Biswas, S. 2016. Enhanced room-temperature magnetoresistance in self-assembled Ag-coated multiphasic chromium oxide nanocomposites. *Phys. Chem. Chem. Phys.* 18, 23879–23887.
13. Dwivedi, S., Biswas, S. 2018. Enhanced magnetoresistance in pulsed laser deposited stable chromium oxide thin films. *Thin Solid Films*, 655, 13–21.
14. Dwivedi, S., Biswas, S. 2014. Pulsed laser deposited ferromagnetic chromium dioxide thin films for applications in spintronics. *Phys. Proc.* 54, 62–69.
15. Ando, T., Kwon, U., Krishnan, S., Frank, M. M., Narayanan, V. 2016. Chapter 7: High-k oxides on Si : MOSFET gate dielectrics, In: *Thin Films on Silicon, Electronic and Photonic Applications, Materials and Energy*, Eds. Narayanan, V. 8, 323–367.

16. Robertson, J. 2008. Maximizing performance for higher *k* gate dielectrics. *J. Appl. Phys.* 104, 124111 (1–7).
17. Robertson, J. 2009. Model of Interface States at III-V Oxide Interfaces. *Appl. Phys. Lett.* 94, 152104(1–3).
18. Greene, B., Liang, Q., Amarnath, K., et al. 2009. High performance 32 nm SOI CMOS with high-k/metal gate and 0.149 µm2 SRAM and ultra low-k back end with eleven levels of copper. *VLSI Symp. Tech. Dig.* 140–141.
19. Muneshwar, T., Miao, M., Borujeny, R. E., Cadien, K. 2018. Atomic layer deposition : Fundamentals, Practice and Challenges. In: *Handbook of Thin Film Deposition*, Eds. Seshan, K., Schepis, D. Elsevier Inc. 359–377.
20. George, M. S. 2010. Atomic layer deposition : An overview. *Chem. Rev.* 110, 111–131.
21. Ritala, M., Leskelä, M. 2002. Atomic layer deposition. In: *Handbook of Thin Film Materials*, Ed. Nalwa, H. S., Vol. 1: Deposition and Processing of Thin Films. Academic Press. 103–159.
22. Zhang, X.-Y., Han, J., Peng, D.-C., et al. 2022. Crystallinity effect on electrical properties of PEALD-HfO_2 thin films prepared by different substrate temperatures. *Nanomat.* 12, 3890(1–14).
23. Yin, Y.-T., Jiang, Y.-S., Lin, Y.-T., Chang, T.-J., Lin, H.-C., Chen, M. J. 2020. Enhancement of dielectric properties of nanoscale HfO_2 thin films via atomic layer bombardment. *Appl. Electron. Mater.* 2, 2440–2448.
24. Ota, Y. 1983. Silicon molecular beam epitaxy. *Thin Solid Films* 106, 3–136.
25. Joyce, B. A. 1985. Molecular beam epitaxy. *Rep. Prog. Phys.* 48, 1637–1697.
26. Dimoulas, A. (2005). Molecular-beam deposition of high-*k*gate dielectrics for advanced CMOS. In: *Materials for Information Technology. Engineering Materials and Processes*, Ed. Zschech, E., Whelan, C., Mikolajick, T. Springer.
27. Niu, G., Saint-Girons, G., Vilquin, B. 2018. Epitaxial systems combining oxides and semiconductors. In: *Molecular Beam Epitaxy*, Ed. Henini, M. Elsevier Inc. 377–402.
28. Fissel, A., Czernohorsky M., Osten, H. J. 2006. Characterization of crystalline rare-earth oxide high-k dielectrics grown by molecular beam epitaxy on silicon carbide. *J. Vac. Sci. Technol.* 24, 2115–2118.
29. Wang, J., Hao, J., Zhang, Y., Wei, H., Mu, J. 2016. Molecular beam epitaxy deposition of Gd_2O_3 thin films on $SrTiO_3$ (100) substrate. *Physica E* 80, 180–195.
30. Su, C.-Y., Frederickx, M., Menghini, M., et al. 2012. Deposition and characterization of MgO/Si gate stacks grown by molecular beam epitaxy. *Thin Solid Films* 520, 4508–4511.
31. Müller, P., Kern, R. 2000. Equilibrium nano-shape changes induced by epitaxial stress (generalised Wulf-Kaishew theorem). *Surf. Sci.* 457, 229–253.
32. Matthews, J.W., 1989. *Dislocations in Solids* Vol. 2. Academic Press.
33. Sarkar, K. D., Brassard, D., Khakani, El A. M., Ouellet, L. 2007. Dielectric properties of sol-gel derived high-k titanium silicate thin films. *Thin Solid Films* 515, 4788–4793.
34. Sareni, B., Krähenbühl, L., Beroual, A. 1996. Effective dielectric constant of periodic composite materials. *J. Appl. Phys.* 80, 1688–1696.
35. Tetzner, K., Schroder, A. K., Bock, K. 2014. Photonic curing of sol-gel derived HfO_2 dielectrics for organic field effect transistors. *Ceram. Int.* 40, 15753–15761.

5 Tuning of Electrical and Optical Properties of Metal Oxide-based High-K Dielectrics

S. Mishra and Santosh Kumar Parida

5.1 INTRODUCTION

5.1.1 OVERVIEW

Condensed matter physics researchers are focusing on adaptable elements that are expected to garner significant interest due to their potential in solid-state electronic systems. Transition-metal elements draw the attention of researchers due to their fundamental electrical and magnetic properties as well as a wide range of unique and intriguing physical features with compositional variations, which hold significant technological promise [1–2]. Construction criteria for ordering elements at their respective sites include ionic radius and formal charge threshold discrepancies, as well as a phase diagram [3]. The symmetry of the structure to show the dielectric characteristics is determined by the degree of ordering and ionic size. The presence of mutually coupled magnetic and dielectric properties is of current interest in various transition oxides because it allows for the development of new materials with significant scientific and technological effects [4–5]. Since they are utilized in catalysis, magnetic media, electrical conductors, and gas sensors, among other things, there has been a lot of scholarly interest in these materials, with both the content and structural distortions at the lattice sites impacting the physical behavior [6]. As a result, a substantial amount of work on lead-based double perovskite has been described [7–10]. Extensive research has recently been conducted to develop novel lead-free alternatives to various lead-based ceramics with enhanced multiferroic characteristics for environmental protection.

Ceramic materials are typically made up of metallic and nonmetallic components. Ceramics' numerous applications are unavoidable. They are utilized in loudspeakers, environmental sensors, fuel cells, bulletproof vests, high-voltage insulators, and medical diagnostic equipment, all of which have an impact on our daily lives. Understanding the physics and chemistry of these materials in the recent period of new technology has resulted in tremendous advances in current studies. The ability of systems to perceive changes in their environs (i.e., temperature, pressure) to optimize the functioning (self-regulation) in a dominating manner (like live organisms) is a common feature of modern technology. Ceramics are characterized by transition metal oxides (TMO) as compounds with a high Coulomb interaction that lies between metallic and nonmetallic elements. Ceramic materials, namely ferroelectrics, are a subclass of dielectrics. Natural materials have been classified as conductors, insulators, and semiconductors based on their conductivity. Due to their many applications in appliances and devices for the improvement of human welfare, semiconductor-based technologies stand out among the rest as a subject of study attention. Any discipline's progress demands a solid base in terms of its physical, chemical, and biological advancements. Unwittingly or deliberately, the products of semiconductor technology (microelectronics), such as smartphones, television, cars, air conditioners, refrigerators, etc., have become a part of human civilization [11]. Transistors generated an amazing output for the global economy as a result of digitalization in the semiconductor

DOI: 10.1201/9781003352365-5

research ecosystem (from vacuum tubes to integrated circuits, junction diodes, optoelectronic, magneto-electronic coupling, etc.) and appropriate advancement in system integration and system-on-chip (SOC) demand [12]. Complementary metal oxide semiconductor (CMOS) and field effect transistor (FET), which were built on silicon-based oxides, were also developed during the electronic age (in particular SiO_2). Silicon has a greater application than the other materials when it comes to semiconductors due to its low power consumption and improved performance that complies with Moore's law of scaling (see Section 5.2 for further discussion). According to Moore's law, the number of devices with integrated circuits (IC) increases continuously, doubling exponentially over two to three years, in turn leading to an exponential decrease in transistor size (per year). However, there were restrictions due to silica (SiO_2) thinness, most likely over a range of 20 nm. As a result, materials with lower dielectric constants are utilized to replace SiO_2-based dielectrics, which cause the phenomena of gate leakage current and resistive-capacitive (RC) delay [13]. Direct tunneling of electron species through a very thin layer (SiO_2) of gate dielectrics results in a significant fraction of gate leakage current, which increases the dissipation energy [14–20]. Therefore, looking for a substitute for SiO_2 is necessary. Again, based on the formula $I = c{*}V{*}e^{-kd}$, where k and c are constants and V is the bias voltage, an exponential decrease like the current is seen as the distance increases. The well-known relationship $C = \frac{\varepsilon_0 KA}{t}$, which is framed by the variable's relative permittivity, area, and oxide thickness, plays a key role in the FET technology. Technologists are now concentrating on the development of materials with high dielectric constants (permittivity) (K). The high-K dielectrics continue to maintain capacitance while supplying a reduction in tunneling current. Since there were signs of gate leakage in the late 1990s, choosing new oxides to replace the silicon oxides that were already in use led to the appearance of these novel oxides. HfO_2 is an important oxide in this race. According to Murto and colleagues, some issues have been linked to high-K dielectrics, including lower equivalent oxide thickness (EOT), density flaws, loss of mobile charge carriers in the Si channel (when using high-K oxides), and shifting of the gate voltage threshold (possibly requiring metal gates) [21]. As a result, high-K dielectrics are needed to maintain Moore's scaling law, thermal and kinetic stability, and potential electrical interface qualities with the Si substrate, among other things. For this reason, numerous combinations from the periodic table have been used, such as $SrTiO_3$ used in dynamic random-access memory (DRAM) capacitors [22].

5.2 DIELECTRICS

Dielectrics are insulators that store electrical energy with little power loss because electrons are securely linked to atom nuclei and hence have no free charge. When a potential difference is placed across their ends, these are utilized to oppose the passage of current through them. All crystalline dielectrics in which polarization or an electric dipole moment can be created by applying an electric field are classified as (1) polar or (2) nonpolar. If the centers of the positive and negative ions do not coincide, the dielectric is said to be polar, and hence a permanent polarization, known as spontaneous polarization (P_s), persists even in the absence of an applied electric field. However, dielectrics are said to be nonpolar if there is no such permanent polarization because the positive and negative ion centers coincide with each other. The action of an external electric field on a dielectric substance produces dipoles, resulting in four types of polarization. Electronic or induced polarization, atomic or ionic polarization, orientation or dipolar polarization, and interfacial or space charge polarization are the four types of polarization.

5.2.1 Electronic or Induced Polarization

Electronic or induced polarizations are caused by interactions between the individual dielectric atoms and the local electric field that exists within the atom. Initialization of the dipole moment occurs in this regard as a result of the displacement of the atom's electronic charge cloud and concerning the positioning of the nucleus (under the effect of the localized electric field).

5.2.2 Atomic or Ionic Polarization

Due to localized field ions, shifting of anions and cations in dielectrics happens when the external AC field is present. Here, the motion of the corresponding positive and negative ions is in opposite directions, creating a dipole moment in the materials used to make the dielectrics. While this kind of polarization is known as ionic or atomic polarization, the induced dipole moment per unit localized field is known as ionic polarizability.

5.2.3 Orientation or Dipolar Polarization

These terms refer to the portion of the polarization in polar dielectrics that result from the orientation of the molecular dipoles. Each atom or molecule in a dielectric material experiences a local or microscopic field when exposed to an external electric field. The strength of this field depends on the size of the dipole moment and how the dipoles are organized around the atoms.

5.2.4 Interfacial or Space Charge Polarization

Accumulation of charges at the interface of two regions or materials produces the effect of interfacial or space charge polarization. The frequency range for the space charge polarization is shown in Figure 5.1. The charge accumulation at the interface of two materials with dissimilar permittivities and conductivities causes interfacial polarization. If the inclusion has higher conductivity and the matrix phase is insulating, it becomes more obvious. A bilayer dielectric is the most straightforward model; however, formulas have been created for more complex geometries as well. Conductivity of the inclusion has a major impact on the frequency of the interfacial (or Maxwell-Wagner-Sillars) polarization, while the inclusion's size, concentration, and orientation have an impact on the change in apparent permittivity that results. Therefore, mobile charge carriers in a heterogeneous material move a substantial distance through the dielectric before being trapped or hindered by physical barriers in interfacial or space-charge polarization. The polarization of the substance is determined by this accumulation of charge. Common barriers include free surface and grain/phase boundaries.

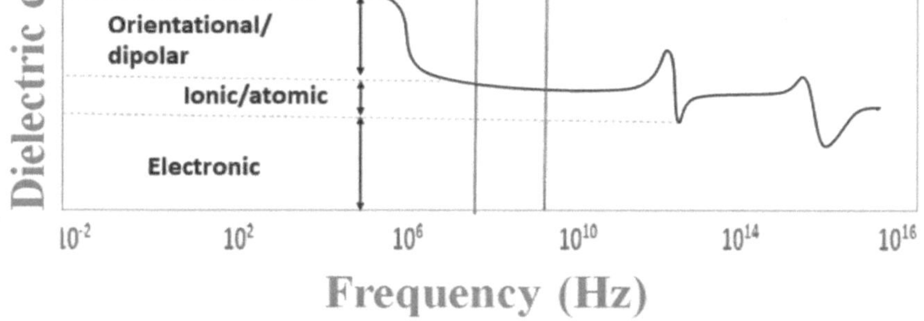

FIGURE 5.1 Schematic diagram that illustrates static dielectric constant on frequency.

5.3 CLASSIFICATION OF DIELECTRIC MATERIALS

According to the state of polarization, the dielectric materials are classified into different groups and subgroups, which are shown in Figure 5.2.

5.3.1 Centro- and Non-centrosymmetric Ceramics

Centrosymmetric has to do with the site symmetry in which an atom is located. An inversion center is one of the symmetry elements of such site symmetry. Certain crystal characteristics may be hindered by the inversion center. There are several methods for determining whether a material is centrosymmetric or non-centrosymmetric. By using X-ray powder diffraction, which will reveal the crystal system's point group, it is possible to examine the general structure. To check for potential disorders in your point symmetry, one can employ extended X-ray absorption fine structure (EXAFS) analysis for more detailed information. Another effective method is to examine the photoluminescent spectroscopy of materials that have been doped with Eu^{3+} to ascertain the site symmetry of a given system. The electric dipole transitions in the Eu^{3+} are suppressed in a centrosymmetric environment, making it a spectroscopic probe. Vibrational spectroscopy is another possibility. Because of this, infrared spectroscopy (IR)-active vibrations are not Raman-active in centrosymmetric crystals and molecules, and vice versa (mutual exclusion principle). However, not all Raman-active vibrations are also IR-active. In contrast, IR-active vibrations are also Raman-active if there is no inversion center. It can be inferred that the structure in question is a point group with an inversion center as one of its symmetry components. That indicates that for every atomic species represented by the R vector in the unit cell there is an identical atomic species stationed there signified by the R vector (negative). A non-centrosymmetric point group is a point group that lacks inversion symmetry.

5.4 FERROELECTRICS AND RELATED MATERIALS

Ferroelectric materials experience spontaneous polarization when no external electric field is present. The first experiment on this was conducted by Pierre Seignette in 1665. The reversal of spontaneous polarization is caused by the application of an electric field for a brief period in certain polar materials. This was originally seen in Rochelle salt ($NaKC_2H_4O_{6.4}H_2O$) by Valasek [23]. Rochelle salt crystals are widely utilized in piezoelectric devices like crystal microphones and phonograph

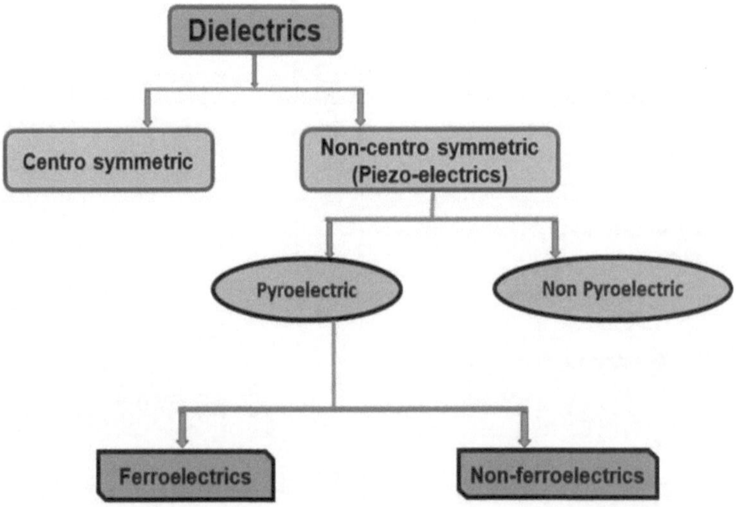

FIGURE 5.2 Dielectrics types.

pickup cartridges because they are easily manufactured. After the discovery of piezoelectricity and pyroelectricity in barium titanate ($BaTiO_3$), Wul and Goldman [24] were the first to identify ferroelectricity in it. The discovery of this substance sparked intense interest in the development of fresh ferroelectric ceramics. The nonlinear connection between polarization and the applied electric field is the distinguishing characteristic of ferroelectric materials [25–27]. The parameters hysteresis loop, coercive field E_c, remanent polarization P_r, saturation polarization P_s, etc. are of relevance in identifying and classifying ferroelectric materials. Different forms of loops are seen depending on the material's starting state of polarization, frequency amplitude, and applied field intensity [28–36].

Busch explored how symmetry affects the characteristics of crystals [37]. It is essential, but not sufficient, for the center of symmetry to be absent for a solid to be ferroelectric [38]. The symmetry around a point in space that corresponds to the center of a unit cell is used by crystallographers in the form of (1) rotational axes, (2) centers of symmetry, (3) mirror planes, and (4) combinations of the aforementioned. Depending on the symmetrical components of the crystal, there are 32-point groupings, 11 of which have a center of symmetry. By rising symmetry, these 32-point groups are separated into seven distinct crystal systems: triclinic, monoclinic, orthorhombic, tetragonal, rhombohedral (trigonal), hexagonal, and cubic. There are 20 piezoelectric and 21 non-centrosymmetric 32-point groups. The piezoelectric effect, which emerges when these 20 groups are subjected to stress, permits electrical polarity [39]. Only 10 out of 20 piezoelectric groups are pyroelectric, because the center of symmetry is devoid of pyroelectricity and there is either no or only one non-trivial axis of appropriate rotational symmetry. Even so, spontaneous polarization occurs in these crystals in the absence of an external electric field. A substance's interatomic location or interatomic bonding may vary as a result of a change in temperature, which alters the dipole's strength. The surface charges once more redistribute themselves to account for the interior polarization. The pyroelectric effect is the name of this phenomenon [40]. In a typical ferroelectric, it has been found that spontaneous polarization, P_s, decreases with an increase in temperature and becomes zero at Curie temperature, T_c [41]. Also, at Curie temperature, the ferroelectric material undergoes a transition from the ferroelectric to the paraelectric phase

Relative permittivity (ε_r), which is the ratio of the permittivity of the medium to that of open space, is one of the properties of ferroelectric materials. ε_r typically relies on the applied field's frequency, stress, temperature, and other factors [42]. At the transition temperature, Tc, the value of the dielectric constant reaches its maximum in double perovskite ferroelectric. The Curie Weiss law [43] states that $\varepsilon = \varepsilon_0 + C/(T - T_c)$, where ε_0 is the permittivity of free space, T_c is the para-electric Curie temperature or projected Curie point, and C is the Curie constant, and it governs the thermal fluctuation of ε_r above T_c. In single-crystal ferroelectric or other polar materials, the value of the dielectric constant, P_s, and other parameters are particularly sensitive to dopants and defects [44–47]. The polar dielectrics that now exist in ceramics are relatively strong and are thus easily changed by factors like porosity, particle size, sintering time, temperature, etc. [48]. The behavior of ceramics is relatively influenced by the domain wall's contribution [49–51]. At T_c, these ferroelectrics have strong piezoelectric characteristics. These crystals have no center of symmetry above T_c [52–54]. In this study, spontaneous polarization occurs in a polar class of materials because of internal symmetry within the unit cell of the crystal, which is where our main interest in the ferroelectric characteristics of materials lies.

5.5 MOSFET: A BRIEF INTRODUCTION

Ewald Georg von Kleist from Germany experimented with an electrostatic generator, wire, and water (an assemblage kept in a glass jar) in Germany back in 1745. A wire has been used to link the generator to a container of water, which demonstrated the storage of charge. Pieter van Musschenbroek, who created the Leyden jar and was the first to demonstrate a novel material property, put this concept into practice. Dielectricity is a property of materials that demonstrates a hybridization of conductor and insulator characteristics. Dielectrics often lack free electrons (like insulators), but they do have significant capacity for atoms to become polarized in the presence of an applied field (same as conductors). The concepts "dia" and "electric" were combined to form the phrase "dielectric," which was first used by William Whewell (with the support of Michael Faraday). It was discovered that a perfect dielectric had no conductivity. When the dielectric loss (amount of heat loss) is very modest, the dielectric—a material that retains energy better known as the dielectric constant—is regarded to be good. In the late 1980s, the development of CMOS and ICs transformed the industrial era [55]. Silicon-based units have advanced the microelectronic industry and astounded the research community with their results. In the later half of the 20th century, gate oxides were employed to isolate the metal-oxide-semiconductor field-effect transistor (MOSFET) gate terminal and resembled a dielectric layer. The nature of the metal oxide materials is explored by a characteristic study on metal oxide–based semiconductors (MOS). Gate dielectrics are inserted into transistors in the form of a thin layer and carry a heavy load, hence they are meant to be delicate. This section of the dielectric gets biased during operation and is intended to be very thin. Lilienfeld was the first to propose the field effect transistor, or FET, in the 1930s [56]. After nearly 30 years, SiO_2 was employed to create MOSFETs in the 1960s [57]. These transistors serve as the core component of logic-based ICs. In both p and n channels, complementary MOSFETs are specially created to link the gates with the drains. These devices' low power requirements make them useful for a variety of static applications, regardless of whether they are in a state of 1 or 0. Since the MOSFET state is "off," there is no energy loss other than when the switch is turned on (known as zero static power consumption). Once more, the clever idea of shrinking the size and density of the prepared devices—better known as scaling of transistors [58–60]—improves the practicality of ICs.

5.6 MOORE'S LAW: A HISTORY OF SMART TECHNOLOGY RENAISSANCE

Gordon Moore, a cofounder of the Intel Corporation, produced a remarkable hypothesis in 1965. The exponential rise of transistors per square inch on integrated circuits was given a new definition by Moore, who had a background in both physics and chemistry [61]. Based on technological advancement, Moore's law declares that the total number of transistors on ICs doubles roughly every 18 months (in terms of price), which is thought to be the explanation for the outputs' low cost. This has led engineering firms to abandon the notion of functional concentration in favor of management focus. The wide acceptance of Moore's theory allowed the creation of new hardware and software systems, including computers, smartphones, and numerous technical devices on the platform of power, vision, memory, communications, etc. Transistors are essentially silicon and carbon molecule–based devices designed to transmit electricity through the diffusion of mobile ions. The speed at which processes are completed in computers depends on how quickly the ICs react. Hardware needed for the reasonably priced creation of enterprise resource planning (ERP) systems is present in modern PCs.

5.6.1 Moore's Law and the Development of Enterprise Resource Planning

ERP systems have brought remarkable changes including architectural mainframe orientation and functionally evolved warehouses and industries. Details of an ERP system are given in Figure 5.3. For transistors smaller than 22 nm, a reduction in gate length is probably not conceivable. According to certain assumptions (including Moore's remark), this law will likely terminate by 2025 [61–64]. Misconceptions surrounding Moore's law have been long-lasting. Intel recently created a transistor with a gate length of less than 22 nm. However, progress and current trends indicate that this novel approach will endure as long as there is demand for technologically oriented research.

5.7 TRANSITION METAL OXIDES

These metals contain 5d ions, which are particularly intriguing due to their reported diversity in magnetic characteristics. The strong spin-orbit coupling and broad spatial range of the d-orbitals are

FIGURE 5.3 Various systems included in the ERP database.

assumed to be responsible for these peculiar magnetic properties of 5d transition metal oxides. Susceptibility measurements at various temperatures are the standard method for researching magnetism. The Curie-Weiss law can be used to determine more details about the magnetic behavior by examining the paramagnetic area of these measurements and can be written as $\chi = CT - \theta$, where χ is the magnetic susceptibility, C is a material-specific Curie constant, T is the temperature, and θ is the Weiss constant, which measures the strength of the coupling between magnetic ions. An antiferromagnetic coupling is indicated by a negative Weiss constant, whereas a ferromagnetic coupling is shown by a positive Weiss constant. The Curie constant can be related to the effective moment $(\mu_{eff}) = 2.84\sqrt{C}$. There have been discoveries of magnetic properties such as ferromagnetism, antiferromagnetism, and spin glass for the $5d^1$ and $5d^2$ electron configurations. There is just one unpaired electron in the d-orbitals in d^1 electron configurations. These compounds' magnetic properties are mostly a result of their structure and magnetic interactions with nearby atoms. Inorganic solids known as transition metal oxides (TMOs) have proved a fertile ground for many lucrative and frequently used phenomena, such as high-temperature superconductivity [28] and enormous magnetoresistance [29, 30]. Certain phenomena, such as a change in lattice orientation, a shift in composition, and an electrical effect in the space between crystallite surfaces, can occasionally occur in TMOs. It is quite difficult to conduct a situational analysis of ceramic characteristics. However, research into the characteristics of a single crystal's main constituents has provided important insights into the behavior of ceramics. Again, single-crystal formation is typically time consuming and challenging, while the complexity of ceramic microstructures makes it difficult to predict qualities from those of the matching single crystal. The transition elements are consistently notable in the history of technological advancement among the numerous elements contained in the periodic table. The IV-B group elements, such as oxides of hafnium ($_{72}Hf^{178}$), zirconium ($_{40}Zr^{91}$), and titanium ($_{22}Ti^{47}$), are among these. Hafnium has been identified as the most viable contender out of them due to its strong breakdown field (3.9–6.7 MV/cm), band gap energy (5.5–6.0 eV), dielectric constant (22–25), thermal stability, and heat of production (–271 kcal/mol).

5.8 SILICON-BASED TECHNOLOGY

It is well recognized in the scientific world that silicon dioxide (SiO_2) performs potentially suitably, serves as the building block for gate oxide dielectrics, and is a promising contender for CMOS technology. It is also a noteworthy means of trapping electrons and holes. The MOS-based applications have seen noticeably better results thanks to good compatibility (especially at high temperatures), fascinating intrinsic features, availability, etc. But for technology researchers, it is especially difficult to reduce transistor size and establish the infrastructure needed to design and produce ultra-thin chips to meet rising demand. There are two main issues. One is reducing the size of gate oxide dielectrics to manage the MOSFET channel's electrostatics issue. To prepare for a 22-nm node, the thickness was lowered from 1 nm to 0.35 nm in 2007, according to a report from the International Technology Roadmap for Semiconductors. The other is that quantum mechanical current leakage via gate dielectrics is comparably significant when the thickness is reduced at the same cost. As a result, power is lost while a chip's standby mode, which has billions of individual transistors, is active.

5.9 ELECTRICAL AND OPTICAL PROPERTIES OF HIGH-K DIELECTRICS BASED ON METAL OXIDES

5.9.1 Hafnium Oxides

With a band gap of 5.7 eV and a dielectric constant of 25, hafnium oxide effectively demonstrated its impact on production technology in 2007 [65]. In comparison to SiO_2 (218 kcal/mol), the heat emission was found to be around 271 kcal/mol. Again, with an operational voltage of 1–1.5 Volts, these hafnium oxides were reported to have leakage currents less than SiO_2 and an EOT of 0.9–2 nm [66–69]. Hafnium-based oxides, such as hafnium oxide (HfO_2), hafnia alumina (HfAlO), and hafnium-lanthanum oxide (HfLaO), among others, have properties that make them better suited for use in CMOS technology. HfO_2 is found in a variety of structures, including cubic, tetragonal, and monoclinic phases. The tetragonal phase is assumed to shift from the monoclinic phase to the cubic phase at a temperature of 2,422°C while the monoclinic phase is observed at the given ambient pressure and temperature [70]. The dielectric constant and other properties change depending on whether a structure is present. The dielectric constant for the monoclinic phase is 18 [59], that for the tetragonal structure is 28 [69, 71–72], and that for the cubic phase is 50 [73–74], which is thought to be the highest among the others. According to the dopants and annealing temperature, these values should vary. With some proportional doping of Si to HfO_2 at 700°C, Tomida and colleagues were able to achieve a dielectric constant of 36 [75]. When the parent HfO_2 has been mixed with some amount of lanthanum, Yamamoto and coworkers reported essentially the same result [76–78]. HfO_2 and Si together provide a band offset that is more than 1 eV, which appears to lessen gate leakage current [77]. Calcination temperature, annealing circumstances, structure, and the presence of flaws all appear to have an impact on how these illnesses are prepared [78–79]. Theoretical HfO_2/Si analysis was explored by Hwang, who suggested that these sources might be thermodynamically stable [80]. It is believed that the $HfO_2/Hf_{1-x}Si_xO_2$/Si heterostructure will perform well in its optical field. The reported offset values were discovered to be 4.75 eV for SiO_2 and 2.53 eV for HfO_2, according to Xie et al. [81]. HfO_2 and Si were stacked in a manner that resulted in the production of an oxygen interlayer, which led to the growth of an ultrathin EOT that is predicted to have high dielectric properties [82–86]. There are several disadvantages to the hafnium-silicon–based technology, despite all these benefits. The stability of the thermal interface between the HfO_2/Si substrate, as well as its thickness and value control, are thought to be the main issues in this preparation process. The essential problem for technological advancements is the crystal structure of any material, which controls the dielectric constant, thermal stability, band gap, point defects, energy creation, etc. Under the given pressure and temperature, HfO_2 can exist in a variety of structures, including monoclinic, tetragonal, and cubic phases [87–90]. Under normal circumstances, hafnium bonds with oxygen and achieves sevenfold coordination, of which Baddeleyite has been thought to be a frequent structure. The tetragonal phase is assumed to shift from the monoclinic phase to the cubic phase at a temperature of 2,422 °C, while the monoclinic phase is observed at the given ambient pressure and temperature [91–92]. The dielectric constant and other properties change depending on whether a structure is present. It is 18 for the monoclinic phase [93], 28 for the tetragonal structure [94–98], and 50 for the cubic phase [97], which is considerably greater than the others.

5.9.2 Electrical and Optical Properties

To ascertain the core 4f state contribution to the valence band structure, Losovyi and colleagues used angle-resolved photoemission experimentation to study the hybridization between the 4f and 5d levels of HfO_2 thin films with silicon [99]. The literature demonstrates a symmetric relationship between band dispersion and 4f values (relied on wave vector). As a result, the band dispersion and the HfO_2 crystal structure have been appropriately connected. Three sub-bands (O-2p, Hf-4f, and Hf-5d states) that are isolated from one another by ionic gaps are thought to make up the valence band of HfO_2 [100]. According to electron energy loss spectroscopy, amorphous HfO_2 has a band gap of 5.7 eV. The effective mass of an electron falls between 0.7 and 2.0 m_0, and band calculation findings reveal the presence of light and heavy holes with masses ranging from 0.3 to 8.3 m_0, respectively. In general, vacancies and impurities have an impact on the electrical properties of molecules with an oxide structure. Using a first-principles calculation, Mckenna looked at the diffusion of positive and neutral vacancies (oxygen vacancies) align near grain boundaries in m-HfO_2 [101]. At the grain boundaries, positive vacancies are found to accumulate, although the maximum amount of these vacancies that can accumulate is due to electrostatic repulsions. The vacancies left by HfO_2 at grain boundaries are partially or fully filled in CMOS technology devices by silicon electrons [101]. Once more, significant influences on the band alignments of both HfO_2 and Si are greatly influenced by the presence of defects, the characteristics of crystal structure, and also depend on compositions (parent materials), as investigated by the combination of first principle calculations and XPS study [102–104]. As seen, partial occupancies in the H_f (d orbitals) at the HfO_2/Si interface can occasionally lead to the formation of dangling bonds. Hydrogen passivation and interfacial layer design are two techniques that can be used to reduce such design and difficulties. For H_f atoms in the interface layer, hydrogen passivation and interfacial layers are both recommended to eliminate dangling bonds (bonds formed from immobilized free radicals). Factors influencing band alignment at the semiconductor/dielectric interface are (1) post-deposition annealing, (2) type of chemical at the interface, (3) vacancies, (4) passivation of hydrogen atoms, and (5) breakdown (major reliability issue in the making of gate dielectrics).

The time-dependent dielectric breakdown operation within operation voltage for high-dielectrics must, in general, satisfy 10 years for higher reliability. When trap density approaches the critical breakdown limit, a weakly localized breakdown channel forms between the gate electrode and the substrates. This condition is known as a soft breakdown (SBD). In contrast, a sizable tunnel of current runs through the layer when the joule heating takes place in discrete breakdown channels. Hard breakdown voltage (HBD) is a term used to describe this situation [105–106]. Kim and coworkers observed Weibull slopes, scaling factors area, and lifetime estimates to gain a deeper understanding of the breakdown processes of HfO_2 [107–108]. Gate oxide statistics are typically described in terms of the Weibull distribution using the formula $F(x) = 1 - \exp[-(x/)]$ where the Weibull slope is (β)

and is the characteristic lifetime. For HBD, the value is 2, whereas for SBD, it is 1.4 at EOT = 1.4 nm. It was discovered that a two-step breakdown process happens in the MOSFET stack of HfO_2, with the Weibull slope of the soft breakdown (first SBD) having a lower value than that of the hard breakdown (second SBD). The mechanical strength of brittle structures is observed using a finite-size relatively small link model as a result of Le et al.'s studies of mathematical analogy inside dielectric breakdown, particularly in semiconductor-based electronic devices. Again, according to this hypothesis, the proportion of the Weibull modulus for breakdown also increases from 2 to 4 nm when the gate dielectric thickness grows from 4.8 to 9.7 nm [109–110].

5.10 SYNTHESIS METHODS

There are different types of synthesis methods for high-K dielectric ceramic materials. Some of them are given in Figure 5.4.

5.10.1 Wet-Chemical Method

5.10.1.1 Sol-Gel Method

HfO_2 is created using organometallic precursors including hafnium tetrachloride, hafnium ethoxide [111–112], and hafnium n-butoxide [113]. The monoclinic structure is calcined at 500°C and annealed at 550°C HfO_2. The literature also demonstrates that this phase was present at temperatures up to 900°C [110–112]. Dielectric constant measured at 15–22 [114–116]. Leakage current density ranged 10–8 A/cm^2 at a bias voltage of –1.5 V and an EOT of roughly 0.6 nm [105].

5.10.1.2 Physical Method

(a) **Physical Sputtering**
Here, plasma oxygen and Hf-metal (the goal) are used to synthesize HfO_2. Additional requirements include the need for voltage and an annealing temperature to manage the interfacial layer during the filmmaking process [117–118] as well as gas mixtures (O_2/Ar) [119–120]. Additionally, these samples' interface characteristics and the density of HfO_2/Si are improved by annealing them in an environment where N_2 is being deposited [121].

(b) **Pulsed Laser Deposition**
A powerful laser beam is employed to focus on the HfO_2 ceramic, which serves as the target, producing a high deposition rate. Instead of using thin film deposition techniques, this type of technology enhances layer quality [122–123]. With an EOT of 7.4 nm and a bias voltage of –1.5 V, Wang and coworkers observed a band gap near 5.8 eV and a leakage current density of 10–8 A/cm^2 [124]. Hullavarad reported the capacitance-voltage measurement of the dielectric constant for HfO_2 thin films in the range of 17 to 24 [125]. According to the literature, the growth rate of ultrathin HfO_2 on the Si substrate is rather sluggish. Only a 19–20 dielectric constant, a leakage current density of 10–6 A/cm^2, an EOT of 1.3 nm, and a bias voltage of 1.5 V were reported [126–127].

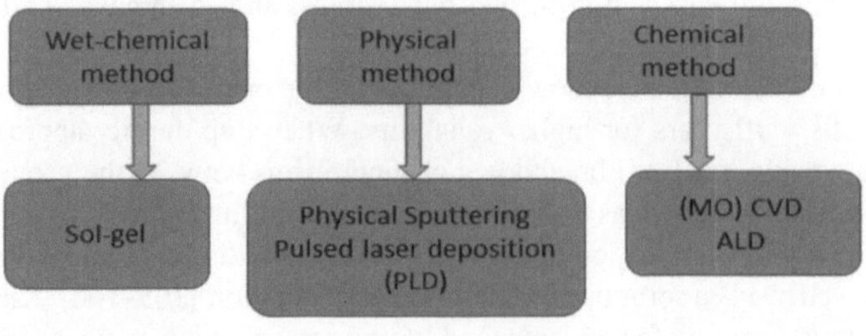

FIGURE 5.4 Method of synthesis.

5.10.1.3 Chemical Method

HfO_2 deposition can be accomplished using the appealing MOCVD process (metalorganic chemical vapor deposition). This deposition approach improves good growth (over a large area), composition control, film uniformity, and conformal step coverage on nonplanar geometries [128]. Tetrakis-diethylamidohafnium (TDEAH) [117], modified amide $Hf(Net_2)_2(guard)_2$ [129–130], and oxidant (where needed) [130–131] are prerequisites for the proper preparation of MOCVD, as are metal halides ($HfCl_4$ and Hf_{I4}) and alkoxide ($Hf(Ot\ Bu)_4$). Alkaline oxides were used to prepare HfO_2, which was discovered to be amorphous below 500°C and monoclinic between 500°C and 750°C [132–134]. Other physical parameters of the above compound are: band gap energy 5.8 eV; leakage current ~10^{-6} A/cm^2 (EOT = 1.4 nm); dielectric constant in a 23–25 range; and a bias voltage of –1.0 V [135–136].

5.11 ZIRCONIUM OXIDES

Hafnium oxide (HfO_2) and zirconium oxide (ZrO_2) are the most suitable substitutes for SiO_2 gate oxides in field-effect transistors made of metal oxide semiconductors (ZrO_2). SiO_2 gate oxide is utilized to reduce leakage current in gate dielectrics; it also uses less power in general [137–139]. For high-K gate dielectrics, these two are thought to be the best options [140–143]. ZrO_2 has a relatively low crystallization temperature of 450°C when compared to HfO_2 [144–146]. Tetragonal ZrO_2 has a dielectric constant of approximately 47 [147], but cubic ZrO_2 has a dielectric constant of 37 according to the literature [148]. Atomic layer deposition (ALD), plasma-enhanced ALD (PE-ALD), and other techniques can be used to deposit these films [149–154].

5.11.1 Optical Properties

Wideband gaps (5.8–7.8 eV), strong thermal stability with silicon, and a good correlation with silicon-based dielectrics are all characteristics of zirconium oxide (ZrO_2) reflection [155]. Additionally, it has a low absorption coefficient, a good refractive index (2.1–2.2), and is opaque in the visible and infrared spectra [156–159]. These compounds also exhibit strong absorption for direct and indirect interband transitions. For dental applications, ZrO_2 has a high hardness of 3.9 MPa m$^{1/2}$, which serves as a protective layer from an optical perspective [160]. Current improvements of ZrO_2 lead to its use in high-K dielectrics, high refractive mirrors, optoelectronic devices, optical interface filters, and transparent components, among others [156, 161]. The optical direct band gap is anticipated to be between 4.83 and 4.74 eV, whereas the indirect band gap may be between 5.6 and 5.73 eV. These values can vary according to the annealing temperature, flaws, impurities, preparation method, etc. [162]. The hydrothermal method of optical absorption and photoluminescence (PL) was the subject of research by Kumari et al. [163]. Zirconia nanoparticles were created by Liang and colleagues using a microwave irradiation technique, and they then used wavelengths of 254 and 412 nm to investigate PL spectra [164]. Various devices generally thought to depend on the luminescent nature of prepared materials are CRT lamps, electroluminescent, plasma display devices, vacuum fluorescent, etc. [165], whereas compact disc heads are thought to be predisposed to PL spectra with short wavelength and band gap [164].

5.11.2 Electrical Properties

Constant technical activity associated with the ongoing scaling of ICs and capacitors has led to a mad scramble for suitably high dielectric materials (to maintain Moore's law–consistent nature) [166]. Due to its good dielectric constant and loss (the nature of these parameters are easily worked out within a large range of temperatures and frequencies due to the polarization effect) employed as gate FinFETS [167–168], zirconium oxide becomes a suitable candidate for electrical device offsets.

5.12 DISCUSSION

Ferroelectric field effect transistors (FeFETs), ferroelectric random-access memory (FeRAM) devices, IC chips, and many more digitally based applications are expected to benefit from high-K dielectric technology. Researchers use different techniques and strategies, such as varying the concentration of dopants and working with various dopants (in particular transition metals), different fabrication methods, manipulating surface interfaces, etc., to create new materials with properties such as high K value, broadband gap, ultra-thin capability, dependability, and strong compatibility with Si-based substrates [169–179]. To encourage the conduction mechanism carried out by the Schottky effect of electrons/holes into the oxide-prone bands, these high-K materials are again expected to function as an insulator for which the height of the barrier wall at the corresponding band is near 1 eV [180]. According to published research, oxides typically exhibit better properties in practical applications when the measured conduction bands are comparably thinner than the valence bands indicated in Table 5.1 [181, 182]. The goal of research is to develop scaling devices that are dependable, take up little space, and operate quickly. These characteristics depend on the rate of mobility's degradation, which is meant to be governed by a variety of mechanisms and gate fields. However, the well-known process as seen in the literature works to restrict mobility. One of its mechanisms, according to Fischetti and coworkers, is remote phonon scattering, which is likely to occur at low-field energy polar phonon modes [183]. Saito and colleagues suggested a different mechanism—distant Coulomb scattering.

5.13 FUTURE PROSPECTS

Extant research makes it abundantly evident that current work has been heavily concentrated on the production of high-K dielectrics, metal gates for CMOS preparation, and storage devices. Current literature, however, shows that research is also being done on next-generation DRAMs. In general, encouraging factors have supported the planning of intelligent technological growth. The global high-K dielectric materials market is anticipated to be driven by an increase in demand for microelectronics. High-K dielectric materials, which are used as gate dielectric materials but have inferior qualities to a standard silicon oxide, are anticipated to restrain market growth over the forecasted period. It is anticipated that increased demand for electric vehicles, which it is utilized to make powerful batteries, will present a significant opportunity for market growth. This timely volume tackles the difficulties associated with high-K gate dielectric materials and will serve as a useful guide to the fundamentals and the effective application of high-K dielectric materials in IC technology in the future.

TABLE 5.1
Select High-K Dielectric Materials

Sl. no.	Materials	Dielectric value (K) [181]	Energy Gap (eV) [181]	Conduction Band Offset (eV) [182]
1	Si		1.1	
2	SiO_2	3.9	9	3.5
3	Al_2O_3	9	8.8	2.8
4	ZrO_2	25	5.8	1.4
5	HfO_2	25	5.8	1.5
6	TiO_2	80	3.5	0
7	$HfSiO_4$	11	6.5	1.5
8	Y_2O_3	15	6	2.3

5.14 CONCLUSION

A discussion of high-K ceramic materials is crucial for numerous practical applications in the modern world. As this chapter has demonstrated, it is essential to create ceramic and polymer composites with the precise K value required for a certain application. As was evident from the previous sections, the behavior of K is determined by the polarization of the material, including its interfacial, ionic, orientation, vibrational, and electronic polarizations. Additional polarization kinds can significantly raise a material's K value; however, each type of polarization has particular frequency-dependent dielectric losses. Understanding how polymers are polarized and how losses are managed by tuning the dielectrics to suitable standards are prerequisites for preventing loss dispersion.

ACKNOWLEDGMENT

The authors would like to thank the host institute for providing the digital library during the collection of the data related to high-K dielectric materials.

REFERENCES

1. Dass. R. I., Goodenough. J. B., Oxygen stoichiometry, ferromagnetism, and transport properties of $La_{2-x}NiMnO_{6+\delta}$, *Phys. Rev. B*, 2003, 67, 014401-1-9.
2. Rogado, N.S., Li, J., Sleight, A.W., Subramanian, M.A., Magnetocapacitance and magnetoresistance near room temperature in a ferromagnetic semiconductor: La_2NiMnO_6, *Adv. Mater.* 2005, 17, 2225–2227.
3. Newnham, R.E., Ruschau, G.R., Smart electroceramics, *J. Am. Ceram. Soc.* 1991, 74, 463–480.
4. Goldschmidt, V.M., Die Gesetze der Krystallochemie, *Naturwissenschaften* 1926, 14, 477–485.
5. Anderson, M.T., Greenwood, K.B., Taylor, G.A., Poppelmeier, K.R., B-cation arrangements in double perovskites, *Prog. Solid State Chem.* 1993, 22, 197–233.
6. Howard, C.J., Kennedy, B.J., Woodward, P.M., Ordered double perovskites – A group-theoretical analysis, *ActaCrystallogr. Sect. B: Struct. Sci.* 2003, 59, 463–471.
7. Zhang, S., Li, F., High performance ferro electric relaxor-$PbTiO_3$ single crystals: Status and perspective. *J.Appl. Phys.* 2012, 111, 031301.
8. Eerenstein, W., Mathur, N.D., and Scott, J.F., Multiferroic and magnetoelectric materials, *Nature* 2006, 442, 759–765.
9. Zhang, S., Li, F., Jiang, X., Kim, J., Luo, J, Geng, X., Advantagesandchallengesofrelaxor-PbTiO3 ferroelectric crystals for electroacoustic transducers—A review. *Prog. Mater. Sci.* 2015, 68, 1–66.
10. Li, F., Jin, L., Xu, Z., Zhang, S., Electrostrictive effect in ferroelectrics: An alternative approach to improve piezoelectricity. *Appl. Phys. Rev.* 2014, 1, 011103.
11. Chakraborty, H., Misra, D., Characterization of High-K Gate Dielectrics using MOS, *Int. J. Sci. Res.* 2013, 3, 1–5.
12. He, G., Sun, Z., Liu, M., and Zhang, L., Scaling and limitation of Si-based CMOS, high-k gate dielectrics for CMOS technology, Chapter 1. *High-k gate dielectrics for CMOS technology*, First ed. Wiley-VCH Verlag GmbH & Co. KGaA., 2012, 1–29.
13. Robertson, J., Wallace, R., High-K materials and metal gates for CMOS applications, *Mater Sci Eng R: Rep.* 2015, 88, 1–41.
14. Wilk, G.D., Wallace, R.M., Anthony, J.M., High-κ gate dielectrics: Current status and materials properties considerations, *J. Appl. Phys.* 2001, 89, 5243–5275.
15. Wallace, R.M., Wilk, G.D., Critical reviews in solid state and materials sciences. *Solid State Mater. Sci.* 2003, 28, 231–243
16. Wallace, R.M., Wilk, G., High dielectric constant gate oxides, *MRS Bull* 2002, 27, 391.
17. Robertson, J., High dielectric constant oxides, *Eur. Phys. J. Appl. Phys.* 2004, 28, 265–291.
18. Huff, H., Gilmer, D., *High K gate dielectrics* (2004) Springer.
19. Houssa, M., *High dielectric constant materials: VLSI MOSFET applications* (2003) IOP.
20. Demkov, A.A., Navrotsky, A., *Materials fundamentals of gate oxides* (2005) Springer).
21. Parida, S.K., Influence of cerium substitution on structural and dielectric properties of the modified $BiFeO_3$-$PbTiO_3$ ceramics, *Ferroelectric* 2021, 583, 19–32.

22. Lue, H.T., Liu, C.Y., Tseng, T.Y., An improved two-frequency method of capacitance measurement for SrTiO3 as high-k gate dielectric, *IEEE Electron. Device Lett.* 2002, 23, 553–555.
23. Valasek, J., Piezoelectric and allied phenomena in Rochelle salt, *Phys. Rev.*, 1920, 15, 537–538.
24. Wul, B., Goldmann, I.M., Dielectric constant of barium titanate as a function of the strength of an alternating field, *C.R. Acad. Sci. USSR.* 1945, 46, 172–179.
25. Parida, S.K., Studies on structural, dielectric, and optical properties of Cu/W double substituted calcium manganite for solar cells and thermistor applications, *Ph. Transit.* 2021, 94, 1033–1052.
26. Mishra, S., Choudhary, R.N.P., Parida, S.K., Structural, dielectric, electrical and optical properties of Li/Fe modified barium tungstate double perovskite for electronic devices, *Ceram. Int.* 2022, 48, 17020–17033.
27. Ivanov, S.A., Nordblad, P., Tellgren, R., Rundlof, H., Andre, F., Bouree, F., Neutron powder diffraction and magnetic study of perovskites Pb(Mn1/2Nb1/2)O3 and Pb(Mn1/4Fe1/4Nb1/2)O3, *Mater. Res. Bull.* 2008, 43, 3074–3087.
28. von Helmolt, R., Wecker, J., Holzapfel, B., Schultz, L., Samwer, K., Giant negative magnetoresistance in perovskite like $La_{2/3}Ba_{1/3}MnO_x$ ferromagnetic films, *Phys. Rev. Lett.* 1993, 71, 2331–2333.
29. Jin, S., Tiefel, T.H., McCormack, M., Fastnacht, R.A., Ramesh, R., Chen, L.H., Thousandfold change in resistivity in magneto resistive La-Ca-Mn-O films, *Science* 1994, 264, 513–515.
30. Troyanchuk, I.O., Lobanovsky, L.S., Kasper, N.V., Hervieu, M., Maignan, A., Michel, C., Szymczak, H., Szewczyk, A., Magnetotransport phenomena in $A(Mn_{3-x}Cu_x)Mn_4O_{12}$ (A=Ca, Tb, Tm) perovskites, *Phys. Rev. B* 1998, 58, 14903.
31. Parida, S.K., Das P.K., Choudhary, R. N. P., Structural and electrical characterization of $SrMn_{0.97}Ce_{0.03}O_3$ ceramics, *Integr. Ferroelectr.* 2021, 221, 215–230.
32. Parida, S.K., Structural, electrical and optical properties of zinc and tungsten modified lead titanate ceramics, *SPIN* 2021, 11, 2150018.
33. Von Hippel, A.R., Ferroelectricity, domain structure, and phase transitions of barium titanate, *Rev. Mod. Phys.* 1950, 22, 221–227.
34. Von Hippel, A.R., *Dielectric materials and applications* (1954) Technology Press, MIT and Wiley.
35. Zheludev, I. S., *Physics of crystalline dielectrics* (1971) Plenum.
36. Megaw, H.D., *Ferroelectricity in crystals* (1957) Methuen and Co.
37. Busch, G., Early history of ferroelectricity, *Ferroelectrics* 1987, 74, 267–284.
38. Kanzig. W., History of ferroelectricity 1938–1955, *Ferroelectrics*, 1987, 74, 285–291.
39. Cross, L.E., Newnham, R. E., *Ceramics and civilization*, Vol. III (1998) Am. Ceram. Soc., 289.
40. Busch, G., How I discovered the ferroelectric properties of KH_2PO_4, *Ferroelectrics* 1987, 71, 43–47.
41. Fousek, J., Ferroelectricity: Remarks on historical aspects and present trends, *Ferroelectrics* 1991 113, 3–20.
42. Megaw, H.D., *Ferroelectricity in crystals* (1957) Methuen.
43. Deri, M., *Ferroelectric ceramics* (1969) Gordon and Breach.
44. Burfoot, C., *Ferroelectrics* (1967) Van Nostrand.
45. Mitsui, T., Tatsuzaki, I., Nakamura, E., *An introduction to the physics of ferroelectrics* (1976) Gordon and Breach.
46. Nye, J. F., *Physical properties of crystals* (1957) Oxford.
47. Bhagavatam, S., *Crystal symmetry and physical properties* (1966) Academic Press.
48. Cady, W.G., *Piezoelectricity* (1946) McGraw Hill.
49. Lang, S.B., *Source book of pyroelectricity* (1974) Gordon and Breach.
50. Lines, M.E., Glass, A.M., *Principles, and applications of ferroelectric and related materials* (1977) Oxford University Press.
51. Tareev, B., *Physics of dielectric materials* (1975) Mir Publication.
52. Burfoot, J.C., Taylor, W., *Polar dielectrics and their applications* (1971) MacMillan Press Ltd.
53. Kittel, C., *Introduction to solid state physics* (1966) John Wiley.
54. Merz, W.J., Double hysteresis loop of $BaTiO_3$ at the Curie point, *Phys. Rev.* 1953, 91, 513–517.
55. Mosfet, M., T. Hori, Gate dielectrics and MOS ULSIs (1997) Springer.
56. Lilienfeld, J.E., Method and apparatus for controlling electric currents, U.S. Patent 1 745 175, 1930.
57. Kahng. D., Atalla, M., Silicon-silicon dioxide field induced surface devices. *IRE Solid-State Device Research Conference*, Pittsburgh, PA, 1960.
58. Hori, T., *Gate dielectrics and MOS ULSIs* (1997) Springer.
59. Critchlow, D.L., Mosfet scaling—The driver of VLSI technology, *Proc. IEEE.* 1999, 87, 659–667.

60. Moore, G.E., Cramming more components onto integrated circuits, *Proc. IEEE* 1998, 86, 82–85.
61. Sze S. M., *Physics of semiconductor devices*, (2nd ed.). (1981) Wiley.
62. Waldrop, M., The chips are down for Moore's law, *Nature* 2016, 530 (7589), 144–147.
63. Moore, G.E., Progress in digital integrated electronics, *IEEE Solid-State Circuits Soc. Newsl.* 2006, 11, 36–37.
64. This. T. N., Wong, H.S.P., The end of Moore's law: A new beginning for information technology, Computing in Science & Engineering; *IEEE* 2017, 19, 41–50.
65. Robertson, J., Interfaces, and defects of high-K oxides on silicon, *Solid State Electron.* 2005, 49, 283–293.
66. Ruh, R., Hollenbers, G.W., Charles, E.G., Patel, V.A., Phase relations and thermal expansion in the system HfO_2-TiO_2, *J. Am. Ceram. Soc.* (1976) 56, 606.
67. Humphrey, G.L., The heats of formation of TiO, Ti_2O_3, Ti_3O_5 and TiO_2 from combustion calorimetry, *J. Am. Chem. Soc.* 1951, 73, 1587–1590.
68. Adam, J., Rogers, M.D., The crystal structure of ZrO_2 and HfO_2, *ActaCrystallogr.* 1959, 12, 951.
69. McClain, L.A., Coppel, C.P., Electrical conductivity studies of tetragonal zirconia, *J. Electrochem. Soc.* 1966, 113, 80
70. Consiglio, S., Papadatos, F., Naczas, S., Skordas, S., Eisenbraun, T.E., Kaloyeros, A.E., Metallorganic chemical vapor deposition of hafnium silicate thin films using a dual source dimethyl-alkyl amido approach, *J. Electrochem. Soc.* 2006, 153, F249–F254.
71. Nakajima, Y., Kita, K., Nishimura, T., Nagashima, K., Toriumi, A., Phase transformation kinetics of HfO2 polymorphs in the ultra-thin region, *2011 Symposium on VLSI Technology Digest of Technical Papers*, 2011, 84–85.
72. Rignanese, G.M., Dielectric properties of crystalline and amorphous transition metal oxides and silicates as potential high-κ candidates: the contribution of density-functional theory, *J. Phys. Condens. Matter* 2005, 17, R357–R379.
73. Zhao, X., Vanderbilt, D., First-principles study of structural, vibrational, and lattice dielectric properties of hafnium oxide, *Phys. Rev. B*,2002, 65, 233106.
74. Neumayer, D.A., E. Cartier, Materials characterization of ZrO2–SiO2 and HfO2–SiO2 binary oxides deposited by chemical solution deposition, *J. Appl. Phys.* 2001, 90, 1801–1808.
75. Tomida, K., Kita, K., Toriumi, A., Dielectric constant enhancement due to Si incorporation into HfO2, *Appl. Phys. Lett.* 2006, 89, 142902–142903.
76. Yamamoto, Y., Kita, K., Kyuno, K., Toriumi, A., Structural and electrical properties of HfLaOx films for an amorphous high-k gate insulator, *Appl. Phys. Lett.* 2006, 89, 032903–032903.
77. Boscke, T.S., Govindarajan, S., Fachmann, C., Heitmann, J., Avellan, A., Schroder, U., Kudelka, S., Kirsch, P.D., Krug, C., Hung, P.Y., Song, S.C., Ju, B.S., Price, J., Pant, G., Gnade, B.E., Krautschneider, W., Lee B.H., Jammy, R., Tetragonal phase stabilization by doping as an enabler of thermally stable HfO2 based MIM and MIS capacitors for sub 50 nm deep trench DRAM, *IEEE International Electron Devices Meeting*, 2006, 1–4.
78. Yamamoto, Y., Kita, K., Kyuno, K., Toriumi, A., Structural and electrical properties of HfLaOx films for an amorphous high-k gate insulator, *Appl. Phys. Lett.* 2006, 89, 032903–032903.
79. Mishra, S., Choudhary, R.N.P., Parida, S.K., Structural, dielectric, electrical and optical properties of a double perovskite: $BaNaFeWO_6$ for some device applications, *J. Mol. Struct.*, 2022, 1265, 133353.
80. Robertson, J., High dielectric constant gate oxides for metal oxide Si transistors, *Rep. Prog. Phys.* 2006, 69, 327–396.
81. Xie, L., Zhao, Y., and White, M.H., Interfacial oxide determination and chemical/ electrical structures of HfO2/SiOx/Si gate dielectrics, *Solid-State Electron.* 2004, 48, 2071–2077.
82. Migita, S., Morita, Y., Mizubayashi, W., Ota, H., Preparation of epitaxial HfO2 film (EOT¼0.5 nm) on Si substrate using atomic-layer deposition of amorphous film and rapid thermal crystallization (RTC) in an abrupt temperature gradient, *IEEE International Electron Devices Meeting* 2010, 11.5.1–11.5.4.
83. Migita, S., Watanabe, Y., Ota, H., Ito, H., Kamimura, Y., Nabatame, T., Toriumi, A., Design, and demonstration of very high-k (k 50) HfO2 for ultra-scaled Si CMOS, *IEEE Symposium on VLSI Technology* 2008, 152–153.
84. Ragnarsson, L.A., Li, Z., Tseng, J., Schram, T., Rohr, E., Cho, M.J., Kauerauf, T., Conard, T., Okuno, Y., Parvais, B., Absil, P., Biesemans, S., Hoffmann, T.Y., Ultra low-EOT (5 A) gate-first and gate-last high-performance CMOS achieved by gate-electrode optimization, *IEEE International Electron Devices Meeting* 2009, 1–4.

85. Ando, T., Frank, M.M., Choi, K., Choi, C., Bruley, J., Hopstaken, M., Copel, M., Cartier, E., Kerber, A., Callegari, A., Lacey, D., Brown, S., Yang, Q., Narayanan, V., Understanding mobility mechanisms in extremely scaled HfO2 (EOT 0.42 nm) using remote interfacial layer scavenging technique and Vt-tuning dipoles with the gate-first process, *IEEE International Electron Devices Meeting*, 2009, 1–4.
86. Changhwan, C., Chang-Yong, K., Se Jong, R., Akbar, M.S., Krishnan, S.A., Manhong, Z., Hyoung-Sub, K., Tackhwi, L., Ok, I., Feng, Z., and Lee, J.C., Aggressively scaled ultra-thin undoped HfO2 gate dielectric (EOT < 0.7 nm) with TaN gate electrode using engineered interface layer, *IEEE Electron Device Lett.* 2005, 26, 454–457.
87. Hwang, W.S., Shen, C., Wang, X., Chan, D.S.H., Cho, B.J., A novel hafnium carbide HfCx metal gate electrode for NMOS device application, *IEEE Symposium on VLSI Technology* 2007, 156–157.
88. He, G., Fang, Q., Zhang, L.D., High-k $HfSi_xO_y$ gate dielectrics grown by solid phase reaction between sputtered Hf layer and SiO_2/Si. *J. Appl. Phys.* 2006, 100, 083517
89. Xie, L., Zhao, Y., White, M.H., Interfacial oxide determination and chemical/ electrical structures of HfO2/SiOx/Si gate dielectrics, *Solid-State Electron.* 2004, 48, 2071–2077.
90. Choi, J.H., Mao, Y., Chang, J.P., Development of hafnium-based high-k materials—a review, *Mater. Sci. Eng. R: Rep.* 2011, 72, 97–136.
91. Desgreniers, S., Lagarec, K., High-density ZrO2 and HfO2: Crystalline structures and equations of state, *Phys. Rev. B* 1999, 59, 8467–8472.
92. Copel, M., Gribelyuk, M., Gusev, E., Structure, and stability of ultrathin zirconium oxide layers on Si(001), *Appl. Phys. Lett.*, 2000, 76 (4), 436–438.
93. Catlow, C.R.A., In: O.T. Sorensen (Ed.), *Nonstoichiometric oxides* (1981) Academic, 61–98.
94. McClaine, L.A., Coppel, C.P., Electrical conductivity studies of tetragonal zirconia *J. Electrochem. Soc.* 1966, 113, 80–85.
95. Kuo, Y., Lin, C.C.; A light emitting device made from thin zirconium-doped hafnium oxide high-k dielectric film with or without an embedded nanocrystal layer, *Appl. Phys. Lett.* 2013, 102, 031117.
96. Ruh, R., Garrett, H.J., Nonstoichiometry of ZrO_2 and its relation to tetragonal-cubic inversion in ZrO 2, *J. Am. Ceram. Soc.* 1967, 50, 257–261.
97. Campbell, S.A., Gilmer, D.C., Wang, X.C., Hsieh, M.T., Kim, H.S., Gladfelter, W.L., Yan, J.H., MOSFET transistors fabricated with high permitivity TiO_2 dielectrics, *IEEE Trans. Electron Dev.* 1997, 44, 104–109.
98. Lowther, J.E., Dewhurst, J.K., Leger, J.M., Haines, J., Relative stability of ZrO_2 and HfO_2 structural phases, *Phys. Rev. B* 1999, 60, 14485–14488.
99. Losovyj, Y.B., Tang, J., Wang, W., Hong, Y., Parshin, V.,Tittsworth, R., 4f–5d hybridization in a high k dielectric; *Phys. Lett. A.* 2006, 357, 240–244.
100. Perevalov, T.V., Gritsenko, V.A., Erenburg, S.B., Badalyan, A.M., Wong, H., Kim, C.W., Atomic and electronic structure of amorphous and crystalline hafnium oxide: X-ray photoelectron spectroscopy and density functional calculations, *J. Appl. Phys.* 2007, 101, 053704–053708.
101. McKenna, K., Shluger, A., Electronic properties of defects in polycrystalline dielectric materials, *Microelectron. Eng.* 2009, 86, 1751–1755.
102. Puthenkovilakam, R., Chang, J.P., An accurate determination of barrier heights at the HfO2⁄SiHfO2⁄Si interfaces; *J. Appl. Phys.* 2004, 96, 2701–2707.
103. Ganga RajuAchary, P., Choudhary, R.N.P., Parida, S.K., Structure, electric and dielectric properties of $PbFe_{1/3}Ti_{1/3}W_{1/3}O_3$ single perovskite compound, *Process. Appl. Ceram.* 2020, 14, 146–153
104. Geppert, I., Lipp, E., Brener, R., Hung, S., and Eizenberg, M., Effect of composition and chemical bonding on the band gap and band offsets to Si of HfxSi1−xO2 (N) films, *J. Appl. Phys.* 2010, 107, 053701–053704
105. Zheng, Y.B., Wang, S.J., Huan, C.H., Microstructure-dependent band structure of HfO_2 thin films; *Thin Solid Films* 2006, 504, 197–200.
106. Chatterjee, S., Kuo, Y., Lu, J.,Tewg, J.Y., Majhi, P., Electrical reliability aspects of HfO2 high-k gate dielectrics with TaN metal gate electrodes under constant voltage stress, *Microelectron. Reliab* 2006, 46, 69–76.
107. Kim, Y.H., Lee, J.C., Hard- and soft-breakdown characteristics of ultrathin HfO2 under dynamic and constant voltage stress. In: *Hf-based high-k dielectrics. Synthesis lectures on solid state materials and devices* (2005) Springer, 19–46
108. Kuo. Y., Plasma enhanced chemical vapor deposited silicon nitride as a gate dielectric film for amorphous silicon thin film transistors—A critical review, *Vacuum.* 1998, 51, 741–745.

109. Le, J.L., Bazant, Z.P., Bazant, M.Z., Lifetime of high-*k* gate dielectrics and analogy with the strength of quasi-brittle structures, *J. Appl. Phys.* 2009, 106, 104119.
110. Nishida, T., Honda, S., Matsuura, M., Ide. M., Surface, structural and optical properties of sol-gel derived HfO2 films, *Thin Solid Films*, 2000, 371, 61–65.
111. Villanueva-Ibanez, M., Le Luyer, C., Marty, O., Mugnier, J., Influence of Sr/Hf ratio and annealing treatment on structural and scintillating properties of sol–gel Ce3+-doped strontium hafnate powders, *Optical Materials*, 2005, 27, 1541–1546.
112. Blanchin, M.G., Canut, B., Lambert, Y., Teodorescu, V.S., Barau, A., Zaharescu M., Structure and dielectric properties of HfO2 films prepared by a sol-gel route, J. Sol–Gel Sci. Technol. 2008, 47, 165–172.
113. Aoki, Y., Kunitake, T., Nakao, A., Sol–Gel fabrication of dielectric HfO2 nano-films; formation of uniform, void-free layers, and their superior electrical properties, *Chem. Mater.* 2005, 17, 450–458.
114. Wilk, G.D., Wallace, R.M., Anthony, J.M., High-k gate dielectrics: current status and materials properties considerations, *J. Appl. Phys.* 2001, 89, 5243.
115. Wang, Z.T., Kumagai, T., Kurokawa, H., Ichiki, M., Maeda, R., Preparation of hafnium oxide thin films by sol-gel method, *J. Electroceramics* 2008, 21, 499–502.
116. Wang, Z.J., Kumagai, T., Kurokawa, H., Ichiki, M., Maeda, R. Preparation of hafnium oxide thin films by sol–gel method, *J. Electroceramics* 2008, 21, 499–502.
117. Acton, O., Ting, G., Ma, H., Ka, J.W., Yip, H.L., Tucker, N.M., Jen, A.K.Y., π-σ-phosphonic acid organic monolayer/Sol–Gel hafnium oxide hybrid dielectrics for low-voltage organic transistors, *Adv. Mater.* 2008, 20, 3697–3701.
118. Nahar, R.K., Sharma, A., Study of the effect of substrate bias on the electrical properties of sputtered HfO2 thin film deposited on the silicon substrate, *Microelectron. Int.* 2007, 24, 46.
119. He, G., Liu, M., Zhu, L.Q., Chang, M., Fang, Q., Zhang, L.D., Effect of postdeposition annealing on the thermal stability and structural characteristics of sputtered HfO2 films on Si (1 0 0), *Surf. Sci.* 2005, 576, 67.
120. Pereira, L., Barquinha, P., Fortunato, E., Martins, R., Influence of the oxygen/argon ratio on the properties of sputtered hafnium oxide, *Mater. Sci. Eng. B* 2005, 118, 210–213.
121. Callegari, A., Cartier, E., Gribelyuk, M., Okorn-Schmidt, H.F., Zabel, T., Physical and electrical characterization of hafnium oxide and Hafnium silicate sputtered films, *J. Appl. Phys.* 2001, 90, 6466–6475.
122. Hu, H., Zhu, C.X., Lu, Y.F., Wu, Y.H., Liew, T., Li, M.F., Cho, B.J., Choi, W.K., Yakovlev, N., Physical and electrical characterization of HfO2HfO2 metal–insulator–metal capacitors for Si analog circuit applications, *J. Appl. Phys.* 2003, 94, 551.
123. Hidenori, H., Ohta, H., Suzuki, T., Honjo, C., Ikuhara, Y., Ueda, K., Kamiya, T., Hirano, M., Hosono, H., Mechanism for heteroepitaxial growth of transparent P-type semiconductor: LaCuOS by reactive solid-phase epitaxy, *Cryst. Growth Des.* 2004, 4, 301–307.
124. Wang, H., Wang, Y., Feng, J., Ye, C., Wang, B.Y., Wang, H.B., Li, Q., Jiang, Y., Huang, A.P., Xiao, Z.S., Structure and electrical properties of HfO2 high-k films prepared by pulsed laser deposition on Si (100), *Appl. Phys. A.* 2008, 93, 681–684.
125. Hullavarad, S.S., Pugel, D.E., Jones, E.B.,Vispute, R.D., Venkatesan, T., Low leakage current transport and high breakdown strength of pulsed laser deposited HfO2/SiC metal-insulator-semiconductor device structures, *J. Electron. Mater.* 2007, 36, 648–653.
126. Parida, S.K., Choudhary, R.N.P., Preparation method and cerium dopant effects on the properties of $BaMnO_3$ single perovskite, *Phase Transitions* 2020, 93, 981–991.
127. Lee, W.C., Lee, Y.J., Wu, Y.D., Chang, P., Huang, Y.L., Hsu, Y.L., Mannaerts, J.P., Lo, R.L., Chen, F.R.,Maikap, S., Lee, L.S., Hsieh, W.Y., Tsai, M.J., Lin, S.Y., Gustffson, T., Hong, M., Kwo, J., MBE-grown high –k gate dielectrics of HfO_2 and $(Hf–Al)O_2$ for Si and III–V semiconductors nano-electronics, *J. Cryst. Growth* 2005, 278, 619–623.
128. Jones, A.C., Chalker, P.R., Some recent developments in the chemical vapor deposition of electroceramic oxides; *J. Phys. D: Appl. Phys.* 2003, 36, R80–R95.
129. Elshocht, S.V., Baklanov, M., Brijs, B., Carter, R., Caymax, M., Carbonell, L., Claes, M., Conard, T., Cosnier, V., Date, L., Gendt, S.D., Kluth, J., Pique, D., Richard, O., Vanhaeren, D., Vereecke, G., Witters, T., Zhao, C., Heyns, M., Bulk properties of MOCVD-deposited HfO2 layers for high k dielectric applications, *J. Electrochem. Soc.* 2004, 151, F228.
130. Baunemann, A., Thomas, R., Becker, R., Winter, M., Fischer, R.A., Ehrhart, P., Waser, R., Devi, A., Mononuclear precursor for MOCVD of HfO2 thin films, *Chem. Commun.* 2004, 14, 1610–1611.

131. Milanov, A., Thomas, R., Hellwig, M., Merz, K., Becker, H.-W., Ehrhart, P., Fischer, R.A., Waser, R., Devi, A., LI-MOCVD of HfO2 thin films using engineered amide based Hf precursors; *Surf. Coat. Technol.* 2007, 201, 9109–9116.
132. Ivanovici, S., Puchberger, M., Fric, H., Kickelbick, G., Coordination behavior of acetoacetate ligands with attached methacrylate groups containing alkyl-spacers of different length to titanium and zirconium alkoxides, *Monatsh. fur Chem.* 2007, 138, 529–539.
133. Loo, Y.F., Kane, R. O., Jones, A.C., Aspinall, H.C., Potter, R.J.,Chlker, P.R., Bickley J.F., Taylor, S., Smith, L.M., Deposition of HfO2 films by liquid injection MOCVD using a new monomeric alkoxide precursor, [Hf(dmop)4], *Chem. Vap. Deposition* 2005, 11 (6–7), 299–305.
134. Thomas, R., Rije, E., Ehrhart, P., Milanov, A., Bhakta, R., Bauneman, A., Devi, A., Fischer, R., Waser, R., Thin films of HfO2 for high-k gate oxide applications from engineered alkoxide- and amide-based MOCVD precursors; *J. Electrochem. Soc.* 2007, 154 (3), G77.
135. Parida, S.K., Choudhary, R.N.P., Achary, P.G.R., Study of structural and electrical properties of polycrystalline $Pb(Cd_{1/3}Ti_{1/3}W_{1/3})O3$ tungsten perovskite, *Int. J. Microstruct. Mater. Prop.* 2020, 15, 107–121
136. Hino, S., Nakayama, M., Takahashi, K., Funakubo, H., Tokumitsu, E., Characterization of hafnium oxide thin films by source gas pulse introduced metalorganic chemical vapor deposition using amino-family Hf precursors, *Jpn. J. Appl. Phys.* 2003, 4, 6015.
137. Yeo, Y.C., King, T.J., Hu, C., Direct tunneling leakage current and scalability of alternative gate dielectrics, *Appl. Phys. Lett.* 2002, 81, 2091–2093.
138. Lo, S.H., Buchanan, D.A., Taur, Y., Modeling and characterization of quantization, polysilicon depletion, and direct tunneling effects in MOSFETs with ultrathin oxides, *IBM J. Res. Dev.* 1999, 43, 327–337.
139. Lo, S.H., Buchanan, D., Taur, Y., Wang, W., Quantum-mechanical modeling of electron tunneling current from the inversion layer of ultra-thin-oxide nMOSFET's, *IEEE Electron. Device Lett.* 1997, 18, 209–211.
140. Lee, B.H., Kang, L., Nieh, R., Qi, W.J., Lee, J.C., Thermal stability and electrical characteristics of ultrathin hafnium oxide gate dielectric reoxidized with rapid thermal annealing, *Appl. Phys. Lett.* 2000, 76, 1926–1928.
141. Robertson, J., Band offsets of wide-band-gap oxides and implications for future electronic devices, *J. Vac. Sci. Technol. B* 2000, 18, 1785–1791.
142. Hubbard, K., Schlom, D., Thermodynamic stability of binary oxides in contact with silicon, *J. Mater. Res.* 1996, 11, 2757–2776.
143. Gutowski, M., Jaffe, J.E., Liu, C.L., Stoker, M., Hegde, R.I., Rai, R.S., Tobin, P.J., Thermodynamic stability of high-K dielectric metal oxides ZrO2 and HfO2 in contact with Si and SiO2, *Appl. Phys. Lett.* 2002, 80, 1897–1899.
144. Huang, J.J., Huang, L.T., Tsai, M.C., Lee, M.H., Chen, M.J., Enhancement of electrical characteristics and reliability in crystallized ZrO2 gate dielectrics treated with in-situ atomic layer doping of nitrogen, *Appl. Surf. Sci.* 2014, 305, 214–220.
145. Huang, J.J., Huang, L.T., Tsai, M.C., Lee, M.H., Chen, M.J., Improvement of capacitance equivalent thickness, leakage current, and interfacial state density based on crystallized high-k dielectrics/nitrided buffer layer gate stacks, *ECS J. Solid State Sci. Technol.* 2013, 2, 524–528.
146. Chen, L.M., Lai, Y.S., Chen, J., Influence of pre-deposition treatments on the interfacial and electrical characteristics of ZrO2 gate dielectrics, *Thin Solid Films*, 2007, 515, 3724–3729.
147. Vanderbilt, D., Zhao, X., Ceresoli, D., Structural and dielectric properties of crystalline and amorphous ZrO2, *Thin Solid Films*, 2005, 486, 125–128.
148. Copel, M., Gribelyuk, M., Gusev, E., Structure, and stability of ultrathin zirconium oxide layers on Si (001), *Appl. Phys. Lett.* 2000, 76, 436–438.
149. George, S.M., Atomic layer deposition: an overview, *Chem. Rev.* 2009, 110, 111–131.
150. Puurunen, R. L., Surface chemistry of atomic layer deposition: a case study for the trimethylaluminum/water process, *J. Appl. Phys.* 2005, 97, 121301.
151. Ten Eyck, G.A., Senkevich, J.J., Tang, F., Liu, D., Pimanpang, S., Kara Back, T., Wang, G.C., Lu, T.M., Jezewski, C., Lanford, W.A., Plasma-assisted atomic layer deposition of palladium, *Chem. Vap. Depos.* 2005, 11, 60–66.
152. Wiedwald, U., Fauth, M., Hessler, M., Boyen, H.G., Weigl, F., Hilgendorff, M., Giersig, M., Schutz, G., Ziemann, P., Farle, M., From colloidal Co/CoO core/shell nanoparticles to arrays of metallic nanomagnets: surface modification and magnetic properties, *ChemPhys Chem.* 2005, 6, 2522–2526.
153. Antoniak, C., Lindner, J., Spasova, M., Sudfeld, D., Acet, M., Farle, M., Fauth, K., Wiedwald, U., Boyen, H.G., Ziemann, P., Wilhelm, F., Rogalev, A., Sun, S., Enhanced orbital magnetism in Fe(50)Pt(50) nanoparticles, *Phys. Rev. Lett.* 2006, 97, 117201.

154. Boyen, H.G., Fauth, K., Stahl, B., Ziemann, P., Kästle, G., Weigl, F., Banhart, F., Hessler, M., Schütz, G., Gajbhiye, N.S., Ellrich, J., Hahn, J., Büttner, M., Garnier, M.G., and Oelhafen, P., Electronic and magnetic properties of ligand-free FePt nanoparticles, *Adv. Mater.* 2005, 17, 574–578.
155. Perkins, C.M., Triplett, B.B., McIntyre, P.C., Saraswat, K.C., Haukka, S., Marko Tuominen, Electrical and materials properties of ZrO 2 gate dielectrics grown by atomic layer chemical vapor deposition, *Appl. Phys. Lett.* 2001, 78, 2357.
156. Balakrishnan, G., Thanigaiarul, K., Sudhakara, P., Song, J., Microstructural and optical properties of nanocrystallineundoped zirconia thin films prepared by pulsed laser deposition, *ApplPhys A* 2013, 110, 427.
157. Andrieux, M., Ribot, P., Gasquères, C., Servet, B., Garry, G., Effect of the oxygen partial pressure on the toughness of tetragonal zirconia thin films for optical applications; *Appl Surf Sci* 2012, 263, 284–290.
158. Ali, N., Bashir, M., Batool, S., Riaz, S., Naseem, S., Structural and optical properties of zirconia thin films, *Mater Today: Proceed.* 2015, 2 (10), 5771–5776.
159. Heshmatpour, F., Aghakhanpour, R.B., Synthesis and characterization of nanocrystalline zirconia powder by simple sol–gel method with glucose and fructose as organic additives, *Powder Technol* 2011, 205, 193–200.
160. Harris, D.C., Durable 3–5 µm transmitting infrared window materials, *Infrared PhysTechnol* 1998, 39 (4), 185–201.
161. Zhao, S., Ma, F., Xu, K.W., Liang, H.F., Optical properties and structural characterization of bias sputtered ZrO_2 films, *J Alloys Compd* 2008, 453 (1–2), 453–457.
162. Huang, M.H., Mao, S., Feick, H., Yan, Y. Wu, Kind, H., Weber, E., Russo, R., Yang, P., Room-temperature ultraviolet nanowire nanolasers, *Science* 2001, 8 (292), 1897–1899.
163. Kumari, L., Li, W.Z., Xu, J.M., Leblanc, R. M., Wang, D.Z., Yi, Li, Haizhong, Guo, Jiandi, Zhang, Controlled hydrothermal synthesis of zirconium oxide nanostructures and their optical properties, *Cryst Growth Des* 2009, 9, 3874–3880.
164. Ma, J., Preparation and characterization of ZrO_2 nanoparticles capped by trioctylphosphine oxide (TOPO), *J. Wuhan Univ. Technol. Mater. Sci. Ed.* 2011, 26, 611–614.
165. Blasse, G., Grabmaier, B.C., *Luminescent Materials* (1994) Springer, (Spriner 1st Edn.).
166. Yalamaç, E., Basturk, S.B., *Ceramic fibers, Fiber Technology for Fiber-Reinforced Composites* (2017) Woodhead Publishing Series in Composites Science and Engineering, (Elsevier), (1st Edn.).
167. Nesamani, F.P., Prabha, V.L., Paul, A., Nirmal, D., Synthesis and dielectric studies of monoclinic nanosized zirconia, *Adv. Condens Matter Phys.* 2014, 1–7.
168. Chintaparty, C.R., Influence of calcination temperature on structural, optical, dielectric properties of nano zirconium oxide, *Optik - International Journal for Light and Electron Optics* 2016, 127(11), 4889–4893.
169. Yim, K., Yong, Y., Lee, J., Lee, K., Ho-Hyun Nahm, Yoo, J., Lee, C., Seong Hwang, C., Han, S., Novel high-K dielectrics for next-generation electronic devices screened by automated ab initio calculations, *NPG Asia Mater.* 2015, 7, 1–6.
170. Tomida, K., Kita, K., Toriumi, A., Dielectric constant enhancement due to Si incorporation into HfO2, *Appl. Phys. Lett.* 2006, 89, 2–5.
171. Wiemer, C., Lamagna, L., Baldovino, S., Perego, M., Schamm-Chardon, S., Coulon, P.E., Salicio, O., Congedo, G., Spiga, S., Fanciulli, M., Dielectric properties of Er-doped HfO_2 (Er∼15%) grown by atomic layer deposition for high- κ gate stacks, *Appl. Phys. Lett.* 2010, 96, 1–4.
172. Das, D., S. Jeon, High-k $Hf_xZr_{1-x}O_2$ ferroelectric insulator by utilizing high-pressure anneal, *IEEE Trans. Electron Dev.* 2020, 67, 2489–2494.
173. Böscke, T.S., Hung, P.Y., Kirsch, P.D., Quevedo-Lopez, M.A., Ramírez-Bon, R., Increasing permittivity in HfZrO thin films by surface manipulation, *Appl. Phys. Lett.* 2009, 95, 2–5.
174. Zhang, X.Y., Hsu, C.H., Lien, S.Y., Wu, W.Y., Ou, S.L., Chen, S.Y., Huang, W., Zhu, W.Z., Xiong F.B., Zhang, S., Temperature-dependent HfO2/Si interface structural evolution and its mechanism, *Nanoscale Res. Lett.* 2019, 14, 1–7.
175. Choi, K.J., Park, J.B., Yoon, S.G., Control of the interfacial layer thickness in hafnium oxide gate dielectric grown by PECVD, *J. Electrochem. Soc.* 2003, 150, F75.
176. Cho, Ho J., Kim, Y.D., Park, D. Su, Lee, E., Park, C.H., Jang, J.S., Lee, K.B., Kim, H.W., Ki, Y.J., Yong, K.H., Song, W., New TIT capacitor with ZrO2/Al2O3/ZrO2 dielectrics for 60 nm and below DRAMs, *Solid-State Electron.* 2007, 51, 1529–1533.

177. Kil, D.S., Song, H.S., Lee, K.J., Hong, K., Kim, J.H., Par, K.S., Yeom, S.J., Roh, J.S., Kwak, N.J., Sohn, H.C., Kim, J.W., and Park, S.W., Development of new TiN/ZrO2/Al2O3/ZrO2/ TiN capacitors extendable to 45 nm generation DRAMs replacing HfO2 based dielectrics, *Symp. VLSI Technol., Dig. Tech. Pap.*, 2006, 65, 38–39.
178. Kim, S.K., Choi, G.J., Lee, S.Y., Seo, M., Lee, S.W., Han, J.H., Ahn, H.S., Han, S., Hwang, C.S., Al-doped TiO2 films with ultralow leakage currents for next generation DRAM capacitors, *Adv. Mater.* 2008, 20, 1429–1435.
179. Lee, S.W., Han, J.H., Han, S., Lee, W., Jang, J.H., Seo, M., Kim, S.K., Dussarrat, C., Gatineau, J., Min, Y.S., and Hwang, C.S., Atomic layer deposition of SrTiO3 thin films with the highly enhanced growth rate for ultrahigh density capacitors, *Chem. Mater.* 2011, 23, 2227–2236.
180. Robertson, J., High dielectric constant gate oxides for metal oxide Si transistors, *Rep. Prog. Phys.* 2006, 69, 327.
181. Robertson, J., Wallace, R., High-K materials and metal gates for CMOS applications, *Mater Sci Eng R: Rep.* 2015, 88, 1–41.
182. Robertson, J., Band offsets of wide-band-gap oxides and implications for future electronic devices, *J. Vac. Sci. Technol.* 2000, B18, 1785.
183. Fischetti, M.V., Neumayer, D.A., Cartier, E.A., Effective electron mobility in Si inversion layers in metal–oxide–semiconductor systems with a high-κ insulator: The role of remote phonon scattering, *J. Appl. Phys.* 2001, 90, 4587.

6 Silicon-based High-K Dielectrics

Debajani Tripathy, Bibhuti B. Sahu, and Srikanta Moharana

6.1 INTRODUCTION

The initial hypothesis underlying the effectiveness of Moore's law in enhancing transistor performance in complementary metal-oxide semiconductor (CMOS) technology was that scaling silicon dioxide dielectrics would be an effective technique [1]. So, in recent years, size reduction in Metal-Oxide-Semiconductor Field-Effect Transistor (MOSFET) has allowed for the packing of more transistors onto a chip while keeping costs low and enhancing circuit performance. Effective oxide thickness (EOT) of a typical SiO_2 dielectric material, however, must be less than 1 nm when devices lead to the sub-45 nm measure; this is near the physical limit and amounts to around three monolayers. Because of the unmistakable quantum tunneling effect at this size, leakage currents in the gate become too large. In the past two decades, many researchers have expended considerable time and effort to understand and select high-K gate dielectrics, and to improve our ability to integrate CMOS technology. Due to its superior compatibility with polysilicon gate electrodes and its thermal stability with silicon substrates, Hf-based dielectrics stand out as a top candidate among the multiple high-K choices [2, 3].

Since the late 1980s, CMOS (complementary metal-oxide-semiconductor) IC (integrated circuit) technology has advanced rapidly, allowing the silicon-based microelectronics sector to meet numerous technology-related demands quickly and to promote marketplace growth. Specific industry-related demands include high speed, low static power, and adaptability to a broad array of input and output voltages. Learning to scale the field effect transistor (FET), the circuit's major active component, has allowed for this [3, 4]. Because of this, the markets for electronics and communication have grown substantially. For example, low-power applications like wireless systems are in great demand, as are high-performance microprocessors. It has been stated that silicon dioxide, the dielectric used for decades in CMOS devices for isolating the transistor gate out of the Si channel, is the critical element permitting the scaling of silicon-based MOSFETs [5–8]. Using amorphous, thermal energy–produced SiO_2 as a gate dielectric in CMOS processing delivers advantages such as a stable state at a range of temperatures and electric current, superior-quality Si–SiO_2 boundary, and strong electric insulation. Defect charge densities of 1,010/cm^2, mid-gap boundary state densities of 1,010/cm^2 eV, and strong breakdown fields of 15 MV/cm are routinely achieved in modern CMOS processing, regardless of device size. Because of SiO_2 outstanding electrical properties [9, 10], other gate dielectric solutions face an uphill challenge to effectively replace it.

6.2 THERMODYNAMIC STABILITY OF SILICON

Contact through silicon is the most crucial factor in determining the ultimate electrical characteristics of practically all thin gate dielectrics. Several high-K metal oxide systems have been demonstrated to respond through silicon via equilibrium environments, forming an undesirable interfacial layer; this is known as an unstable interface. As previously mentioned, this necessitates the presence of an interfacial reaction barrier between the two substances. Problems with quality, consistency, and dependability are present in this thickness regime for any ultra-thin interfacial reaction barrier, not just SiO_2 [11]. It is specifically the case as the interfacial region takes a critical part in

DOI: 10.1201/9781003352365-6

determining the resulting electrical characteristics. In order to control the Si interface, it is essential to understand the thermodynamics of these systems. Envisaging and understanding the relative stability of a certain three-component system for device applications are made easier with the help of ternary-phase diagrams. The Gibbs free energies regulating the relevant chemical processes show that neither Ta_2O_5 nor TiO_2 or mixtures comprising Si are stable to SiO_2 formation for the ternary systems Ta-Si-O and Ti-Si-O. Yet the a and b tie lines show that when Ta_2O_5 and TiO_2 are placed on Si, they will tend to phase-split into SiO_2, metal oxide M_xO_y, metal M_5, and perhaps silicide (M_xSi_y) phases. Representations of phase in Figure 6.1(a) and (b) have been exhibited in the 700–900°C range, although analogous relations hold for considerably lower temperatures as well. These metal oxides are experimentally unstable with respect to SiO_2 formation; therefore, an additional interfacial layer is required [12, 13]. As shown by the lines in the representation of phase for the Zr-Si-O system, unlike the Ta and Ti systems, the metal oxide–based ZrO_2 as well as $ZrSiO_4$ composite will become constant through interaction with Si at elevated temperatures. As shown in Figure 6.1(c), a huge field of phase of $(ZrO_2)_x(SiO_2)_{1-x}$ configurations projected to become steady on silicon through higher temperatures is depicted by the gray shaded area. Film conformations within the gray range may be preferable, however, as it is best to prevent any Zr-Si bonding. Inside the sheltered area, configurations having higher O stages closer to $ZrSiO_4$ are preferable, as they might be extra prospective to avert the M-Si bond and phase development of silicide. Because of the large range of stable $(ZrO_2)_x(SiO_2)_{1-x}$ compositions, it may be possible to increase the quantity of zirconium or hafnium absorbed through the film of silicon progressively as one advances to higher technological

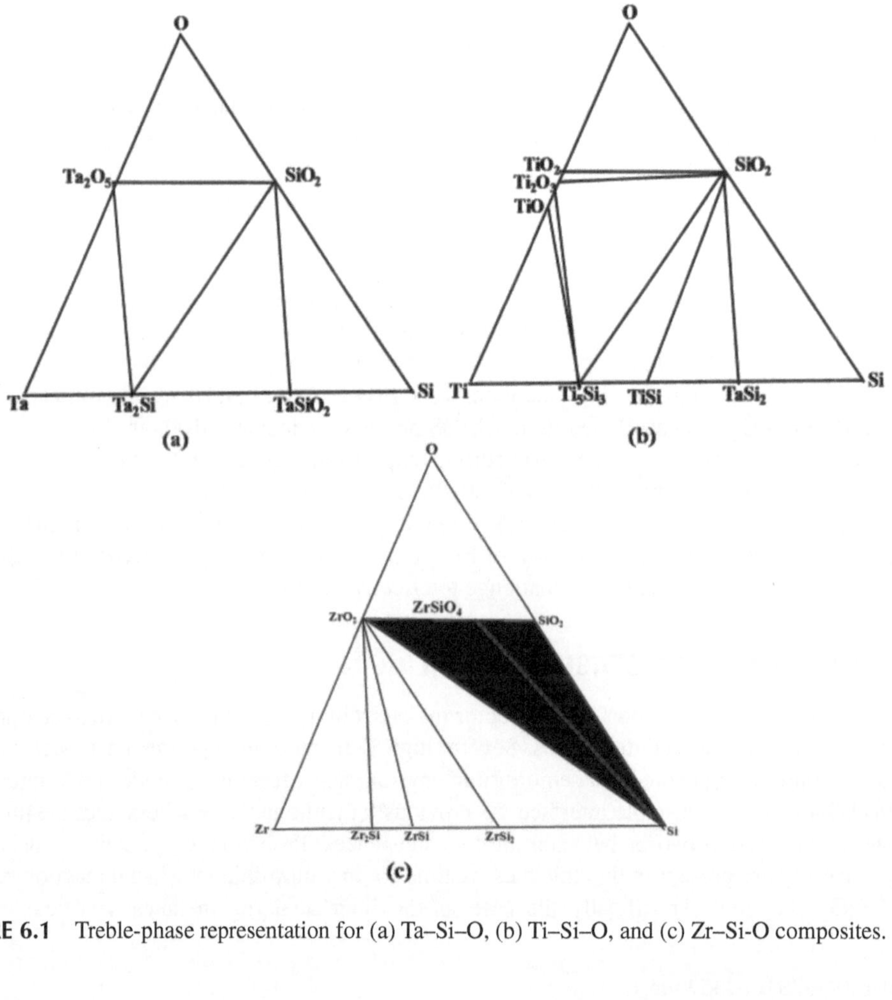

FIGURE 6.1 Treble-phase representation for (a) Ta–Si–O, (b) Ti–Si–O, and (c) Zr–Si-O composites.

nodes [14, 15]. This steady alteration in film composition has the potential to provide a rising value of K through a certain limit, which is necessary for meeting device presentation necessities.

The configuration of film between the gray ranges might be desirable; however, it is best to avoid any Zr-Si attachment. Inside the darkened region, compositions having more O levels nearer to $ZrSiO_4$ are preferable, as they make it possible to avert the M-Si bonding and creation of the silicide phase. Another possible advantage of this large phase field of stable $(ZrO_2)_x(SiO_2)_{1-x}$ conformations is that the amount of zirconium or hafnium absorbed into the silicate layer might be gradually increased as technological nodes progress. The slow shifting in the configuration of the film has the potential to provide a rising K value up to a certain limit, which is necessary to address device recital necessities. Coordination chemistry theory suggests that Hf, Si, and O systems will all have the same characteristics [16]. For the system of Hf-Si-O as a whole, comprehensive thermodynamic data are lacking; nevertheless, the data we do have propose that HfO_2 and $HfSiO_4$, and a wide array of $(HfO)_x(SiO_2)_{1-x}$ precursors, would be steady in a straight interaction with silicon at high temperatures. This fundamental divergence between the Ta and Ti systems is crucial because it suggests the dielectric-Si contact may be manipulated. Despite the fact that all of the deposition procedures of importance are finished in non-equilibrium circumstances, it is doubtful that a desired phase of less stability, such as amorphous Ta_2O_5, will be maintained for all of the thermal cycling necessary for CMOS dispensation. With process augmentations like an auxiliary gate flow, which enables the high-temperature processing to be done before deposition of the final gate dielectric, continued heat cycling subsequently may result in unacceptable electrical properties. Because of this, choosing a material system in which the desired state is innately stable is crucial. The use of $(ZrO_2)_x(SiO_2)_{1-x}$ and, by extension, $(HfO_2)_x(SiO_2)_{1-x}$ may provide a workable answer to a significant problem with high-K gate dielectric material techniques. As the K values of $(HfO_2)_x(SiO_2)_{1-x}$ and $(ZrO_2)_x(SiO_2)_{1-x}$ are significantly lower than those of pure HfO_2 and ZrO_2, respectively, this compromise of control in the boundary region would be tolerable provided that the accompanying leak currents are negligible.

6.3 REQUIREMENTS AND CONSIDERATION IN HIGH-K CANDIDATES

The primary reason why silicon is used rather than other semiconductors in microelectronics is because of silicon dioxide. Despite the fact that SiO_2 is often a good insulator, Si performs badly when compared to a semiconductor. SiO_2 may be produced from silicon by oxidation (related to heat energy), unlike any additional semiconducting material (Ge, GaAs, GaN, SiC, etc.). SiO_2 forms a great contact with silicon because it is amorphous, has few electrical flaws, and is easy to work with. Nanoscale etching and patterning are technically possible. The main issue is that when it is extremely thin, tunneling current flow through it is possible. Thus, the need has arisen to switch from employing SiO_2 to a new high-K oxide. There are a number of considerations that must be made during the development of high-K gate dielectrics on silicon to ensure their compatibility with the Si substrate [17].

Here are some specifications for several elevated permittivity-valued gate dielectrics.

- A higher enough value of K to be utilized due to scaling for a sufficient number of years.
- Good thermal stability when in straight interaction with the Si channel.
- Kinetic steadiness and ability to work by processing at high temperatures (up to 1,000°C).
- Having band offsets with Si of more than 1 eV and being an insulator.
- Having a strong electrical connection to silicon.
- Possessing a few electrically active flaws in bulk.

The aforementioned are just a few of the many potential requirements when choosing high-K materials for gate dielectric applications. As will be covered in the following sections, several crucial aspects require attention.

6.4 CRYSTALLIZATION TEMPERATURE

Amorphous composites are favored over crystal symmetric ones for usage as high-K gate dielectrics because of their high diffusion barrier and absence of grains. Among the grains that make up crystalline formations, the pathways for the diffusion and breakdown of dopants may often be located. High-K oxides, in contrast to SiO_2, often have low crystalline temperatures and are rapidly formed upon exposure to rapid thermal annealing, resulting in incomparable electric properties. When the leak current increases, the grain boundaries of crystalline films also tend to rise. Because of their low crystallization temperatures, upsurged valued permittivity composites are prone to powder-phase growth during modest temperature annealing [18]. Hence, it is important to remember that increasing the temperature during crystallization of elevated-K gate dielectrics is now a major challenge.

6.5 INTERFACE QUALITY AND STABILITY

The ultimate aim of any probable high-K gate dielectric is to produce a boundary with the Si channel that is as good as that of SiO_2. It is difficult to think of any material creating an interface that is better than that of SiO_2, as SiO_2 gate dielectrics used in manufacturing sometimes have a middle-gap boundary state density of about 2X 1,010/cm^{2}eV. Almost all of the high-K composites described also have interface trap density (Dit 1,011–1,012/cm^2 eV), and a large flat band voltage shift (VFB > 300 mV) (perhaps from fixed charge density > 1,012/cm^2 in the film or at the interface). Optimal high-K/Si interface performance may be achieved with any high-K gate dielectric by first understanding the factors that contribute to the characteristics of the interface. Numerous monolayers of a Si-O material, such as nitride silicate, may make up the "planned" interface at the channel of the ideal gate stacks. This thin layer may aid in preserving the vital, superior quality of the SiO_2 boundary while also giving a higher permittivity (K) value for that layer. Either a distinct elevated-K composite is cast off at the topmost layer between the interfaces or the same pseudo-binary material is used and continues beyond the interface. To meet scaling performance requirements, the particular composition of the gate stack structure is probably to be altered. As an obvious example, consider the presence of gradients in the dielectric layer's N concentration or any other component's concentration. The superior-K gate dielectrics would have thermodynamic stability, unreactive to the Si substratum or the gate probe under any dispensation conditions, and the boundaries should be free of stress. Hence, defects (dislocations, point defects) or even new phases like crystalline oxides would occur as a result of the large interface constraints, leading to poor performance. The amorphous condition of a dielectric must be preserved from the time of its creation until the end of its service life. For transferring ions, the dielectric has to be of less diffusivity for extra protection against contamination of the channel of the transistor. Moreover, most high-K dielectrics show flat-band variations that are either positive or negative, possibly due to fixed charges, as well as consistently high interface charge density and potential.

6.6 K VALUE, BAND GAP, AND BAND ALIGNMENT

To replace SiO_2 in CMOS devices, forthcoming superior-K alternatives will need to process K values over 3.9. Attention must be paid when choosing these high-K dielectric materials, as a K value that is too high will result in a film thickness that is greater than the size of the junction transistor conduit. The supposed fringing field that stretches from the gate to the host and drain can speed up short channel effects and reduce gate control over the channel. As stated in [19], K values below 25 are preferred for gate dielectric applications involving high-K composites to diminish the fringing effect. As most available data on values of K are based on observations from thickened bulk composites, there are some discrepancies between bulk and ultra-thin layers. Hence, we need additional investigational data to back up specific K values in this width domain [20]. Although Table 6.1 and Figure 6.2(a) show that permittivity of appropriate oxides frequently changes contrariwise with the

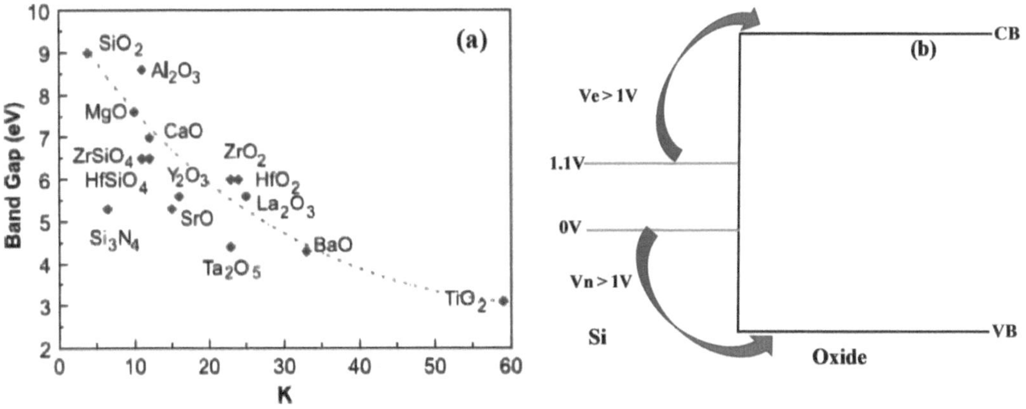

FIGURE 6.2 (a) The influence of carrier injection on band offsets in oxide band states and the (b) relationship between the static dielectric constant and band gap for suitable gate oxides.

TABLE 6.1
Candidate Gate Dielectrics Experimental Band Gap, Conduction Band Offset, and Static Dielectric Constant on Si

Dielectric Materials	K	Band Gap (eV)	CB Offset (eV)
Si		1.1	
SiO_2	3.9	9	3.2
Si_3N_4	7	5.3	2.4
Al_2O_3	9	8.8	2.8
Ta_2O_5	22	4.4	0.35
TiO_2	80	3.5	0
$SrTiO_3$	2,000	3.2	0
ZrO_2	25	5.8	1.5
HfO_2	25	5.8	1.4
$HfSiO_4$	11	6.5	1.8
La_2O_3	30	6	2.3
Y_2O_3	15	6	2.3

energy gap [21–23], a comparatively low K value is adequate. It is true that some oxides, such as the ferroelectric $BaTiO_3$, have extremely high K values, but their band gaps are too small to be practical. For obvious reasons related to their physical thickness, thicker films with larger K values would offer fewer straight tunneling currents through the insulating layer. Nonetheless, Schottky release of agitated electrons or holes is still possible in the conveyance or valence band of high-K composites. Hence, high-K materials with a greater conveyance band offset are the best option.

It has been proposed that prospective elevated dielectric constant composites would have a band offset bigger than 1 eV for their valence band (VB) and conduction band (CB) with reverence to Si to restrict Fowler-Nordheim (FN) and straight tunneling and achieve reduced leakage current at projected operating voltage [24, 25]. Figure 6.2(b) shows the VB and CB offsets of the materials must be more than 1 eV. As the conduction band offset is less than the VB offset, oxides of band gaps larger than 5.7 eV are viable candidates for use as gate dielectrics. If the CB offsets or VB offsets for oxides with small band gaps are less than 1 eV, then the materials available to you are limited.

6.7 FERMI-LEVEL PINNING

For metal gate electrodes to perform as expected they need to have the appropriate work functions; however, this is hindered by Fermi-level pinning. Fermi pinning occurs when a metal and a dielectric (or a semiconducting material) are brought together. The formation of a boundary "pins" the operative metal work function to a lower energy than its vacuum work function. This occurs as a result of charge transference between the metallic Fermi level and gap states at the metal–dielectric boundary, which results in a movement of the metal Fermi level with respect to its unpinned site. As maximum Fermi-level pinning research is concerned with the metallic semiconducting material boundary and is unresponsive to the metallic work function [26], the Schottky-Mott rapport is unsupported via experimental investigation. One method for adjusting the blockade height in Schottky contacts is the introduction of an induced gap state by a meal (MIGS). The MIGS are inherent conditions within the semiconductor's band gap because the wave functions of the metal electrons penetrate semiconductors in the energy range, where the metal conduction band overlaps the semiconductor band gap [27–29]. As a rule of thumb, conditions near the semiconducting band are of acceptor type, and the ones near to the valence band are of donor type. At the energy of the charge neutrality level (ECNL), a change from the donor type to the acceptor type occurs. Charged states produce a dipole crossways the interface, which causes the Fermi level to pin itself, weakening the dipole. Depending on the semiconductor band gap and the position of the ECNL, the derived work function can either increase or decrease. Recent research has postulated that Fermi-level pinning and convergence of the Schottky barrier height toward one-half of the band gap might be caused by polarization of chemical bonds at the metal–semiconductor interface. There is still some debate about this hypothesis. The Fermi level appears to be pinned at the metal–dielectric and poly-Si–dielectric interfaces in MOS devices [30]. However, the mechanism behind this phenomenon is likely different from that at metal–semiconductor interfaces. Many researchers have found that the calculated work function or threshold voltage changes diverge from anticipated values, with varying reports arriving at different conclusions. Poly-Si gates on high-K dielectrics, on the other hand, may be less vulnerable to Fermi-level pinning, according to a mode based on the interface dipole theory. Fermi-level pinning of poly-Si gates on HfO_2 and Al_2O_3 gate dielectrics, for instance, has been observed. In the subsequent paper, we offer a model for the pinning of the metal work function on high-K dielectric toward the ECNL at the Fermi level due to MIGS. The model predicts that the efficiency of Fermi-level pinning will increase as the K value of the dielectric material increases [31]. Mid-gap values attributed to extrinsic state in the interface have also been observed during thermal annealing of metal gates at temperatures above 800°C; this will be discussed in depth in the sections that follow. The effect of Fermi-level pinning on the electrical performance is essentially the result of previously choosing high-K gate dielectrics as a replacement for SiO_2 in forthcoming CMOS devices.

6.8 SILICON-BASED MICROELECTRONICS

By decreasing the size of the MOSFET, the number of transistors that could be packed onto a single chip increases, which in turn increases processing speed and indirectly boosts the semiconductor industry's business. The early success (which lasted until 1990) was a direct result of the scalability of the SiO_2 ($\varepsilon_r = 3.9$) gate dielectric. One might classify this as the very first gate dielectric generation. Since then, Si and gate oxides have been achieved in a variety of dielectric materials, including SiO_2, Si-O-N, and Hf-Si-ON, with the thickness as low as 1.2 nm as measured by conventional SiO_2-Si-O-N thickness (three to four monolayers) [32]. The key aspects of 45-nm and 32-nm technologies are high interface defect density and high inherent leakage. New dielectric materials with suitably large dielectric constants are needed to replace SiO_2-Si-O-N in future CMOS technology with an EOT in a sub-nanometer range. Depositing N_2 on the silicon film prevents the creation of a SiOx layer at the silicon interfaces [32, 33]. The *ab initio* calculation also demonstrated that N_2 has no detrimental effect on the conduction band offset. It was discovered that optimizing the metal oxide–SiO_2 phase separation led to significantly reduced component diffusivities in these materials. Si-Hf-ON and its offspring offer an alternative to Si-ON for use as a gate oxide (second generation). It has been demonstrated that these materials can scale to EOTs smaller than 1 nm. Substituting N_2 for O_2 during silicate film deposition prevents the production of a SiOx-rich layer at the silicate–silicon point of contact [33, 34]. This class of compounds, now in its third generation, is being mass-produced in tandem with metal as an alternate gate material to polysilicon in advance of the 45-nm technology node. Over the four decades of development of CMOS, gate oxides with K values between 3.9 and 14 have been effectively implemented at a variety of node levels. Research and production of fourth-generation gate-oxide logic devices is of great interest to engineers and physicists. A high-K gate oxide in the 20–40 K range is needed for this, ideally with low fixed-charge densities, attractive interface properties, and sufficient band offset with Si [33, 34].

6.9 RECENT HIGH-K DIELECTRIC DEVELOPMENT: PROGRESS AND CHALLENGES

Due to the possible challenges and shortcomings of the SiO_2 film in a dielectric stack, it is highly desired to hire a modern gate dielectric that is steady on Si as well as shows an interfacial superiority to Si in contrast to SiO_2. By adding nitrogen to SiO_2, the need for a supplementary interfacial film can be avoided. Dielectric constant may be increased, and the resistance to boron can be strengthened by introducing nitrogen into SiO_2. Advantages of good contact between the oxide and substrate are preserved in a nitride or oxide stack structure [35, 36]. While advancements in SiO_2, oxynitrides, and high-K gate dielectric materials are promising, the thin gate oxide poses a significant threat to the scaling of all gate dielectrics. In order to build a thicker dielectric to reduce current leakage and boost gate capacitance, several investigations have been undertaken on high-K metal oxides as an alternative to SiO_2 or SiO_xN_y. Motivated by applicability in memory capacitor and the subsequent infrastructure for the development of semiconductor manufacturing techniques [37–40], materials including Ta_2O_5, TiO_2, $SrTiO_3$, and Al_2O_3 were first identified from this group as promising replacement gate dielectric options. Except for Al_2O_3, none of these materials are thermodynamically stable when placed in close proximity to Si. Particular examples of interfacial reaction are Ta_2O and TiO_2 on Si. Certain metal oxides necessitate an additional barrier layer, which may increase production costs and put constraints on the maximum thickness that can be achieved. Nonetheless, Al_2O_3 is a popular option due to its many necessary characteristics, such as its broad band gap, excellent thermal stability on a Si substrate, and ability to retain its amorphous form even at high annealing temperatures. Its high threshold voltage and relatively low permittivity mean it can only be used as a stopgap measure. Recently, high-K gate oxides based on Hf (Zr) are being discussed as a potential replacement for SiO_2 in future devices' gate dielectric due to their comparatively higher permittivity, comparatively large band gap, higher temperature synthesis, and greater thermodynamic consistency. As HfO_2's Gibbs energy of generation is lower than that of ZrO_2 (1,088 kJ/mol vs. 1,040 kJ/mol) [41–45], it may have superior stability in interaction through Si. HfO_2 has been revealed to be well matched with a polysilicon gate; ZrO_2, on the other hand, has been shown to exhibit significant mismatches. Owing to the desirable properties, thin HfO_2 films or Hf-based high-K dielectrics have remained the subject of much study as a possible upcoming-generation gate dielectric composites. Nevertheless, HfO_2 has a tendency to crystallize at temperatures as low as 400°C, which increases leaking current and creates trails for dopant dissemination and dielectric breakdown. At this point, researchers from a wide range of institutions have explored the possibility of elevating the crystallizing temperature of the oxides by adding other materials including N, Si, Al, Ta, and La. The advantages and potential applications of Hf-based high-K dielectrics in CMOS devices are more significant than those of gate dielectrics based on Zr. Thus, this chapter discusses only the most recent state-of-the-art studies that have focused on Hf-based high-K gate dielectrics as an alternative to silicon oxide–based ones [46].

The challenges of finding a suitable replacement for SiO_2 as the gate dielectric for CMOS technologies are well known in the scientific community. SiO_2's remarkable intrinsic properties, compatibility with high-temperature production methods, and silicon's inherent abundance have all contributed to the growth of the $200-billion silicon-based manufacturing business. On the other hand, thinner gate oxides are required when the MOS transistor and its supporting architecture on the chip are getting progressively smaller [47]. While some argue that the rate of technological advancement outpaces the ability of circuit originators to effectively exploit the merits of any assumed technological node, the reality is that technological advancement and consumer demand for ever-increasing processing speeds drive the paradigm. In order to regulate the electrostatics of the MOSFET conduit and overcome the drain voltage incursion in the struggle to lessen undesirable short channel effects, the gate oxide must be made extremely thin. Comparable oxide thicknesses, however, were expected to decrease to 0.35 nm by the 22-nm node in 2007 in accordance with International Technological Roadmap for Semiconductors (ITRS). The native oxide has a lower limit of around 0.7 nm, since it requires at least three monolayers of SiO_2 to obtain "bulk"-like features [48]. The cost of reducing oxide thickness in devices with a billion transistors is the excessively high quantum mechanical current leaking across the gate as well as the consequent standby power consumption. To reduce leakage, it is desirable to use a thicker oxide to maintain the similar gate capacitance, and composites with elevated dielectric constants are being investigated for this purpose. It is essential to reduce gate leakage deprived of compensating the transistor's current drive (ION). SiO_2 is a hard act to follow, and any potential rivals will need to fulfill a number of prerequisites [49].

In a nutshell, these prerequisites are thermodynamic consistency when in interaction with Si; a value of K high enough to permit the investment in R&D; scalability; stability during the high-temperature CMOS fabrication process; and band offsets of electrons and holes > 1 eV that translate to band gap energies (E_g) > 5 eV considering the inverse connection between E_g and K. The periodic table reveals (maybe not unexpectedly) that there aren't many possibilities when these constraints are taken into account. Taking into account the ITRS recital necessities, the metal oxide of HfO_2 is the best option for the near and medium term. At temperatures above 450°C, its silicates and aluminates could diminish the risk of crystallization, with a slight decrease in the value of K [50]. When compared to the necessities for the 22-nm node, candidates like Pr and La seems to have potential, while Gd, Ce, and Sm oxides are also deserving of attention. Several different metrological methods have been developed throughout the years for characterizing SiO_2 from the most fundamental physics and engineering perspectives. There is a lot of trust in this technology because the procedures constitute a self-reliable practice for manufacturing extremely dependable gate oxides in a gross manufacture setting. It quickly became apparent that the novel dielectrics had features that call for a reexamination and recalibration of the experimental methods. Consider the example of HfO_2, where the construction of higher-bandwidth measurement instruments with data detention periods on the order of microseconds or less was necessary due to the quick electron trapping, resulting in large threshold instability [51]. In terms of physical depiction, spectro-ellipsometry (SE) is a very effective implement to gain basic parameters and characteristics such as oxide thickness, dielectric constant, and energy gap. The multifaceted portion of the dielectric constant or absorption coefficient, known as the losses, can be studied to learn more about the defect levels in the oxide band gap and, more importantly, at the band edges. The SE can offer effective, self-reliable, nondestructive strategies for depicting composites when combined with conventional C-V and I-V measurements.

6.10 LATEST DEVELOPMENTS IN SI-BASED HIGH-K DIELECTRICS

HfO_2's basic crystalline phase has the lattice constants of a = 5.116, b = 5.172, and c = 5.294. Some research suggests that, under the right processing circumstances, HfO_2 can exist in an amorphous state at temperatures lower than 973 K. HfO_2 is mostly ionic and has a low ionization potential and electron affinity [52, 53]. However, problems with EOT scaling and consistency arise while Hf-based high-K gate oxides are integrated into traditional CMOS devices because of pure HfO_2's lower crystallization temperature and limited resistance to O and B by way of superior thermal dispensation. Crystallized HfO_2 in Hf silicates reduces electron mobility by inducing Coulomb scattering in addition to substrate impurity scattering, as reported in a recent study by Yamaguchi et al. [54]. In addition, Liu's modeling outcomes evidently demonstrate that boron may diffuse through a thin HfO_2 polycrystalline layer only via diffusing grain boundary and not in bulk [55]. The verdict underlines the ongoing need for the amorphous film to prevent the diffusion of impurities. Some promising electrical characteristics have been recorded despite the high-K layer having solidified. Crystallization of HfO_2, as reported by Kim et al. [56], has negligible influence on gate leak current and the supplementary trap conditions familiarized by the creation of grain boundaries. Crystalline ZrO_2 has leakage current that is equal to or lower than that of amorphous high-K dielectric (EOT) as reported by Perkin et al. [57]. The results of both studies show the SiO_2 layer in the interface with the Si substratum is at least 15, which would restrict the downscaling of EOT. Thus, the impact of high-K film crystallization on leakage current remains a contentious topic of discussion. HfO_2's electric properties and crystallinity are greatly enhanced when alloyed with Al, Si, or N, as has been established by multiple research groups [58–60]. The increased thermal stability of the HfAlO with a 67% HfO_2 composition, or $(HfO_2)_{0.67}(Al_2O_3)_{0.33}$, can be seen in the material's strong thermal stability, which is equivalent to that of an Al_2O_3 film, and a much lower leakage current compared to pure HfO_2 film. While HfO_2 is crystallized during production, HfO_xN_y remains amorphous even after toughening at 900°C, as reported by Choi et al. [61]. When comparing the evolution of EOT and leakage current with increasing annealing temperature, HfO_xN_y shows a more modest increase than does HfO_2. Since N incorporation allows the film to remain amorphous even after high-temperature annealing, oxygen transport through the film and leakage current are both reduced. Pulse laser deposition [62], annealing of PVD-based HfO_2-Al_2O_3 nanolaminate gate dielectrics [63], and reoxidation or annealing of PVD-HfN [64] are just a few of the methods used to create Hf-based gate dielectrics. These dielectrics have a suitable K value and band-offset value to Si, making them an excellent choice for high-K gate dielectrics application.

6.11 CONCLUSIONS AND FUTURE DIRECTIONS

The study of silicon-based high-K dielectrics can take a variety of potential directions. One of the most serious issues is how to improve the efficiency and reliability of high-K dielectrics used in cutting-edge electronic systems. Improved dielectric constants and leakage currents with less impact from defects and interface states need the development of new materials and deposition techniques. Incorporation of high-K dielectrics through state-of-the-art semiconductor manufacturing processes will also be an important field of research. New surface passivation techniques and interface engineering strategies will need to be developed to ensure sure the high-K dielectric layer is compatible with the silicon substratum and other device components. There is a growing interest in exploring the probable applications of high-K dielectrics in emerging fields like spintronics, neuromorphic computing, and quantum information processing, which highlights the importance of not only improving the performance of high-K dielectrics but also learning more about these technologies. High-K dielectrics are appealing for these uses because of their large permittivity, low phonon energy, and tunable band gap, but unconventional approaches to material and device design are required to fully exploit their promise. Finally, in the development of silicon-based high-K dielectrics, sustainability and environmental challenges, including waste minimization and reduced reliance on potentially toxic precursors, will need to be taken into account. To make significant strides in the progress of silicon-based high-K dielectrics, experts from a wide range of fields and industries will need to collaborate. This is because the future of silicon-based high-K dielectrics will be influenced by technical progress, scientific discoveries, and social requirements.

ACKNOWLEDGMENTS

The author Srikanta Moharana acknowledges Centurion University of Technology and Management, Odisha, for the support and facilities.

REFERENCES

1. Moore, G.E. Progress in digital integrated electronics, *IEDM Tech. Dig*., 1975, 11–13.
2. Chau, R., Datta, S., Doczy, M., Doyle, B., Kavalieros, J. and Metz, M. High-k/Metal–Gate Stack and Its MOSFET Characteristics. *IEEE Electron Dev. Lett*., 2004, *25*(6), 408–410.
3. Wilk, G. D., Wallace, R. M., and Anthony, J. High-κ gate dielectrics: Current status and materials properties considerations. *J. Appl. Phys*., 2001, *89*(10), 5243–5275.
4. He, G., Zhang, L. D., Liu, M., Zhang, J. P., Wang, X. J. and Zhen, C. M. Thickness-modulated optical dielectric constants and band alignments of HfOx Ny gate dielectrics. *J. Appl. Phys*., 2009, *105*(1), 014109.
5. He, G., Zhang, L. D., Meng, G. W., Li, G. H., Fei, G. T., Wang, X. J. and Boyd, I. W. Composition dependence of electronic structure and optical properties of Hf $_{1-x}$ Six O y gate dielectrics. *J. Appl. Phys*., 2008, *04*(10), 104116.
6. He, G., Meng, G. W., Zhang, L. D. and Liu, M.Temperature-dependent interfacial chemical bonding states and band alignment of HfOxNy/Si O_2/ Si gate stacks. *Appl. Phys. Lett*., 2007, *91*(23), 232910.
7. Choi, J. H., Mao, Y., & Chang and J. P. Development of hafnium based high-k materials—A review. *Mater. Sci. Eng.: R: Report*, 2011, *72*(6), 97–136.
8. Gang, H. E. and Zhang, L. Current progress of Hf (Zr)-based high-k gate dielectric thin films. *J. Mater. Sci. Technol*., 2007, *23*(04), 433.
9. He, G., Zhang, L. D. and Fang, Q. Silicate layer formation at $HfO_2/SiO_2/Si$ interface determined by x-ray photoelectron spectroscopy and infrared spectroscopy. *J. Appl. Phys*. 2006, *100*(8), 083517.
10. He, G., Liu, M., Zhu, L. Q., Chang, M., Fang, Q. and Zhang, L. D. Effect of postdeposition annealing on the thermal stability and structural characteristics of sputtered HfO_2 films on Si (1 0 0). *Surf. Sci*., 2005, *576*(1–3), 67–75.
11. Beyers, R. Thermodynamic considerations in refractory metal-silicon-oxygen systems. *J. appl. Phys*., 1984, *56*(1), 147–152.
12. Hubbard, K. J. and Schlom, D. G. Thermodynamic stability of binary oxides in contact with silicon. *J. Mater. Res*., 1996, *11*(11), 2757–2776.
13. Wang, S. Q. and Mayer, J. W. Reactions of Zr thin films with SiO_2 substrates. *J. appl. Phys*., 1998, *64*(9), 4711–4716.
14. Barin, I., Knacke, O. and Kubaschewski, O. *Thermochemical properties of inorganic substances: supplement*. Springer Science & Business Media, 2013.
15. Jacob, K.T., Lwin, K. T. and Waseda, Y. System Pr–Pd–O: Phase diagram and thermodynamic properties of ternary oxides using solid-state cells with special features. *Int. J. Mater. Res*., 2022, *92*(7), 731–739.
16. He, G., Zhu, L., Sun, Z., Wan, Q. and Zhang, L. Integrations and challenges of novel high-k gate stacks in advanced CMOS technology. *Prog. Mater. Sc*., 2011, *56*(5), 475–572.
17. He, G., Fang, Q., Liu, M., Zhu, L. Q. and Zhang, L. D. The structural and interfacial properties of HfO_2/Si by the plasma oxidation of sputtered metallic Hf thin films. *J. Cryst. Growth*, 2004, *268*(1–2), 155–162.
18. Frank, D. J., Taur, Y. and Wong, H. S. Generalized scale length for two-dimensional effects in MOSFETs. *IEEE Electron Dev. Lett*., 1998, *19*(10), 385–387.
19. Green, M. L., Gusev, E. P., Degraeve, R. and Garfunkel, E. L. Ultrathin (< 4 nm) SiO_2 and Si–O–N gate dielectric layers for silicon microelectronics: Understanding the processing, structure, and physical and electrical limits. *J. Appl. Phys*., 2001, *90*(5), 2057–2121.
20. Robertson, J. Band offsets of wide-band-gap oxides and implications for future electronic devices. *J. Vac. Sci. Technol. B*, 2000, 1*8*(3), 1785–1791.
21. Robertson, J. and Chen, C. W. (1999). Schottky barrier heights of tantalum oxide, barium strontium titanate, lead titanate, and strontium bismuth tantalate. *Appl. Phys. Lett*., 1999, *74*(8), 1168–1170.
22. Tung, R. T. Recent advances in Schottky barrier concepts. *Mater. Sci. Eng.: R: Rep*., 2001, *35*(1–3), 1–138.
23. Wang, B., Huang, W., Chi, L., Al-Hashimi, M., Marks, T. J. and Facchetti, A. High-k gate dielectrics for emerging flexible and stretchable electronics. *Chem. Rev*., 2018, *118*(11), 5690–5754.
24. Heine, V. Theory of surface states. *Phys. Rev*., 1965, *138*(6A), A1689.

25. J. Tersoff, Theory of semiconductor heterojunction: the role of quantum dipoles, *Phys. Rev. B*, 1984, 30, 4874–4877.
26. Mönch, W. Barrier heights of real Schottky contacts explained by metal-induced gap states and lateral inhomogeneities. *J. Vac. Sci. Technol. B*, 1999, 17(4), 1867–1876.
27. Tung, R. T. Chemical bonding and Fermi level pinning at metal-semiconductor interfaces. *Phys. Rev. lett.*, 2000, *84*(26), 6078.
28. Hobbs, C. C., Fonseca, L. R., Knizhnik, A., Dhandapani, V., Samavedam, S. B., Taylor, W. J. and Tobin, P. J. Fermi-level pinning at the polysilicon/metal-oxide interface-Part II. *IEEE Trans. Electron Devices*, 2004, *51*(6), 978–984.
29. Hobbs, C. C., Fonseca, L. R., Knizhnik, A., Dhandapani, V. and Samavedam, S. B. Fermi level pinning at the polysilicon/metal oxide interface-Part I, *IEEE Trans Electron Dev*., 2004, 51, 971–977.
30. Yeo, Y. C., King, T. J. and Hu, C. Metal-dielectric band alignment and its implications for metal gate complementary metal-oxide-semiconductor technology. *J. Appl. Phys*., 2002, *92*(12), 7266–7271.
31. Yu, H. Y., Ren, C., Yeo, Y. C., Kang, J. F., Wang, X. P., Ma, H. H. H. and Kwong, D. L. Fermi pinning-induced thermal instability of metal-gate work functions. *IEEE Electron Dev. Lett.*, 2004, *25*(5), 337–339.
32. Raghavan, N., Pey, K. L. and Shubhakar, K. (2014). High-κ dielectric breakdown in nanoscale logic devices–Scientific insight and technology impact. *Microelectronics Reliab*., 2014, *54*(5), 847–860.
33. Sharma, U., Kumar, G., Mishra, S. and Thomas, R. Advancement of gate oxides from SiO_2 to high-k dielectrics in microprocessor and memory. In *Journal of Physics: Conference Series*, IOP Publishing, 2022, (Vol. 2267, No. 1, p. 012142).
34. Abuwasib, M., Lu, H., Li, T., Buragohain, P., Lee, H., Eom, C. B. and Singisetti, U. Scaling of electro-resistance effect in fully integrated ferroelectric tunnel junctions. *Appl. Phys. Lett*., 2016, *108*(15), 152904.
35. Ellis, K. A. and Buhrman, R. A. Time-dependent diffusivity of boron in silicon oxide and oxynitride. *Appl. Phys. Lett*., 1999, 74(7), 967–969.
36. Kumar, K., Chou, A. I., Lin, C., Choudhury, P., Lee, J. C. and Lowell, J. K. Optimization of sub 3 nm gate dielectrics grown by rapid thermal oxidation in a nitric oxide ambient. *Appl. Phys. Lett*., 1997, *70*(3), 384–386.
37. Campbell, S. A., Gilmer, D. C., Wang, X. C., Hsieh, M. T., Kim, H. S., Gladfelter, W. L. and Yan, J. MOSFET transistors fabricated with high permitivity TiO/sub 2/dielectrics. *IEEE Trans. Electron Dev*., 1997, *44*(1), 104–109.
38. Chaneliere, C., Autran, J. L., Devine, R. A. B. and Balland, B. Tantalum pentoxide (Ta_2O_5) thin films for advanced dielectric applications. *Mater. Sci. Eng: R: Reports*, 1998, *22*(6), 269–322.
39. Eisenbeiser, K., Finder, J. M., Yu, Z., Ramdani, J., Curless, J. A., Hallmark, J. A. and Overgaard, C. D. Field effect transistors with $SrTiO_3$ gate dielectric on Si. *Appl. Phys. Lett*., 2000, *76*(10), 1324–1326.
40. Park, D. G., Cho, H. J., Lim, K. Y., Lim, C., Yeo, I. S., Roh, J. S. and Park, J. W. (2001). Characteristics of n+ polycrystalline-Si/Al 2 O 3/Si metal–oxide–semiconductor structures prepared by atomic layer chemical vapor deposition using Al $(CH_3)_3$ and H_2O vapor. *J. Appl. Phys*., 2001, *89*(11), 6275–6280.
41. He, G., Zhu, L. Q., Liu, M., Fang, Q. and Zhang, L. D. Optical and electrical properties of plasma-oxidation derived HfO_2 gate dielectric films. *Appl. Surf. Sci*., 2007, *253*(7), 3413–3418.
42. Pant, G., Gnade, A., Kim, M. J., Wallace, R. M., Gnade, B. E., Quevedo-Lopez, M. A., & Kirsch, P. D.Effect of thickness on the crystallization of ultrathin HfSiON gate dielectrics. *Appl. Phys. Lett*., 2006, *88*(3), 032901.
43. Zhu, L. Q., Zhang, L. D., Li, G. H., He, G., Liu, M. and Fang, Q. Thermal stability and energy-band alignment of nitrogen-incorporated Zr O 2 films on Si (100). *Appl. Phys. Lett*., 2006, *88*(23), 232901.
44. Zhu, L. Q., Fang, Q., Wang, X. J., Zhang, J. P., Liu, M., He, G. and Zhang, L. D. Structural, optical properties and band gap alignments of ZrOxNy thin films on Si (1 0 0) by radio frequency sputtering at different deposition temperatures. *Appl. Surf. Sc*., 2008, *254*(17), 5439–5444.
45. Monaghan, S., Greer, J. C. and Elliott, S. D. Atomic scale model interfaces between high-k hafnium silicates and silicon. *Phys. Rev. B*, 2007, *75*(24), 245304.
46. Wilk, G. D., Wallace, R. M. and Anthony, J. M. Hafnium and zirconium silicates for advanced gate dielectrics. *J. Appl. Phys*., 2000, *87*(1), 484–492.
47. Hossain, M. K., Ahmed, M. H., Khan, M. I., Miah, M. S. and Hossain, S. Recent progress of rare earth oxides for sensor, detector, and electronic device applications: A review. *ACS Appl. Electron. Mater*., 2021, *3*(10), 4255–4283.

48. D. A. Muller, T. Sorsch, S. Mocclo, F. H. Baumann, K. Evans-Lutterodt, and G. Timp, The electronic structure at the atomic scale of ultrathin gate oxides. *Nature*, 1999, 399, 758–761.
49. Robertson, J. Interfaces and defects of high-k oxides on silicon. *Solid-State Electron.*, 2005, 49, 283–293.
50. Wilk, G. D., Wallace, R. M. and Anthony, J. (2001). High-κ gate dielectrics: Current status and materials properties considerations. *J. Appl. Phys.*, 2001, *89*(10), 5243–5275.
51. Rignanese, G. M. Dielectric properties of crystalline and amorphous transition metal oxides and silicates as potential high-κ candidates: The contribution of density-functional theory. *J. Phys. Cond. Mat.*, 2005, *17*(7), R357.
52. Lee, B. H., Kang, L., Nieh, R., Qi, W. J. and Lee, J. C. Thermal stability and electrical characteristics of ultrathin hafnium oxide gate dielectric reoxidized with rapid thermal annealing. *Appl. Phys. Lett.*, 2000, *76*(14), 1926–1928.
53. Feng, A., McCoy, B. J., Munir, Z. A. and Cagliostro, D. Wettability of transition metal oxide surfaces. *Mater. Sci. Eng.: A*, 1998, *242*(1–2), 50–56.
54. Yamaguchi, T., Iijima, R., Ino, T., Nishiyama, A., Satake, H. and Fukushima, N. Additional scattering effects for mobility degradation in Hf-silicate gate MISFETs. In *Digest. Int. Electron Device Meeting*, 2002, (pp. 621–624). IEEE.
55. Liu, C. L., Jiang, Z. X., Hegde, R. I., Sieloff, D. D., Rai, R. S., Gilmer, D. C. and Lu, S. Theoretical and experimental investigation of boron diffusion in polycrystalline HfO_2 films. *Appl. Phys. Lett.*, 2002, *81*(8), 1441–1443.
56. Kim, H., McIntyre, P. C. and Saraswat, K. C. Effects of crystallization on the electrical properties of ultrathin HfO_2 dielectrics grown by atomic layer deposition. *Appl. Phys. Lett.*, 2003, *82*(1), 106–108.
57. Perkins, C. M., Triplett, B. B., McIntyre, P. C., Saraswat, K. C., Haukka, S. and Tuominen, M. Electrical and materials properties of ZrO_2 gate dielectrics grown by atomic layer chemical vapor deposition. *Appl. Phys. Lett.*, 2001,*78*(16), 2357–2359.
58. Yu, H. Y., Li, M. F., Cho, B. J., Yeo, C. C., Joo, M. S., Kwong, D. L. and Ramanathan, S. Energy gap and band alignment for (HfO_2) x $(Al_2 O_3)_{1-x}$ on (100) Si. *Applied Phys. Lett.*, 2002, *81*(2), 376–378.
59. He, G., Fang, Q., & Zhang, L. D. Structural and interfacial properties of high-k HfOxNy gate dielectric films. *Mater. Sci. semicon. Proc.*, 2006, *9*(6), 870–875.
60. Callegari, A., Cartier, E., Gribelyuk, M., Okorn-Schmidt, H. F. and Zabel, T. Physical and electrical characterization of Hafnium oxide and Hafnium silicate sputtered films. *J. Appl. Phys.*, 2001, *90*(12), 6466–6475.
61. Choi, C. H., Rhee, S. J., Jeon, T. S., Lu, N., Sim, J. H., Clark, R. and Kwong, D. L. Thermally stable CVD HfO/sub x/N/sub y/advanced gate dielectrics with poly-Si gate electrode. In *Digest. Int, Electron Devices Meeting* 2002, (pp. 857–860). IEEE.
62. Zhu, J., Liu, Z. G. and Feng, Y. Thermal stability and electrical properties of pulsed laser-deposited Hf-silicate thin films for high-k gate dielectric applications. *J. Phys. D: Appl. Phys.*, 2003, *36*(23), 3051.
63. Liu, M., He, G., Zhu, L. Q., Fang, Q., Li, G. H. and Zhang, L. D. Microstructure and interfacial properties of HfO2–Al2O3 nanolaminate films. *Appl. Surf. Sci.*, 2006, *252*(18), 6206–6211.
64. Liu, M., Fang, Q., He, G., Zhu, L. Q. and Zhang, L. D. Characterization of HfOxNy gate dielectrics using a hafnium oxide as target. *Appl. Surf. Sci.*, 2006, *252*(24), 8673–8676.

7 Rare-Earth Oxide-based High-K Dielectrics

Shohaib Abass and Khalid Sultan

7.1 INTRODUCTION

Rare-earth elements are a group of elements comprising lanthanides, yttrium, and scandium. These elements are of particular interest nowadays because of their use in modern technologies and the promotion of clean energy technologies [1, 2, 3]. These rare-earth elements have been found to occur in low concentrations and dispersed form, and they are highly sought after due to their extraordinary magnetic and luminescent behaviors. Rare-earth elements have applications in aerospace, high-intensity street lamps, cancer treatments, strength enhancement of alloys, high-temperature superconductors, camera lenses, battery electrodes, permanent magnets, electric motors, lasers, nuclear reactors, fuel cells, sonar systems, hard disk devices, glass coloring, integrated circuits, signal amplifications, glass microphones, catalysis, steel production, and metallurgy, among many others [4].

As far as the oxides of rare-earth elements are concerned, these oxides have a pivotal role in systems like portable compact disc players and supercomputers. Rare-earth oxides are chemically active and thermodynamically stable. In general, rare-earth ions are in a trivalent state, so the formula for rare-earth oxides is R_2O_3; however, in some cases, other oxidation states are also possible. The structure of R_2O_3 is stable in a hexagonal, monoclinic, or cubic form depending on the ionic radius of the rare-earth element in the structure. Rare-earth oxides as well as transition metal oxides are associated with defects because the oxygen present in these oxides is generally volatile and is able to escape from materials, and transport properties like the ion's diffusion are accordingly controlled [5]. Studies also suggest that rare-earth oxides that are sesquioxides exist in different structures depending on temperature conditions. Some oxides of rare earth such as CeO_2, PrO_2, and TbO_2 with an oxidation structure other than the trivalent one have been found to occur in cubic fluorite configurations [6].

The conductivity of sesquioxides as well as nonstoichiometric oxides of rare earth lies in the range 10^{-9} to $10^{-1}\Omega^{-1}$ cm-^{1}. Studies have also concluded that in these compounds, the conductivity receives contributions from ionic as well as electronic parts, and the mechanism of conduction is best explained using the hopping mechanism or polaron mechanism [7]. Rare-earth oxides have the ability to provide the information of varying properties of materials containing these rare earths, and accordingly they can be utilized for testing theories. In addition to considering these rare-earth oxides as materials of research interest, they have applications for practical use. At some moderate temperatures, the oxides of rare earth are solid, and the capacity of oxygen to combine with a rare-earth element depends on the electronic configuration. In case of such element belonging to a lanthanide series, the valence electrons ($5d^1$ and $6s^2$) from the outer orbitals are shielded from the rising nuclear charge as the electrons are added to 4f orbitals. Due to this, the charge state of all lanthanides subjected to a slow change in energy across the series is trivalent [8].

The study of dielectric properties of rare-earth oxides by resonance cavity perturbation method shows that Pr_5O_{11} and Tb_4O_7 are excellent candidates for microwaves, as they both appear to gush at temperatures above 500°C; for example, at 800°C, the dielectric constant is 10.814(F/M) for Pr_5O_{11} and 10.157(F/M) for Tb_4O_7. Studies show that these rare-earth oxides can be heated from 150°C to

DOI: 10.1201/9781003352365-7

800°C in just 8.8 min [9]. The dielectric constant of binary earth oxides can be studied through linear correlation between the composition of crystal and dielectric constant. This shows that the dielectric constant of a material can be modified by changing the composition of material. It has been found that in a series of rare-earth oxides having similar structure the dielectric constant depends on the atomic number of rare-earth ions, which varies across the series [10].

In this chapter we will discuss the dielectric properties of rare-earth–based oxides. Rare-earth–based oxides of the RBO_3 structure (also called oxide perovskite) and $R_2BB'O_6$ (also called double perovskite) structure have been found to show a high dielectric constant. We will begin with a discussion of the high-K dielectric properties of oxide perovskites and then extend the concept to double perovskites. We will support our discussion with some examples and applications from day-to-day applications.

7.2 RARE-EARTH–BASED OXIDE PEROVSKITES AND THEIR HIGH DIELECTRIC CONSTANT

Oxide perovskites, which originate from the discovery of $CaTiO_3$, are the compounds having the ABO_3 structural formula, where A and B are the cations. In the case of rare-earth–based oxide perovskites, A is usually the rare-earth element while B can be any other element such as transition elements. These oxide perovskites have been found to be insulators with a high dielectric constant. Considering the B cation at the center of cube, we see that this B cation is shared with six oxygen anions forming a BO_6 octahedron. Cations A lie at the corners of the cube and within the interstitial sites created by BO_6 octahedra. The ideal structure of oxide perovskite is shown in Figure 7.1. The stability and structure of oxide perovskites have been determined through the famous tolerance factor t given by Equation (7.1) [11].

$$t = \frac{r_A + r_O}{\sqrt{2}\left(r_B + r_O\right)} \tag{7.1}$$

where r_A, r_B, and r_O are the ionic radii corresponding to A, B, and O ions, respectively. Depending on the value of t, these perovskites can have different structures. For example, if the value of t is 1, then the structure is cubic and perovskite has an ideal structure. On the other hand, if the value of t deviates from 1, then the structure is nonideal, and this usually happens due to the ionic size of A and B present in the structure of a particular compound. The distortion and tilting of BO_6 octahedra are generally responsible for changes in structure from the ideal form. Depending on the different possible combinations of A and B cations, a large number of oxide perovskites exist. This means that oxide perovskites have high structural flexibility. The possibility of doping at the A or B site is also

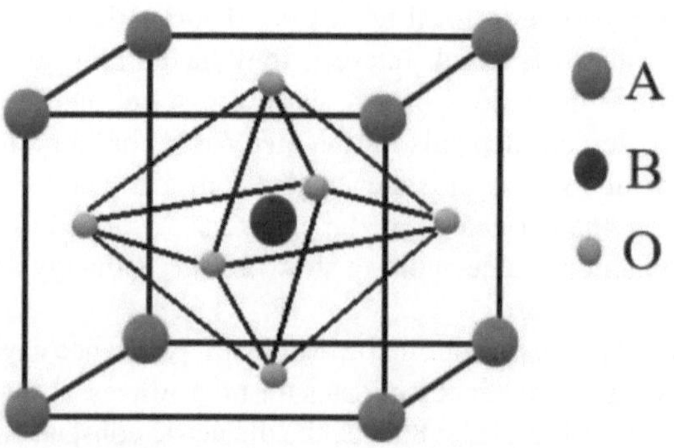

FIGURE 7.1 Structure of ideal simple perovskite.

a way to tune the physical properties of these compounds [12]. In this section we will discuss the high dielectric properties of rare-earth–based oxide perovskites.

Orthoferrites of the form $RFeO_3$, with R representing the rare-earth element, are considerable in replacing the lead-based perovskites that are used as relaxors, capacitors, sensors, and other energy storage devices [13]. These $RFeO_3$ are from the perovskite family having distorted orthorhombic structure and *pbnm* space group. It is to be noted that in these orthoferrites the distortion is largely due to polyhedra formed by rare-earth cation R as compared to octahedra formed by Fe ion, and this distortion rises with an increase in ionic radius of rare-earth R [14, 15]. Orthoferrites like $PrFeO_3$ and $EuFeO_3$ are the known compounds having a high dielectric constant and low dielectric loss, as a result of which they are potential candidates for electric devices. These orthoferrites have distorted orthorhombic structures with mixed valency and show remarkable electrical and magnetic properties [16, 17].

In $PrFeO_3$ it has been found that both the dielectric constant as well as dielectric loss measured in the frequency range of 20 Hz to 1 MHz show a decreasing trend with an increase in frequency, and this is the most usual property of polar dielectrics. In the high-frequency range, the dielectric constant shows a constant nature because the electric dipoles are unable to follow the varying frequency of the applied alternating electric field in that high-frequency region. Another possible reason is that the Fe ions exists in $Fe^{2+}$and Fe^{3+} charge states, and the electronic exchange between these two charge states would not be following the applied alternate electric field. The dependence of dielectric constants as well as dielectric loss on temperature shows that both entities do not vary at low temperatures but both experience an increase (for all frequencies) at high temperatures. This is due to the thermal effect leading to a rise in mobility of the carriers within the system, thereby increasing the chances of hopping. When doped with different concentrations of Mn in place of Fe, $PrFeO_3$ shows a considerable change in a dielectric constant, and even relaxation behavior comes into play at higher concentrations of Mn. This behavior is due to chemical pressure occurring due to Mn doping. The value of dielectric constant measured at room temperature and at 20 Hz is 103 for $PrFeO_3$, 6.68×10^3 for $PrFe_{0.9}Mn_{0.1}O_3$, 6.22×10^4 for $PrFe_{0.7}Mn_{0.3}O_3$, and 9.32×10^6 for $PrFe_{0.5}Mn_{0.5}O_3$. These values clearly indicate the high dielectric constant of this Mn-doped material, which can thus be considered for device miniaturization [18]. A similar study on the dielectric properties of Mn-doped $EuFeO_3$ explores the high dielectric constant of compounds. Polycrystalline $EuFeO_3$ has been considered as the best dielectric material due to its high dielectric constant and low dielectric loss. The behavior of dielectric constant as well as dielectric loss is the same as that of $PrFeO_3$. At room temperature and at frequency of 1 kHz the dielectric constant of $EuFeO_3$ is 330; this value has been found to increase with an increase in concentration of Mn [19].

The effect of Ni doping in $HoFeO_3$ on the dielectric properties suggests an increase in the dielectric constant with an increase in Ni concentration, and both the dielectric constant and loss have been found to exhibit different behavior with varying frequency and temperature. The behavior with frequency is similar to that for $PrFeO_3$, and there is a decrease in dielectric constant with increases in frequency in the low frequency range and a steady behavior in the high frequency range. The behavior is determined by the formation of new charge states of cations and creation of space charge polarization due to formation of dipoles. It has been assumed that the heterogeneous conduction in such materials is responsible for high dielectric constant in these materials in low frequency range. Moreover, Ni doping at the Fe site in $HoFeO_3$ results in higher values of dielectric constant. Actually, the Fe ion exists in two charge states, Fe^{2+} and Fe^{3+}, and there is electron hopping between these ions that contribute to the electrical nature of this $HoFeO_3$ compound. Ni doping induces chemical pressure in the system, which leads to a change of Ni from Ni^{2+} to Ni^{3+} and accordingly a change of Fe from Fe^{3+} to Fe^{2+}, thereby causing a net decrease in resistance of grains. This leads to an increased chance of electrons reaching grain boundaries and increasing polarization [20]. Similarly, the effect of Ni doping in $SmFeO_3$ suggests the colossal dielectric constant in the compound. Ni-doped samples show a dielectric constant of the order of 10^3–10^4. The dielectric constant of materials increases with an increase in Ni concentration that can be explained on the basis of Maxwell-Wagner mechanism. The rise in dielectric constant is associated with electron hopping and space charge polarization. Space charge polarization is due to grains and grain boundaries leading to local accumulation of charge and causing polarization. Electron hopping is, as explained above, the transfer of electrons between the different charge states of Fe. The values of dielectric constant measured at room temperature and a frequency of 1 MHz is 12,205 for $SmFeO_3$, 15,639 and for $SmFe_{0.7}Ni_{0.3}O_3$. These values give the idea that a high dielectric constant occurs due to Ni doping in $SmFeO_3$ [21].

The mixed valency of an A site doped with rare-earth R-based perovskites of the form $R_{1-x}A_xMnO_3$ has been under investigation for many decades [22, 23]. The substitution of R by some divalent cation such as Ca, Sr, Ba, etc. results in hole-type doping, and there is every possibility that divalent doping will enhance the conversion of Mn^{3+} to Mn^{4+}. The latter, being Jahn-Teller inactive, causes a decrease in distortion of MnO_6 octahedra with an increase in divalent dopant. However, it should be noted that the range of doping has always been determined by the tolerance factor. Consider the case of $LaMnO_3$ where doping has been done at La site by Sr. Both La and Sr have approximately similar sizes, so there would be random substitution at R site. Moreover, there would be tilting and distortion of MnO_3 octahedra depending on the concentration and ionic size of the dopant [24]. Studies have shown that $La_{0.55}Sr_{0.45}MnO_3$ is associated with high dielectric constant at room temperature and low frequency. At low frequency, the value of the dielectric constant is high because all four types of polarization are contributing in this low frequency range. The dielectric constant decreases with a rise in frequency up to 5 MHz and finally achieves a constant behavior [25]. $GdMnO_3$ ceramic sintered at 1,350°C depicts a higher value of dielectric constant and a smaller value of dielectric loss. In this compound, at higher frequencies, the relaxation process comes into play. The study shows that the sintering temperature affects the dielectric constant of the material under investigation. The dielectric constant of this ceramic measured in the high frequency range is best when sintered at 1,350°C as compared to the same ceramic when sintered at temperatures of 1,300°C and 1,400°C. Similarly, the dielectric loss is minimal for the sample sintered at 1,350°C [26].

A study of the dielectric properties of Bi and Fe-doped $TbMnO_3$ perovskite has been conducted [27]. In this study, the dielectric constant of $TbMnO_3$, $Tb_{0.95}Bi_{0.05}MnO_3$, and $Tb_{0.9}Bi_{0.1}Mn_{0.95}Fe_{0.05}O_3$ has been determined in the frequency range from 100 Hz to 1 MHz and the temperature range from 80 K to 400 K. It has been observed that Bi doping results in a decrease in dielectric constant with the disappearance of dissipation peaks in high temperature range. On the other hand, Fe doping again results in a high dielectric constant of the order of 10^3 along with reappearance of dissipation peaks in the high temperature range. This indicates the effect of doping on the dielectric properties, especially the dielectric constant. We have summarized the value of the dielectric constant of some perovskites in Table 7.1.

7.3 RARE-EARTH–BASED OXIDE DOUBLE PEROVSKITES AND THEIR HIGH DIELECTRIC CONSTANT

An active class of materials under investigation for energy storage applications is double perovskites. The general formula of oxide double perovskites is $A_2BB'O_6$, where A is generally rare earth or alkaline earth metal while B and B′ are transition metals. As it is evident from the studies, double perovskites are the derivatives of single perovskites that are known for their extraordinary and multifunctional properties. Perovskites have such compositional and structural freedom that they can accommodate a large number of elements. In comparison, double perovskites have the freedom to accommodate two different types of cations at A sites as well as at B sites. Due to this, double perovskites can offer more flexibility to alter the properties. A double perovskite of the form $AA'BB'O_6$ is called doubly ordered perovskite or double-double perovskite. About a thousand double perovskites have been studied, with divalent alkaline earth metals or trivalent rare earth elements at the A site and transition metals at the B site. These divalent and trivalent cations leave the possibility of different charge states of transition metals at B sites [28]. Since in this chapter we are

TABLE 7.1
Dielectric Constant of Some Perovskites and Double Perovskites

Rare-Earth–Based Oxide Perovskite/ Double Perovskite	Dielectric Constant at a Given Frequency and Room Temperature	Reference
$PrFeO_3$	10^3 at 20 Hz	[18]
$PrFe_{0.9}Mn_{0.1}O_3$	6.68×10^3 at 20 Hz	[18]
$PrFe_{0.7}Mn_{0.3}O_3$	6.22×10^4 at 20 Hz	[18]
$PrFe_{0.5}Mn_{0.5}O_3$	9.32×10^6 at 20 Hz	[18]
$EuFeO_3$	330 at 1 kHz	[19]
$SmFeO_3$	12205 at 1 MHz	[21]
$SmFe_{0.7}Ni_{0.3}O_3$	15639 at 1 MHz	[21]
$SmFe_{0.5}Ni_{0.5}O_3$	17940 at 1 MHz	[21]
Gd_2NiMnO_6	595 at 100 kHz	[30]
$Gd_{1.9}Sr_{0.1}NiMnO_6$	805 at 100 kHz	[30]
$Gd_{1.7}Sr_{0.3}NiMnO_6$	1055 at 100 kHz	[30]
Nd_2NiMnO_6	4471.8 at 1 kHz	[32]
$Nd_{1.9}Sr_{0.1}NiMnO_6$	7502.9 at 1 kHz	[32]
$Nd_{1.7}Sr_{0.3}NiMnO_6$	8205.7 at 1 kHz	[32]
Dy_2NiMnO_6	10^3 at 100 Hz	[35]
La_2FeMnO_6	7×10^4 at 1 Hz	[39]

concerned with the rare-earth–based oxides, we will limit our discussion to the high dielectric status of the rare-earth–based oxide double perovskite.

We will begin with the double perovskites of the form R_2NiMnO_6 (R = rare earth). The extensive literature pertaining to the exploration of dielectric properties of this type of double perovskites reveals the very high dielectric constant in the double perovskites of the form R_2NiMnO_6. La_2NiMnO_6 double perovskite doped with Ba has been investigated for the dielectric properties. Studies show that the dielectric constant decreases in value with increasing frequency, but the decreasing trend is comparatively smaller in doped samples than in the pristine ones, in which the decrease in value is drastic. The behavior is best understood in terms of the Maxwell-Wagner model. Moreover, the grain boundaries have a remarkable impact on the dielectric response of materials [29]. Gd_2NiMnO_6 double perovskite shows a dielectric constant of 595 at a frequency of 100 kHz, which has been further improved by doping a divalent Sr at the site of Gd, where the dielectric constant has been found to increase up to 1,055 at same frequency for a doping concentration of 0.3. Dielectric loss in this Gd_2NiMnO_6 has a value of 98 at 100 kHz for the pristine sample, which decreases to 4 for the doped sample of $Gd_{1.7}Sr_{0.3}NiMnO_6$. Sr doping also results in a decrease in grain size, thereby causing an increase in density of grain boundary and thus an increase in free charge accumulation at grain boundaries. This in turn leads to increases in interfacial polarization and consequently higher dielectric constant. With increases in temperature, the dielectric constant first remains constant at low temperature, then starts increasing slowly and finally achieves a very high value. These features make Sr-doped Gd_2NiMnO_6 suitable for electronic devices operating in a chosen range of frequency and temperature [30]. A similar study of dielectric constant of Gd_2NiMnO_6 investigated through Ca doping suggests high dielectric constant of the material, where the grains, grain boundaries, and transition metal ions have been found to influence the dielectric nature of samples. Ca doping at the Gd site results in a large increase in the dielectric constant and a corresponding decrease in the dielectric loss of the material [31].

Nd_2NiMnO_6 is also a promising candidate for high dielectric materials. At room temperature and at a frequency of 1 kHz, the material shows a dielectric constant of 4,471.8 and dielectric loss of 119.43. At same temperature and frequency but when doped with Sr at the Nd site, the dielectric constant becomes 7,502.9 and dielectric loss becomes 30.2 for $Nd_{1.9}Sr_{0.1}NiMnO_6$ while the dielectric constant becomes 8,205.7 and dielectric loss becomes 158.2 for $Nd_{1.7}Sr_{0.3}NiMnO_6$. The same measurements when performed at low temperature of 200 K and frequency of 1 kHz results in values of 1,840.5 for dielectric constant and 12.612 for dielectric loss in case of pristine sample Nd_2NiMnO_6. Similarly, the values become 2,080.9 for dielectric constant while 3.1831 for dielectric loss in case of $Nd_{1.9}Sr_{0.1}NiMnO_6$ and 4,640.4 for dielectric constant while 15.583 for dielectric loss in case of $Nd_{1.7}Sr_{0.3}NiMnO_6$. At both temperatures the dielectric constant has been observed to increase with an increase in Sr concentration, which is attributed to a decrease in grain size causing an increase in grain boundary density and hence net polarization. The values of dielectric loss do not show a regular trend with increasing concentration of Sr, as in this compound the dielectric loss is also determined by presence of oxygen vacancies. The trends are in accordance with the quantity of oxygen vacancies in three samples of $Nd_{2-x}Sr_xNiMnO_6$ ($x = 0.0, 0.1, 0.3$) [32].

In Sm_2NiMnO_6, it has been observed that the dielectric constant remains almost constant up to the temperature of 60°C and then starts increasing up to about 200°C at which point the dielectric constant decreases abruptly for all the frequencies ranging from 50 kHz to 1 MHz. This means that the peak in dielectric constant value can be fairly accurately determined for all the selected frequencies, and this peak has been found to shift toward the higher temperature region as the frequency at which the data are collected increases. This phenomenon is known as dielectric relaxation. The relaxor behavior in this compound can be confirmed through Vogel-Fulcher fitting, and the temperature-dependent effect on grain and grain boundary can be confirmed through Nyquist plots. For dielectric loss expressed in terms of tan δ, the value is low at low temperatures but rises for higher temperatures. This increase in dielectric loss can generally be attributed to the thermally activated carriers, defects, and conductivity. In this compound one can achieve a dielectric constant as high as 1,500 at a frequency of 50 kHz. Moreover, the dielectric loss at room temperature expressed in terms of tanδ is less than 10 [33]. Studies also suggest the nonlinear dielectric behavior of Sm_2NiMnO_6, revealing that the compound possesses small hysteretic and large nonlinearity properties. Moreover, the nonlinear dependence of dielectric constant on field reveals an ability toward saturation for high fields of approximately 30 kV/cm[34].

In Dy_2NiMnO_6 double perovskite, the dielectric relaxation can be seen by observing the dielectric constant of 10^3 at a frequency of 100 Hz to a dielectric constant of 50 at a frequency of 1 MHz. The relaxation frequency has been observed to shift with changing temperature. The dielectric relaxation also exhibits a loss peak that is positioned at relaxation frequency. The frequency dependence of dielectric constant as well as dielectric loss shows that the relaxation in Dy_2NiMnO_6 is non-Debye type of relaxation [35]. The study of dielectric measurements on Tb_2NiMnO_6 in terms of real and imaginary part of dielectric permittivity reveal three plateaus. These plateaus clearly predict that the dielectric relaxation receives contributions from bulk, grain boundaries, and sample electrode interfaces. The compound has a very high dielectric constant depending on the suitable frequency and temperature at which the compound is being investigated [36]. Co-doping in these rare-earth–based oxide double perovskites has produced tremendous increases in dielectric constant values. Sr and Sb co-doped samples of La_2NiMnO_6 synthesized through gel combustion techniques have been investigated for dielectric constant. The temperature dependence of dielectric constant in a pristine sample measured at a frequency of 2 kHz shows a plateau at high temperatures where the dielectric constant has value of the order of 10^6. When the temperature is decreased below 175 K, the dielectric constant also starts decreasing. However, the increase in Sr and Sb concentration in the compound causes a decrease in the plateau region, and the drop-off temperature moves toward the higher-temperature region. The dielectric loss has been associated with a peak in both the pure and doped samples, and this peak moves toward the higher-temperature region with an increase in Sr and Sb concentration. This study clearly shows the effects of co-doping-based modification in dielectric constant and dielectric loss of La_2NiMnO_6 [37].

Investigations of of electrical properties for the compounds of the form R_2CoMnO_6 (R = Ho, Tm, Yb, Lu) show the semiconducting behavior. Studies show that these compounds have a high dielectric constant at room temperature, which decreases abruptly to a value between 10 and 26 when the temperature is lowered below 50 K. They also show that at low temperatures, there is very little or no frequency dependence compared to the dielectric constant at high temperatures where a considerable dynamic nature can be seen. The dielectric constant in Lu_2CoMnO_6 rises with the increasing temperature in a step-like fashion, with the K value increasing from approximately 10 in the low-temperature region to about 10^3 at room temperature. However, this step-like rise has been found to shift toward the high-temperature side with the rise in frequency at which the measurements are being made. The dielectric loss in Lu_2CoMnO_6 expressed in terms of tanδ is associated with loss peak. For low frequencies, there is a rise in dielectric loss above the peak, which suggests contribution from the electronic part at room temperature. Actually, these phenomena come from the Maxwell-Wagner contribution. The effect of rare-earth element on the dielectric properties has also been investigated. All compounds of the form R_2CoMnO_6 (R = Ho, Tm, Yb) behave similarly to Lu_2CoMnO_6. However, the step-like increases in dielectric constant and loss peak are both observed to shift toward the low-temperature side as the size of rare-earth element present in the material decreases [38].

In double perovskite La_2FeMnO_6, the dielectric constant has a value of 7×10^4 at room temperature and at frequency of 1 Hz while the value of dielectric constant becomes 0.5 at room temperature and frequency of 10 MHz. The high value of dielectric constant at low frequency is because of various contributions to polarization like charge accumulation at grain boundaries, electronic polarization, polaronic contribution, and ionic contribution. Polaronic contribution comes from the mixed valence of ions and distortions of the lattice. On the other hand, dielectric loss has been found to increase with an increase in frequency up to a maximum value, known as loss peak. This loss peak has been observed when there is matching of applied frequency and hopping frequency. Loss peak shifts toward the high-frequency end when the temperature at which the measurements are being carried out is increased [39].

In ceramic materials the dielectric properties come from two types of contributions: intrinsic contribution and extrinsic. Intrinsic contribution is decided by the structure of the system under investigation while the extrinsic contribution is determined by defects, dopants, microstructure, etc. The intrinsic contribution can be achieved through the B site ordering while the extrinsic contribution is achieved through calcination temperature. It is observed that the calcination temperature has a greater impact on the dielectric constant of a material under investigation. In La_2CoMnO_6 the static dielectric constant has been observed to increase with an increase in the calcination temperature [40]. Studies have also shown that the sample of La_2CoMnO_6 calcinated at 1,000°C is perfectly ordered, and in this ordered compound the colossal dielectric effect has been observed that is due solely to extrinsic contribution [41]. Dy_2CoMnO_6 is also a potential candidate for high dielectric properties. This compound shows the same features of decreasing dielectric constant with increasing frequency, which appears to be the normal behavior of a dielectric material. Moreover, the dielectric behavior can be described in terms of Maxwell-Wagner phenomenon according to which the dielectric material consists of grains and grain boundaries. Grains are more conducting as compared to the grain boundaries. Grains are effective at high frequencies, and the low frequencies make the grain boundaries sensitive. The decrease in dielectric constant that occurs with an increase in frequency due to the hopping of electrons, which makes them incapable of following the applied electric field. Due to this the electrons can travel across the grains and grain boundaries. Since it has been assumed by the Maxwell-Wagner theory that grain boundaries are less conductive, the carriers thus get accumulated at grain boundaries, leading to space charge polarization. On the other hand, the loss tangent also shows the decreasing behavior with increasing frequency. That means the dielectric loss is high at low frequency compared to that at high frequency. The loss may be due to different reasons depending on the frequency. The movement of ions may be responsible for dielectric loss at low frequencies; jumping of ions, loss in conduction, or ion polarization may be responsible for dielectric loss at medium frequencies; and the vibration of the ion may be the reason for dielectric loss at high frequencies [42]. Apart from the compounds discussed in this section, many other oxide double perovskite have been investigated. All this research points at the high dielectric constant in oxide double perovskites. We have summarized the value of the dielectric constant of some double perovskites in Table 7.1.

7.4 ILLUSTRATION AND EXAMPLES

7.4.1 Dielectric Studies of $PrFe_{1-x}Mn_xO_3$ ($x = 0, 0.1, 0.5$)

The discussion in the previous sections indicates that ceramic materials are associated with a very good value static dielectric constant, and thus are an important part of electronic applications for technological use. We have also understood that the dielectric constant has a direct dependence on the polarizability. There are various approaches by which the dielectric constant of a material can be modified. One approach, as discussed in the preceding section, is by doping at a particular size of the system under investigation. Another way is by irradiation. The ionizing type of radiation, such as gamma radiation, has been used mostly for the creation of structural defects in materials subjected to them to alter the physical properties of solids. Gamma radiation is important with respect to other irradiation effects, as gamma rays can get through the shielding and penetrate the system, thereby changing the properties of the system. Since the change in the physical system depends heavily on the internal characteristics of the absorbed substance, investigations have shown that structural defects are the main factors created in the materials on their exposure to ionizing gamma radiation. Gamma ray irradiation has been found to reduce polarization of ceramic materials. Based on technical use, it is very important to know the effects of gamma ray irradiation on structural and dielectric properties of materials. Here we will discuss the dielectric constant of $PrFe_{1-x}Mn_xO_3$ ($x = 0, 0.1, 0.5$) system that is being subjected to gamma ray irradiation. The measurements have been done with the gamma radiation dose from the Co^{60} isotope having an average dose rate of 6.4 KGy/hr.

The dielectric constant has been measured before and after the sample exposure to an absorbed gamma radiation dose of 625 KGy. As evident from Figure 7.2, the variation of dielectric constant with frequency for all the samples of $PrFe_{1-x}Mn_xO_3$ ($x = 0, 0.1, 0.5$) shows a familiar trend of decreasing value with increasing frequency until eventual saturation, for both the non-irradiated and irradiated samples. However, we observe a net decrease in the dielectric constant of all the samples after irradiation. The purpose of irradiation is to create defects in the samples. When a sample is irradiated, defects are formed in the sample that may obstruct the flow of space charge carriers. This obstruction is, in turn, reducing space charge buildup and thus space charge polarization. Hence, there would be a net decrease in the values of the dielectric constant for irradiated samples. It can also be assumed that when the doses of gamma radiation are high, they can lead to atomic rearrangement and the creation of primary defects. The atomic rearrangement would cause the production of stable defects because the gamma radiation energy is redistributed over the sample. This finally causes the reduction in polarization and the suppression of defects migration. Ternary and binary ceramic compounds are more prone to such type of effects. However, it can be noticed that although gamma ray irradiation has an impact in modifying the dielectric properties of ceramic materials, this dependence on gamma ray irradiation is not stronger, as the dielectric properties depend as much on the pre-exposure state of the material under study as on the radiation dose [12].

7.4.2 Dielectric Studies of $Nd_{2-x}Sr_xNiMnO_6$ ($x = 0, 0.1, 0.3$)

7.4.2.1 Frequency Dependence of Dielectric Constant and Dielectric Loss

The frequency dependence of the dielectric constant as well as dielectric loss of the three samples of $Nd_{2-x}Sr_xNiMnO_6$ ($x = 0, 0.1, 0.3$) measured at a temperature of 100 K and in the frequency range from 1 Hz to 1MHz are depicted in Figure 7.3. The plots indicate the usual behavior of polar dielectrics. Both the parameters—dielectric constant and dielectric loss—show a decreasing trend with

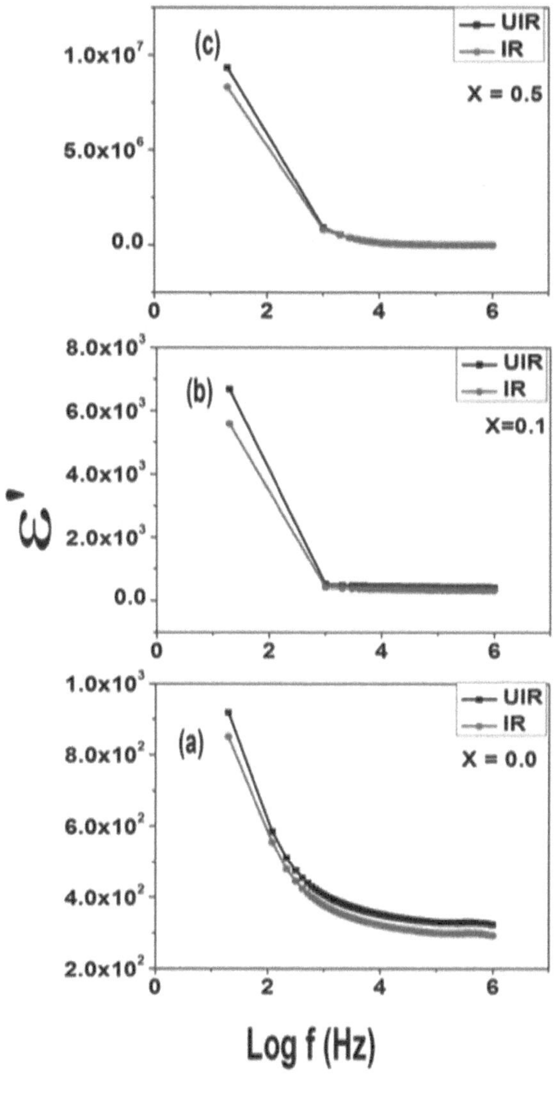

FIGURE 7.2 Room temperature dependence of dielectric constant on frequency for the samples of $PrFe_{1-x}Mn_xO_3$ (x = 0.0, 0.1, 0.5) system measured before and after gamma irradiation.

increasing frequency. As discussed earlier, at low frequencies, all types of polarizations are active and all are contributing to the high dielectric constant. As the frequency is increased, all the types of polarizations are no longer contributing, and thus a decrease in dielectric constant can be observed that finally leads to a consistent behavior at higher frequencies. Moreover, at higher frequencies, induced dipoles do not follow the applied frequency. These are some of the reasons for the decreasing trend of dielectric constant with increasing frequency.

On the other hand, the dielectric loss is also high at low frequency, and decreases with an increase in frequency before finally achieving a constant-like behavior. All these effects have been described in terms of surface charge effects and the motion induced in the different charged ions present in the sample [32]. Studies also suggest that the dielectric loss is associated with electron hopping and oxygen vacancies. Grains and grain boundaries also play a crucial role in the determination of the dielectric parameters of these oxide perovskites [43]. As far as the effect of Sr doping is concerned, we can see in Figure 7.3 a remarkable influence on dielectric constant as well as dielectric loss on changing the concentration of Sr dopant. The dielectric constant increases with increases in Sr concentration, which is due to Sr having an ionic radius that differs from the exact Nd ion radius. This in turn causes strain and stress within the system and hampers the grain growth process. A decrease in grain size leads to an increase in grain boundary density and thus a rise in polarization and dielectric constant.

In terms of the dielectric loss, Sr doping does not give a regular trend. When the sample is doped with a Sr concentration of $x = 0.1$, the dielectric loss decreases compared to that of the pristine sample, while for the Sr concentration of $x = 0.3$, the dielectric loss is greater than that of the pristine sample. This may be due to the oxygen vacancies present in the pristine and doped samples [32]. At 100 K temperature and a frequency of 1 Hz, the pristine sample Nd_2NiMnO_6 shows a dielectric constant of 572.84 and a dielectric loss of 13.906. At the same temperature and frequency, the doped sample $Nd_{1.9}Sr_{0.1}NiMnO_6$ shows a dielectric constant of 1,294.3 and a dielectric loss of 4.4134 while the doped sample $Nd_{1.7}Sr_{0.3}NiMnO_6$ shows a dielectric constant of 2,110.8 and a dielectric loss of 52.368.

7.4.2.2 Temperature Dependence of Dielectric Constant and Dielectric Loss

In the temperature range from 80 K to 300 K variations in dielectric constant as well as dielectric loss have been observed as shown in Figure 7.4 for all samples of $Nd_{2-x}Sr_xNiMnO_6$ (x = 0, 0.1, 0.3). Moreover, the temperature dependence has been observed at three different frequencies, specifically 3.162×10^5 Hz, 3.162×10^4 Hz, and 3.162×10^3Hz. From the graphs it is obvious that the dielectric constant remains unchanged for all three frequencies and for all three samples in the low-temperature range. This is because at low temperatures, the dipoles cannot align along the field applied to them and transition metal ions of, e.g., Ni and Mn are unable to move due to less thermal energy. With temperature increases, thermal energy rises, allowing more motion to transition metal ions in the samples and a net increase in the dielectric constant of the material. On the other hand, dielectric loss, which is usually expressed in terms of loss tangent, is also constant in the low-temperature range for all frequencies and all three samples of $Nd_{2-x}Sr_xNiMnO_6$ (x = 0, 0.1, 0.3) (as shown in Figure 7.4) This is again due very little motion among the dipoles. At higher temperatures, motion of the transition metal ions of Ni and Mn increases along with the induced motion in the dipoles, thereby producing a higher loss tangent value [32]. We can also observe a strong low-frequency dependence of loss tangent in the low-temperature range. However, at higher temperatures, both dielectric constant as well as dielectric loss show a dispersive behavior. The value of dielectric constant and loss tangent decreases with increasing frequency from 3.162×10^3 Hz to 3.162×10^5 Hz, which follows the familiar behavior of decreasing dielectric properties with increasing frequency in these perovskite materials.

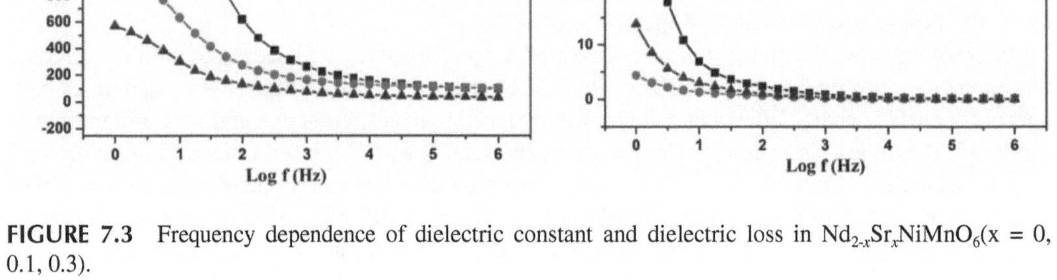

FIGURE 7.3 Frequency dependence of dielectric constant and dielectric loss in $Nd_{2-x}Sr_xNiMnO_6$(x = 0, 0.1, 0.3).

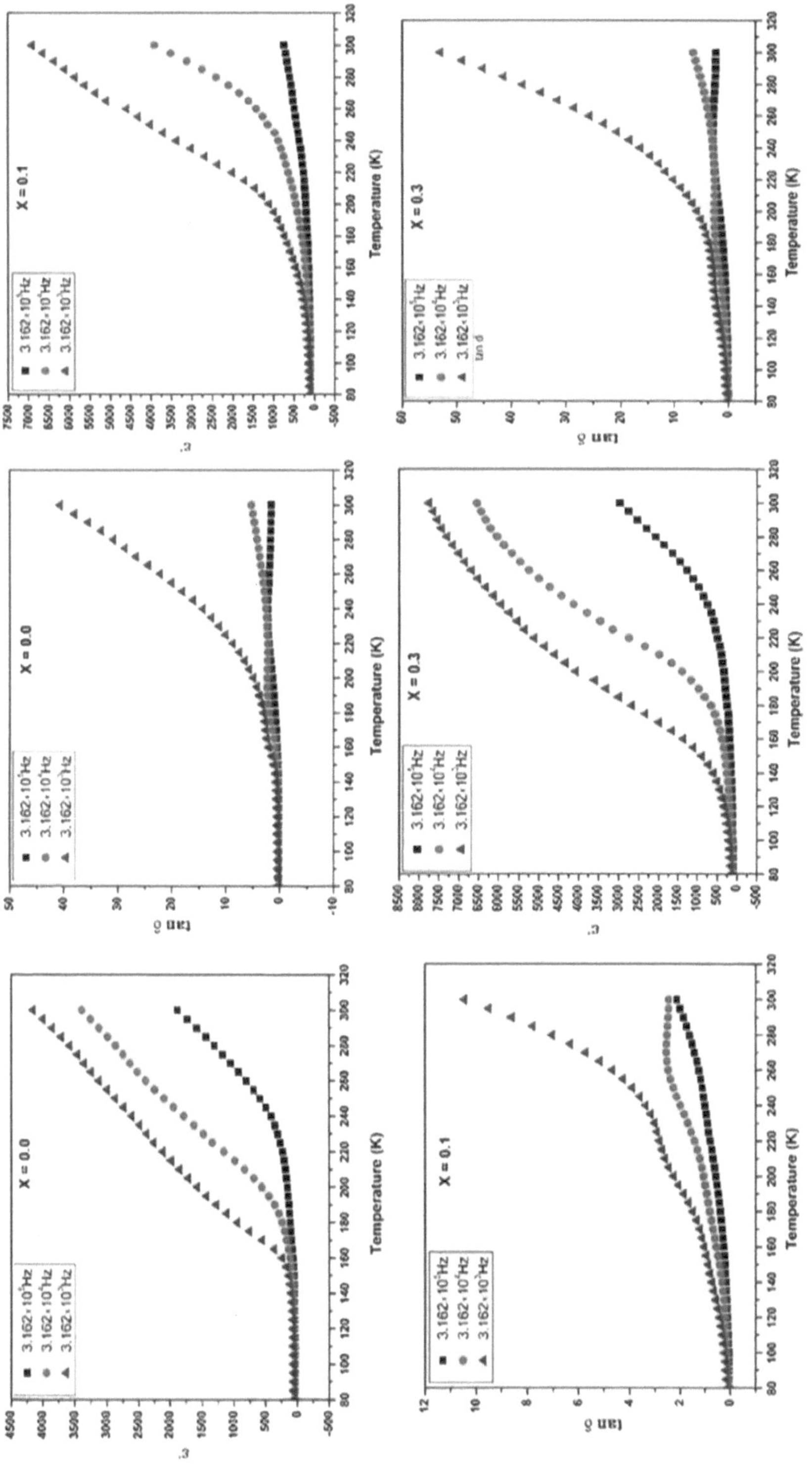

FIGURE 7.4 Temperature dependence of dielectric constant as well as dielectric loss in samples of $Nd_{2-x}Sr_xNiMnO_6$ (x = 0, 0.1, 0.3).

Let us now give an idea of the high dielectric constant in these materials by listing and comparing the values of dielectric constant and dielectric loss at different frequencies and at different temperatures. Starting with the pristine sample Nd_2NiMnO_6, at a temperature of 300 K and frequency of 3.162×10^5 Hz, the dielectric constant is 1,879.9 and the dielectric loss is 1.4832. At a same temperature of 300 K and frequency of 3.162×10^4 Hz, the value of the dielectric constant is 3,388.9 and the value of loss tangent is 5.18. Again, at the same temperature of 300 K and frequency of 3.162×10^3 Hz, this dielectric constant and dielectric loss become 4,153.5 and 40.669 respectively. For the doped sample $Nd_{1.9}Sr_{0.1}NiMnO_6$, at a temperature of 300 K and frequency of 3.162×10^5 Hz, the dielectric constant is 738.12 and the dielectric loss is 2.1358. At a frequency of 3.162×10^4 Hz, the value of the dielectric constant is 3,912.9 and of the dielectric loss is 2.474, and at a frequency of 3.162×10^3 Hz, this dielectric constant and dielectric loss become 6,916.6 and 10.462, respectively. Similarly, for the doped sample $Nd_{1.7}Sr_{0.3}NiMnO_6$, at a sample temperature of 300 K and frequency of 3.162×10^5 Hz, the dielectric constant is 2,983 and the dielectric loss is 2.2987. At 3.162×10^4 Hz, the value of the dielectric constant is 6,539.5 and of the dielectric loss is 6.4801. Finally, at a frequency of 3.162×10^3 Hz, the dielectric constant and dielectric loss become 7,750.5 and 52.962, respectively.

Similarly, we can compare the data at a very different temperature of 100 K and frequency of 3.162×10^3 Hz. We observe a dielectric constant of 101.33 and dielectric loss of 0.67533 for a pristine sample, dielectric constant of 143.12 and dielectric loss of 0.22101 for $Nd_{1.9}Sr_{0.1}NiMnO_6$, and dielectric constant of 197.61 and dielectric loss of 0.59772 for $Nd_{1.7}Sr_{0.3}NiMnO_6$. From the above values we observe that depending on frequency, temperature, dopant concentration, grain size, and oxygen vacancies, these compounds can have very high dielectric constant values combined with a very low dielectric loss tangent. The above-discussed examples clearly illustrate the high dielectric constant of rare-earth–based oxide perovskites and double perovskites. In the next section we will cast some light on applications of these rare-earth–based oxides.

7.5 APPLICATIONS

Dielectric materials can store charge and as a result are most commonly used to store energy and for the construction of transmission lines. There are various other applications for high dielectric materials, such as in liquid crystal displays, as dielectric resonators, oscillators, and in enhancing the performance semiconductors. Rare-earth elements have a wide range of applications in technology, and their alloys, ceramics, and oxides have been proven to imbue materials to which they are added with special useful properties [44]. Rare-earth oxides have been found to have high dielectric constants and are thus very good candidates for gate dielectrics. La_2O_3 synthesized under proper calcination and sintering conditions has emerged as the best candidate for metal-oxide-semiconductor field-effect transistor (MOSFET) due to its suitable electrical properties [45]. These rare-earth–based oxides have a great role in electronic devices, particularly in memory devices, dielectric materials, capacitors, and other energy storage devices. Rare-earth–based oxides are the focus of considerable current research for industrial and other applications [46].

In making electronic components, the value of dielectric constant determines the degree of miniaturization of electronic circuits. Most of the rare-earth–based oxide perovskites show potential for use in electronic devices because of their high dielectric constant values [18, 19]. Investigation of dielectric tunable materials is also under way because of their potential applications in capacitors, communication devices involving microwaves, oscillators, phase shifters, and filters. La_2NiMnO_6 in bulk form has emerged as a compound showing large dielectric tunability [47]. There is always a demand to exceed the available abilities of dielectric materials so that they can be even more useful for electronic devices for day-to-day human activities. Device miniaturization has always been a concept of great perspective in the research field of material science. In memory devices based on the concept of capacitance and random-access memory, the extent of miniaturization is determined by the static dielectric constant of the material [48]. Miniaturization of devices and improvement of their features can be achieved by engineering the dielectric properties of materials. One such method of altering the dielectric constant of materials is by doping. Doping always leads to structural, morphological, and electrical changes in a material and thereby causes changes in the dielectric properties. We have illustrated in the above example of Nd_2NiMnO_6 how the dielectric constant as well as dielectric loss can be altered by doping this material with Sr. Moreover, it should be noted that all rare-earth–based oxide perovskites and oxide double perovskites can modify the dielectric properties by inducing some structural or electrical changes in their system. Based on this, we can conclude that rare-earth–based oxide perovskites are the potential candidates for energy storage devices and particularly for device miniaturization.

7.6 FUTURE PROSPECTS AND CHALLENGES

We have observed that rare-earth–based oxide perovskites and double perovskites can be used as dielectric materials. They show a high dielectric constant even in pristine form, and this dielectric constant can be altered by doping or by irradiation. Doping in oxide perovskites and double perovskites can be carried out in different ways. In oxide perovskites of the form ABO_3, doping can be done at the A site or B site and even at both sites simultaneously. Similarly, in double perovskites of the form $A_2B_2O_6$ or $AA'BB'O_6$, doping can be done at any of the sites A, A′, B, or B′. We have observed that doping has a large influence in enhancing the dielectric constant of the rare-earth–based oxides discussed in this chapter. A new study about the dielectric constant can be carried out by doping at another site or doping with different dopants to make a comparison. Future investigations of the dielectric properties of these rare-earth–based oxides, by depositing their thin films, synthesizing their nanoparticles, or irradiating the deposited thin films, could provide more information for expanded technological applications.

7.7 SUMMARY

In this chapter, the dielectric constant of rare-earth–based oxide perovskites of the form ABO_3 and double perovskites of the form $A_2BB'O_6$ (where A is a rare-earth element and B or B' is a transition metal) has been discussed, with such examples as $PrFeO_3$, $EuFeO_3$, $SmFeO_3$, Gd_2NiMnO_6, and La_2NiMnO_6. We have also discussed the frequency dependence and temperature dependence of the dielectric constant as well as the dielectric loss of Sr-doped Nd_2NiMnO_6. Rare-earth oxides have been found to have a high dielectric constant, and their derivatives like oxide perovskites and double perovskites also show high values of dielectric constant at different frequencies and temperatures. We have found that the K value of these rare-earth–based oxides can be enhanced with doping. The dielectric properties of these compounds predict that they can be used in device miniaturization, memory devices, and many other technological applications. Finally, we presented the scope and future prospects of the study discussed in this chapter.

REFERENCES

[1] Humphries, M. Rare Earth Elements: The Global Supply Chain; Report No. 41347; Congressional Research Service: Washington D.C., 2013.

[2] Gschneidner, Karl A. Industrial Applications of Rare Earth Elements; *American Chemical Society*, Washington, DC, 1981, *164*, 326.

[3] Hoenderdaal, S., Tercero Espinoza, L., Marscheider-Weidemann, F., Graus, W. Can a dysprosium shortage threaten green energy technologies? *Energy*, 2013, *49*, 344.

[4] Zaimes, George G., Hubler, Berlyn J., Wang, S., and Khanna, V. Environmental life cycle perspective on rare earth oxide production, *ACS Sustainable Chemistry & Engineering*, 2015, *3*, 237.

[5] Adachi, G., Imanaka, N. and Kang, Z. C. *Binary Rare earth Oxides*, Kluwer Academic Publishers, 2004.

[6] Sato, S., Takahashi, R., Kobune, M., and Gotoh, H. Basic properties of rare earth oxides. *Applied Catalysis A: General*, 2009, *356*, 57.

[7] SubbaRao, G. V., Ramdas, S., Mehrotra, P. N., and Rao, C. N. R. Electrical transport in rare-earth oxides. *Journal of Solid StateChemistry*,1970, *2*, 377.
[8] Eyring, L. The Binary Rare Earth oxides, in *Handbook on the Physics and Chemistry of Rare Earths*, North-Holland Publishing Company, 1979, 337–339.
[9] Lv, P., Zhang, L., Li, S., Chen, K., Peng, J., Yin, S., Libo Zhang. Measurement of dielectric properties of rare earth oxides by resonance cavity perturbation method, *Journal of Physics and Chemistry of Solids*, 2022, *160*, 110362.
[10] Xue, D., Betzler, K., and Hesse, H. Dielectric constants of binary rare-earth compounds, *Journal of Physics. Condensed Matter*, 2000, *12*, 3113.
[11] Goldschmidt, V.M., et al. "Geochemische Verteilungsgesetz der elemente (VII)," Skr Nor Vidensk Akad Oslo I, *Matematisk-Naturvitenskapelig Klasse*, 1926, *2*, 1.
[12] Sultan, K., *Ph.D Thesis*, National Instuite of Technology Srinagar, India, 2015.
[13] Samara, G.A. The relaxational properties of compositionally disordered ABO_3 perovskites, *Journal of Physics. Condensed Matter*. 2003, *15*, 367.
[14] Marezio, M., Remeiko, J. P. and Dernier, P. D. The crystal chemistry of the rare earth orthoferrites, *Acta Crystallographica. Section B*, 1970, *26*, 2008.
[15] Marezio, M. and Dernier, P.D. The bond lengths in LaFeO3, *Materials Research Bulletin*, 1971, *6*, 23.
[16] Trukhanov, S.V., Investigation of stability of ordered manganites, *JETP*, 2005, *101*, 513.
[17] Trukhanov, S.V., Lobanovski, L.S., Businsky, M.V., et al. Study of A-site ordered $PrBaMn_2O_{6-\delta}$ manganite properties depending on the treatment conditions, *Journal of Physics. Condensed Matter*,2005, *17*, 6495.
[18] Sultan, K.,Ikram, M., Asokan, K.Structural, optical and dielectric study of Mn doped $PrFeO_3$ ceramics, *Vacuum*, 2014,*99*,251.
[19] Sultan, K., Ikram, M., and Asokan, K. Effect of Mn doping on structural, morphological and dielectric properties of $EuFeO_3$ ceramics, *RSC Advances*, 2015, *5*, 93867.
[20] Habib, Z.,Majid, K., Ikram, M., and Asokan, K. Structural analysis and dielectric properties of $HoFe_{1-x}Ni_xO_3(0 \leq x \leq 0.5)$, *Journal of Electronic Materials*, 2015, *44*, 1044.
[21] Mir, S.A., Ikram, M., Sultan, K., Habib, Z., Kausar, H., Asokan, K. Correlative exploration of structural, optical and electric properties of colossal dielectric Ni doped Sm orthoferrites, *Advanced Materials Letters*, 2015, *6*, 1081.
[22] Jonker, G. H., and Santen, J. H. V. Ferromagnetic compounds of manganese with perovskite structure, *Physica*, 1950, *16*, 337.
[23] Wollan, E. O., and Koehler, W. C. Neutron diffraction study of the magnetic properties of the series of perovskite-type Compounds $[(1-x)La, x Ca]MnO_3$, *Physics Review*, 1955, *100*, 545.
[24] Sultan, K., Ikram, M. An investigation of electrical, magnetic and optical properties of $La_{1-x}Ca_xMnO3$ ($x = 0.0, 0.3, 0.5$ and 0.7) system, *Advanced Materials Letters*, 2015, *6*, 749.
[25] Bally, M.A.A., and Khan, F.A. Structural, dielectric and magnetic properties of $La_{0.55}Sr_{0.45}MnO_3$ polycrystalline perovskite, *Journal of Magnetism and Magnetic Materials*, 2020, *509*, 166897.
[26] Ye, F., Dai, H., Wang, M., et al. The structural, dielectric, and magnetic properties of $GdMnO_3$ multiferroic ceramics. *Journal of Materials Science: Materials in Electronics*, 2020, *31*, 3590.
[27] Xu, J., Cui, Y., and Xu, H. Dielectric properties of $TbMnO_3$ ceramic doped with Bi and Fe ions, *Results in Physics*, 2016, *6*, 811.
[28] Saha-Dasgupta, T. Double perovskites with 3d and 4d/5d transition metals: Compounds with promises, *Materials Research Express*, 2020, *7*, 014003.
[29] Islam, S.A., Andrabi, F.A., Mohmed, F., Sultan, K., Ikram, M., and Asokan, K. Ba doping induced modifications in the structural, morphological and dielectric properties of double perovskite La_2NiMnO_6 ceramics, *Journal of Solid State Chemistry*, 2020, *290*, 121597.
[30] Nazir, N., and Ikram, M. Structural, dielectric, and conductivity studies of strontium-doped Gd_2NiMnO_6 perovskite, *Journal of Materials Science: Materials in Electronics*, 2020, *31*, 23002.
[31] Nazir, N., and Ikram, M. Tuning of the structural, morphological, dielectric, and magnetoresistance properties of Gd_2NiMnO_6 double perovskite by Ca doping, *Physica B: Condensed Matter*, 2022, *632*, 413734.
[32] Abass, S., Bagri, A., and Sultan, K. Modifications induced in structural, electronic and dielectric properties of Nd_2NiMnO_6 double perovskite by Sr doping, *Journal ofAlloys and Compounds*,2023, *930*, 167463.
[33] Das, R., and Choudhary, R.N.P. Dielectric relaxation and magneto-electric characteristics of lead-free double perovskite: Sm_2NiMnO_6, *Journal of Advanced Ceramics*, 2019, *8*, 174.

[34] Gheorghiu, F., Curecheriu, L., Lisiecki, I., et al., Functional properties of Sm_2NiMnO_6 multiferroic ceramics preparedby spark plasma sintering, *Journal of Alloys and Compounds*, 2015, *649*, 151.

[35] Masud, M.G., Dey, K., Ghosh, A., Majumdar, S., and Giria, S. Occurrence of magnetoelectric effect correlated to the Dy orderin Dy_2NiMnO_6 double perovskite, *Journal of Applied Physics*, 2015, *118*, 064104.

[36] Nhalil, H., Nair, H.S., Bhat, H. L., and Elizabeth, S. Glassy dielectric response in Tb_2NiMnO_6 double perovskite with similarities to a Griffiths phase, *EPL*, 2013, *104*, 67002.

[37] Cao, Z., Li, Z., Gao, Y., Liu, J., Ruan, X., and Fang, M. Giant dielectric tunability properties of Sr and Sb co-doped La_2NiMnO_6 ceramicsinduced by extrinsic contribution, *Physica Status Solidi A: Applications and Materials Science*, 2014, *211*, 1207.

[38] Blasco, J., García-Muñoz, J.L., García, J., et al. Magnetic order and magnetoelectric properties of R_2CoMnO_6 perovskites (R = Ho, Tm, Yb, and Lu), *Physical Review B*, 2017, 96, 024409.

[39] Kumar, G. J., Kumar, E. S., Navaneethan, M., and Bharathi, K.K. Exchange bias, magnetic, and dielectric properties of La_2FeMnO_6 nanoparticles, *Journal of Materials Science: Materials in Electronics*, 2022, *33*, 8467.

[40] Silva, R. X., de Menezes, S.A., Almeida, R.M., et al. Structural order, magnetic and intrinsic dielectric properties of magnetoelectric La_2CoMnO_6, *Journal of Alloys and Compounds*, 2016, *661*, 541.

[41] Silva, R.X., Moreira, R.L., Almeida, R.M.,Paniago, R.,Paschoal, C.W.A. Intrinsic dielectric properties of magnetodielectric La_2CoMnO_6, *Journal of Applied Physics*, 2015, *117*, 214105.

[42] Das, R., and Choudhary, R.N.P. Investigation of electrical and magnetic properties of rare-earth based double perovskite: Dy_2CoMnO_6, *Ferroelectrics*, 2022, *588*, 125.

[43] Das, R., and Choudhary, R.N.P. Studies of structural, dielectric relax or and electrical characteristics of lead-free double Perovskite: Gd_2NiMnO_6, *Solid State Sciences*, 2019, *87*, 1.

[44] Preinfalk, C.,Morteani, G. The industrial applications of rare earth elements. In: Möller, P., Černý, P., Saupé, F., (Eds.), Lanthanides, Tantalum and Niobium, Special Publication No. 7 of *the Society for Geology Applied to Mineral Deposits*, Springer, 1987, *7*, 359.

[45] Kakushima, K., Tsutsui, K., Ohmi, S-I., Ahmet, P., Rao, V.R., and Iwai, H. Rare Earth oxides in microelectronics. In: Fanciulli, M., and Scarel, G. (Eds), *Rare Earth oxide thin films. Topics in applied physics*, Springer, 2006, 106, 345.

[46] Hossain, M.K., Ahmed, M.H., Khan, M. H., Miah, M. S., and Hossain, S. Recent progress of rare earth oxides for sensor, detector, and electronic device applications: A review. *ACS Applied Electronic Materials*, 2021, *3*(10), 4255.

[47] Hossain, A., AtiqueUllah, A.K.M., Guin, P.S., and Roy, S. An overview of La_2NiMnO_6 double perovskites: Synthesis, structure, properties, and applications, *Journal of Sol-Gel Science and Technology*, 2019, *93*, 479.

[48] Jha, P., and Ganguli, A.L. New perovskite-related oxides having high dielectric constant: $Ln_2Ba_2CaZn_2Ti_3O_{14}$ (Ln = La and Pr). *Proceedings - Indian Academy of Sciences, Chemical Sciences*, 2003, *115*, 431.

8 High-K Dielectric and Related Traits of Some Transition Metal- and Non-Metal-based Double Perovskites

Santanu Sen, Rimpi Chakravarty, B. N. Parida, and R. K. Parida

8.1 INTRODUCTION AND BRIEF HISTORY OF HIGH-K DOUBLE PEROVSKITES

Ceramics materials of perovskite-type, particularly transition, metal oxides are in absolute demand, owing to their intriguing properties. These materials have a similar structure to that of the $CaTiO_3$ mineral and exhibit properties like ferroelectricity and piezoelectricity, which brought an evolutionary change in the utility starting from hospitality to electronic components. Moreover, these materials are used in communication, information technology, aeronautics, military, and many other engineering industries. A search for novel ceramics is desired to meet the technological demands, as well as to make huge benefits of production technology. For this, the significant factors concerned are the availability of materials and their properties like dielectric, optical, and piezoelectricity. In addition to perovskites, many double perovskites have also been synthesized and looked over for any probable utilities in the last few decades. In these materials, the variation of octahedral tilting or changing the orbital overlapping, the magnetic, electrical, and dielectric properties are affected. Therefore, these properties can be optimized by controlling the degree of tilting in the double perovskite structure. Despite this, until 1980 the structure of double perovskite materials was not well characterized. In many cases, double perovskite samples have only been synthesized in polycrystalline form but, due to the lack of availability in synchrotron, X-ray, and neutron diffractometers, the structural information regarding complex polycrystalline samples has been sparse. The configuration of double perovskites may be $AA'B_2O_6$ (e.g., $CaFeTi_2O_6$ and $NdAgTi_2O_6$), $A_2BB'O_6$ (e.g., Ba_2YRuO_6 and La_2LiRuO_6), or $AA'BB'O_6$ (e.g., $NaLaMgWO_6$). Apart from this, different perovskites can be synthesized by adopting doping or substitution of A or B cations. The replacement might revamp the crystal anatomy because of variation of cation sizes, while the replacements with non-isovalent +ve ions enhance the concentration of multiple valency in the materials. Hence multiple possessions of the moderator composites and novel enhanced behavior might be drawn out due to the substitution.

In late 1950, the work in the field of double perovskite oxides (DPO) expanded after the discovery of ferromagnetism in some rare-earth-based DPO at room temperature reported by Longo and Ward [1]. In 1951, Steward and Rocksin together discovered Ba_2CaWO_6 and reported many fascinating results [2]. Subsequently, Paterson et al. revealed the existence of ferromagnetism in Mo and W-based DPO materials above room temperature in 1961 [3]. In the meantime, some Mo and Re-based DPO materials were found to exhibit very high conductivity comparable to that of metals [4–6]. Thereafter, the development of DPO slowed down until the discovery of the half-metallic behavior of Sr_2FeMoO_6 in 1998 by Kobayashi et al [7]. Besides, it is reported to display ferromagnetism above ambient temperature. Due to this, the material is a potential contender for spintronics

 DOI: 10.1201/9781003352365-8

based applications, and this discovery enhances the demand for DPO materials in scientific and technological applications [8–10]. Later, half metallicity was also found in Sr_2FeReO_6 by the first principle density of state calculation [11]. In 1999, Yuan et al. reported the correlation of magneto-resistance features with the inter-grain size of Sr_2FeMoO_6. It was found that, by reducing the inter-grain size up to the nanometer range, the tunneling magneto-resistançe of Sr_2FeMoO_6 can be enhanced quite significantly over a wide range of temperatures in a weak magnetic field [12]. From the technological point of view, the most significant features are colossal magneto-resistance, magnetoelectric coupling, magneto-dielectric features, and high ionic conductivity. The presence of such interesting features in DPO materials makes them a potential candidate in various aspects like magnetic recording material, spin filtering tunneling junctions, weak field magneto-resistive sensors, multi-media storage, multilayer capacitors, etc.

8.2 PREPARATION APPROACH OF HIGH-K DOUBLE PEROVSKITES

8.2.1 Solid-state Reaction Path

Solid solutions prepared following the solid-state reaction path, where the end products are in the solid complex state, resulting from the thermal fusion process of participating precursors in the exact stoichiometry as to end in final products such as oxides or halides, are said to have followed the solid-state path synthesis mechanism. DPOs can be prepared through the solid-state reactions route by fusing various carbide or oxide-based precursors at the A/B location as per the ABO_3 perovskite structure in the needed balance amounts to achieve the final product with the needed components. This process is usually carried out via mortar pestle/ball milling in the proper dry and wet media, such as methanol or isopropanol. The so obtained mix is dried to remove any moisture at around 100°C and calcined in the presence of air, at the desired calcination temperatures, and for the required time interval. Following that, the calcined composite was dry ground again and put again under thermal treatment following a repetitive heating methodology until a solid solution formed.

8.2.2 Gas Phase Deposition

Gas phase transport deposition-based DPO films with a desired composition and thickness can be obtained. Laser ablation, DC sputtering, magnetron sputtering, molecular beam epitaxy [MBE], thermal evaporation, and electron beam evaporation are the well-known traditional techniques developed for carrying out such an approach for gas phase deposition. This deposition approach is divided into three parts: (a) deposition at low substrate temperature, following post-annealing at high thermal energies, (b) deposition at mid sweep of temperature from 873 to 1073 K, followed by post-annealing treatment, and lastly, (c) deposition at the phase formation temperature and desired environment.

8.2.3 Solution-based Wet Chemical Paths

These approaches comprise the sol-gel synthesis, co-precipitation of transition ions, with the help of precipitating helpers such as cyanide, oxalate, citrate, and many more, and after that thermal processing, after which the solid solution forms in a single-phase material high degree of homogeneity. The particular low-temperature approach was found to be more advantageous compared to the solid-state approach in terms of parameters such as isotropy, enhanced uniformity while fabricating thin-film, and enhanced reactivity, together with improved compositional concentration with uniform grain dimensions. Hence, it created new pathways for the molecular framework in the preparation of perovskites. Further, depending upon the means utilized for solvent removal, solution-based methods are classified into two classes: (i) precipitation, filtration, centrifugation, and other steps needed for the extraction of the solid from the liquid phases (ii) thermal processing, such as evaporation, sublimation, and combustion, necessary for reduction of the solvent.

8.3 LITERATURE REVIEW

Synthesis of novel-derived composites, such as double perovskites or higher complex oxides, is now an emerging thrust area of research in the mineralogists and material science research community. The technological field is facing ever-inflating demands of novel, efficient but sustainable routes of deriving renewable energy to meet the energy crisis demands of humanity in the future. Double perovskites or similarly derived oxides are one such promising candidate that comes to the rescue under such circumstances. Various material scientists in this quest studied and analyzed numerous possibilities and derivatives of DPOs and looked for their enhanced physiochemical capacities for various application purposes. One such composite, namely Sr_2NiMO_6, with M as Te/W, was formed along the solution-based citrate sol-gel methodology and analyzed. It was noted that SN-TeO_6, as well as SN-WO_6, inhibits elevated values of dielectric permittivity as 341 and 308, possessing weak tangent delta loss of 0.06 for SN-WO_6 at a definite frequency 1 kHz. These numerical dielectric parameters saturate to low magnitudes exponentially relative to the increment of frequency up to 10^3 kHz, where the relative permittivity constant gets fixed to 260 values for both SN-TeO_6 & SN-WO_6 and tan δ loss reduces to 0.01 for SN-WO_6 and 0.04 for SN-TeO_6. The Cole-Cole arcs for all the complexes assure the departure from a debye kind of relaxation; hence, the ascendancy of low-sweep hopping mechanisms was verified [13]. The complex of Ba_2PrRuO_6 was synthesized by employing the SS reaction mechanism, and the dielectric response was investigated along the scale of 20 Hz ~1 MHz, in the temperature extent 10 Kelvin-300 Kelvin. Enhanced disbandment concerning the frequency and monotonic growth in the real resolution of permittivity was spotted when the temperature rose above the 150 Kelvin mark. The nature of the temperature-supported behavioral curves of the real dielectric permittivity was related to the maxima of the temperature-reliant tanδ undulates, where it was inferred that the maxima values discretely shift higher magnitudes with the frequency enhancement [14]. A novel double PO composite, namely Ho2CdTiO6 was grown via a solid state path, ensuing in an outstanding dielectric relaxation particularly at mid-frequency sweep, due to the frequency and inverse relaxation time resonance. The frequency-dependent modulus studies revealed temperature-dependent capacitance fall, suggesting long sweep conductance dominance by thermally enhanced charge carrier density [15]. The permittivity relaxation and charge flow operation of $Cs_2AgBiBr_6$ single phase crystallites following a temperature-relying C-IS analysis approach concerning the CMS was looked for. A possible contribution from the ionic movement and space-charge polarization toward the dielectric response was postulated. A well-defined transition value of frequency ~ 191 Hz at 273°K & 492 Hz at 263°K was estimated from the ε' vs T ^{0}C plot, which revealed that the temperature-reliant gradient of ε' points toward a +ve and -ve coupling coefficient beneath and above the transition frequency values, respectively. The binding energy and polaron hopping were evaluated to be $E_P = 0.34$ eV and $W_H = 0.17$ eV, respectively. Since the BE of the polaron was observed to be more than the R-T thermal energy, we may theorize the dominance of polaron hopping short-path conductions, relative to free-charge carrier transport. In addition, the imaginary reactance -Z'' and capacitive parameter M'' delineates the existence of thermally generated polaron relaxation, following a transition from non-debye to debye deportment [16]. Sr_2CoNbO_6 ceramic DPO was cast via the solid-state path. A temperature variant study of the dielectric characteristics of Sr_2CoNbO_6 was undertaken in the thermal scale 90~330 ^{0}K and in the frequency domain from 20 Hz~10 Mega Hz. The conductivity was seen to be dominated by short-sweep hopping carriers. It also enhances the dipolar effects heading towards the relaxation of polarons in the mid-temperature domain. At sufficiently low temperatures, frozen hopping conductions following the Maxwell-Wagner freezes out relaxation were identified. Hence a multi-relaxation mechanism over single relaxation was held responsible for the elevated dielectric parameters [17]. The ferroelectric, dielectric and transport characteristics of the polycrystalline $PbSrBiNbO_6$ double perovskite was performed, which resulted in enhancement of the aforementioned physical quantities with potential applicability [18]. From the dielectric study performed by S. Bhattacharjee et.al. on $Ba_{0.5}Sr_{1.5}FeVO_6$, the numerical value of the ambient condition relative permittivity parameter was extracted to be $\varepsilon_r = 250.7$, which directly implies that the complex has promising layered dielectric storage potential. The co-existence of ferroelectric and magnetic orders in the same phase was also identified at R.T. [19]. The dielectric relaxor, impedance, AC conductivity, and CMS studies were undertaken on the DP oxide Nd_2NiMnO_6, prepared following the SS reaction way. During the inspection of the temperature-dependent behavioral pattern of the dielectric constant, the relaxor response of the composite was noted. Such a type of response might be elaborated following the modified Vogel-Fulcher law and Curie-Weiss. The M" curves reach the maxima at the same frequency; however, C(T) values decrease as M" maxima amplitudes enhance with thermal energy enhancement [20]. Hence, from the detailed survey of the present literature on rare earth, transition metal-based double POs, we synthesized our understanding regarding them as the most reliant, dependable in terms of stability, and possessing high dielectric permittivity for their direct use in storage device manufacturing. They were also seen to exhibit enhanced long-range conductivities or stabilized short-range conductivity mechanisms with thermal energy enhancement for capacitive device utilities, according to the requirement. Excellent remnant values of ferroelectric polarizabilities open up the door for electrical memory storage applications, together with sensing abilities.

8.4 SOME USEFUL RESULTS AND DISCUSSION ON THE REPORTED DIELECTRIC RELATED FINDINGS

Dielectric spectroscopy is a vital tool that helps in analyzing the influence of various types of polarization mediated relaxations, dislocations, oxygen vacancies on the dielectric facet of a composition. For this purpose, significant parameters like dielectric constant (ε_r) and loss tangent ($\tan\delta$) are investigated over a broad spectrum of frequency and temperature. The dielectric response of a material is a frequency-reliant complex parameter and can be expressed by the following relation given by Debye:

$$\varepsilon_r = \varepsilon_\infty + \left(\varepsilon_s - \varepsilon_\infty\right) / 1 + i\omega\tau$$

The real and imaginary resolutions are segregated and expressed as follows:

$$\varepsilon' = \varepsilon_\infty + \left(\varepsilon_s - \varepsilon_\infty\right) / 1 + \left(\omega\tau\right)^2$$
$$\varepsilon'' = \left(\varepsilon_s - \varepsilon_\infty\right)\omega\tau / 1 + \left(\omega\tau\right)^2$$

Here ε_s and ε_∞ are the dielectric parameters at the invariant frequency ($\omega = 0$) and high-frequency limit ($\omega \rightarrow \infty$) respectively and τ being the relaxation time.

Additionally, the loss tangent is expressed as:

$$\tan\delta = \frac{\varepsilon''}{\varepsilon'}$$

To measure ε_r and $\tan\delta$, several methods like the microwave method (distributed circuit method), R F method (lump circuit method), and impedance method may be used, depending upon the nature of the sample and selected frequency range for investigation. In this particular work, the impedance method was incorporated, which measures the capacitance of the sample by utilizing impedance bridges like the Schering bridge, transformer ratio bridge, etc., depending on the frequency. Subsequently, the ε_r value can be estimated from the measured capacitance and the geometry of the sample.

8.5 DIELECTRIC STRENGTH AND LOSS FINDINGS OF Sr_2Co NIOBATES

Figure 8.1(a) and (b) displays the outcomes in changes of real dielectric values (ε') and permittivity loss tangent ($\tan\delta$) relative to temperature for a Sr_2Co-based double niobate. At the first appearance, the $\varepsilon'(T)$ curves, specifically belonging to 300 Hz instance elevate in a step-like manner, increasing from a lower magnitude up to 100 until a saturated magnitude of 6000, as the thermal energy rises from 90 up until 330 Kelvin. The step-like enhancement moves to enhanced temperatures and enhanced perturbation ω, which is brought together by maxima relative to the $\tan\delta$ curve. This points out that there exists a thermal-mediated dielectric relaxation. Additionally, a conscientious inspection delineates that farther group of bumps [(glared with help of arrow in Figure 8.1(b)] shows up at the low-temperature arm of the $\tan\delta$ maxima. Hence this is a direct indication of two types of thermally mediated relaxations in Sr_2Co Niobates. For compendiousness, these relaxation maxima were assigned to R1 and R2 in the sequence of enhancing temperature. In order to point out their peak locations more precisely, two Gaussian deconvolutes were fitted to the plot of tangentδ. The fitted Gaussian peaks and the relation among the fitting outcome and the investigational results got at 1 Mega Hz were cited in Figure 8.1(c). The fitting result was found to be ideal. The resultant

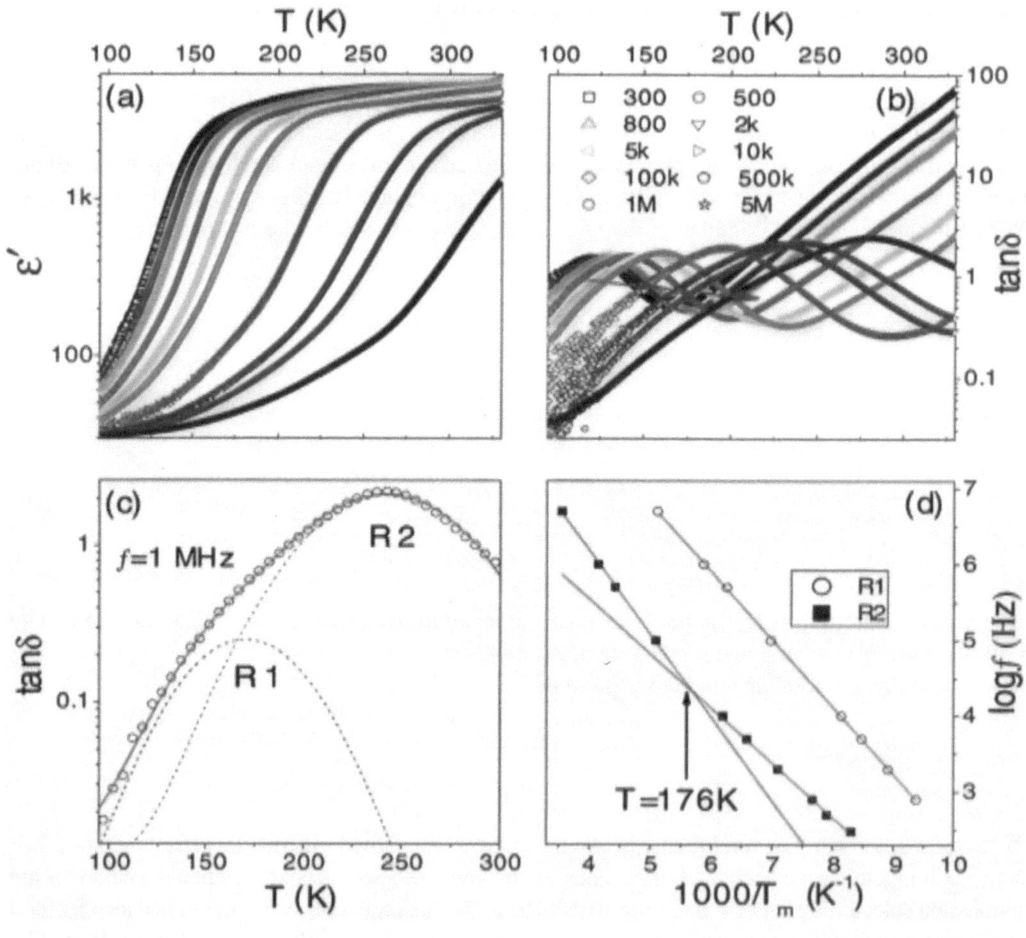

FIGURE 8.1 Temperature reliances of real dielectric part (a), tangentδ & (b) of SC-NO read at various frequencies. (c) relates the investigated data and fitting curve of tangent δ evaluated at 1 Mega Hz. The dotted curves are the resultant maximums of R1 and R2. (d) shows the Arrhenius plots for R1 and R2.

maxima location at T_m was traced with frequency, following the relation $f = f_0 e^{E_a (k_b T)^{-1}}$, where f is the instantaneous and f_0 is a proportionate factor, E_a E_a being the activation energy.

Figure 8.1(d) shows R1 and R2 Arrhenius plots. It was observed that R1 obeys the Arrhenius condition quite well. The activation energy as well as the relaxation time, $1/2\pi f_0$, for R1, were estimated to be lying around 0.177 eVolts & 8.54×10^{-13} secs. However, in case of R2, the results mark at low temperatures separately curve out from the linear response, heading toward two Arrhenius parts, with the limiting temperature at 176 K. The relaxation times and activation energies were extracted to be 6.2×10^{-10} s, 0.14 eV, & 4.30×10^{-12} s, 0.22 eV for the lower and enhanced thermal energy sweeps. The activation for the low-temperature domain (0.14 eV) matches well with that of the DPO compound Ca_2CoNbO_6 [21]. The low-temperature relaxation mechanism belongs to high K behavior in Ca_2CoNbO_6, which was identified to be coming from the dipolar/orientational phenomena due to the short sweep hopping electrons, jumping back and forth in sync with the field. It is a fact that both the +ve oxygen vacancies and −ve Co vacancies exist simultaneously in the Co-dependent material [22]. Therefore, the dielectric traits of SC Niobates should be examined from the standpoint of rivalry among the e-s and holes formed by the V^0 and Co vacancies, respectively. Delineated dielectric traits are anticipated to be subtle to annealing processes in degeneration and diminutive conditions.

Figure 8.2 displays the temperature variance of dielectric constant (left windows) and loss tangent (right widows) post-annealing in a N_2 and O_2 environment at 1073 K for 2 hrs for the same composite utilized in Figure 8.1. Here, two prime characters deserve to be highlighted. First, is the annealing done in N_2 conditions, which shows considerable decay in dielectric K numeric and loss tan delta as compared with the outcomes enumerated in (Figure 8.1(a)). The K value of dielectric at a temperature <150 K starts to become like a plateau with a value around ~15. An enhanced hole-to-electron recombination is expected due to nitrogen annealing, due to enhanced O vacancy concentration. Hence, few resultants may contribute toward conductivity, thereby decreasing it in the background of the tangent delta. Moreover, the permittivity plateau was designated to be independent of frequency as well as temperature, indicating the intrinsic dielectric behavior without dipolar relaxation. Additionally, it implies too that the electron-hole pairs are the ones responsible for the relaxation at the lowest. Secondly, a thermally generated relaxation at the elevated temperatures (for $T > 200$ K) becomes identifiable; whose temperature domain is more compared to R2. This relaxation appears to be a unique one. The inset panel of Figure 8.2 traces the Arrhenius relaxation behavior. The E_a and pre-functional multiplier corresponding to the relaxation were found out to be 0.30 eVolts and 9.31×10^{10} Hz, respectively. These numerics are quite new and contrasting relative to those of the R1 and R2, suggesting that the observed relaxation is implying some new trend in the behavioral aspects. Hence, we may now appoint it as R3. It may be recognized as a puny stridden enlargement in $\varepsilon'(T)$ at the soaring temperatures, as depicted in Figure 8.1(a), if the trace was graphed relative to a linear scale. Hence, altogether three relaxation processes were observed in SC-NO.

8.6 DIELECTRIC STRENGTH AND LOSS FINDINGS OF $SrLa_{1-x}Nd_xLiTeO_6$

Figure 8.3(a) depicts the ε' variations against frequency in the interval 50 Hz~1 MHz for the $SrLa_{1-x}Nd_xLiTeO_6$ (for x as 0.00, 0.25, 0.50, 0.75, 1.00) complexes at 298 K [23]. The reaction of ε' to frequency is separate for an individual sample relative to the amount of Nd doping. For frequencies beneath 100 Hz, the complexes displayed a comparatively equivalent reaction of ε' for the frequency, when ε' suddenly made a low-value transition relative to the frequency from an elevated epoch value. Nevertheless, the particulate reaction for every sample exhibited different behaviors at frequencies >100 Hz. The pure Nd-containing composite manifested as a weak decay in ε' versus frequencies. A relatively steeper decreasing gradient of dielectric K value was noted for enhanced doping concentration Nd from 0.25 to 1.00, until 1 MHz. However, the sample with Nd 0.50 was found to be an exception, where real dielectric values (ε') were seen to be high in magnitude and

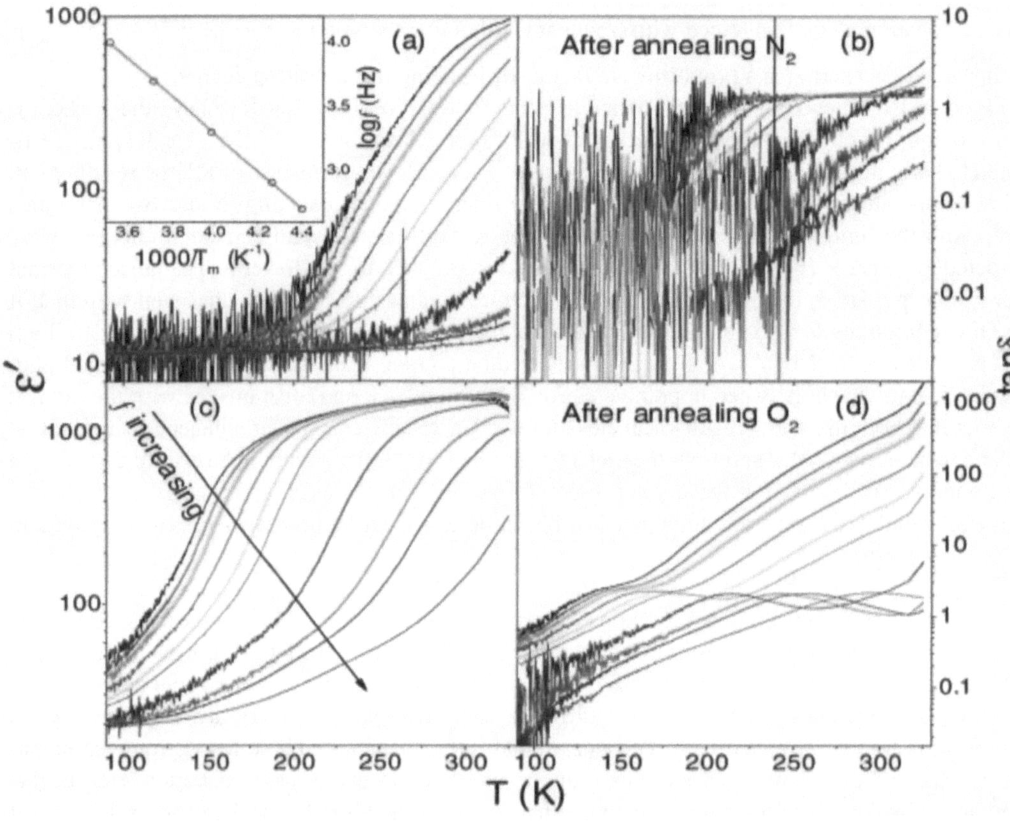

FIGURE 8.2 Temperature reliant contrast in dielectric parameter (right panels) and dielectric loss tan delta (left panels) for SCNO complexes post-annealing in N_2 & O_2 at the probing frequencies of 300 Hz, 500 Hz, 800 Hz, 2 kHz, 5 kHz 10 kHz, 100 kHz, 500 kHz, 1 MHz, and 5 MHz (from top to bottom). The inset depicts the Arrhenius plot for the high-temperature relaxation. (Figures from G. J. Wang et.al. AIP Advances 3, 022109 (2013); https://doi.org/10.1063/1.4791763)

constant beyond 100 Hz. The inset of Figure 8.3(a) displays the changes in dielectric parameter ε′, at frequencies of 250–750 kHz. Amazingly, these fluctuations showed a monotonically up trend in ε′ up to a peak magnitude in Nd0.50, followed by a steep decay with further enhanced Nd dopant.

Figure 8.3(b) is the *tan δ* fluctuations for $SrLa_{1-x}Nd_xLiTeO_6$ (with *x* as 0.00, 0.25, 0.50, 0.75, 1.00) at 298 K. Dielectric loss (*tangent δ*) corresponds to the energy dissipation due to the forced idleness of the space charges while short-range polarizability is dominant. The plot exhibits a drop in *tangent δ* at frequencies less than 500 Hz, continued by a commensurate saturated behavior in the range of frequencies (600~10 kHz). For composites such as Nd-0.00 & 0.50, the dissipation feebly enhanced at around 100 kHz and beyond. Although for the complexes such as Nd 0.25, Nd 0.75, and Nd 1.00, a considerably higher enhancement in *tangent δ* magnitude were seen at the same frequency domain. This enhancement implies the existence of a relaxation bump beyond 1 Mega Hz.

The starting jump of ε′ to lower magnitudes, particularly at lower frequencies (f < 100 Hz) for every composite in Figure 8.3(a) can be designated to the relaxor dipoles/space charge orientational moments. The space-charge dipoles were possibly due to the vacancies in grain walls, which cannot cope with the flipping of the electric field direction as the frequencies enhanced. This decrease was backed by the large magnitude of *tangent δ* at the same frequency domain in all complexes, which implies that loss of energy is prevalent. High-temperature annealing oxygen vacancies creation might be the separate cause legitimizing the hand of immobile dipoles in this frequency domain. The existence of oxygen vacancies in the dielectric composites can lower the phonon modes, resulting in

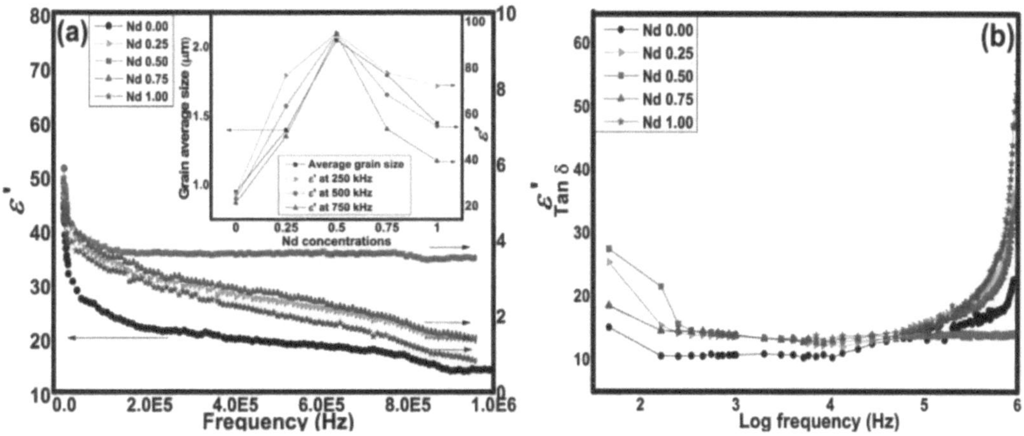

FIGURE 8.3 (a) ***ε'*** of $SrLa_{1-x}Nd_xLiTeO_6$ (x = 0.00, 0.25, 0.50, 0.75, 1.00) within 50–1 MHz, (inset) average grain sizes with ***ε'*** at selected frequencies for different Nd concentration and (b) Tan ***δ*** of $SrLa_{1-x}Nd_xLiTeO_6$ (x = 0.00, 0.25, 0.50, 0.75, 1.00) within 50–1 MHz. (Figure from M. Z. M. Halizan et al. 2020 *Mater. Res. Express* 7 086301, DOI 10.1088/2053-1591/abaa8e.)

space charge polarization [24]. This low-value transition might also be due to the direct conduction loss. Although such an assumption is negated since the trace of *ln* ε' ver. *ln ω* doesn't exhibit a slope of value comparable to (−1).

At elevated frequencies (i.e., greater than 100 Hz), the reaction of ε' may be thought to be coming from the probable combined sources namely the orientational/dipolar, ionic, and electronic moments. In the Nd-0.00 composite, the monotonic decay process of ε' were related to the increase of *Tangent δ*, which were particularly observed at 10 kHz and above. The resultant influence of the Nd mixing can be observed in the frequency range of 100 kHz to 1 MHz, where, for solid solutions with Nd 0.25, Nd 0.75, and Nd 1.00 amounts, enhanced Nd mixing resulted in a sharper declination in ε', which was again correlated to the greater increase in *Tangent δ* for frequencies above 100 kHz. Beyond 100 kHz, the moderate-dimensional dipoles in Nd 0.25, N0.75, and Nd 1.00 were not keeping up with the fast phase alternations of the applied electric component of the field, until 1 Mega Hz, thus resulting in a low polarization contribution. On the other hand, the near saturated feature of ε' for Nd 0.50 was quite fascinating, although it can be seen relative to the almost saturated nature of *Tangent δ* for the said solid-sample. The truancy of such of conduct in *Tangent δ* relative to frequency for these solid-oxide points toward the profusion of middle-sized dipoles that can easily correspond to high frequencies, and thus can generate enhanced polarizations.

The variability of ε' can be well recognized by taking into consideration the inclination of ε' in the inset of Figure 8.3(a), persistent relative to grain dimension and enhancing Nd concentration. Bigger grains permit the genesis of larger and medium dimensional dipoles with a enhanced ε' compared to the miniature grains. Also, larger sized grains are anticipated to hold an enhanced density of intermediate-sized dipoles compared to smaller sized grains. This process illustrates the steadiness of the variance in grain size relative ε'. A similar description of the consequences of grain size on ε' is reported in the literature, where ε' is linearly related to the grain size [25]. Figure 8.4 exemplifies the variance of the ε' for $SrLa_{1-x}Nd_xLiTeO_6$ as a temperature dependent function, in the sweep of 298 Kelvin~343 Kelvin. The ε' variability reaction follows the same fashion in nearly all the samples. All composites were seen to exhibit a declination trend in ε' at low frequencies until 300 Hz, emanating from R-T up to 323 K, with Nd 0.25 as an exception that initiated from room temperature to 333 Kelvin & for Nd 1.00, commencing from R-T until 313 K. The numerics show an enhancing ε' trend at the exact frequencies, at higher temperatures. On the other hand, the ε' value at the mid and high frequencies describes a declining trend for all temperatures, and the ε' value increases with the temperature, especially at high frequencies.

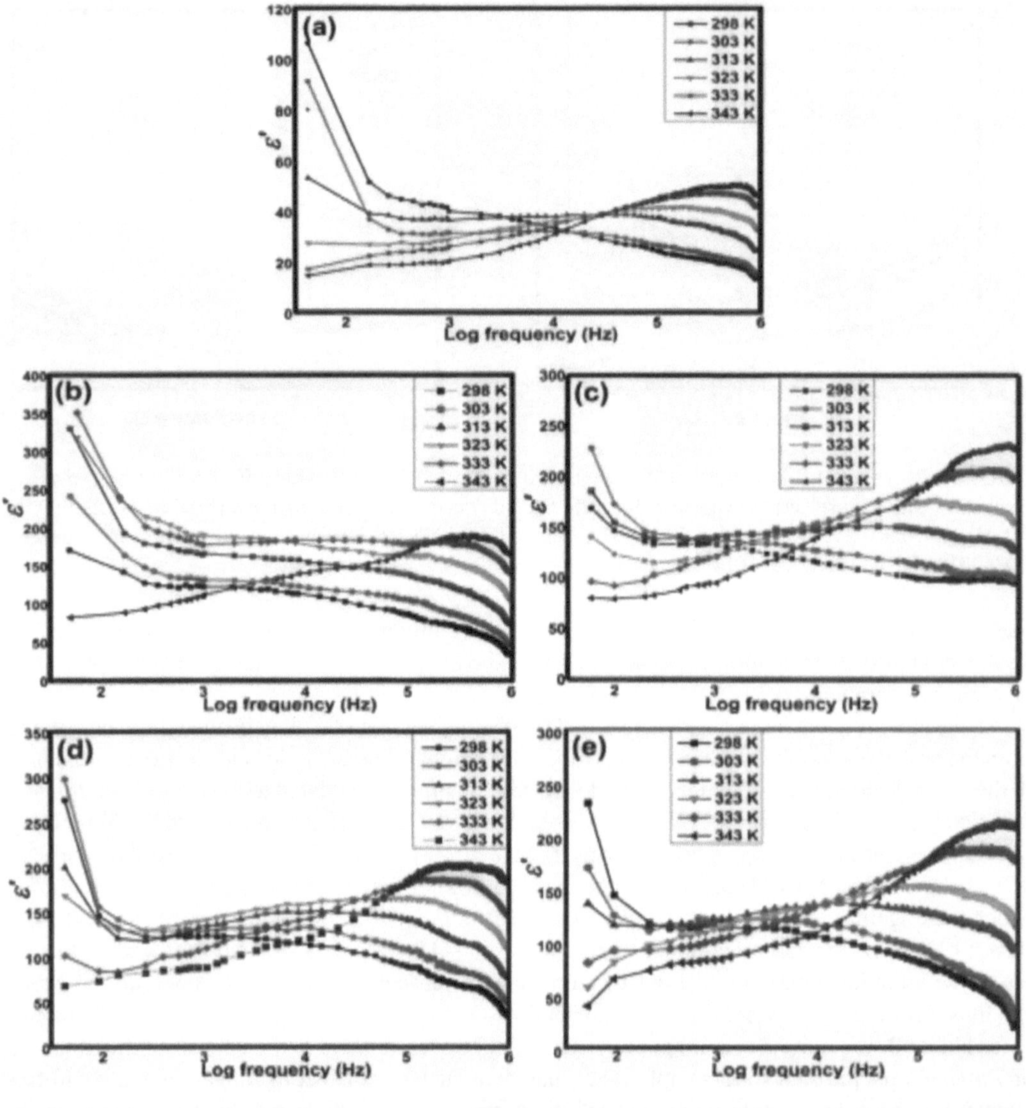

FIGURE 8.4 *ε'* of $SrLa_{1-x}Nd_xLiTeO_6$ (x = (a) 0.00, (b) 0.25, (c) 0.50, (d) 0.75, (e) 1.00) at different temperatures. (Figure from M. Z. M. Halizan et al. 2020 *Mater. Res. Express* 7 086301, DOI 10.1088/2053-1591/abaa8e.)

8.7 DIELECTRIC STRENGTH AND LOSS FINDINGS OF Sr_2FeVO_6

The dielectric features of the solid-state derived Sr_2FeVO_6 double perovskite compound have been reported previously by S. Bhattacharjee et al [26]. In their work, they analyzed the frequency-dependent variation of dielectric constant (ε_r) and loss tangent (tan δ) of the material at different temperatures, as depicted in Figure 8.5(a) and (b). From the figures, it is evident that both ε_r and tan δ decline with the enhancement of frequency for all the selected temperatures. This pattern, especially in the lower frequency domain, signifies the impact of various polarization effects like electronic, ionic, orientation, and space charge. Among them the impact of orientation polarization, as well the space charge polarization, is quite significant in the lower frequency domain. The dipoles Sr-O, Fe-O, and V-O contribute to orientation polarization, whereas the presence of multivalent element Fe enhances the space-charge polarization by an electron hopping mechanism between Fe^{2+}-Fe^{3+}. These dipoles and hopping electrons remain aligned with the applied field at low

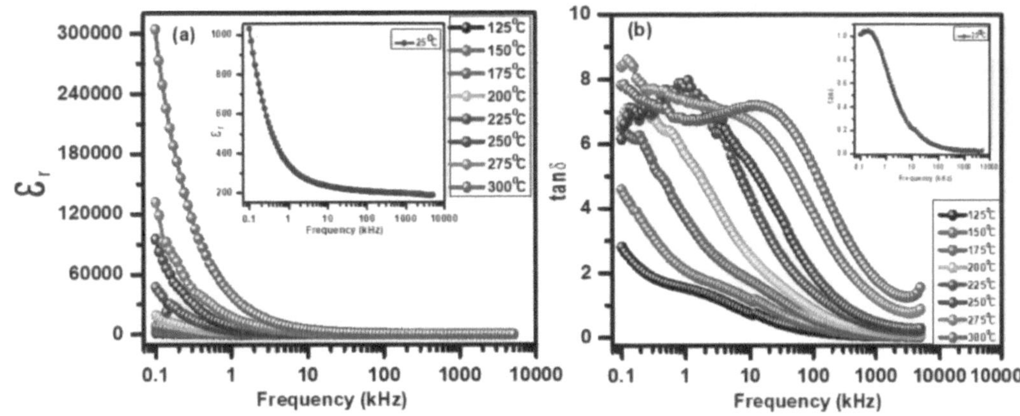

FIGURE 8.5 (a) and (b) Depict the variation of ε_r with frequency and variation of tanδ with frequency (Figure from S. Bhattacharjee et al. *Mat. Chem. Phys.* https://doi.org/10.1016/j.matchemphys.2021.125254)

frequency, resulting in a high ε_r value. However, as frequency increases, they are unable to cope, and this leads to a decline in ε_r value. The dielectric anomaly can also be elucidated through the Maxwell-Wagner model. According to this model, the dielectric ceramics that consist of resisting grain boundaries offer high impedance to the flow of charge carriers at lower frequencies. As a result, the carrier concentration at the grain boundaries gets enhanced; thus, polarization of space charges increases. With increasing frequency, the trapped charges at the grain boundaries get scattered, and consequently, the carrier density starts to fall, which results in a decrease in ε_r value. Moreover the value of ε_r increases with the rise in temperature, which is due to the enhancement of space charge polarization through the thermally activated hopping conduction mechanism. Similar to the frequency-dependent variation of ε_r, the noticed anomaly in the tan δ plot can be explained through Koop's phenomenological theory. The high value of tan δ in the lower frequency domain is due to the high opposing nature of grain boundaries toward the movement of charge carriers. The charge carriers get trapped inside the grain boundaries. To move them from one site to another within the crystal lattice, more energy is required, which results in a high tan δ value. However, at high frequencies, these carriers can easily transport through the active conducting/semiconducting grains without consuming much energy. Hence, tan δ is significantly low. Additionally, the tan δ value increases sharply with temperature which may be due to the intensification of oxygen vacancies and domination of conductivity.

8.8 DIELECTRIC STRENGTH AND LOSS FINDINGS OF Pb_2BiNbO_6

The dielectric properties of the solid-state derived Pb_2BiNbO_6 double perovskite have been reported earlier by Parida et al. [27]. In their work, they have mainly analyzed the frequency and temperature-dependent variations of both ε_r and tan δ. It is evident from Figure 8.6 that the value of ε_r sharply decreases as frequency increases, and almost turns invariant in the higher frequency range, indicating a frequency-independent outcome. This type of dielectric response can be expounded by Koop's phenomenological theory and Maxwell-Wagner model which states that a dielectric material comprises poorly conducting grain boundaries and good conducting grains. Accordingly, the grain boundaries are effective in the lower frequencies, whereas the grains are in the higher frequencies. Moreover, as the investigated material is Nb based, which is a multivalent element, there may be hopping of charge carriers between the Nb^{5+} and Nb^{4+} ions present at the crystallographic equivalent sites or oxygen octahedral sites. The charge carriers have to cross both conducting grains, as well as poorly conducting grain boundaries. Because of the opposing nature of the grain boundaries to the charge carriers, they tend to accumulate together and induce substantial space charge polarization. This results in the high value of ε_r at lower frequencies. With further increase in frequency, the

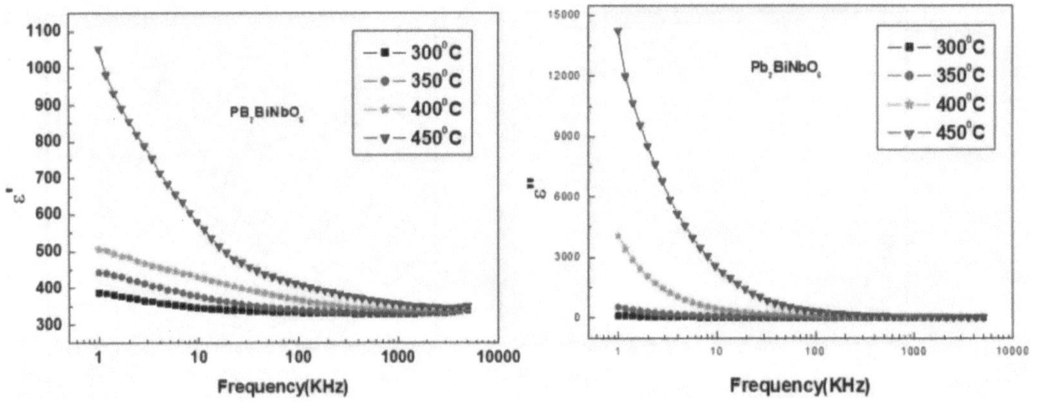

FIGURE 8.6 Variation of real and imaginary relative dielectric constant (ε_r) with frequency at various temperatures of Pb_2BiNbO_6.

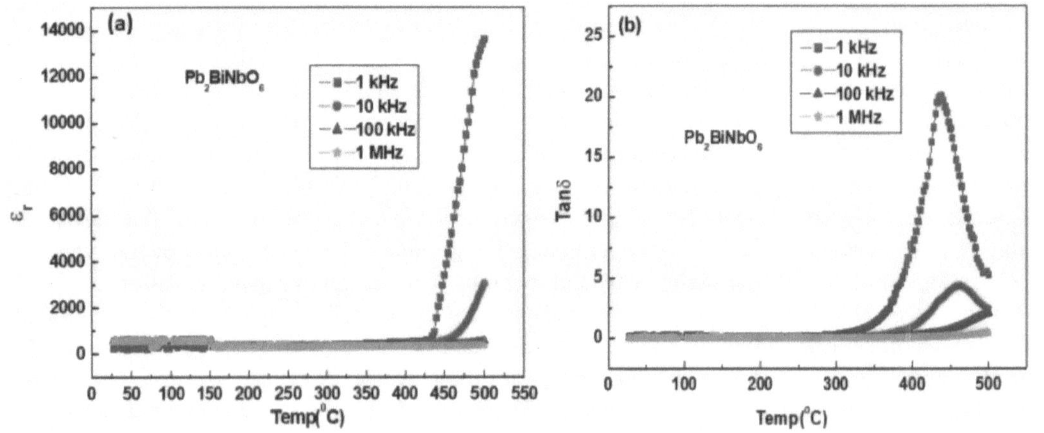

FIGURE 8.7 Temperature dependence of ε_r and tanδ of Pb_2BiNbO_6 at different frequencies.

accumulated charge carriers at the grain boundaries get scattered, and consequently, their density gets reduced. This results in the decrease of space charge polarization and thus leads to the fall in ε_r value. It is also observed that, at lower frequencies, the value of ε_r increases with temperature but is nearly stable at the higher frequencies. This dispersion in the lower frequency domain is due to the thermally activated hopping mechanism of charge carriers between Nb^{5+} and Nb^{4+} ions at octahedral sites. This hopping mechanism leads to a local displacement of charges in the direction of the applied electric field and increases dielectric polarization. Thus, ε_r increases with the temperature rise.

Figure 8.7 reflects the response of ε_r and tan δ to temperature variance at specified frequencies. The ε_r value increases gradually with the rise in temperature for all frequencies and can be ascribed to the electron-phonon interaction as noticed usually in Nb-based perovskites. The absence of transition temperature (T_c) in the investigated temperature domain indicates that T_c lies past the experimental limitation temperature. The higher value of ε_r noticed at a lower frequency (1 kHz) may be because of the accumulation of charge carriers at the grain boundaries. Similarly, for all selected frequencies, the variation of tan δ with temperature follows an analogous pattern. At lower temperatures, the value of tan δ is almost constant but rises rapidly with the increase in temperature. This rise can be ascribed to the existing oxygen vacancies and Nb^{4+} as well as Bi^{3+} ions, which act as donor centers initiating extrinsic n-type conduction in the investigated material. Also at higher temperatures, the conductivity begins to dominate, which in turn, is responsible for the rise in tan δ value.

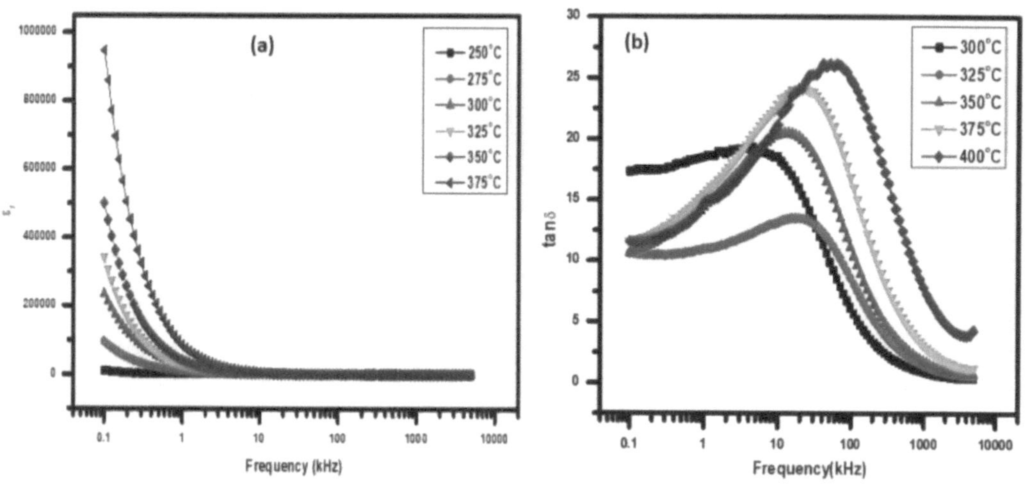

FIGURE 8.8 (a) and (b) Variation for ε_r and $\tan\delta$ for Ca_2BiNbO_6. (From Panda et al. *Appl. Phys. A* 127, 950 (2021). https://doi.org/10.1007/s00339-021-05105-4)

8.9 DIELECTRIC STRENGTH AND LOSS FINDINGS OF Ca_2BiNbO_6

Panda et al [28] have reported the dielectric properties of the solid-state derived Ca_2BiNbO_6 double perovskite compound in their work. In this aspect, the variance of ε_r and tan δ with frequency at selected temperatures are analyzed as depicted in Figure 8.8(a) and (b), respectively. Both ε_r and tan δ depict a decreasing trend, with an increase in frequency at selected temperatures. In their work, they have mentioned the influence of various polarizations on the net ε_r value, as well as the governance of space charge effects in the lower frequencies. The existence of both Bi^{3+} and Nb^{5+} cations in the material generates oxygen vacancies, which in turn increases the space charge polarization effect and causes the hopping of charge carriers. This appearance of a high ε_r value at the lower frequency domain can be well explained through the Maxwell–Wagner theory. Accordingly, a dielectric material comprises good-conducting grains and poorly conducting grain boundaries. The grains offer low resistance and low activation energy, whereas the grain boundaries offer high resistance and high activation energy.

As a consequence, the charge carriers gather near the grain boundaries and space charge polarization dominates. Thus, the material exhibits a high ε_r value at the lower frequency domain. However, as the frequency of the applied field increases, the space charge polarization at the grain boundary breaks and reduces the ε_r value. The response of ε_r and tan δ with temperature variation at selected frequencies for the material is illustrated in Figure 8.9(a) and (b). The ε_r value for 1 kHz is almost constant at lower temperatures (i.e., < 300 °C) and then gradually increases with the temperature rise. A phase transition from ferroelectric to paraelectric is observed at a temperature above 350°C. The observed variation in tan δ value at a higher temperature for all frequencies may be because of possible reasons like the formation of oxygen vacancies, thermally excited charge carriers, or any inherent lattice defects, etc.

8.10 SOME UNREPORTED COMPARATIVE RESULTS ON HIGH-K DPO DIELECTRIC

8.10.1 Comparative Dielectric K Strength and Loss Analysis

Figure 8.10(a) and (b) presents the response of ε′ and ε″ to frequency variation for $Pb_{1.5}Sr_{0.5}BiNbO_6$ and $Pb_{0.5}Sr_{1.5}BiNbO_6$ double perovskites samples at specified temperatures. It is evident from the plots that the value ε' and ε" of both the samples sharply decreases as frequency increases and almost turns invariant in the higher frequency range, indicating a frequency-independent outcome.

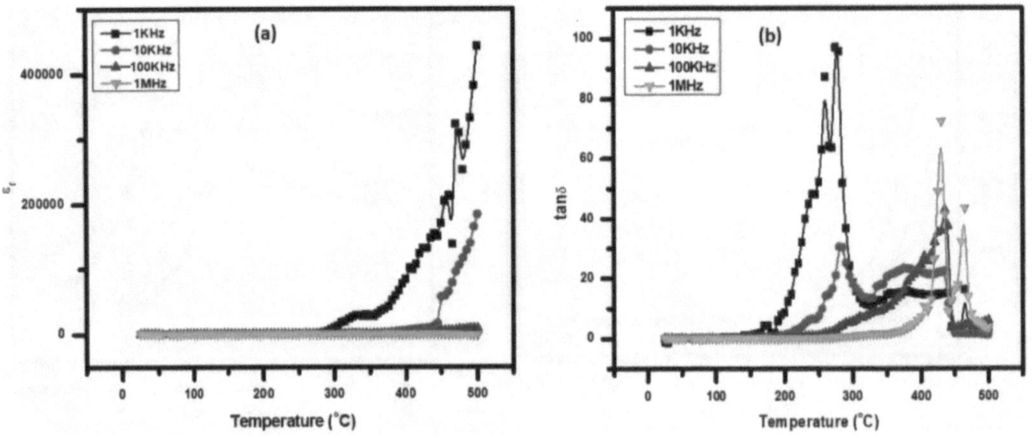

FIGURE 8.9 (a) and (b). ε_r and tan δ variance with temperature. (Panda et al. *Appl. Phys. A* 127, 950 (2021). https://doi.org/10.1007/s00339-021-05105-4)

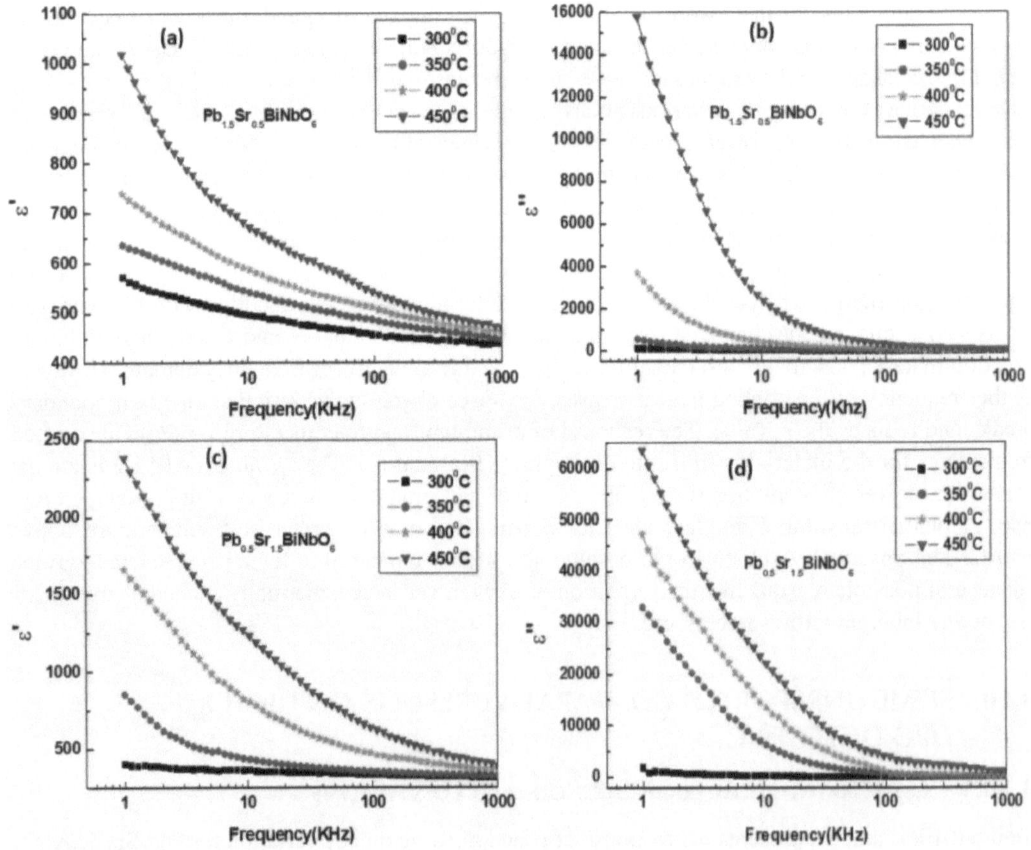

FIGURE 8.10 (a) ε′ and ε″ as a function of frequency for $Pb_{1.5}Sr_{0.5}BiNbO_6$. (b) ε′ and ε″ as a function of frequency for $Pb_{0.5}Sr_{1.5}BiNbO_6$.

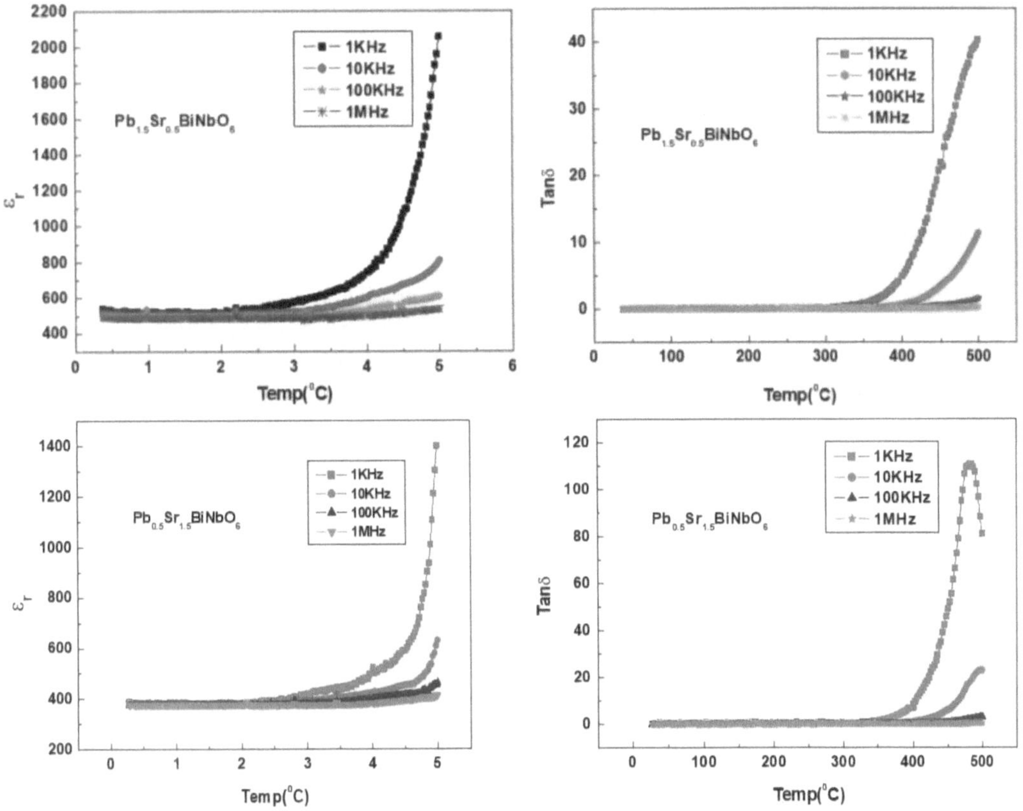

FIGURE 8.11 (a) Relative K(ε_r) as a function of frequency for $Pb_{1.5}Sr_{0.5}BiNbO_6$. (b) Relative K(ε_r) as a function of frequency for $Pb_{0.5}Sr_{1.5}BiNbO_6$.

This type of dielectric response can be expounded by Koop's and Maxwell-Wagner's phenomenological theory, which states that a dielectric material comprises poorly conducting grain boundaries and good conducting grains [29]. Accordingly, the grain boundaries are effective in the lower frequencies, whereas the grains are in the higher frequencies. As the investigated samples are Nb based, which is a multivalent element, there may be a hopping of charge carriers between the Nb^{5+} and Nb^{4+} ions present at the crystallographic equivalent sites or oxygen octahedral sites. The charge carriers have to cross both conducting grains, as well as poorly conducting grain boundaries. Because of the opposing nature of the grain boundaries to the charge carriers, they tend to accumulate together and induce substantial space charge polarization. This results in the high value of ε' and ε" at lower frequencies. And a comparative analysis suggests that the dielectric K values enhance with enhanced Sr modifications at low frequencies, the behavior of which is repeated also at enhanced temperatures. With further increase in frequency, the accumulated charge carriers at the grain boundaries get scattered. Consequently, their density gets reduced. This results in the decrease of space charge polarization, thus leading to the fall in ε' and ε" values. It is also observed that at lower frequencies, the values of both ε' and ε" increase with temperature but are nearly stable at the higher frequencies. This dispersion in the lower frequency domain is due to the thermally activated hopping mechanism of charge carriers between Nb^{5+} and Nb^{4+} ions at octahedral sites. This hopping mechanism leads to a local displacement of charges in the direction of the applied electric field and increases dielectric polarization. Thus, both ε' and ε" increases with the temperature rise.

Figure 8.11(a) and (b) reflects the response of ε_r and tan δ to temperature variance for $Pb_{1.5}Sr_{0.5}BiNbO_6$ and $Pb_{0.5}Sr_{1.5}BiNbO_6$ double perovskite samples at specified frequencies. It is noticed that up to 200° C, the values of both ε_r and tan δ are almost constant, which may be because of the moisture present in the samples. After approximately 200° C the ε_r value gradually increases, along with temperature for all selected frequencies, and may be attributed to the electron-phonon interaction as usually noticed in Nb-based perovskites [30]. A ferroelectric to paraelectric transition is expected past the experimental limitation temperature (~550° C) in the samples. This transition beyond 550° C may be due to the occurrence of space charge polarization in the samples. In this aspect, the oxygen vacancies created due to the evaporation of Pb during the sintering of the samples also contribute toward space charge polarization as they enhance the hopping mechanism of charge carriers between Nb^{5+} and Nb^{4+} [31–33]. Moreover, in both samples, higher values of ε_r are noticed at low frequencies (1 kHz) which may be because of the accumulation of charge carriers at the grain boundaries [34]. The response of tan δ to temperature variance also follows a similar pattern to ε_r. At lower temperatures, the value of tan δ is almost constant but rises rapidly with the increase in temperature. This rise can be ascribed to the existing oxygen vacancies and Nb^{4+} ions which behave as donor centers, initiating extrinsic n-type conduction in the investigated perovskite samples [35].

8.10.2 Complex Modulus Spectroscopy (CMS)

The CMS is a useful approach to drawing out useful information regarding the suppressed capacitive effects resulting from the relaxed space charge and dipolar orientational effects in the dielectric complex, together with any possible electrode-to-solid solution interface effects. It is generally associated with the real and imaginary dielectric loss permittivity as [36]

$$M^*(2\pi f) = \varepsilon_{real}\left(\varepsilon_{real}^2 + \varepsilon_{img}^2\right)^{-1} + i\varepsilon_{real}\left(\varepsilon_{real}^2 + \varepsilon_{img}^2\right)^{-1} = M' + iM''$$

Figure 8.12(a) and (b) frequency versus M' plot for Sr doped $Pb_{1.5}Sr_{0.5}BiNbO_6$ and $Pb_{0.5}Sr_{1.5}BiNbO_6$ double perovskite. In both the solid oxides electrode polarizations were not identified since curve variation is negligible at lower perturbation frequency. The curves appear to be relatively more disbanded in the studied compounds concerning pure lead-based and Ba-doped complexes [37, 38]. This immediately implies that the Sr addition enhances the short-span conductions, due to grain-to-grain interface effects, short-range polaron hopping, and exchange interaction dominance relative to Pb-only and Ba-modified solid solutions. Relatively enhanced disbandment is observed with the Sr^{2+} amount enhancement when we do a comparative analysis of Figure 8.12(a) and 8.12(b). This behavior can be accurately assumed to be coming from enhanced defect densities, existence, and elevation in the concentration of dual valencies of Sr, which in turn help the hopping mechanisms a lot through singly ionized oxygen vacancies. The enhancement in the saturation values of all M' curves enhances with Sr content, thereby implying that Sr has played a crucial role toward dielectric relaxation in the composite and hence toward enhancement in the values of dielectric parameters. The frequency-reliant variations in M" are depicted in Figure 8.12(c) and (d). The peaks in both plots exhibit negligible dispersion in width as temperature values rise. However, relaxation time for $Sr_{0.5}$ modified ceramic, with respect to $Sr_{1.5}$ content is comparatively weakly defined. This allows the $Pb_{0.5}Sr_{1.5}BiNbO_6$ complex to show enhanced dielectric relaxation [39]. Some extensive comparative studies also reveal that the peak heights belonging to $Pb_{1.5}Sr_{0.5}BiNbO_6$ slowly rise, indicating temperature-dependent decrement in capacitive relaxation mechanisms. However, the peak heights of the $Pb_{0.5}Sr_{1.5}BiNbO_6$ compound rather maintain a constant height; hence, improved temperature stability of capacitance is ensured.

8.10.3 Temperature-Dependent P-E Loop Trace

The existence electric ferroic phase, both at the ambient environment and at some higher temperatures, was performed on the two DPOs, and their results were compared. Ferroelectricity is usually

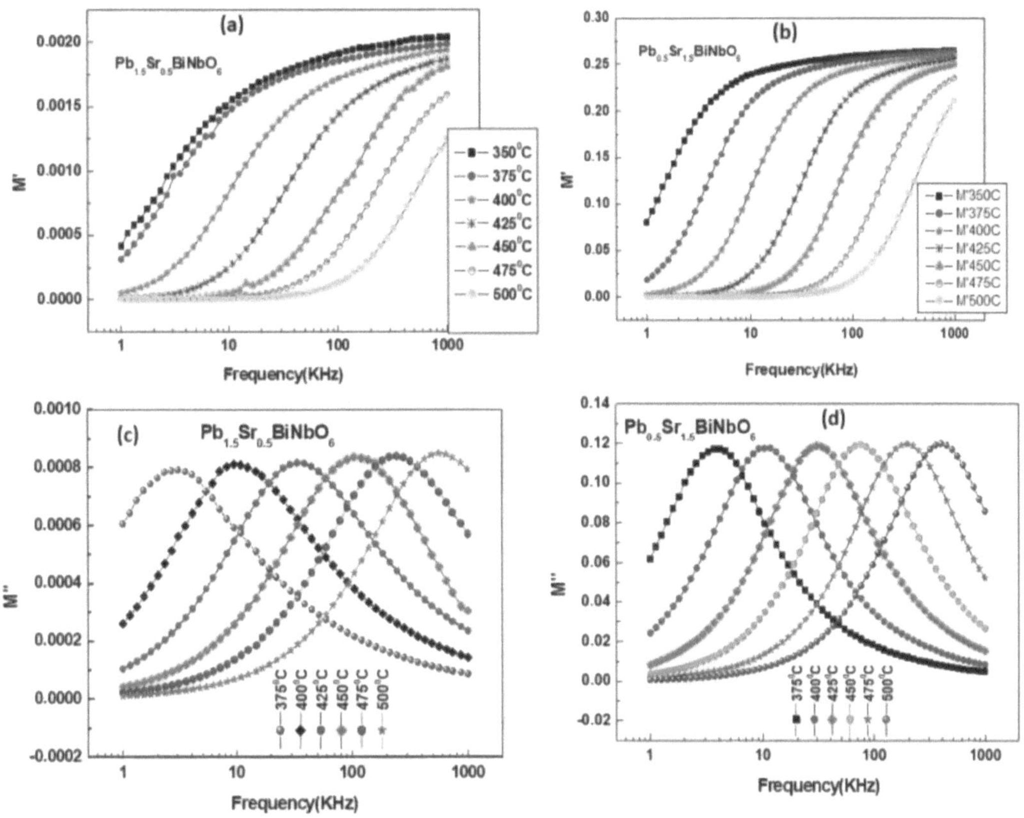

FIGURE 8.12 (a) and (b) M' against frequency at selected temperatures. (c) and (d) M'' against frequency at selected temperatures.

contributed by the lone pair A site cationic moments and their spontaneous relative orientations in response to the applied electric field perturbations. These instantaneous values of polarizability vary as $P\left(\mu Ccm^{-2}\right) = P_{sat} \tanh\left[\{\pm E - E_c\}(2\delta)^{-1}\right]$ [40–42], with polarizability being connected to the frequency-dependent dielectric capacitance as $P\left(\mu Ccm^{-2}\right) = Voltage_{x-axis} \times 10^6 \times C(F) / \pi r^2$ [43], where, the symbols have their universal implication. However, some lossy nature is exhibited by most of the solid oxide route synthesized DPOs [44], which is characterized by the blunt corners of the loops in Figure 8.13(a) and (b). The above-cited phenomenon is mostly contributed by the leakage of charges, induced by defect carriers [45].

Similar nature was also seen in our experimental curves, irrespective of temperature. However, a temperature enhancement results in a reduction in coercive field and remnant polarization values, indicating temperature-enhanced long-range conductivities. Also, the comparative analysis, from Table 8.1 revealed that there is a substantial amplification in room temperature and high-temperature $2E_c$ and $2P_r$ magnitudes, with the increase in Sr content. This explicitly signifies the role of Sr doping toward polarizability improvements in the host, by reducing the disorderliness contributions in orientational polarizability brought in by thermal energy enhancement. The improved ferroelectricity by higher $SrCO_3$ doping is accredited to the lessening proportion of oxygen defects and hence debilitated defect mobility, which is usually considered to be the contributive factor towards domain pinning [46, 47]. Thus, it directly infers the usefulness of strontium modifications toward high storage electrical capacity in transition metal-based ceramics.

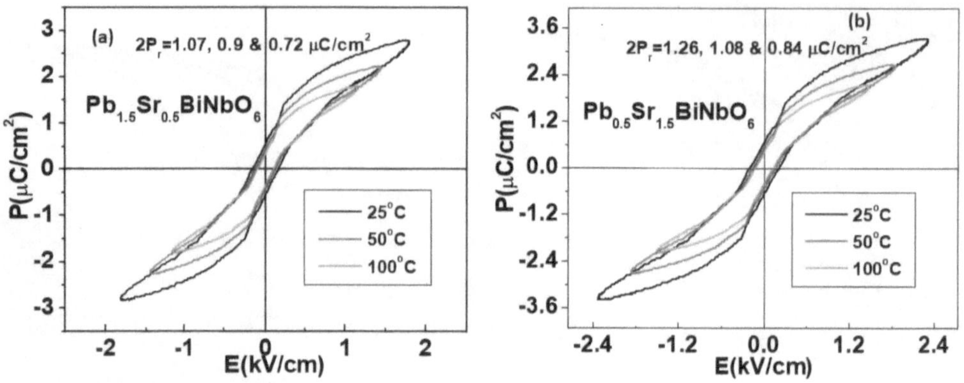

FIGURE 8.13 (a) and (b) The temperature vs spontaneous P-E loops.

TABLE 8.1
Showing the Variations of Various Ferroelectric Parameters for the Compounds under Comparison

Name of the samples	$2E_c$ (kV/cm) 25°C	$2E_c$ (kV/cm) 50°C	$2E_c$ (kV/cm) 100°C	$2P_r$ (μC/cm²) 25°C	$2P_r$ (μC/cm²) 50°C	$2P_r$ (μC/cm²) 100°C
$Pb_{1.5}Sr_{0.5}BiNbO_6$	0.32	0.27	0.21	1.07	0.90	0.72
$Pb_{0.5}Sr_{1.5}BiNbO_6$	0.40	0.34	0.26	1.26	1.08	0.84

8.11 THE FREQUENCY-BASED AC CONDUCTIVITY AND A VS N PLOT

To understand more regarding the flow or interaction among the charge carriers, a frequency variant analysis of ac conductivity at discrete temperature values is vital. A. K. Jonscher for the very first time undertook the task and put forward the universal power law [48], a relation governing the conduction mechanism in ceramic-based dielectrics. In his relation, two conductive parameters found their place, one being the DC part and another being the AC part, which is a function of the nth power of angular field frequency, where n is a parameter that determines the dielectric strength. It is expressed mathematically as $\sigma(T,\omega) = \sigma_{dc}(T) + A(T)\omega^n$, where $\sigma_{dc}(T)$ is a temperature-dependent but frequency-independent direct conductivity parameter and A(T) is a constant, which relies on temperature. Such variations of AC conduction are a function of frequency at selected temperatures, for both samples are shown in Figure 8.14(a). For low concentrations of Sr, better DC conductivities are observed, compared to the higher Sr content samples. In the higher Sr content sample ($Pb_{0.5}Sr_{1.5}BiNbO_6$), a better continuous increase of slope was observed, indicating improved AC conductivities at higher frequencies. It also suggests that the value of 'n' is comparatively higher in $Pb_{0.5}Sr_{1.5}BiNbO_6$ solid oxide, suggesting better dielectric strength. Figure 8.14(b) exhibits the thermal variation of $\sigma(T,\omega)$ fitting parameters (A and n) for both composites. It was noted that 'n' decreases with temperature increment, while the values of pre-frequency exponent factor 'A' undergoes a reverse pattern for all the compounds. The intersection point between the 'A' and 'n' curves indicates the curie dielectric transition point, which also corresponds to the dielectric relaxation strength of the composite [49]. This value was observed to be higher in the case of the complex having more Sr content.

8.12 FUTURE SCOPE AND CHALLENGES

Apart from the limited domain discussion, many of such physiological applicable potentialities of double Pos or similar complexes remain to be investigated, specifically the application part, since

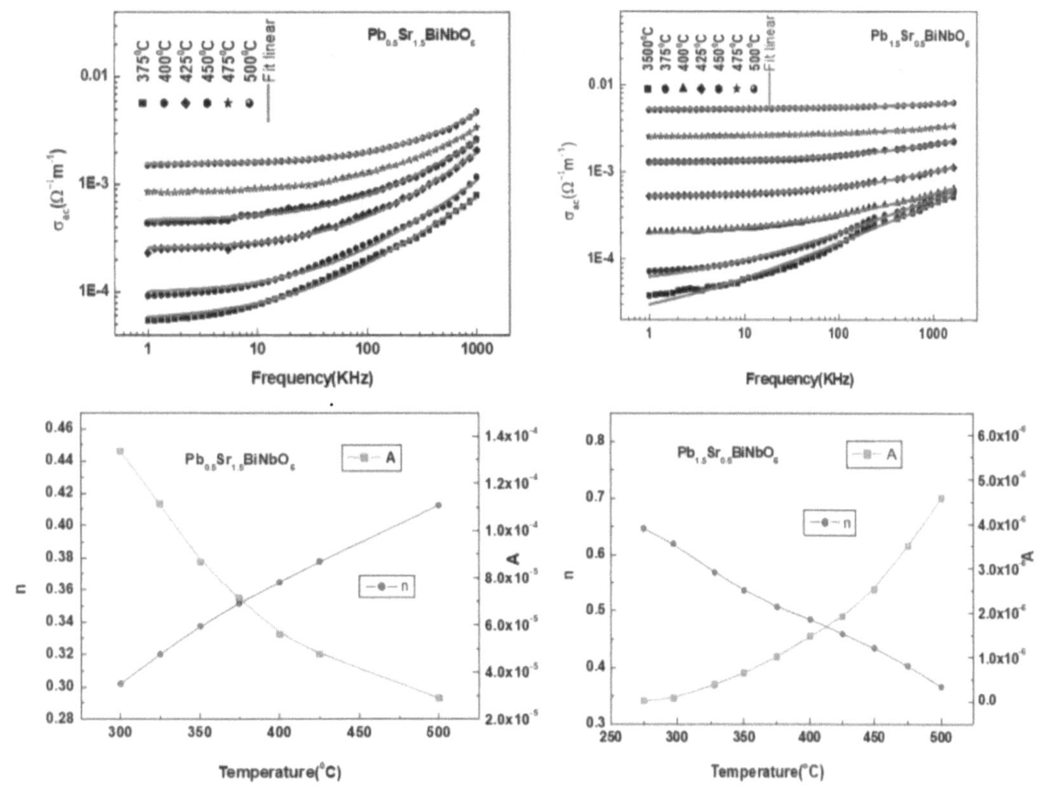

FIGURE 8.14 (a) $\sigma_{ac}(T)$ vs frequency for $Pb_{1.5}Sr_{0.5}BiNbO_6$ and $Pb_{0.5}Sr_{1.5}BiNbO_6$. (b) A vs n with temperature variations for $Pb_{1.5}Sr_{0.5}BiNbO_6$ and $Pb_{0.5}Sr_{1.5}BiNbO_6$.

almost no work has been undertaken to fabricate perovskite-based cells following the thin film heterojunction formation, oxide-based fuel cells, and electrodes, storage/conversion devices. Various unexplored physical characterizations/investigations, such as magnetoelectric, piezoelectric, DTA, TGA, stress-field response, magnetoresistive, and magnetocaloric, have hardly been deployed until now over DPOs. The main challenge which the material researcher community commonly encounters with DPOs is what is the most perfect fabrication approach, which must be chosen carefully in order to form industry-grade phase-purity DPO high-K ceramics.

8.13 CONCLUSION

Double perovskite ceramics thus proves itself as a potential candidate because of its high K dielectric ability for electrical storage applicability, together with microwave applications for loss device fabrication. Greater ambient condition high K values were achieved, with enhanced modifier content, particularly at low frequencies, and improvement in its response is directly evident as a function of thermal treatment for all-purpose high dielectric storage utilities. The lower loss tangent signifies the samples' microwave device fabricability. The enhancement in hopping mechanisms, higher saturation values of real modulus, stabilized capacitive relaxations and well-defined relation frequency are some of the remarkable achievements, achieved by enhancing Sr mixer content, as evident from modulus spectra. The dielectric strength was found to be enhanced in the composite with higher dopant content, as divulged from the graphical analysis of the dielectric strength parameters, extracted from the conductivity plots. An elevation in ferroelectric remnant values, together with the saturation magnitudes, was seen in the higher modifier concentration complex, with minimal entropy in the orientational polarizability. Since ferroelectric polarizations are inherently related to dielectric polarizability, we conclude that the above-discussed composite may be considered as the best choice to go to for dielectric storage and similar device applications, as compared to the former.

REFERENCES

1. Long, J., and Ward, R. *Magnetic compounds of hexavalent rhenium with the perovskite-type structure.* JACS, 1961, 83, p. 2816–2818.
2. Capece, A.M., Polk, J.E., and Shepherd, J.E. *X-ray photoelectron spectroscopy study of* $BaWO_4$ *and* Ba_2CaWO_6. J. Electron Spectros. Relat. Phenomena, 2014, 197, p. 102–105.
3. Patterson, F.K., Moeller, C.W., and Ward, R. *Magnetic oxides of molybdenum (V) and tungsten (V) with the ordered perovskite structure.* Inorg. Chem., 1963, 2, p. 196–198.
4. Goodenough, J.B. *Theory of the role of covalence in the perovskite-type manganites [La, M (II)] Mn* O_3. Physiol. Rev., 1955, 100, p. 564.
5. Anderson, P.W. *Antiferromagnetism. Theory of superexchange interaction.* Phys. Rev., 1950, 79, p. 350.
6. Kramers, H.A. *L'interaction entre les atomes magnétogènes dans un cristal paramagnétique* Physica, 1934, 1, p. 182–192.
7. Kuepper, K., Balasz, I., Hesse, H., Winiarski, A., Prince, K.C., Matteucci, M., Wett, D., Szargan, R., Burzo, E., and Neumann, M. *Electronic and magnetic properties of highly ordered Sr2FeMoO6.* Phys. Status Solidi., 2004, 201, p. 3252–3256.
8. Garcıa-Landa, B., Ritter, C., Ibarra, M.R., Blasco, J., Algarabel, P.A., Mahendiran, R., and Garcıa, J. *Magnetic and magnetotransport properties of the ordered perovskite* Sr_2FeMoO_6. Solid State Commun., 1999, 110, p. 435–438.
9. Sarma, D.D. *A new class of magnetic materials: Sr2FeMoO6 and related compounds.* Curr. Opin. Solid State Mater. Sci., 2001, 5, p. 261–268.
10. Sarma, D.D., Sampathkumaran, E. V., Ray, S., Nagarajan, R., Majumdar, S., Kumar, A., Nalini, G., and Row, T.N.G. *Magnetoresistance in ordered and disordered double perovskite oxide* Sr_2FeMoO_6. Solid State Commun., 2000, 114, p. 465–468.
11. Wang, J., Zhang, J.M., Wang, S.F., and Xu, K.W. *First-principles study on the half metallic Tc-doped* Sr_2FeReO_6, J. Magn. Magn. Mater. 2013, 329, p. 30–33.
12. Yuan, C.L., Wang, S.G., Song, W.H., Yu, T., Dai, J.M., Ye, S.L., and Sun, Y.P. *Enhanced intergrain tunneling magnetoresistance in perovskite* Sr_2FeMoO_6 *polycrystals with nanometer-scale particles.* Appl. Phys. Lett., 1999, 75, p. 3853.
13. Bijelić, J., Tatar, D., Hajra, S., Sahu, M., Kim, S.J., Jagličić, Z., and Djerdj, I. *Nanocrystalline antiferromagnetic High-κ Dielectric* Sr_2NiMO_6 *(M = Te, W) with Double Perovskite Structure.* Molecules, 2020, 25(17), p. 3996.
14. Chen, Jenq-Wei., Chiou, Kuan Ru., Hsueh, An-Chih., and Chang, Ching-Ray. *Dielectric relaxation of the double perovskite oxide Ba2PrRuO6.* RSC Adv., 2019, 9, p. 12319.
15. Mahato, Dev K., Dutta, A., and Sinha, T.P. *Dielectric relaxation in double perovskite oxide, Ho2CdTiO6.* Bull. Mater. Sci. Ind. Acad. Sci., 2011, 34, 3, p. 455–462.
16. Tailor, Naveen Kumar, Parikh, Nishi, Yadav, Pankaj., and Satapathi, Soumitra. *Dielectric Relaxation and Polaron Hopping in* $Cs_2AgBiBr_6$ *Halide Double Perovskites.* J. Phys. Chem. C, 2022, 126 (24), p. 10199–10208.
17. Wang, G. J., Wang, C. C., Huang, S. G., Sun, X. H., Lei, C. M., Li, T., and Liu, L.N. *Origin of the colossal dielectric properties in double-perovskite* Sr_2CoNbO_6. AIP Adv. 2013, 3, p. 022109.
18. Parida, B.N., Panda, N., Sen, S., Padhee, R., and Parida, R.K. *Investigation of multifunctional features in new double perovskite* $PbSrBiNbO_6$ *for possible devices.* Inorg. Chem. Commun., 2021, 134, p. 109074.
19. Bhattacharjee, S., Brahma, F., Parida, R.K. and Parida, B.N. *Synthesis and characterization of revived double perovskite* $Ba_{0.5}Sr_{1.5}FeVO_6$. J. Mater. Sci.: Mater. Electron., 2020, 31, p. 2789–2798.
20. Das, Rutuparna, Choudhary, Ram Naresh. *Structure, dielectric and electrical properties of relaxor lead-free double perovskite:* Nd_2NiMnO_6. Process. Appl. Ceram. 2019, 13. p. 1–11.
21. Wang, G. J., Wang, C. C., Huang, S. G., Sun, X. H., Lei, C. M., Li, T., and Mei, J. Y. *Low-temperature dielectric properties of double-perovskite* Ca_2CoNbO_6. J. Electroceram. 2012, 28, p. 172
22. Wang, C. C., and Zhang, L. W., *Dielectric behaviour of cobalt titanium oxide.* J. Phys. D: Appl. Phys., 2007, 40, p. 6834.

23. Halizan, M Z M et al *Simultaneously improved dielectric, optical and conductivity properties of $SrLa_{1-x}Nd_xLiTeO_6$ double perovskites*. Mater. Res. Express 2020, 7, p. 086301.
24. Mostafa, Y., Van Vliet, K.J., and Yildiz, B. Polarizing *Oxygen Vacancies in Insulating Metal Oxides under a High Electric Field*. Phys. Rev. Lett., 2017 119.12, p. 126002.
25. Badapanda, T., Singh, A.K., and Panigrahi, S. *An approach for correlating the structural and electrical properties of Zr^{4+}-modified $SrBi_4Ti_4O_{15}$/SBT ceramics*. RSC Adv., 2017, 7, p. 16319–16331.
26. Bhattacharjee, S., Mohanty, B., Parida, R.K., and Parida, B.N. *Multifunctional feature of double perovskite strontium iron vanadate for storage device*. Mater. Chem. Phys., 2022, 275, p. 125254.
27. Parida, B.N., Panda, N., Padhee, R., Das R. P., Choudhary, R.N.P. *Dielectric relaxation and impedance analysis of ferroelectric double perovskite Pb_2BiNbO_6*. J. Mater. Sci.: Mater. Electron., 2017, 28, p. 1824–1831.
28. Panda, A.K., Sahoo, L., Chakravarty, R., Nayakm N.C., Parida, R., Parida, B.N., Dutta, R. *Transport and semiconducting behavior of Ca_2BiNbO_6 new inorganic double perovskite*. Appl. Phys. A, 2021, 127, p. 950.
29. Koops, C. G. On the *Dispersion of Resistivity and Dielectric Constant of Some Semiconductors at Audiofrequencies*. Physiol. Rev.,1951, 83, p. 121–124.
30. Kityk, I.V., Makowska-Janusik, M., Fontana, M. D., Aillerie, M., and Fahmi, A. *Band Structure treatment of the influence of nonstoichiometric defects on optical properties of $LiNbO_3$*. J. Appl. Phys., 2001, 90, p. 5542.
31. Garcia, J. E., Gomis, V., Perez, R., Albareda, A., and Eiras, J.A. *Unexpected dielectric response in lead zirconate titanate ceramics: The role of ferroelectric domain wall pinning effects*. Appl. Phys. Lett., 2007, 91, p. 042902.
32. Wu, D. Li, A., and Ming, N. *Dielectric characterization of $Bi_{3.25}La_{0.75}Ti_3O_{12}$ thin films*. Appl. Phys. Lett., 2004, 84, p. 4505.
33. Dai, Z., Akishige, Y. *Electrical properties of multiferroic $BiFeO_3$ ceramics synthesized by spark plasma sintering*. J. Phys. D: Appl. Phys., 2010, 43, p. 445403.
34. Jawahar, K., Choudhary, R.N.P. *Structural and dielectric properties of Y3/2Bi3/2Fe5O12*. Mater. Lett., 2008, 62 (6–7), p. 911–913.
35. Cross, L. E. *Relaxor ferroelectrics*, Ferroelectrics, 1987, 76, p. 241.
36. Bharti, C., Dutta, A., and Sinha, T.P., *Modulus spectroscopy study of a double perovskite oxide Sr2CeNbO6: Frequency and time domain analyses*. Solid State Sci., 2012, 14, 7, p. 920–925.
37. Parida, B.N., Panda, N., Padhee, R., Das, P.R., Choudhary, R.N.P. Mater Sci: Mater Electron., 2017, **28**, p. 1824–1831.
38. Parida, B. N., Panda, N., Padhee, R., and Parida, R. K. *Ferroelectric and optical behavior of $Pb_{0.5}Ba_{1.5}BiNbO_6$ double perovskite*. Ferroelectrics, 2019, 540:1, p. 18–28.
39. Mahato, Dev K., Rudra, M., and Sinha, T.P. *Structural and electrical features of rare earth based double perovskite oxide: Pr2NiZrO6*. J. Alloys Compd., 2016, 689, p. 617–624,
40. Smith, R.C., and Hatch, A.G. *Parameter estimation techniques for a polarization hysteresis model*. Smart Structures and Materials 2004, *Proceedings of SPIE, Vol. 5383, Bellingham*, p. 155–163.
41. Lupascu, D.C. *Fatigue in Ferroelectric Ceramics and Related Issues*. Series in Materials Science, Springer Book, 61, Berlin.
42. Sheikholeslami, A., Gulak, P.G. *A survey of behavioural modelling of Ferrolecetric capacitors*. IEEE Trans. Ultrason. Ferroelectr. Freq. Control, 1997, 44(4), p. 917–924.
43. Mahmood, N.B., and Al-Shakarchi, E.K. *Simulation and Experimental Data of P-E Hysteresis Loop in BNT and BKT*. J. Mod. Phys., (2017), 8, p. 844–851.
44. Sahoo, L., Parida, B.N., Parida, R.K., Padhee, R., Mahapatra, A.K. *Structural, optical dielectric and ferroelectric properties of double perovskite $BaBiFeTiO_6$*. Inorg. Chem. Commun., 2022, 146, p. 110102.
45. Zhou, C., Feteira, A., Shan, X., Yang, H., Zhou, Q., Cheng, J., Li, W., and Wang, H. *Remarkably high-temperature stable piezoelectric properties of $Bi(Mg_{0.5}Ti_{0.5})O_3$ modified $BiFeO_3$–$BaTiO_3$ ceramics*. Appl. Phys. Lett., 2012, 101, p. 032901.
46. Matthias, B., and Hippel, Von A. *Domain Structure and Dielectric Response of Barium Titanate Single Crystals*. Physiol. Rev., 1948, 73, p. 1378–1384.
47. Hippel, Von A. *Ferroelectricity, Domain Structure, and Phase Transitions of Barium Titanate*. Rev. Mod. Phys., 1950, 22, p. 221–227.
48. Jonscher, A. K., *The Universal Dielectric Response*. Nature, 1977, 267, p. 673–679.
49. Nobre, M. A. L., and Lanfredi, S. *Ferroelectric state analysis in grain boundary of $Na_{0.85}Li_{0.15}NbO_3$ ceramic*. J. Appl. Phys., 2003, 93, p. 5557.

9 Interfacial Engineering of Atomic Layer Deposited High-K Dielectrics on High-Mobility Ge/GaAs Semiconductors

Yan-Qiang Cao and Ai-Dong Li

9.1 INTRODUCTION

With the rapid development of microelectronics technology, the feature size of complementary metal oxide semiconductor (CMOS) devices is shrinking. Therefore, the traditional silicon-based devices will reach their limits of theory and technology [1]. On the one hand, a gate dielectric layer of SiO_2 becomes thinner and thinner, which leads to the very huge leakage current caused by the quantum tunneling effect. As a result, the devices fail to function normally. Therefore, it is necessary to replace SiO_2 with high dielectric constant (high-K) materials, among which HfO_2 is considered to be one of the most promising candidates [2]. On the other hand, deterioration occurs at the interface of gate dielectrics material/channel material, resulting in the significant degradation of channel mobility. This greatly hinders the improvement of speed for CMOS logic devices [3–5]. An effective way to solve this problem is via adopting the channel materials with high mobility. In recent years, it has attracted wide attentions to use high mobility semiconductor substrates for CMOS devices, such as carbon nanotubes [6, 7], Graphene [8, 9], Ge [10, 11] and III-V compound semiconductors [12, 13]. Compared to traditional silicon-based devices, these semiconductor materials with higher carrier mobility can provide a larger drive current, which are the promising channel materials for the manufacture of next-generation high-speed, low-power electronic devices.

Among them, Ge is an ideal material to replace Si substrate. Ge exhibits both high electron and hole mobility with low doping activation temperatures [14]. In addition, Ge is the first successful substrate to fabricate transistors and integrated circuits. Nevertheless, the properties of Germanium oxides are poor. Studies have shown that the natural oxide on the surface of Ge substrate is GeO_2, which is not only water-soluble, but thermal decomposition occurs at 400°C and above [15, 16]. Therefore, the existance of native oxides seriously damages the interface quality of the Ge substrate. In addition to its thermodynamic instability, intrinsic characteristics of GeO_2 are not conducive to further scaling down of MOS devices. First, its dielectric constant is small at only ~ 6. Moreover, the energy band of GeO_2 is 5.1 eV, and the valence band offset of GeO_2/Ge is 3.8 eV. Also, the conduction band offset of GeO_2/Ge is only 0.6 eV, which is too small to block the injection of electrons. Thus, electrons can directly tunnel through the GeO_2 layer, causing a very large gate current. Therefore, interface passivation layer is indispensable and necessary to enhance the properties of Ge based MOS devices [17–19].

GaAs has high electron mobility with a large band gap and high breakdown field intensity [20], which is why it is the other promising substrate to replace Si. However, the chemical properties of GaAs surface are also very unstable, thus it is easy to form a natural oxide layer on the surface. More seriously, the natural oxide layer is of poor quality and loose structure, making it difficult to obtain

DOI: 10.1201/9781003352365-9

a high-quality gate dielectrics/GaAs interface during the preparation of GaAs-based MOS devices. As a result, a serious Fermi surface pinning effect occurs, which is detrimental to the normal operation of devices [21]. Thus, in order to obtain high performance GaAs based MOS devices, it is essential to engineer the high-K dielectric/GaAs interface.

In this chapter, aimed at above problems, several strategies based on our works for interface engineering between high-K dielectrics and Ge/GaAs are discussed, including chemical pretreatment, plasma passivation, self-cleaning during ALD, and interfacial passivation layer, focusing on their interface structures and electrical performance.

9.2 HIGH-K DIELECTRICS ON GE

9.2.1 CHEMICAL PRETREATMENTS

The effects of chemical pretreatments on the interface and electrical properties of ALD deposited Al_2O_3 gate dielectric on Ge were studied. Ge substrates were conventionally cleaned by HF and then passivated by $(NH_4)_2S$. A new combination pretreatment of HBr + $(NH_4)_2S$ was adopted for comparison [22]. X-ray photoelectron spectroscopy (XPS) was adopted for surface chemistry analysis. XPS results show HBr + $(NH_4)_2S$ treated sample has less GeO_x (9.25%) than that of 16.45% on an HF + $(NH_4)_2S$ treated one, as shown in Figure 9.1(a). It implies that the combined treatment of HBr and $(NH4)_2S$ is more effective than the conventional HF and $(NH_4)_2S$ pretreatment, where more native Ge oxides can be removed. In addition, Ge pretreated by HBr + $(NH_4)_2S$ exhibits more Ge-S (7.40%) than the HF + $(NH_4)_2S$ pretreated sample with 3.37%. More importantly, the signal of Ge-S bonds on Ge is very stable in the ambient for 300 min, indicating that G-S bonds can effectively stabilize the Ge surface. To evaluate the electrical properties, 10 nm ALD Al_2O_3 were deposited on Ge surfaces after different chemical pretreatments. Capacitance-voltage (*C–V*) characteristic curves of 10 nm Al_2O_3 on Ge were measured at 1 MHz, as displayed in Figure 9.1(b). From the above the C-V curves, the value of capacitance equivalent thickness (CET) of the MOS capacitor can be determined by the equation of $\varepsilon_0\varepsilon_r A/C_{acc}$ [23]. Therefore, HF + $(NH_4)_2S$ and HBr + $(NH_4)_2S$ pretreated samples exhibit the CET values of 5.50 nm and 5.06 nm, respectively. It can be ascribed to the fact that HBr + $(NH_4)_2S$ pretreatment can remove more of the defective GeO_x layer between Al_2O_3 and Ge, as illustrated from XPS results, resulting in the decrease of CET value. Moreover, by carefully comparing the *C-V* curves of the above two samples, it can be seen that the HF + $(NH_4)_2S$ pretreated sample shows a more serious stretch-out than the HBr + $(NH_4)_2S$ pretreated one. It implies that higher interface trap densities are formed in the HF + $(NH_4)_2S$ pretreated sample [24]. In addition, the HF + $(NH_4)_2S$ pretreated sample exhibits a hysteresis of 498 mV at the flat-band voltage (V_{FB}), which is larger than that of the HBr + $(NH_4)_2S$ pretreated one (335 mV at V_{FB}), which also implies higher density of defects. Furthermore, the leakage current density of HF + $(NH_4)_2S$ pretreated sample is 7.24×10^{-4} A/cm^2 (at V_{FB} + 1 V), which is much larger than that of 8.41×10^{-6} A/cm^2 for the HBr + $(NH_4)_2S$ pretreated sample, as illustrated in the insert of Figure 9.1(b). The significant decrease in leakage current density for HBr + $(NH_4)_2S$ pretreated Al_2O_3/Ge stacks is due to the presence of fewer defective Ge oxides at the interface of Al_2O_3 and Ge substrate [25]. Figure 9.1(c), (d) display the *C-V* curves at various frequencies for the samples with different chemical pretreatments. It can be seen that the HBr + $(NH_4)_2S$ pretreated sample can exhibit well the frequency-dependence characteristics, implying its better interface quality. By contrast, the HF + $(NH_4)_2S$ pretreated sample cannot detect the low frequency *C-V* curves below 200 kHz, as presented in Figure 9.1(d), which may be caused by its poorer interface quality.

Based on the above analysis, HBr + $(NH_4)_2S$ pretreatment seems to be better than the combination of HF + $(NH_4)_2S$ for a Ge based MOS capacitor. HBr + $(NH_4)_2S$ pretreatment can not only effectively remove the defective GeO_x, but also passivate the Ge surface with stable *Ge-S bonding*. As a result, the HBr + $(NH_4)_2S$ pretreated Al_2O_3/Ge stack exhibits better electrical properties, in the term of larger accumulation capacitance, smaller C-V hysteresis, and significant decreased leakage

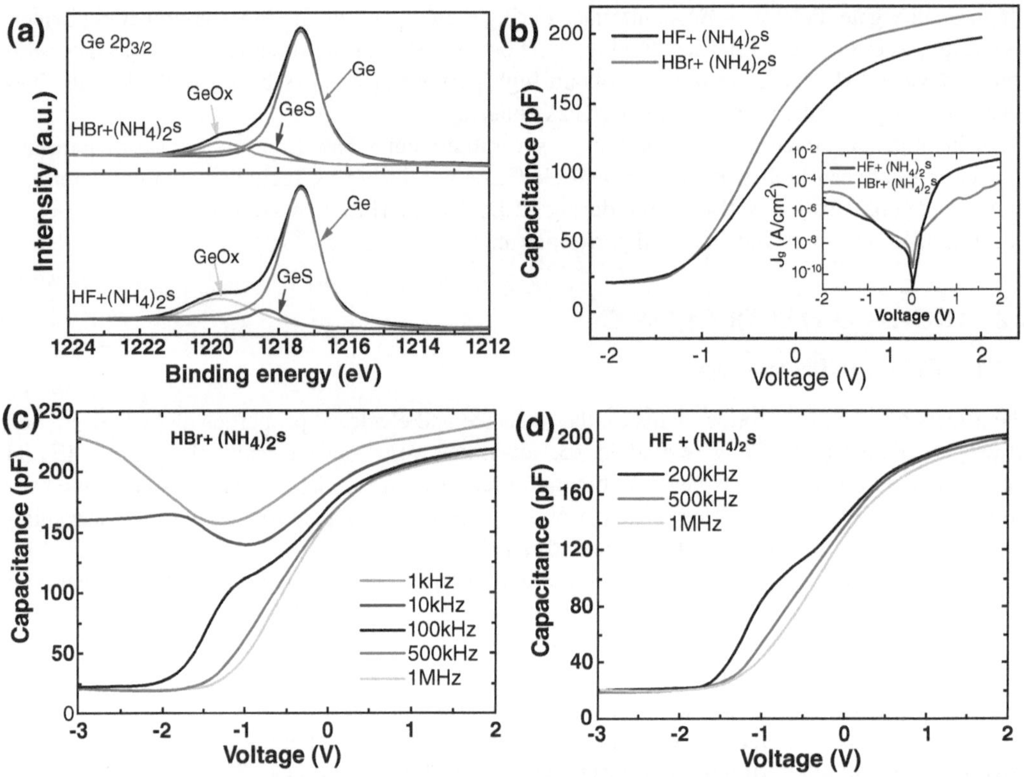

FIGURE 9.1 Interfacial and electrical properties of the Al_2O_3/Ge with different chemical pretreatments [22]. The comparison of (a) Ge 2p XPS spectra and (b) C-V curves for the samples after HBr + $(NH_4)_2S$ and HF + $(NH_4)_2S$ pretreatments. Frequency dispersion characteristics of Al_2O_3/Ge with (c) HBr + $(NH_4)_2S$ and (d) HF + $(NH_4)_2S$ pretreatments. The insert of (b) is the leakage current density versus voltage (J-V) curves. (Copyright 2011. Reproduced with permission from Elsevier.)

current density. However, kinks in the *C-V* curves still can be observed in the HBr + $(NH_4)_2S$ pretreated Ge MOS capacitor, which may be caused by the interface states, as presented in Figure 9.1(c). Therefore, further surface passivation optimization for Ge is still needed after HBr + $(NH_4)_2S$ chemical pretreatment, such as introducing an interfacial passivation layer.

9.2.2 Plasma Passivation

It has been demonstrated that the formation of GeO_xN_y via incorporating nitrogen into GeO_2 can enhance its both thermal and chemical stability [26]. It was also demonstrated that it is an effective way to improve the performance of Ge MOS devices via surface nitriding on Ge substrates prior to deposition of high-K dielectrics films, which can suppress Ge out-diffusion [27]. Therefore, besides chemical pretreatments, *in situ* NH_3 plasma in a plasma-enhanced atomic layer deposition (PEALD) chamber was used to further passivate the Ge surface [28]. After that, *in situ* thermal ALD-HfO_2 was deposited without breaking the vacuum. As respected, XPS results reveal that NH_3 plasma pretreatment is able to turn O-Ge bonds into N-Ge bonds, forming an ultrathin GeO_xN_y layer on a Ge surface. Figure 9.2(a), (b) show the cross-sectional high-resolution transmission electron microscope (HRTEM) images of HfO_2/Ge and HfO_2/GeO_xN_y/Ge, which are treated with 400°C post deposition annealing (PDA) for 180 s. For a sample with HfO_2 directly deposited on Ge, a ~0.9-nm nonuniform and ambiguous layer can be observed at the interface of HfO_2 and Ge, as marked in Figure 9.2(a). This interfacial layer should be the defective $HfGeO_x$, formed by the diffusion of Ge into HfO_2. On

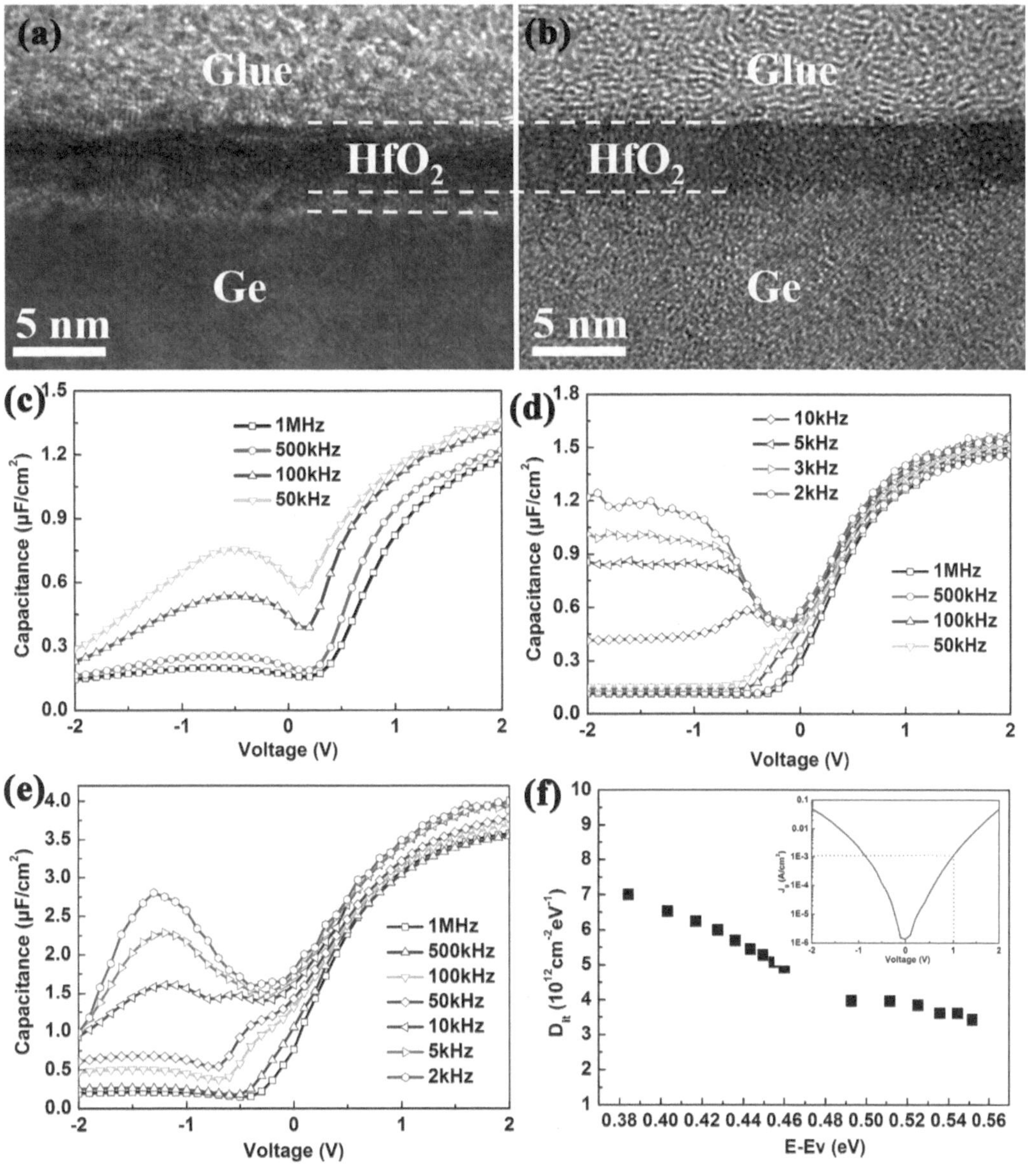

FIGURE 9.2 Interfacial and electrical properties of the HfO_2/Ge without and with NH_3 plasma pretreatments [28]. HRTEM images of (a) HfO_2/Ge and (b) HfO_2/GeO_xN_y/Ge. Frequency dispersion characteristics of (c) HfO_2(7 nm)/Ge, (d) HfO_2(7 nm)/GeO_xN_y/Ge, and (e) HfO_2(3 nm)/GeO_xN_y/Ge. (f) Distribution of D_{it} for HfO_2(3 nm)/GeO_xN_y/Ge, the insert shows the leakage current density properties. (Copyright 2015. Reproduced with permission from Elsevier.)

the contrary, HfO_2/GeO_xN_y/Ge with NH_3 plasma pretreatment after 400°C PDA exhibits a sharp and flat interface between HfO_2 and Ge. No obvious interfacial layer can be observed, as illustrated in Figure 9.2(b). Clearly, the GeO_xN_y interfacial layer formed via *in situ* NH_3 plasma pretreatment is very thermal stable, which can inhibit the Ge out-diffusion to generate a defective interfacial layer during the PDA process, improving the interface quality of high-K dielectrics/Ge.

First, a thick HfO_2 of 7 nm was deposited on both pristine Ge and NH_3 plasma pretreated Ge, where Figure 9.2(c), (d) show the C-V curves measured at 1 MHz after PDA. HfO_2/GeO_xN_y/Ge exhibits a smaller CET of 2.33 nm than that of 2.93 nm for HfO_2/Ge. The decrease in CET value for HfO_2/GeO_xN_y/Ge should be ascribed to the absence of a defective interfacial layer $HfGeO_x$, as illustrated in the above TEM images. Furthermore, HfO_2/Ge stacks show a more serious stretch-out than HfO_2/GeO_xN_y/Ge in C-V curves. It implies that interface trap density is well reduced for in HfO_2/GeO_xN_y/Ge [24]. Furthermore, the V_{FB} is determined to be 0.59 V and 1.01 V for HfO_2/Ge and HfO_2/GeO_xN_y/Ge, respectively. Based on the previous report [29], an interfacial fixed charge caused by poorer interface quality will lead to a larger positive shift of V_{FB}, which also demonstrates the good interface quality for HfO_2/GeO_xN_y/Ge without the defective $HfGeO_x$ layer. Moreover, the HfO_2/GeO_xN_y/Ge also exhibits the smaller leakage current density of 1.51×10^{-4} A cm^{-2} (@ +1 V) than that of 2.12×10^{-3} A cm^{-2} for HfO_2/Ge. There is a high density of interface defects for HfO_2 on pristine Ge, therefore, C-V curves below 50 kHz cannot be measured, as displayed in Figure 9.2(c). At the same time, a large frequency dispersion of 15.6% between 50 kHz to 1 MHz is presented for HfO_2/Ge. On the contrary, a much smaller frequency dispersion (3.5%) is observed for HfO_2/GeO_xN_y/Ge. Moreover, well C-V curves without stretch-out or bumps (Figure 9.2(d)) can be observed for HfO_2/GeO_xN_y/Ge, indicating a significant improvement for interface quality after NH_3 plasma pretreatment [30]. The conductance method has been widely used to quantitatively evaluate the interface quality of high-K film and semiconductors [31], where the density of interface trap (D_{it}) can be calculated by the formula of $D_{it} = 2.5 \times (G_p/\omega)_{max}/Aq$. In the above equation, $(G_p/\omega)_{max}$ is the peak value of conductance-voltage curves, ω is the angular frequency of the measurement, q is the electron charge, and A is the electrode area. Therefore, HfO_2 on pristine Ge exhibits the D_{it} of 7.59×10^{12} cm^{-2} eV^{-1}, whereas, a much smaller D_{it} value of 3.68×10^{12} $cm^{-2}eV^{-1}$ can be detected for HfO_2 on NH_3 plasma pretreated Ge. The above electrical results demonstrate that a much-improved interface quality can be achieved for HfO_2/Ge with NH_3 plasma pretreatment. The thickness of gate dielectric must be reduced in order to achieve sub-nanometer CET for MOS capacitors, therefore, 3 nm ALD HfO_2 was deposited on NH_3 plasma pretreated Ge. Figure 9.2(e) shows the corresponding C-V curves at various frequencies from 1 MHz to 2 kHz, it can be seen that well behaved C-V characteristics can still be presented, with the frequency dispersion of only 6.6% from 1 MHz to 50 kHz. With a much decreased gate dielectric thickness, a much increased accumulation capacitance of 3.58 $\mu F/cm^2$ at 1 MHz can be achieved. Therefore, a sub-nanometer CET value of 0.96 nm can be successfully gained. Moreover, this device still exhibits a moderate D_{it} of 3.42×10^{12} cm^{-2} eV^{-1} near the bottom of the conduction band, as plotted in Figure 9.2(f). Besides, the 3 nm HfO_2 on GeO_xN_y/Ge possesses a relatively small leakage current density of 1.12×10^{-3} A cm^{-2} (@ +1 V), as presented in the inset of Figure 9.2(f). Compared to the literatures with ~1 nm CET, the leakage current densities are normally ~2×10^{-2} A cm^{-2} [32], which are much larger than the value for HfO_2/GeO_xN_y/Ge.

Based on the above discussion, it can be seen that a stable interface layer (GeO_xN_y) with wonderful properties can be formed on the Ge surface by NH_3 plasma pretreatment, at the same time, the native Ge oxides are consumed. Hence, HfO_2/GeO_xN_y/Ge stacks can obtain excellent performance, including small CET value, little leakage current, less frequency dispersion, and a moderate D_{it} value. Fortunately, the device with both small CET (< 1 nm) and leakage current density can be obtained in HfO_2(3 nm)/GeO_xN_y/Ge. In summary, *in situ* NH_3 plasma pretreatment is believed to be a promising strategy to achieve high performance Ge-based MOS devices with perfect interface.

9.2.3 ALD Interfacial Passivation Layer

Although chemical pretreatments can remove native oxides and passivate the Ge surface, it is still not enough to deposit high-K dielectric films on Ge directly because an unwanted and harmful interfacial layer and high interface trap density are generated at the interface of high-K/Ge [33]. Thus, it is essential to prepare a large high-K/Ge interface to fabricate high-speed and low-power Ge-based MOS devices. Therefore, various interfacial passivation layers have been adopted to improve the quality of high-K/Ge interface. Next, we will focus on the effect of interfacial passivation layers prepared by ALD for high-K/Ge stacks.

9.2.3.1 SiO_2

First, an ultrathin SiO_2 layer prepared by *in situ* PEALD was utilized as an interfacial passivation layer for HfO_2 on Ge [34]. After the deposition process of PEALD SiO_2, it is found that an interface layer of Si-O-Ge is generated on Ge. Fortunately, a Si-O-Ge interlayer is extremely stable against thermal annealing. Thus, no formation of defective Hf-silicates and Ge-O degradation can be observed, as shown in Figure 9.3(a), (b). Figure 9.3(c) presents the C-V curves of HfO_2/SiO_2/Ge stacks at 1 MHz before and after PDA process. The PDA process usually can densify and thin the high- K dielectric films. Therefore, the annealed HfO_2/SiO_2/Ge stacks exhibit an evident smaller CET of 1.53 nm with 2.25 μF cm^{-2} in accumulation capacitance compared to the *as prepared* sample. At the same time, the leakage current density also increases from 2.2×10^{-4} A cm^{-2} to 2.1×10^{-3} A cm^{-2} (@ V_{FB+}1 V) after PDA due the decrease of HfO_2/SiO_2 thickness, as plotted in Figure 9.3(d). In addition, a small D_{it} value of 2.67×10^{12} cm^{-2} eV^{-1} can be achieved for the Pt/HfO_2/SiO_2/Ge stacks after PDA. More importantly, the electrical performance of Pt/HfO_2/SiO_2/Ge stacks with *in situ* PEALD SiO_2 is compared to the sample using *ex situ* SiO_2 interlayer formed by metal organic chemical vapor deposition (MOCVD) [35]. It can be seen that the sample using *in situ* SiO_2 via PEALD exhibits better electrical properties, including higher accumulation capacitance and smaller leakage current density. This can be attributed such that PEALD can deposit more uniform film than MOCVD, especially at the scale of nanoscale thickness. Therefore, compared to other technologies, the *in situ* ALD formed interfacial passivation layers may be more promising for the realization of high-quality Ge-based transistor devices.

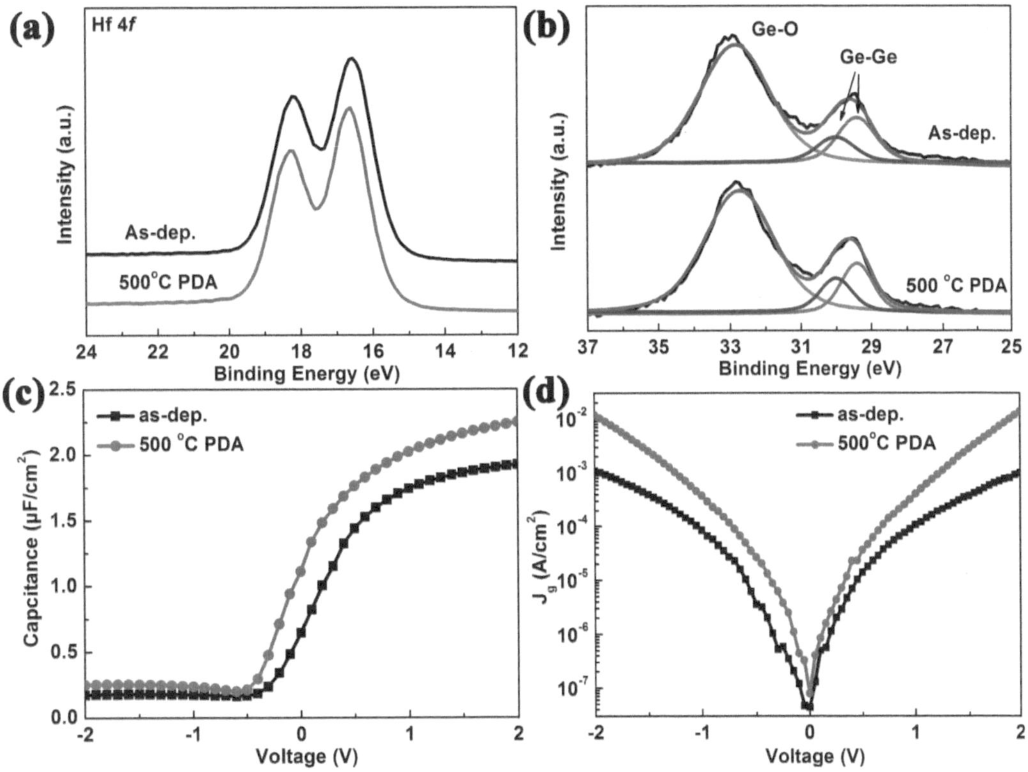

FIGURE 9.3 Interfacial and electrical properties of the HfO_2/Ge with in situ PEALD SiO_2 interfacial passivation layer before and after 500 °C PDA [34]. Comparison of XPS spectra for (a) Hf 4f and (b) Ge 3d, and (c) high-frequency C-V and (d) leakage current curves.

9.2.3.2 La_2O_3

Rare earth oxides exhibit wonderful affinity toward Ge, leading to the formation of rare earth germanate, which can passivate the volatile GeO and other oxide defects [36]. Therefore, various rare earth oxides, such as La_2O_3, Y_2O_3, CeO_2, Gd_2O_3, and Tm_2O_3, have been extensively researched to engineer the interface of high-K/Ge [37–40]. Thus, ALD deposited La_2O_3 was used as interfacial passivation layer for HfO_2/Ge [41]. As discussed above, the interface of HfO_2 and Ge is not stable, where ~1 nm defective $HfGeO_x$ is formed at its interface after 500 °C PDA, as shown in Figure 9.4(a). The ultrathin La_2O_3 can inhibit the Ge out-diffusion and the formation of $HfGeO_x$ layer at the interface. The cross-sectional HRTEM image of the sample with La_2O_3 interfacial passivation layer is displayed in Figure 9.4(b), a uniform interfacial $LaGeO_x$ layer (~1.5 nm) can be recognized at the interface of HfO_2/Ge, implying the improved interfacial quality with the assistance of La_2O_3 interfacial passivation layer.

Figure 9.4(c) shows *C-V* curves at 1 MHz for HfO_2/Ge and HfO_2/La_2O_3/Ge. The Pt/HfO_2/Ge MOS capacitor without La_2O_3 interfacial passivation layer exhibits serious stretch-out and large hysteresis of ~281 mV. This is ascribed to the unstable of HfO_2/Ge interface, where defective $HfGeO_x$ would be formed during PDA process, leading to worse interface quality and high interface state density. Thus, it is essential to introduce the interfacial passivation layer at the interface of high-K/Ge. After adopting the La_2O_3 interfacial passivation layer, well behaved C-V curves with higher accumulation capacitance and smaller hysteresis can be achieved for HfO_2/La_2O_3/Ge.

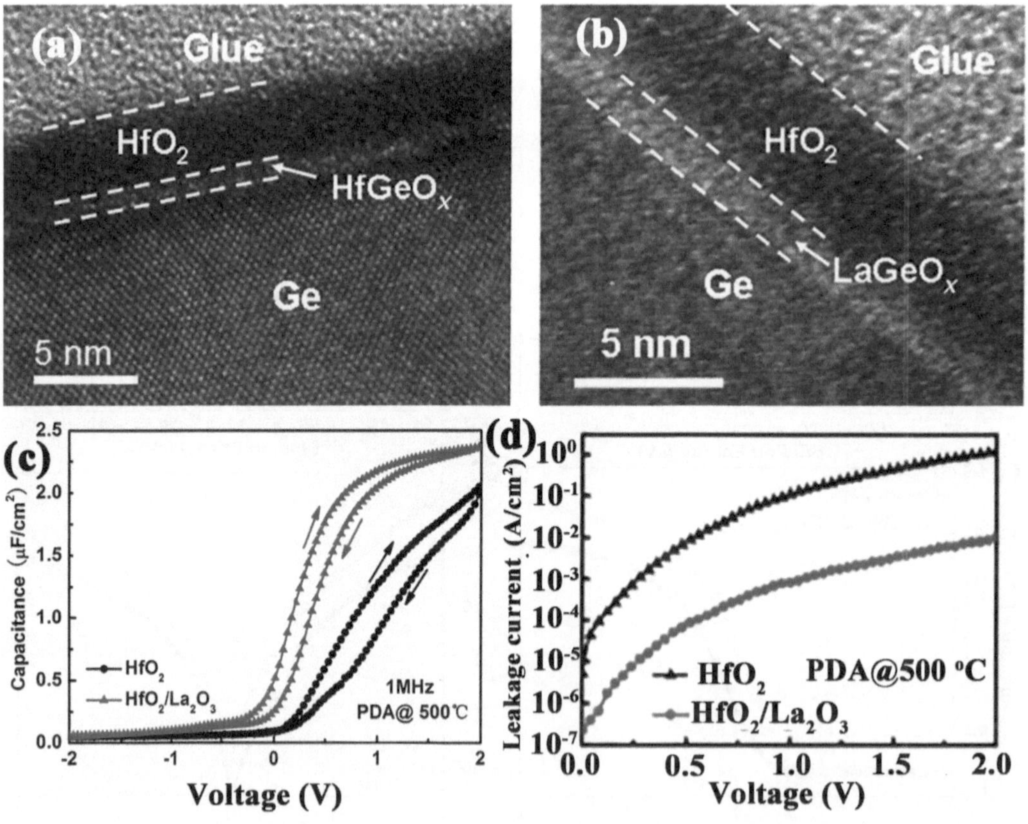

FIGURE 9.4 Interfacial and electrical properties of the HfO_2/Ge without and with La_2O_3 interfacial passivation layer [41]. HRTEM images of (a) HfO_2/Ge and (b) HfO_2/La_2O_3/Ge. Comparison of (c) high frequency C-V and (d) J-V curves for HfO_2/Ge and HfO_2/La_2O_3/Ge. (Copyright 2013. Reproduced with permission from Elsevier.)

HfO_2/Ge without La_2O_3 interfacial passivation layer shows unsaturated *C-V* curves, which is caused by high interface trap density. The CET value of HfO_2/La_2O_3/Ge is about 1.35 nm, which is much smaller than that of 1.67 nm for HfO_2/Ge without passivation. More importantly, HfO_2/La_2O_3/Ge also shows a much smaller leakage current density of 8.3×10^{-4} A cm^{-2} than HfO_2/Ge (1.1×10^{-1} A cm^{-2}). Normally, the samples with smaller CET value exhibit a larger leakage current. However, the MOS capacitors with La_2O_3 interfacial passivation layer possess both smaller CET and much lower leakage current. This is attributed to the improved interfacial quality of Ge/HfO_2 after introducing an La_2O_3 interfacial passivation layer. The samples without an La_2O_3 interfacial passivation layer exhibit poor interface with the presence of defective interfacial oxide of $HfGeO_x$, lowering the conduction-band offset between HfO_2 and Ge [42]. Therefore, the La_2O_3 interfacial passivation layer greatly improves the interface quality of HfO_2/Ge by inhibiting Ge out-diffusion and formation of interfacial defects.

9.2.3.3 Al_2O_3

Although various strategies have been applied to improve the interfacial properties of Ge based devices, it is still challenging to obtain small CET with low D_{it}. For the above goal, Al_2O_3 with wonderful thermal stability and large bandgap was used as the interfacial passivation layer for HfO_2/Ge. A thin interfacial passivation layer of Al_2O_3 (1.2 nm) prepared by ALD was deposited on $(NH_4)_2S$ passivated Ge prior to *in situ* ALD-HfO_2 dielectrics deposition [43]. Figure 9.5(a) shows the Ge 3d XPS spectra without deposition and with deposition of 1.2 nm Al_2O_3 and HfO_2 thin films on the Ge substrates. It can be seen that no Ge oxide signal can be detected on as-cleaned Ge substrate after $(NH_4)_2S$ chemical passivation. After ALD Al_2O_3 process, there is still no signal for Ge oxides. However, there is an obvious signal for Ge oxide that can be observed after an ALD HfO_2 process on Ge substrate. This result indicates that direct deposition of HfO_2 would results in the formation of defective Ge oxides at interface, while Al_2O_3 layer can efficiently suppress the above phenomenon, thus improving the interface quality of HfO_2/Ge.

As discussed above, a nonuniform interfacial layer $HfGeO_x$ will be formed at the interface of HfO_2/Ge after PDA. Thus, adopting ultrathin Al_2O_3 as the interfacial passivation layer can improve the interfacial quality of Al_2O_3/Ge, as presented in Figure 9.5(b), where a flat and sharp interface can be achieved. The HRTEM image of HfO_2/Al_2O_3/Ge demonstrated that Al_2O_3 passivation layer can prevent the formation of low-*k* and defective interfacial layer at the interface of HfO_2/Ge. In order to achieve small CET value, an ultrathin HfO_2 of 3 nm was *in situ* deposited in ALD chamber after Al_2O_3. 4.2 nm HfO_2 on Ge was also prepared for comparison. Figure 9.5(c) compares high frequency (1 MHz) C-V curves for the samples with and without an Al_2O_3 interfacial passivation layer. HfO_2/Al_2O_3/Ge samples exhibit significantly improved C-V shape compared to the HfO_2/Ge sample, such as less stretched-out and higher accumulation capacitance, implying lower D_{it}. Moreover, the leakage current density of the HfO_2/Al_2O_3/Ge sample is only 1.22×10^{-3} A cm^{-2} at +1 V, which is much lower than that of 3.06×10^{-1} A cm^{-2} for the HfO_2/Ge sample, as plotted in the inset of Figure 9.5(c).

The D_{it} values of HfO_2/Ge and HfO_2/Al_2O_3/Ge were measured to quantitatively explore the influence of an Al_2O_3 interfacial passivation layer on improving the interface quality of HfO_2/Ge, as shown in Figure 9.5(d). The D_{it} values of HfO_2/Ge and HfO_2/Al_2O_3/Ge are determined to be 8.61×10^{12} and 2.83×10^{12} eV^{-1} cm^{-2}, respectively. Furthermore, the V_{FB} hysteresis of HfO_2/Ge and HfO_2/Al_2O_3/Ge is found to be 577 mV and 189 mV, respectively, which also confirms the better interface quality that exists in the capacitor with the Al_2O_3 interfacial passivation layer. Larger V_{FB} hysteresis indicates more bulk traps at the interface between HfO_2 and Ge, which is caused by Ge out-diffusion and the defective interfacial layer of $HfGeO_x$.

Above results demonstrate that ultrathin Al_2O_3 interfacial passivation layer between HfO_2 gate dielectrics and Ge substrates can improve the interface quality and electrical properties. It can be ascribed to the wonderful passivation effect of the Al_2O_3 interfacial layer, suppressing the Ge out-diffusion and defective interfacial layer formation. As a result, the double layer dielectrics of HfO_2/Al_2O_3 on $(NH_4)_2S$ passivated Ge can achieve the CET value as low as 1.2 nm, with small leakage.

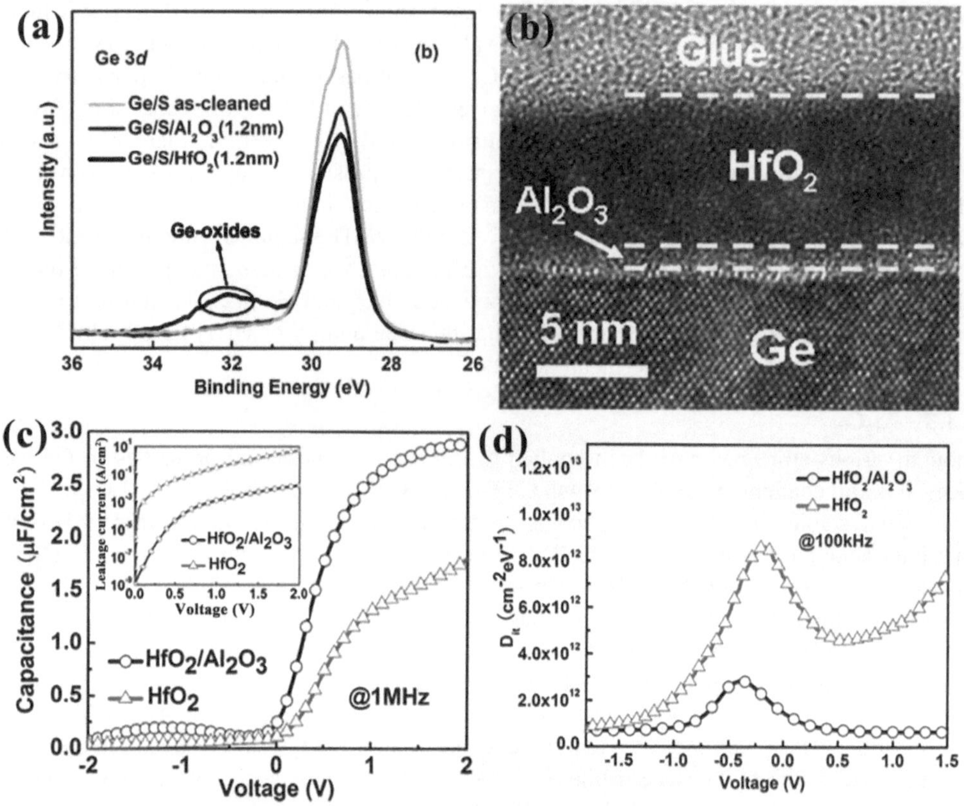

FIGURE 9.5 Interfacial and electrical properties of the HfO_2/Ge without and with Al_2O_3 interfacial passivation layer [43]. (a) Ge 3d XPS spectra of S passivated Ge without and with 1.2 nm Al_2O_3 or HfO_2. (b) HRTEM image of HfO_2/Al_2O_3/Ge. Comparison of (c) high-frequency C-V and (d) Dit for HfO_2/Ge and HfO_2/Al_2O_3/Ge. The insert of (c) is the J-V curves. (Copyright 2012. Reproduced with permission from IOP.)

9.3 HIGH-K DIELECTRICS ON GaAs

9.3.1 Chemical Passivation

Two different chemical passivation pretreatments using NH_4OH and $(NH_4)_2S$ were used to passivate GaAs substrates [44]. GaAs was firstly cleaned by diluted HCl aqueous solution. Then, the cleaned GaAs substrates were soaked in $(NH_4)_2S$ and NH_4OH aqueous solution for 30 minutes for passivation. Figure 9.6(a) shows the XPS spectra of Ga $2p_{3/2}$ and As $2p_{3/2}$ on the surface of GaAs before and after various chemical passivations. For the untreated GaAs, naturally formed gallium oxides can be detected at 1118.2 eV, which may contain various oxidation state information, such as Ga^{5+}, Ga^{3+}, and Ga low oxide. Similarly, in the XPS spectra of As $2p_{3/2}$, As oxides on the untreated GaAs surface are mainly As_2O_3 and As_2O_5. After the sample was cleaned with HCl, the intensity of Ga oxides on the GaAs surface was obviously weakened. In addition, the peak of Ga-O bonding moved to the low oxidation state. At the same time, As oxide was also reduced and As^{5+} oxide was completely removed. Then, the passivation effect of NH_4OH and $(NH_4)_2S$ is compared. For NH_4OH passivated samples, there is no obvious change for As^{3+} oxide. However, for samples pretreated with $(NH_4)_2S$, As^{3+} oxide can be further reduced. It is noteworthy that the peak of original As-As bonding moves to a position about 1.5 eV larger than the As-Ga bonding, indicating the formation of As-S bonding on the surface of GaAs after $(NH_4)_2S$ pretreatment. Moreover, the signal of Ga-O bonds after passivation of $(NH_4)_2S$ is slightly enhanced, which can indirectly prove the formation of Ga-S

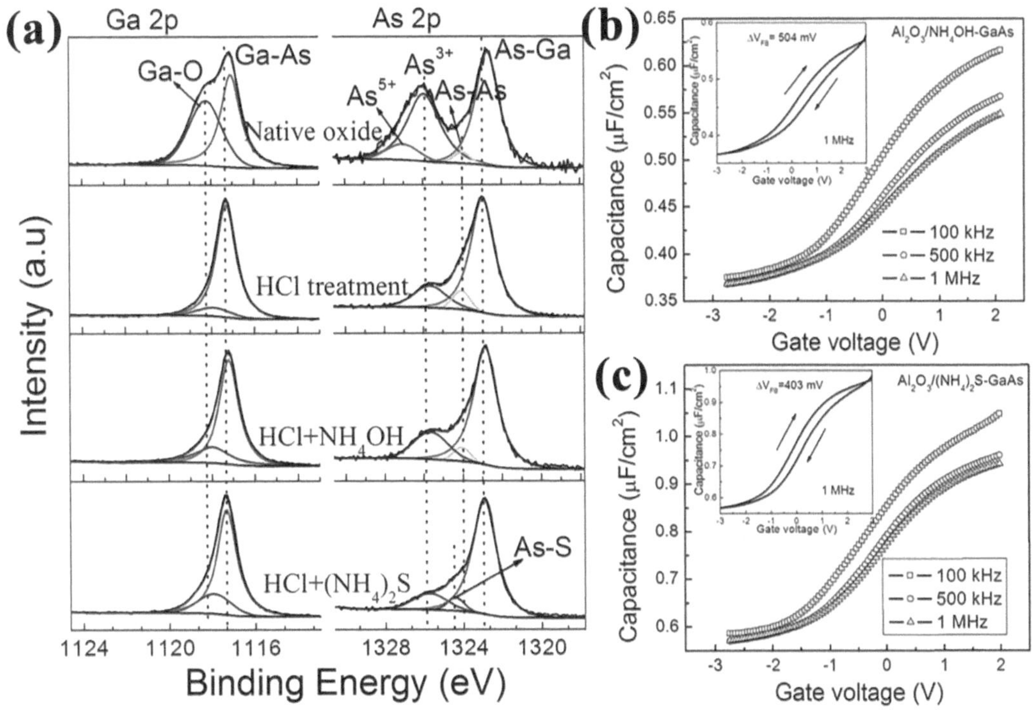

FIGURE 9.6 Interfacial and electrical properties of the GaAs with NH_4OH or $(NH_4)_2S$ passivation [44]. (a) Ga 2p and As 2p XPS spectra of GaAs surface with various chemical pretreatments. C-V curves obtained at various frequencies for (b) NH_4OH and (c) $(NH_4)_2S$ passivated GaAs. The inserts show the capacitor hysteresis. (Copyright 2011. Reproduced with permission from Wiley.)

bonding. Therefore, $(NH_4)_2S$ pretreatment can passivate the GaAs surface with Ga-S and As-S bonding.

Figure 9.6(b), (c) compares the C-V curves of Pt/Al_2O_3/GaAs MOS devices after two different chemical passivation. The insertion exhibits the hysteresis characteristics of C-V curves at 1 MHz. Compared to a NH_4OH passivated sample, the $(NH_4)_2S$ pretreated sample possesses higher accumulation capacitance and lower-frequency dispersion. The ΔV_{FB} of NH_4OH and $(NH_4)_2S$ passivated samples are 504 mV and 403 mV, respectively, indicating that the trap between dielectrics and $(NH_4)_2S$ pretreated GaAs is reduced. Moreover, the leakage current of the $(NH_4)_2S$ passivated sample can be slightly reduced. Compared with NH_4OH, $(NH_4)_2S$ can passivate the GaAs surface with Ga-S and As-S bonding, which can inhibit the mutual diffusion between Al_2O_3 film and GaAs substrate. The quality of the dielectrics/GaAs interface can be significantly improved. As a result, Pt/Al_2O_3/GaAs capacitor shows better electrical characteristics, including higher accumulation capacitance, smaller frequency dispersion, and smaller capacitor hysteresis.

9.3.2 Self-Cleaning during ALD

For GaAs based MOS devices, it is important to get a clean interface without any native oxides. There was an interesting finding that some ALD processes can get rid of the native oxides of GaAs, which is called as the self-cleaning effect. This effect has attracted the researchers' attention. It has been demonstrated that Al_2O_3 and HfO_2 ALD processes can exhibit the self-cleaning effect for GaAs and InGaAs, where As and Ga oxides can be effectively reduced. For example, Ye *et al.* found that Al_2O_3 ALD process using TAM as a metal organic precursor can reduce the interfacial oxides for GaAs [45]. Huang et al. [46] and Chang et al. [47] have discovered TMA and [tetrakis (ethylmethyl-amino) hafnium] (TEMAH) in ALD process can remove the As oxides on InGaAs, respectively. Interestingly, Hinkle et al. have discovered that different metal organic precursors can remove different native oxides with a different oxidation state. It was demonstrated that TEMAH of Hf^{4+} can remove native oxides with higher oxidation states than TMA of Al^{3+} [48]. Therefore, the combination of Al^{3+} and Hf^{4+} metal organic precursors might have a better self-cleaning effect than alone. With this purpose, two different self-cleaning pretreatments using ALD precursors, TMA alone and TMA + [tetrakis (dimethyl-amino) hafnium] (TDMAH) combination, were conducted before HfO_2 deposition [49].

Figure 9.7(a) displays the XPS spectra of Ga 2*p* for 1-nm-thick Al_2O_3 on GaAs with various previous pretreatments. The XPS spectra reveal that untreated GaAs surface has the highest amount of 9.88% Ga-O and 7.83% Ga-O_x/Ga-S, and a TMA pretreated one has a moderate amount of 8.47% Ga-O and 5.0% Ga-O_x/Ga-S, as expected, TMA + TDMAH combination pretreated one exhibits the least amount of 7.12% Ga-O and 3.96% Ga-O_x/Ga-S. Above XPS results indicate that both TMA and TDMAH metal organic precursors can remove the Ga oxides. Moreover, the combination of TMA+TDMAH pretreatment is more effective than the TMA alone. Similarly, XPS spectra of As 2p in Figure 9.7(b) also indicates that TMA + TDMAH combination exhibits a better self-cleaning effect in removing As oxides than TMA. Thus, a TMA + TDMAH combination pretreatment can achieve the cleanest GaAs surface, which is beneficial to obtain the high-quality interface of high-K/GaAs. The self-cleaning phenomenon can be explained using the ligand-exchange reaction mechanism, where both TMA and TDMAH can react with As or Ga oxides (such as As_2O_3 and Ga_2O_3) to generate volatile products (e.g. $As(CH_3)_3$ and $Ga(CH_3)_3$), getting rid of these defective oxides.

The HRTEM images of HfO_2/Al_2O_3/GaAs stacks with different ALD precursor pretreatments prior to HfO_2/Al_2O_3 deposition are presented in Figure 9.7(c–e). All the three samples exhibit the

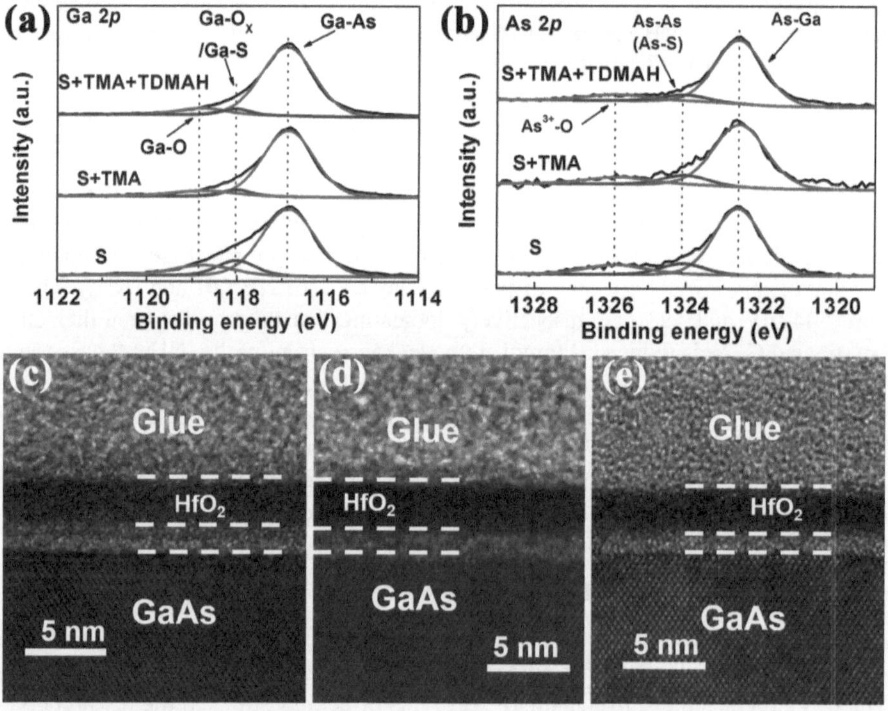

FIGURE 9.7 Interfacial properties of the GaAs with different pretreatments using ALD precursors [49]. (a) Ga 2p and (b) As 2p XPS spectra of GaAs surface with various chemical pretreatments. HRTEM image of HfO_2/Al_2O_3 on GaAs: (c) untreated, (d) TMA pretreatment, and (e) TMA+TDMAH combination pretreatment. (Copyright 2012. Reproduced with permission from Elsevier.)

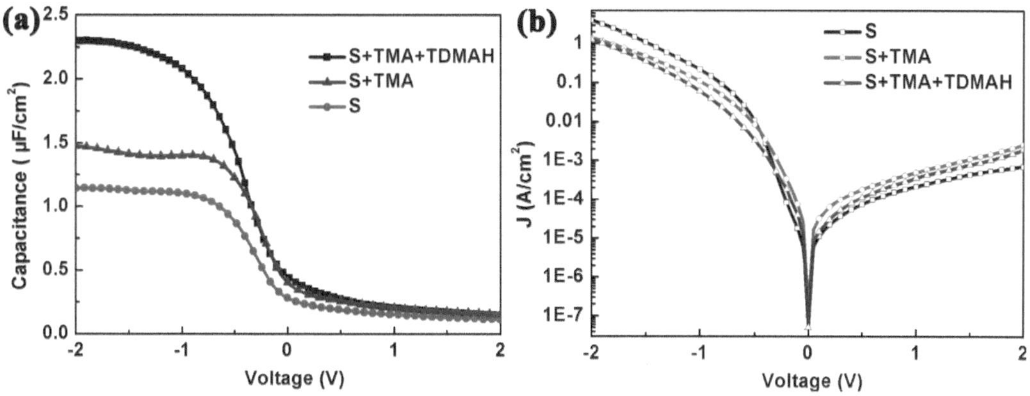

FIGURE 9.8 Electrical properties of the $HfO_2/Al_2O_3/GaAs$ with different pretreatments using ALD precursors [49]. (a) High-frequency C-V curves and (b) J-V curves. (Copyright 2012. Reproduced with permission from Elsevier.)

smooth interface with the thin films that are amorphous. It should be noted that HfO_2 is about 2.8 nm for all the three samples. Nevertheless, there is a 1.9 nm, 1.5 nm, and 1.2 nm interfacial layer existing between HfO_2 and GaAs, for the untreated, TMA pretreated, and TMA +TDMAH pretreated sample, respectively. Assuming that ALD deposited Al_2O_3 is ~1 nm for all the samples, therefore, the corresponding native oxides at the interface are around 0.9 nm, 0.5 nm, and 0.2 nm. These results are consistent with the above XPS data, TMA + TDMAH combination pretreatment that is the most effective to remove the native Ga/As oxide via the self-cleaning effect, achieving the thinnest interfacial layer on GaAs.

Figure 9.8(a) presents C-V curves of GaAs MOS capacitors at 1 MHz, with 1-nm-thick Al_2O_3/2.8-nm-thick HfO_2 gate stacks dielectrics. The accumulation capacitance for the TMA + TDMAH pretreated sample, TMA pretreated one, and untreated one is 2.29 µF cm^{-2}, 1.48 µF cm^{-2}, and 1.15 µF cm^{-2}, respectively. The untreated sample exhibits the lowest accumulation capacitance due to the existence of the thickest As and Ga native oxides layer, with low permittivity and poor properties. After TMA or TMA + TDMAH combination pretreatments, the accumulation capacitance increases, due to consuming these defective native oxides. Moreover, compared to TMA or TMA+TDMAH pretreated samples, the untreated one exhibits more serious stretch-out in C-V curves. This is attributed to poor interface quality for the untreated sample with a more defective interfacial layer. In the other words, the TMA + TDMAH combination treatment can result in the best interface quality for HfO_2/GaAs. Figure 9.8(b) plots the leakage current characteristic curves for all the samples, where the values at -1 V are almost same. Considering that the thickness of gate dielectrics is the thinnest for the TMA + TDMAH pretreated sample, the combination of metal organic precursors pretreatment should be more effective than TMA alone.

It is demonstrated that the pretreatment with the combination of TMA and TDMAH is more effective in "self-cleaning" a GaAs surface than alone, which can effectively reduce the As-O and Ga-O bonds on the GaAs surface. Thus, the interface quality of the pretreated sample is better than the TMA pretreated one and untreated one. Therefore, a much better electrical performance can be achieved. It can be assumed that proper combination of different ALD precursors may result in an even better self-cleaning effect.

9.3.3 Interfacial Passivation Layer

9.3.3.1 Gd_2O_3

ZrO_2 is the other promising gate dielectrics candidate to replace SiO_2, which exhibits the large band gap and high dielectric constant. However, similar to HfO_2, poor interface quality with high

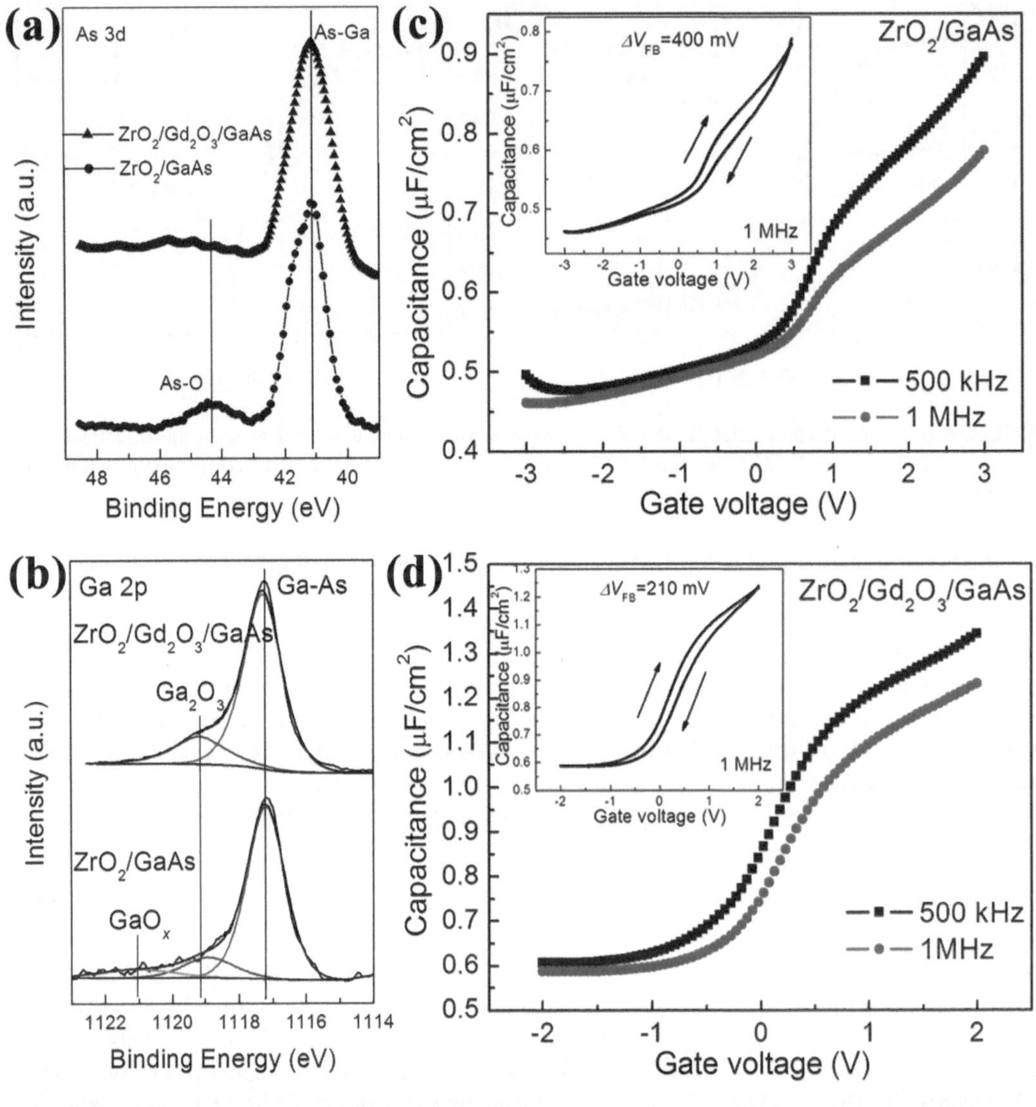

FIGURE 9.9 Interfacial and electrical properties of the GaAs with or without Gd_2O_3 interfacial passivation layer [52]. (a) As 3d and (b) Ga 2p XPS spectra of ZrO_2/GaAs and ZrO_2/Gd_2O_3/GaAs. C-V curves obtained at 1 MHz and 500 kHz for (c) ZrO_2/GaAs and (d) ZrO_2/Gd_2O_3/GaAs. The inserts show the capacitor hysteresis. (Copyright 2014. Reproduced with permission from Elsevier.)

interface traps density is the problem when depositing ZrO_2 directly on GaAs substrate [50, 51]. Thus, an ultrathin Gd_2O_3 film prepared by MOCVD was adopted as the interfacial passivation layer for the ZrO_2/GaAs interface [52]. Figure 9.9(a), (b) display XPS spectra of As 3d and Ga 2p, where As-O bonding located at 44.3 eV can be detected in ZrO_2/GaAs. After introducing the Gd_2O_3 passivation layer, no As-O signal can be observed at the same position in ZrO_2/Gd_2O_3/GaAs. Similarly, the high valence GaO_x signal is also not detected in ZrO_2/Gd_2O_3/GaAs, as illustrated in Figure 9.9(b). XPS results imply that an ultrathin Gd_2O_3 interfacial passivation layer is able to inhibit the generation of both As and Ga oxides at the interface of ZrO_2/GaAs.

Figure 9.9(c), (d) display the C-V curves (@ 1 MHz and 500 kHz) of ZrO_2/GaAs and ZrO_2/Gd_2O_3/GaAs capacitors. Serious stretched-out can be observed in the C-V curves for sample ZrO_2/GaAs. This is mainly ascribed to the presence of native As and Ga oxides, with poor quality at the interface of ZrO_2/GaAs. On the contrary, ZrO_2/Gd_2O_3/GaAs stacks exhibit dramatically improved electrical performances, including higher accumulation capacitance and smaller C-V hysteresis. Thus, Gd_2O_3 incorporation is able to suppress the trap defect and movable charge at the interface. In addition, the leakage current density of ZrO_2/Gd_2O_3/GaAs is much lower than decreased ZrO_2/GaAs, at least by an order of magnitude.

9.3.3.2 Comparison of Al_2O_3 and AlN Interfacial Passivation Layer

It has been demonstrated above that Al_2O_3 is a wonderful interfacial passivation layer for dielectrics on Ge. Herein, Al_2O_3 was also used as an interfacial passivation layer for GaAs. In addition, PEALD AlN was also used as interfacial passivation layer in the HfO_2/GaAs stack for comparison. Both HfO_2/AlN and HfO_2/Al_2O_3 stack dielectrics on GaAs were explored to compare their thermal stability and electrical performance [53]. PDA processes at different temperatures (400°C, 500°C, and 600°C) were applied for both Al_2O_3/GaAs and AlN/GaAs. Figure 9.10(a)(b) illustrate the composition change of the As element after different PDA temperatures. For Al_2O_3/GaAs stack, there is an obvious peak for As-O bonding. Moreover, this peak dramatically increases after 500°C PDA, implying that plenty of As oxides formed during the 500°C PDA process. Nevertheless, this signal corresponding to As-O bonding significantly decreases at the higher temperature of 600°C. This is attributed to As_2O_3 not being stable, which would react with GaAs substrate during annealing as follows: $As_2O_3 + 2GaAs \rightarrow Ga_2O_3 + 4As$. The above phenomenon indicates that the Al_2O_3/GaAs interface is not thermal stable, where not only As oxides generate during PDA process, but also As oxides would decompose at higher temperature. Both cases will impair the performance of GaAs based capacitors. On the contrary, the AlN/GaAs interface is much more stable, as presented in Figure 9.10(b). There is only one peak that can be detected for both as-deposited and 500°C PDA treated samples, indicating a very clean GaAs surface without any As oxides. It also demonstrates that the AlN/GaAs interface is very stable at 500°C. Further increasing the PDA temperature to

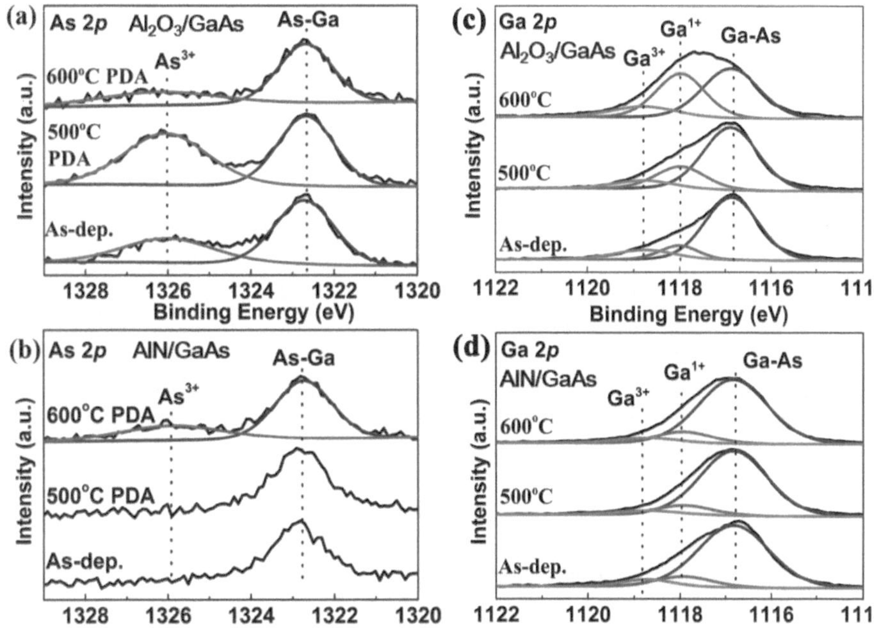

FIGURE 9.10 The thermal stability of the interface for Al_2O_3/GaAs and AlN/GaAs after 500°C and 600°C PDA [53]. As 2p XPS spectra for (a) Al_2O_3/GaAs and (b) AlN/GaAs, Ga 2p XPS spectra for (c) Al_2O_3/GaAs and (d) AlN/GaAs. (Copyright 2015. Reproduced with permission from American Vacuum Society.)

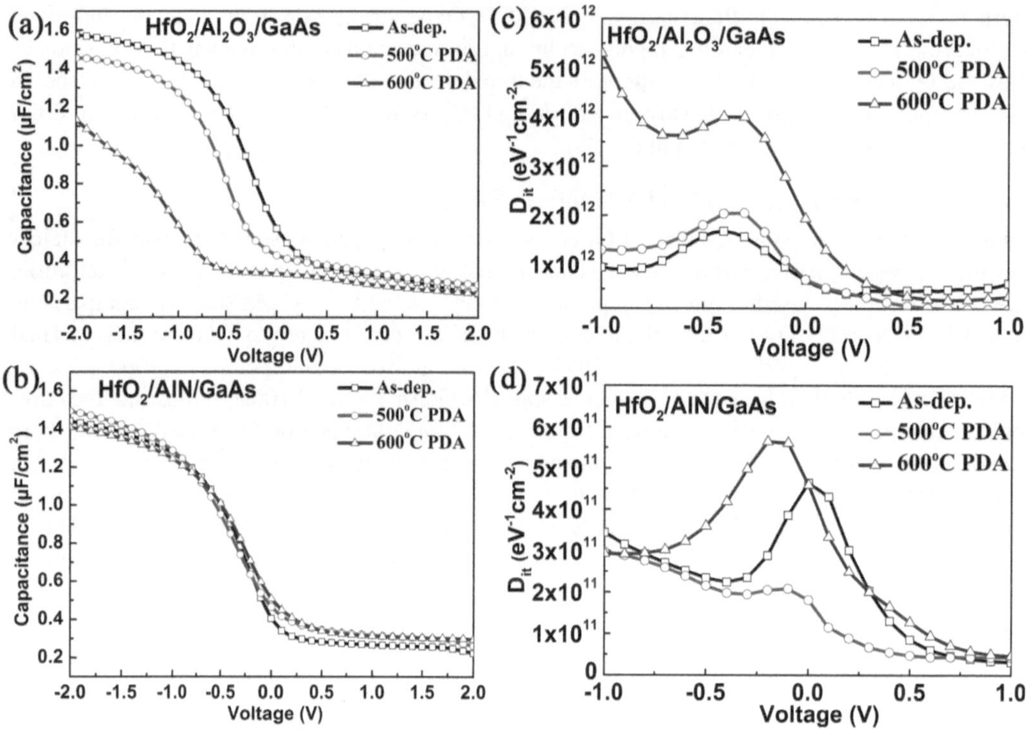

FIGURE 9.11 The thermal stability of electrical properties for $HfO_2/Al_2O_3/GaAs$ and $HfO_2/AlN/GaAs$ after 500°C and 600°C PDA [53]. High-frequency C-V curves for (a) $HfO_2/Al_2O_3/GaAs$ and (b) $HfO_2/AlN/GaAs$, D_{it} distribution (c) $HfO_2/Al_2O_3/GaAs$ and (d) $HfO_2/AlN/GaAs$. (Copyright 2015. Reproduced with permission from American Vacuum Society.)

600°C, there appears a weak signal of As oxides. It could be ascribed to the fact that GaAs reacts with oxygen at extremely high temperatures, which is the oxidizing impurities in annealing ambient, resulting in the formation of As-O products. XPS spectra of Ga 2p also demonstrate that the thermal stability of the $Al_2O_3/GaAs$ interface is weak, as displayed in Figure 9.10(c), (d), plenty of Ga oxides formed during both the 500°C and 600°C PDA processes. However, the AlN/GaAs has good thermal stability without obvious change in the intensity of Ga oxide peaks after PDA.

According to the above discussion in 9.3.2, native oxides on GaAs can be consumed by TMA during ALD Al_2O_3 process, which is called a self-cleaning effect. Further comparing the surface chemical composition for the as-deposited $Al_2O_3/GaAs$ and AlN/GaAs, the interface of AlN/GaAs is much cleaner than $Al_2O_3/GaAs$. Therefore, it can be assumed that NH_3 plasma may also exhibit the self-cleaning effect for GaAs. Thus, GaAs was treated by NH_3 plasma alone, where XPS results reveal that only Ga-N signal can be detected. Thus, the following reactions can be used to explain the self-cleaning effect of NH_3 plasma for GaAs [54].

$$12\,H + As_2O_3 \rightarrow 3\,H_2O + 2\,AsH_3 \tag{9.1}$$

$$2\,NH_3 + Ga_2O_3 \rightarrow 6\,H_2O + 2\,GaN \tag{9.2}$$

$$NH_3 + GaAs \rightarrow AsH_3 + GaN \tag{9.3}$$

Figure 9.11(a) displays the C-V curves at 1 MHz for $HfO_2/Al_2O_3/GaAs$ after different PDA temperatures. For $HfO_2/Al_2O_3/GaAs$ stacks, the C-V curves degrade dramatically, along with increasing the PDA temperature. At the same time, the corresponding CET values of $HfO_2/Al_2O_3/GaAs$ increases from 2.19 nm (as-deposited) to 2.36 nm (500°C PDA) and 3.03 nm (600°C PDA). In comparison, the C-V curves of HfO_2/AlN/GaAs samples are much more stable during PDA processes, as presented in Figure 9.11(b), where the corresponding CET values exhibit almost no change with 2.39 nm (as-deposited), 2.31 (500°C PDA) nm, and 2.45 nm (600°C PDA). Furthermore, after 500°C PDA, the sample with the AlN interfacial passivation layer exhibits a much lower leakage current of 1.72×10^{-4} A cm^{-2} than the one of 1.46×10^{-1} A cm^{-2} with Al_2O_3. Therefore, AlN interfacial layers can dramatically suppress the degradation of electrical properties of a GaAs based MOS capacitor, due to its wonderful stability.

The interface trap density was measured to quantitatively explore the effect of PDA processes on interface quality for both HfO_2/Al_2O_3/GaAs and HfO_2/AlN/GaAs stacks. Figure 9.11(c), (d) show the normalized conductance-voltage curves measured at 100 kHz for HfO_2/Al_2O_3/GaAs and HfO_2/AlN/GaAs stacks after various PDA processes. For HfO_2/Al_2O_3/GaAs, the D_{it} value significantly increases after PDA processes. On the contrary, the interface quality of HfO_2/AlN/GaAs stacks do not degrade after PDA treatments. Moreover, HfO_2/AlN/GaAs after 500°C PDA process exhibits the lowest D_{it} of 2.11×10^{11} $eV^{-1}cm^{-2}$, which is much lower than that of 2.08×10^{12} $eV^{-1}cm^{-2}$ for 500°C annealed HfO_2/Al_2O_3/GaAs. It indicates that HfO_2/AlN/GaAs exhibits a much improved interface quality than HfO_2/Al_2O_3/GaAs.

The above results demonstrate that AlN is a better interfacial passivation layer for GaAs than Al_2O_3. HfO_2/AlN stacks exhibit wonderful thermal stability in suppressing the formation of interfacial oxide and inhibiting the degradation of electrical performances. As a result, HfO_2/AlN/GaAs stacks after 500°C PDA exhibit the smallest CET and the lowest D_{it} value. The above results demonstrate that AlN is a promising interfacial passivation layer for GaAs, which can not only enhance the thermal stability but also improve the electrical performance.

9.4 CONCLUSIONS AND DISCUSSIONS

Tables 9.1 and 9.2 summarize the electrical properties of our Ge and GaAs based MOS capacitors with different interface engineering strategies, respectively. For chemical pretreatments, $(NH_4)_2S$ solution is effective to form S-bonding on both Ge and GaAs, passivating the surface. However, the interface between high-K dielectrics (HfO_2 or ZrO_2) and S passivated Ge or GaAs is not stable, a defective interfacial layer can still form at the interface, degrading the electrical performances. Therefore, the chemical passivation should be combined with the interfacial passivation layer. For HfO_2/Ge, it has been demonstrated that SiO_2, La_2O_3, and Al_2O_3 interfacial passivation layers can effectively suppress the Ge out-diffusion and the formation of a defective interfacial layer. Herein, the *in situ* ALD formed interfacial passivation layers are much more promising than other technologies. It can be ascribed to the fact that ALD derived interfacial passivation layers are more uniform, especially at ultrathin thickness. Besides, an ultrathin N incorporated Ge oxides (GeO_xN_y) as an

TABLE 9.1
Summary of Interfacial Engineering of High-K Dielectrics on Ge

Strategies	Dielectrics	CET (nm)	Leakage@+1V (A cm^{-2})
HBr+$(NH_4)_2S$	10 nm Al_2O_3	5.06	8.41×10^{-6}
HF+$(NH_4)_2S$	10 nm Al_2O_3	5.50	7.14×10^{-6}
NH_3 plasma	3 nm HfO_2	0.96	1.12×10^{-3}
0.7 nm SiO_2(PECVD)	4 nm HfO_2	1.53	2.1×10^{-3} [a]
1 nm SiO_2(MOCVD)	4 nm HfO_2	1.75	3.9×10^{-3} [a]
1.5 nm La_2O_3	4 nm HfO_2	1.35	8.3×10^{-4} [a]
1.2 nm Al_2O_3	3 nm HfO_2	1.2	1.22×10^{-3}

[a] The leakage current was obtained at $V_{FB+}1$ V.

TABLE 9.2
Summary of Interfacial Engineering of High-K Dielectrics on GaAs

Strategies	Dielectrics	CET (nm)	Leakage@-1V (A cm^{-2})
HCl+NH_4OH	10 nm Al_2O_3	6.27	8.41×10^{-6}
HCl+$(NH_4)_2S$	10 nm Al_2O_3	3.63	7.14×10^{-6}
TMA	1 nm $Al_2O_3$2.8 nm HfO_2	2.33	1.12×10^{-1}
TMA+TDMAH	1 nm $Al_2O_3$2.8 nm HfO_2	1.51	6.2×10^{-2}
1 nm Al_2O_3	5 nm HfO_2	2.36	1.46×10^{-1}
1 nm AlN	5 nm HfO_2	2.31	1.72×10^{-4}
1 nm Gd_2O_3	5 nm HfO_2	2.76	1.2×10^{-4}

interfacial passivation layer can be achieved via an *in-situ* NH_3 plasma treatment in PEALD, which exhibits good thermal stability and wonderful interface quality. Therefore, we successfully obtained the Ge based MOS capacitor with sub nanometer CET of 0.96 nm, small current density of 1.12×10^{-3} A cm^{-2}, and moderate D_{it} of 3.42×10^{12} cm^{-2} eV^{-1}.

Compared to Ge based MOS capacitors, it is challenging to achieve small CET for GaAs based MOS capacitors, where the interfacial chemistry between high-K dielectrics and GaAs should be well understood. Among various passivation strategies, the self-cleaning effect deserves more attentions. Metal precursors used in ALD processes can consume the native oxides of GaAs, moreover, metal precursors with different valence states can remove oxides with different valence states. Besides, plasma in PEALD (e.g. NH_3) can also exhibit a self-cleaning effect. Therefore, the pretreatment using the proper combination of ALD dosing (metal organic precursors or plasma) may be hopeful in removing the native oxides of GaAs. Furthermore, AlN is much better than Al_2O_3 as an interfacial passivation layer for GaAs, where a much better thermal stability can be achieved.

In summary, ALD is considered to be the most promising technology to integrate high-K dielectrics with high mobility channel materials (e.g., Ge, GaAs). First, ALD exhibits outstanding characteristics of excellent thickness control, wonderful three-dimensional conformality, and the low deposition temperature, making it promising in depositing ultrathin high-K dielectrics on Ge and III-V substrates. More fortunately, ALD technology can engineer the interface between high-K dielectrics and Ge/III-V substrates via *in situ* plasma pretreatment, self-cleaning treatment of ALD precursors, and introducing various interfacial passivation layers.

ACKNOWLEDGMENTS

This work was supported in part by the Natural Science Foundation of China (51802150, 52073142, and 51721001) and Jiangsu Province (BK20201252).

REFERENCES

1. Akinwande, D., Huyghebaert, C., Wang, C.-H., Serna, M. I., Goossens, S., Li, L.-J., Wong, H. S. P., and Koppens, F. H. L. Graphene and two-dimensional materials for silicon technology. *Nature* 2019, *573*, 507–518.
2. Robertson, J., and Wallace, R. M. High-K materials and metal gates for CMOS applications. *Materials Science and Engineering: R: Reports* 2015, *88*, 1–41.
3. Wallace, R. M., McIntyre, P. C., Kim, J., and Nishi, Y. Atomic layer deposition of dielectrics on Ge and III-V materials for ultrahigh performance transistors. *MRS Bulletin* 2009, *34*, 493–503.
4. Bennett, B. R., Ancona, M. G., and Boos, J. B. Compound semiconductors for low-power p-channel field-effect transistors. *MRS bulletin* 2009, *34*, 530–536.

5. Hong, M., Kwo, J., Lin, T., and Huang, M. InGaAs metal oxide semiconductor devices with Ga_2O_3 (Gd_2O_3) high-κ dielectrics for science and technology beyond Si CMOS. *MRS Bulletin* 2009, *34*, 514–521.
6. Portilla, L., Zhao, J., Zhao, J., Occhipinti, L. G., and Pecunia, V. Ambipolar carbon nanotube transistors with hybrid nanodielectric for low-voltage CMOS-like electronics. *Nano Futures* 2021, *5*, 025001.
7. Xie, Y., and Zhang, Z. Carbon nanotube-based CMOS transistors and integrated circuits. *Science China Information Sciences* 2021, *64*, 1–16.
8. Geim, A. K. Graphene: status and prospects. *Science* 2009, *324*, 1530–1534.
9. Goossens, S., Navickaite, G., Monasterio, C., Gupta, S., Piqueras, J. J., Pérez, R., Burwell, G., Nikitskiy, I., Lasanta, T., and Galán, T. Broadband image sensor array based on graphene-CMOS integration. *Nature Photonics* 2017, *11*, 366–371.
10. Wan, H.-W., Hong, Y.-J., Cheng, Y.-T., Cheng, C.-K., Hsu, C.-H., Wu, C.-T., Pi, T.-W., Kwo, J., and Hong, M. Low-Temperature-Grown Single-Crystal Si Epitaxially on Ge, Followed by Direct Deposition of High-κ Dielectrics-Attainment of Low Interfacial Traps and Highly Reliable Ge MOS. *ACS Applied Electronic Materials* 2021, *3*, 2164–2169.
11. Onik, T. A. M., Hawari, H. F., Sabri, M. F. M., and Wong, Y. H. Growth mechanisms and characteristics of Sm_2O_3 based on Ge semiconductor through oxidation and nitridation. *Applied Surface Science* 2021, *544*, 148949.
12. Takagi, S., Zhang, R., Suh, J., Kim, S.-H., Yokoyama, M., Nishi, K., and Takenaka, M. III-V/Ge channel MOS device technologies in nano CMOS era. *Japanese Journal of Applied Physics* 2015, *54*, 06FA01.
13. Takagi, S., Zhang, R., Kim, S.-H., Taoka, N., Yokoyama, M., Suh, J.-K., Suzuki, R., and Takenaka, M. In *MOS interface and channel engineering for high-mobility Ge/III-V CMOS*, 2012 International Electron Devices Meeting, IEEE: 2012, pp 23.1. 1–23.1. 4.
14. Wang, D., He, G., Hao, L., Qiao, L., Fang, Z., and Liu, J. Interface Chemistry and Dielectric Optimization of TMA-Passivated high-K/Ge Gate Stacks by ALD-Driven Laminated Interlayers. *ACS Applied Materials & Interfaces* 2020, *12*, 25390–25399.
15. Prabhakaran, K., and Ogino, T. Oxidation of Ge (100) and Ge (111) surfaces: an UPS and XPS study. *Surface Science* 1995, *325*, 263–271.
16. Wang, X., Nishimura, T., Yajima, T., and Toriumi, A. Thermal oxidation kinetics of germanium. *Applied Physics Letters* 2017, *111*, 052101.
17. He, G., Li, W., Wei, H., Jiang, S., Xiao, X., Jin, P., and Gao, J. Modulation of electrical properties and current conduction mechanism of HfAlO/Ge gate stack by ALD-derived Al_2O_3 passivation layer. *Journal of Alloys and Compounds* 2017, *695*, 1591–1599.
18. Huang, Y., Xu, J.-P., Liu, L., Cheng, Z.-X., Lai, P.-T., and Tang, W.-M. Improved interfacial and electrical properties of HfTiON gate-dielectric Ge MOS capacitor by using LaON/Si dual passivation layer and fluorine-plasma treatment. *Applied Surface Science* 2019, *493*, 628–633.
19. Zhao, L., Liu, H., Wang, X., Wang, Y., and Wang, S. Electrical properties and interfacial issues of HfO_2/Ge MIS capacitors characterized by the thickness of La_2O_3 interlayer. *Nanomaterials* 2019, *9*, 697.
20. Li, Q., Fang, S., Liu, S., Xu, L., Xu, L., Yang, C., Yang, J., Shi, B., Ma, J., Yang, J., Quhe, R., and Lu, J. Performance Limit of Ultrathin GaAs Transistors. *ACS Applied Materials & Interfaces* 2022, *14*, 23597–23609.
21. Heyns, M., and Tsai, W. Ultimate scaling of CMOS logic devices with Ge and III–V materials. *Mrs Bulletin* 2009, *34*, 485–492.
22. Li, X., Li, A., Liu, X., Gong, Y., Chen, X., Li, H., and Wu, D. Effect of chemical surface treatments on interfacial and electrical characteristics of atomic-layer-deposited Al_2O_3 films on Ge substrates. *Applied Surface Science* 2011, *257*, 4589–4592.
23. McKee, R. A., Walker, F., and Chisholm, M. Crystalline oxides on silicon: the first five monolayers. *Physical Review Letters* 1998, *81*, 3014.
24. Maeda, T., Nishizawa, M., Morita, Y., and Takagi, S. Role of germanium nitride interfacial layers in HfO_2/germanium nitride/germanium metal-insulator-semiconductor structures. *Applied Physics Letters* 2007, *90*, 072911.
25. Darmawan, P., Chan, M., Zhang, T., Setiawan, Y., Seng, H., Chan, T., Osipowicz, T., and Lee, P. S. Lu_2O_3/Al_2O_3 gate dielectrics for germanium metal-oxide-semiconductor devices. *Applied Physics Letters* 2008, *93*, 062901.
26. Chui, C. O., Ito, F., and Saraswat, K. C. Scalability and electrical properties of germanium oxynitride MOS dielectrics. *IEEE Electron Device Letters* 2004, *25*, 613–615.

27. Lu, N., Bai, W., Ramirez, A., Mouli, C., Ritenour, A., Lee, M., Antoniadis, D., and Kwong, D. Ge diffusion in Ge metal oxide semiconductor with chemical vapor deposition HfO_2 dielectric. *Applied Physics Letters* 2005, *87*, 051922.
28. Cao, Y.-Q., Chen, J., Liu, X.-J., Li, X., Cao, Z.-Y., Ma, Y.-J., Wu, D., and Li, A.-D. $HfO_2/GeO_xN_y/Ge$ gate stacks with sub-nanometer capacitance equivalent thickness and low interface trap density by in situ NH_3 plasma pretreatment. *Applied Surface Science* 2015, *325*, 13–19.
29. Xie, Q., Deduytsche, D., Schaekers, M., Caymax, M., Delabie, A., Qu, X.-P., and Detavernier, C. Effective electrical passivation of Ge (100) for HfO_2 gate dielectric layers using O_2 plasma. *Electrochemical and Solid-State Letters* 2011, *14*, G20.
30. Mavrou, G., Galata, S., Tsipas, P., Sotiropoulos, A., Panayiotatos, Y., Dimoulas, A., Evangelou, E., Seo, J. W., and Dieker, C. Electrical properties of La_2O_3 and HfO_2/La_2O_3 gate dielectrics for germanium metal-oxide-semiconductor devices. *Journal of Applied Physics* 2008, *103*, 014506.
31. Nicollian, E. H., and Brews, J. R. *MOS (metal oxide semiconductor) physics and technology*. John Wiley & Sons: 2002.
32. Lin, C.-C., Wu, Y.-H., Wu, C.-Y., and Lee, C.-W. Surface Passivation of Ge MOS Devices by $SmGeO_x$ With Sub-nm EOT. *IEEE Electron Device Letters* 2014, *35*, 384–386.
33. Caymax, M., Van Elshocht, S., Houssa, M., Delabie, A., Conard, T., Meuris, M., Heyns, M., Dimoulas, A., Spiga, S., and Fanciulli, M. HfO_2 as gate dielectric on Ge: Interfaces and deposition techniques. *Materials Science and Engineering: B* 2006, *135*, 256–260.
34. Cao, Y.-Q., Wu, B., Wu, D., and Li, A.-D. Interfacial, electrical, and band alignment characteristics of HfO_2/Ge stacks with in situ-formed SiO_2 interlayer by plasma-enhanced atomic layer deposition. *Nanoscale Research Letters* 2017, *12*, 1–7.
35. Li, X.-F., Liu, X.-J., Fu, Y.-Y., Li, A.-D., Zhang, W.-Q., Li, H., and Wu, D. Effect of annealing on interfacial and band alignment characteristics of HfO_2/SiO_2 gate stacks on Ge substrates. *Journal of Vacuum Science & Technology B* 2012, *30*, 010602.
36. Zhao, L., Liu, H., Wang, X., Wang, Y., and Wang, S. Improvements on the interfacial properties of High-K/Ge MIS structures by inserting a La_2O_3 passivation layer. *Materials* 2018, *11*, 2333.
37. Onik, T. A. M., Hawari, H. F., Sabri, M. F. M., and Wong, Y. H. Enhancement of electrical performance of Ge-based metal-oxide-semiconductor capacitor via formation of trigonal-Sm_2O_3. *Surfaces and Interfaces* 2021, *25*, 101289.
38. Seo, Y., Lee, T. I., Yoon, C. M., Park, B.-E., Hwang, W. S., Kim, H., Yu, H.-Y., and Cho, B. J.,The impact of an ultrathin Y_2O_3 layer on GeO_2 passivation in Ge MOS gate stacks. *IEEE Transactions on Electron Devices* 2017, *64*, 3303–3307.
39. Zimmermann, C., Bethge, O., Winkler, K., Lutzer, B., and Bertagnolli, E. Improving the ALD-grown Y_2O_3/Ge interface quality by surface and annealing treatments. *Applied Surface Science* 2016, *369*, 377–383.
40. Žurauskaitė, L., Jones, L., Dhanak, V. R., Mitrovic, I. Z., Hellström, P.-E., and Östling, M. Investigation of Tm_2O_3 as a gate dielectric for Ge MOS devices. *ECS Transactions* 2018, *86*, 67.
41. Li, X.-F., Liu, X.-J., Cao, Y.-Q., Li, A.-D., Li, H., and Wu, D. Improved interfacial and electrical properties of atomic layer deposition HfO_2 films on Ge with La_2O_3 passivation. *Applied Surface Science* 2013, *264*, 783–786.
42. Afanas'ev, V., Stesmans, A., Mavrou, G., and Dimoulas, A. Beneficial effect of La on band offsets in Ge/high-κ insulator structures with GeO_2 and La_2O_3 interlayers. *Applied Physics Letters* 2008, *93*, 102115.
43. Li, X.-F., Cao, Y.-Q., Li, A.-D., Li, H., and Wu, D. $HfO_2/Al_2O_3/Ge$ gate stacks with small capacitance equivalent thickness and low interface state density. *ECS Solid State Letters* 2012, *1*, N10.
44. Gong, Y. P., Li, A. D., Liu, X. J., Zhang, W. Q., Li, H., and Wu, D. Effect of surface treatments on interfacial characteristics and band alignments of atomic-layer-deposited Al_2O_3 films on GaAs substrates. *Surface and Interface Analysis* 2011, *43*, 734–737.
45. Ye, P., Wilk, G., Yang, B., Kwo, J., Chu, S., Nakahara, S., Gossmann, H.-J., Mannaerts, J., Hong, M., and Ng, K. GaAs metal-oxide-semiconductor field-effect transistor with nanometer-thin dielectric grown by atomic layer deposition. *Applied Physics Letters* 2003, *83*, 180–182.
46. Huang, M., Chang, Y., Chang, C., Lee, Y., Chang, P., Kwo, J., Wu, T., and Hong, M. Surface passivation of III-V compound semiconductors using atomic-layer-deposition-grown Al_2O_3. *Applied Physics Letters* 2005, *87*, 252104.

47. Chang, C.-H., Chiou, Y.-K., Chang, Y.-C., Lee, K.-Y., Lin, T.-D., Wu, T.-B., Hong, M., and Kwo, J. Interfacial self-cleaning in atomic layer deposition of HfO_2 gate dielectric on $In_{0.15}Ga_{0.85}As$. *Applied Physics Letters* 2006, *89*, 242911.
48. Hinkle, C., Sonnet, A., Vogel, E., McDonnell, S., Hughes, G., Milojevic, M., Lee, B., Aguirre-Tostado, F., Choi, K., and Kim, H. GaAs interfacial self-cleaning by atomic layer deposition. *Applied Physics Letters* 2008, *92*, 071901.
49. Cao, Y.-Q., Li, X.-F., Li, A.-D., Li, H., and Wu, D. The combination self-cleaning effect of trimethylaluminium and tetrakis (dimethyl-amino) hafnium pretreatments on GaAs. *Applied Surface Science* 2012, *263*, 497–501.
50. Sánchez-Ahumada, D., Verastica-Ward, L. J., Orozco, M., Vargas-Hernández, D., Castro-Beltrán, A., Ramirez-Bon, R., and Alvarado-Beltrán, C. G. In-situ low-temperature synthesis of PS-ZrO_2 hybrid films and their characterization for high-K gate dielectric application. *Progress in Organic Coatings* 2021, *154*, 106188.
51. Liu, J., Li, J., Wu, J., and Sun, J. Structure and dielectric property of high-K ZrO_2 films grown by atomic layer deposition using tetrakis (dimethylamido) zirconium and ozone. *Nanoscale Research Letters* 2019, *14*, 1–12.
52. Gong, Y., Zhai, H., Liu, X., Kong, J., Wu, D., and Li, A. Impact of Gd_2O_3 passivation layer on interfacial and electrical properties of atomic-layer-deposited ZrO_2 gate dielectric on GaAs. *Applied Surface Science* 2014, *291*, 35–39.
53. Cao, Y.-Q., Li, X., Zhu, L., Cao, Z.-Y., Wu, D., and Li, A.-D. Improved thermal stability and electrical properties of atomic layer deposited HfO_2/AlN high-K gate dielectric stacks on GaAs. *Journal of Vacuum Science & Technology A: Vacuum, Surfaces, and Films* 2015, *33*, 01A136.
54. Augelli, V., Ligonzo, T., Minafra, A., Schiavulli, L., Capozzi, V., Perna, G., Ambrico, M., and Losurdo, M. Optical and electrical characterization of n-GaAs surfaces passivated by N_2-H_2 plasma. *Journal of Luminescence* 2003, *102*, 519–524.

10 Sol–Gel-Processed Metal-Oxide Films for Large-Area Printed High-K Dielectrics

Won-June Lee, Jun-Gyu Choi, Myung-Han Yoon, and Sungjun Park

10.1 INTRODUCTION

Developing high-performance electronic materials with well-matched electrical, physical, and chemical properties is necessary for advances in the practical electronic applications of large-area flexible electronics, such as displays [1], wearable devices [2, 3], and low-cost disposable circuits [4]. In this regard, various printing techniques [5–8], such as spray, inkjet, screen, flexographic/gravure, and blade/bar methods, have been employed for cost-effective film deposition and fabrication on large-area substrates. Correspondingly, several researchers have widely studied solution-processable electronic materials for their widespread use in device components for versatile electronic applications. Among the device components used in electronics, the dielectric layer plays a vital role as an electrically insulating layer comprising non-conducting materials that are responsive to an externally applied electric field. The dielectric layer provides various functional properties, and these can be categorized and utilized in accordance with the linear/non-linear polarization behaviors presented in response to the specific displacement of dipole moments for functional electronic device applications [9–12], such as thin-film transistors (TFTs), resistive switching memories, energy capacitors, and ferroelectric devices.

The dielectric layer, particularly in TFTs, plays important roles: (1) it efficiently modulates the conductance of the adjacent semiconducting channel layer by accumulating charge carriers, and (2) it minimizes leakage currents near the channel region, thus acting as an insulator for minimal static dissipation. Moreover, note that the following are the two essential features of desirable dielectrics: a high areal capacitance and ideal electrical insulation, which can be essentially realized by utilizing a thin layer of robust material with a high dielectric constant (K). Therefore, the development of a method that exploits dielectric material properties to enable the realization of multi-functional and high-performance electronic devices is crucial.

However, previous solid-state metal oxide inorganic dielectrics (e.g., Al_2O_3, HfO_2, ZrO_2, HfO_2, Yt_2O_3, TiO_2, etc.) with high K values have been universally developed and fabricated by vacuum-based deposition methods (e.g., magnetron sputtering, pulsed laser deposition (PLD), chemical vapor deposition, atomic layer deposition (ALD), etc.), along with high-thermal budget post-treatments, as alternatives to the general SiO_2 (i.e., oxidized silicon dioxide) dielectric layer. Therefore, several technical limitations hinder the production of large-scale flexible electronic devices on plastic substrates via high-throughput production. Note that organic dielectric materials can be readily processed in the form of a well-dissolved solution ink; however, they have relatively low K values and severe leakage currents owing to the high surface roughness and the presence of numerous defects (e.g., pinholes, dangling bonds, interfacial charge-trapping sites, and interstitial defects) on the dielectric film. Moreover, without particular structural network

DOI: 10.1201/9781003352365-10

formations (e.g., chemical cross-linking), several organic dielectrics are not compatible with chemical solvent/gas treatments and high-temperature processes (e.g., low glass-transition temperatures (typically below 250 °C) of general insulating polymers). These limiting factors typically require complex additional processing steps for the fabrication of solution-processable high-performance organic electronic devices.

With potential advancements in thesol–gel synthesis of functional metal oxides, solution-processed metal oxide dielectric materials have found a broad range of electronic applications owing to their high K values, accompanied by high areal capacitances, good optical transparency, new functionalities, and chemical/environmental stability [10–14]. To realize better-performing sol–gel-derived dielectric films with novel electrical functionalities, a facile chemical design engineering, subsequent ink deposition, and (post-)annealing process-based production of highly densified metal oxide films with smooth surfaces with low film roughness values and minimal defects is essential for stable electrical insulation and frequency-dependent dielectric capacitance performance. Moreover, advanced methodologies dedicated to the rapid and low-temperature fabrication have to be further analyzed to realize a high-throughput (e.g., roll-to-roll (R2R), etc.) printing process while maintaining satisfactory performance.

In this chapter, we present details regarding the recent progress in developing sol–gel-processed metal oxide dielectric films from the perspectives of large-area printed electronics comprising TFT device arrays and their advanced electronic applications. Figure 10.1 illustrates the major process steps involved in the production of high-K metal oxide dielectric films in the following sequence: (1) sol–gel chemistry and synthesis design engineering, (2) large-area and patternable printing process, (3) external/internal energy-derived low-temperature processing, and (4) combination approaches for the advanced high-throughput device fabrication process. Notably, herein, we broadly summarize the ongoing research development in sol–gel dielectric thin films and their formation mechanisms for practical applications and commercialization and discuss related copious information presented in several recent review articles [10–14].

10.2 HIGH-K METAL OXIDE DIELECTRIC FILMS

Dielectric materials are electric insulators that are known to become polarized when subjected to an electric field. Notably, free electrical carriers produced under the effect of an applied field can create a slight shift from the equilibrium position for dielectric polarization but not the carrier movements. Generally, charge redistributions in dielectric materials are known to weaken the applied electrical field during dielectric polarizations. Figures 10.2a–d illustrate the polarization mechanism of dielectric materials with three different polar components of a charged species rearranged based on the strength of the electric field.(1) Electronic polarization (Figure 10.2a): the materials have electron-cloud nucleus pairs between positively and negatively charged surrounding electron clouds; thus, the symmetrical orbital paths become distorted and separated for induced dipole formation when placed under an electric field. This polarization dominates the dielectric properties of most atomic crystals, with atoms adequately held in the lattice structure [15]. (2) Ionic polarizations from oppositely charged ion pairs (Figure 10.2b): the displacement of cations and anions produces a net dipole moment under an applied electric field (at frequencies up to ~10^{12} Hz); (c.f. the specific net dipole is the vector summation of each product of charge displacements from the superposition principle) [16]. (3) Molecular polarizations from randomly oriented permanent molecular dipoles without an electric field (Figure 10.2c): the random molecular dipoles in several materials are generally active for rearrangement and alignment along the direction of the applied electric field [16].

The quick reorientation and relaxation of dipoles under a rapidly alternating electric field are significant in the use of gate dielectric insulators for TFTs. Therefore, the choice of dielectric insulating materials has to be based on a suitable frequency region because all these polarization mechanisms do not react equally at the frequencies of the time-varying electric fields (Figure

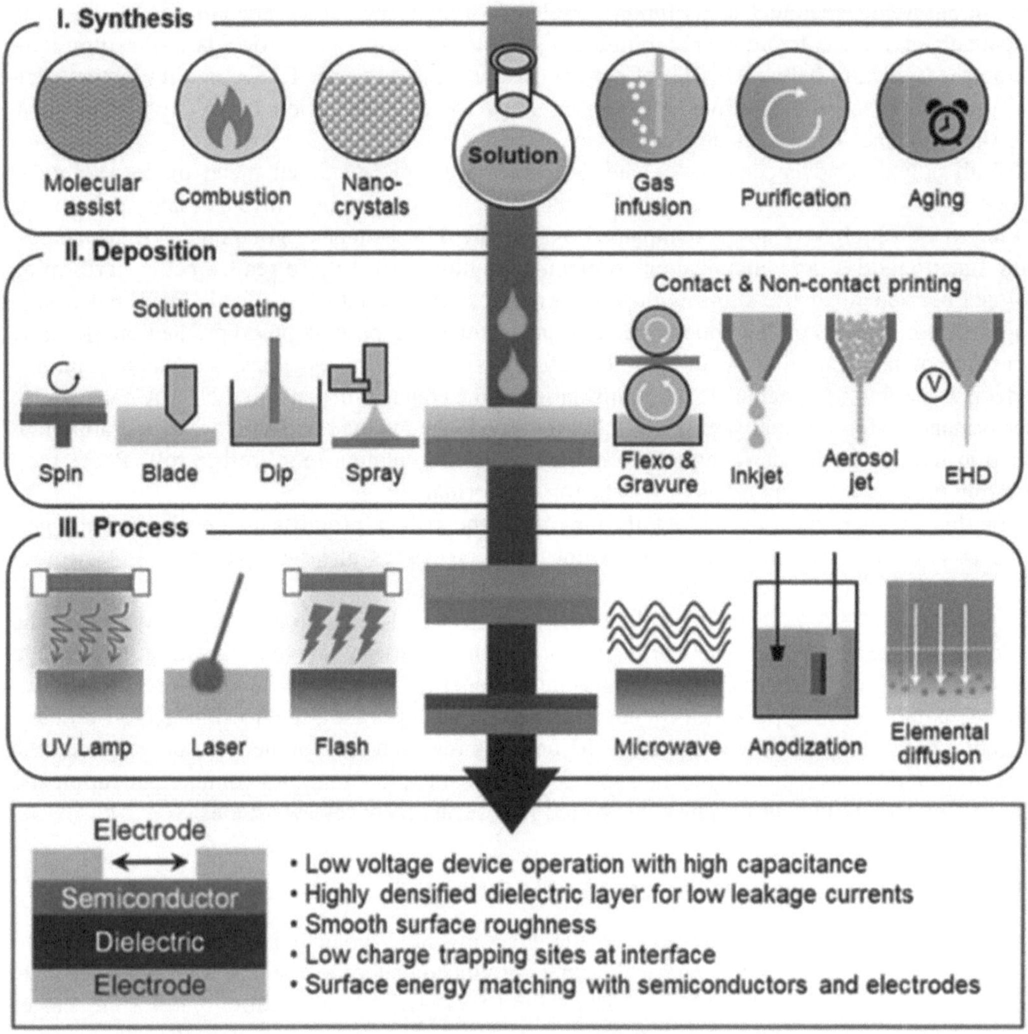

FIGURE 10.1 Scheme of the overall process for sol-gel-processed large-area printed metal oxide-based high-K dielectric films: (I) synthesis design of dielectric precursor solution, (II) solution deposition methods based on both conventional coating process and (non-)contact printing process, and (III) thin film formation process with (post-)treatment methods. The requirements for high-performance sol-gel oxide dielectric layers among the major components of thin film transistor devices. (Reproduced with permission [11] Copyright 2021, AIP Publishing.)

10.2d). For example, electronic polarization (Figure 10.2a) can respond at extremely high frequencies (up to ~10^{15} Hz); however, molecular polarization (Figure 10.2c) can react only at low frequencies, with its cut-off frequency region at~10^{10} Hz.

The illustration in Figures 10.2(e–g) indicates that the polarization behaviors of dielectric materials are responsible for the external electric field based on a model of a two-parallel-plate capacitor. Figure 10.2g (up) illustrates that a dielectric material without an electrical bias has a random or spontaneous charge distribution and no directional effect. Here, when a positive potential is applied to the top electrode (Figure 10.2g (down)), the dielectric material becomes polarized, with the evident appearance of a net dipole moment averaged throughout its volume and surface charge density created on both sides of the dielectric layer boundaries. Typically, the performance of a dielectric capacitor in the form of charged sheets for electrical energy storage

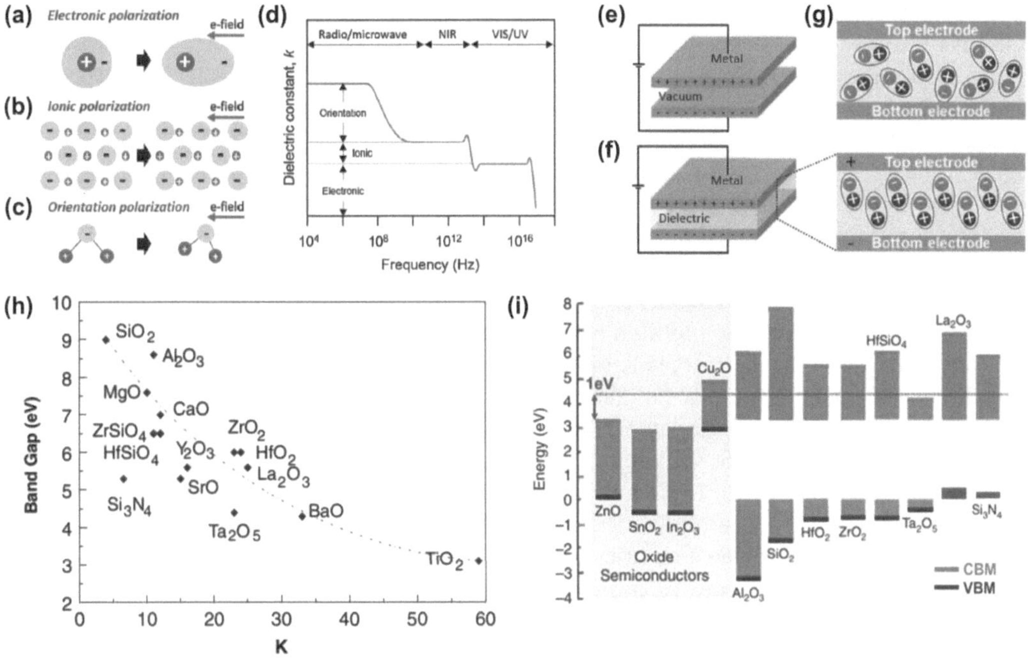

FIGURE 10.2 The basic mechanisms of dielectric polarization with schematics of (a) electronic polarization, (b) ionic polarization, and. (c) orientation polarization. (d) The contribution of electronic, ionic, and orientation polarization to the overall dielectric constant range at the applied different frequencies. (e), (f) Simple images of parallel-plate capacitor model along with dielectric polarization. Schematics of parallel-plate metal-insulator-metal (MIM) capacitors (e) without and (f) with dielectric layer. (g) Illustrated dipole polarization in the absence (top) or presence (bottom) of an external electrical field. Reproduced with permission [10] Copyright 2017, Elsevier. (h) Relationship between dielectric constant and band gap (E_g) for high-K metal oxide dielectrics. (i) Calculated band offsets of the several dielectrics in (h). The values are normalized to the VBM of ZnO (cf. dotted line: the minimum of 1 eV for the conduction band offset). Reproduced with permission [18] Copyright 2012, John Wiley & Sons, Ltd.

corresponds to its capacitance; therefore, the capacitance per area (or areal capacitance), C_i, can be expressed as follows:

$$C_i = \frac{k\varepsilon_0 A}{d} \tag{10.1}$$

where ε_0, A, and d denote the vacuum (free space) permittivity, overlapping area of the two plates, and dielectric film thickness, respectively. Notably, the polarization of dielectric materials and their channel conductance modulations are essential for TFT device operations.

Owing to its abundant reserves in the Earth's crust, conventional amorphous silicon dioxide (SiO_2) (K = 3.9), which has been well-known as an excellent insulator with high electric resistivity (~10^{10} Ω cm), has been extensively employed in the modern complementary metal oxide semiconductor technology for several decades. Moreover, tremendous efforts have been dedicated to reducing the physical dimensions of electrical components in the electronic industry for miniaturization and high integration, following Moore's law. Particularly, scaling down processes for SiO_2 dielectric layers are necessary to fabricate smaller and faster electronic circuits. Unfortunately, the production of a severe gate leakage current is inevitable, owing to the drastic scaling down of SiO_2 dielectric layers for larger capacitance and lower operating voltages.

According to Equation (10.1), metal oxide dielectrics are ideal for integrating high C_i values into the TFT architecture. First, we note that several metal oxide materials with an inherently high-K value are substantially more prominent than amorphous SiO_2. Second, high-quality metal oxide insulating layers exhibit mechanical robustness and minimal defects, thus allowing the downscaling of film thicknesses below 100 nm without serious gate current leakage issues. These two points interactively contribute to overcoming the limitations of the SiO_2 layer and reducing the operating voltages of TFTs owing to the relatively thicker, high-K metal oxide dielectric films with similar high C_i values. Therefore, high-K metal oxide dielectric materials are well suited for the development of advanced integrated electronics, wherein low-power consumption/loss and satisfactory device performance are essential features.

A high band gap energy (E_g) is another important requirement for a reliable high-K dielectric material. This feature allows a dielectric film to further mitigate parasitic electrical conduction by suppressing the charge injection to/from electrodes and by suppressing the generation of intrinsic free carriers based on the thermal/photoexcitation conditions [17, 18]. However, a typical tradeoff exists between E_g and K (Figure 10.2h); thus, to enhance the value of K, dielectric films with a lower E_g are preferable. However, materials with an ideal high dielectric constant (K> ~30) and energy band gap (E_g > ~8 eV) are not yet known [19, 20]. Although some materials with high dielectric constants have K > 20–40 (e.g., Ta_2O_5 and TiO_2), their E_g and band offset energies are negligible and not suitable for alternative gate dielectrics to be used in transistors fabricated with wide band gap semiconductors. Notably, oxide dielectrics with relatively small K values are sometimes employed to avoid current leakages. Figure 10.2i illustrates the energy bands of semiconductors, along with the determined conduction band (CB) and valence band (VB) offsets for all the indicated dielectrics. Note that all the values are normalized to the VBM of zinc oxide (ZnO) (n-type oxide semiconductor), and the dashed line indicates the minimum of the 1 eV requirement for the CB offset. Furthermore, note that the sufficient band offsets of the dielectrics should be > 1 eV, relative to those of semiconductors (i.e., VB maximum (VBM) in a p-type transistor device and the CB minimum (CBM) in an n-type transistor device) to achieve reliable TFT device operations. The minimum 1 eV requirement of the band offsets is set to maintain the Schottky emission leakage at a low enough level [21, 22]. Apart from Ta_2O_5, all the oxide dielectrics are suitable for n-type semiconductors with the minimum 1 eV requirement. Notably, a minimal band offset of less than 1 eV may lead to unacceptable high leakage currents owing to the thermionic emission of electrons or holes into the dielectric energy bands. In addition, the structure-related requirements of a dielectric layer incorporated into a TFT device must also be considered for scaling issues, such as low defect/interfacial trap sites, low fringing capacitive effects, and band(-gap) engineering feasibility [10].

With regard to the film deposition techniques, metal oxide dielectric films have been conventionally fabricated using various vacuum-based deposition methods, such as ALD [23–25], PLD, e-beam sputtering deposition, and magnetron sputtering deposition [26, 27]. These methods have proven useful in the realization of extremely thin films with good mechanical and electrical properties, thus efficiently addressing the limitations on scaling. Nevertheless, most vacuum-based deposition techniques necessitate high initial (e.g., equipment and environment) build-up costs and financial maintenance. They are also commonly accompanied by an additional high-temperature post-annealing treatment; accordingly, they are not well-matched for depositing materials on large-area flexible substrates in a rapid and scalable manner. Therefore, conventional deposition methods need to be replaced in the near future to satisfy the demands of large-area flexible electronic devices and circuits that can be produced using low-cost and mass-production-oriented industrial processes. In the next section, we will discuss the basic principles of solution-based approaches used to produce functional metal oxide materials, including various sol–gel metal oxide dielectrics that can be integrated into high-performance TFT devices.

10.3 SOL–GEL METAL OXIDE PRECURSOR ENGINEERING FOR HIGH-QUALITY HIGH-K DIELECTRIC FILMS

Over the years, intensive research efforts have been dedicated to the cost-effective fabrication of high-quality metal oxide electronic materials with essential considerations, such as raw material and

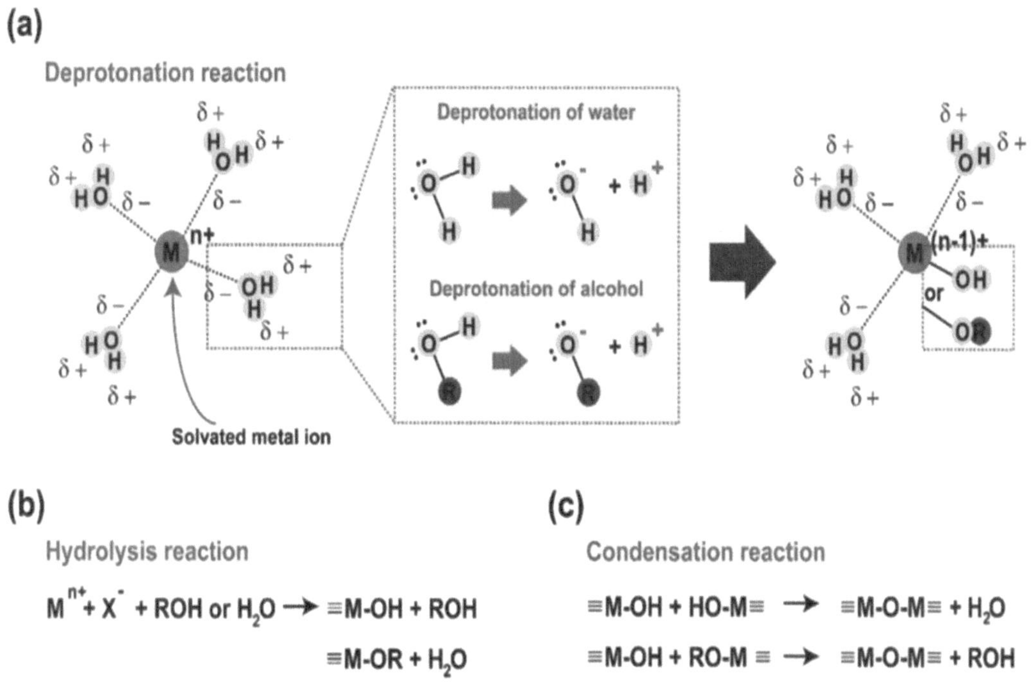

FIGURE 10.3 Basic sol-gel chemistry reactions during the metal oxide film formation from the metal precursor solutions. (a) A schematic of reaction between metal ions and water/alcohol molecules to metal hydroxides. After the deprotonation reactions of water or alcohol molecules, metal and hydroxide ions are bound to each other by electrostatic interaction. (Representative chemical reactions are shown for (b) hydrolysis and (c) condensation. Reproduced with permission [10]. Copyright 2017, Elsevier.)

device manufacturing costs, processing time and complexity, and stability and reliability. Accordingly, sol–gel oxide solution-based methodologies have been considered as the most promising fabrication routes for high-quality metal oxide materials with good uniformity and large scalability. Since the first conceptual study on the production of silica-based oxide materials using silicon alkoxide precursors in the mid-1800s [28], remarkable developments in the sol–gel technology have been made for synthesizing functional oxides with satisfactory performance; these developments include the realization of a broad range of applications and minimization of the issues related to solution-based technologies.

For applications to significant electronics components, typical sol–gel processes are allowed to follow the given reaction pathway: hydrolysis–condensation–gelation from metal precursor solutions to form metal oxide films (i.e., sol colloidal suspension and gel network) [10]. A sol–gel chemical reaction is typically initiated by dissolving precursor compounds (i.e., metal salts or alkoxides) in an organic or aqueous solvent (Figure 10.3a). The solvent molecules (e.g., H_2O or ROH) and solvating metal cations are transformed into hydroxo (OH-) or oxo (O^{2-}) ligands, leading to metal hydroxides (Figure 10.3b). Subsequently, a condensation reaction forms metal–oxide–metal (M–O–M) polymeric frameworks via an oxolation reaction between metal hydroxides, as illustrated in Figure 10.3c. Generally, three reaction product forms (i.e., sol, gel, and nanoparticles) can be obtained by controlling the reaction temperature and duration, agitation, aging time, pH, and additives. Finally, a gel-like oxide layer is formed by depositing a sol–gel precursor solution onto the substrates, followed by thermal (post-)annealing at a high temperature (typically above 400 °C) for an hour-long process time to complete the condensation reaction, eliminate chemical impurities/unreacted chemical residues, and promote the densification (or crystallization) of the resultant oxide film. The desired properties and structures of metal oxide materials can be designed according to the above controllable synthetic conditions in all reaction pathways [29].

Withan emphasis on sol–gel metal oxide dielectrics, the realization of good electrical insulating properties without compromising the high-K values typically depends on the capability to fabricate oxide films with smooth surface morphologies and minimal defect sites. It is noteworthy that solution-processed oxide dielectric layers generally have lower K values than the theoretical value, owing to the associated organic residues and undesired impurities at the surface/interface and in the films during fabrication [30]. Therefore, various synthesis designs of sol–gel-based oxide dielectrics have been developed for broad-impact electronic applications, including the fabrication of metal components (e.g., binary, ternary, and quaternary), ligand/additive engineering, and creation of different film structures (e.g., bilayer or multilayer). Note that all the notations of "MO_x" (M: Al, Zr, Hf, Yt, Gd, Ga, In, Sn, Zn, etc.) relate to a nonspecific chemical formula of the oxide dielectric and semiconductor materials owing to their common non-stoichiometric composition of sol–gel-processed metal oxide compounds.

10.3.1 Single-component Metal Oxide Dielectrics

First, Katz et al. reported a novel approach by including ionic dopants into metal oxide lattices to improve the K value of a host dielectric material [31–33], Figures 10.4a illustrate the structures of ion-doped (e.g., Na^+, K^+, and Li^+) alumina dielectrics, indicating extremely high areal capacitance (200 nF/cm^2 at 50 Hz) values with relatively low leakage currents (<10^{-7} A/cm^2 at 1 MV) for a 75nm insulating film [33]. The fabricated oxide dielectric film presents an extremely high-K value of 170 and a relatively high breakdown electric field of 5.3 MV/cm. This remarkable capacitance can be attributed to the enhanced polarization resulting from small ion displacement, whereas the lack of mobile charges in the bulk dielectric matrix can be deemed responsible for the low leakage current. The electrical characterizations of zinc-tin-oxide (ZTO) TFTs on this dielectric presented a maximum mobility of 28 cm^2/Vs (Figure 10.4b).

Keszleret al. demonstrated the applications of various solution-processed phosphate/sulfate-derivedaluminum, hafnium, and zirconium oxide dielectrics in combination with sputtered ZnO, indium zinc oxide (IZO), and ZTO semiconductors as TFTs [34]. Such unique formulations were intentionally employed to fabricate high-quality amorphous oxide dielectrics by mitigating the undesirable dielectric effects produced by defects, grain boundaries, and charge conduction pathways (i.e., current leakage path). In this pioneering study, the authors provided conceptual inspiration for manipulating dielectric properties via sandwiched multilayer architectures or atomic dopant implantation. The resultant aluminum oxide phosphate ($AlPO_x$) dielectric films exhibited a high breakdown electric field of up to 5 MV/cm with K ~ 4.8. Nevertheless, this relatively low-K value and the AlPO film thickness of approximately 150 nm required a relatively high voltage (~40 V) for the consequent TFT operation, suggesting further dielectric optimization in terms of the K value and film thickness remained an issue.

In parallel, several research groups dedicated considerable efforts to the development of single-component high-performance sol–gel metal oxide dielectrics by avoiding additives in precursor solutions. Previously, sol–gel-based AlO_x dielectrics for oxide electronics have been prepared using various aluminum sources such as aluminum nitrate [31–33, 35–43], aluminum acetyl acetonate [44, 45], aluminum chloride [35, 46], and aluminum alkoxide [32, 47]. Aluminum nitrate, in particular, has been most broadly utilized for sol–gel-based AlO_x preparation with minimal impurities owing to its commercial availability, environmental stability, and low decomposition temperature [48]. It is also remarkable that sol–gel processed AlO_x films are known to maintain their amorphous phases even at relatively high temperatures (~600 °C) [49], unlike other metal oxides, which develop various crystalline states, inducing leakage current pathways.

Several research groups have recently explored other sol–gel-based high-K metal oxide dielectric materials. Consequently, they have reported a variety of oxide semiconductor and dielectric

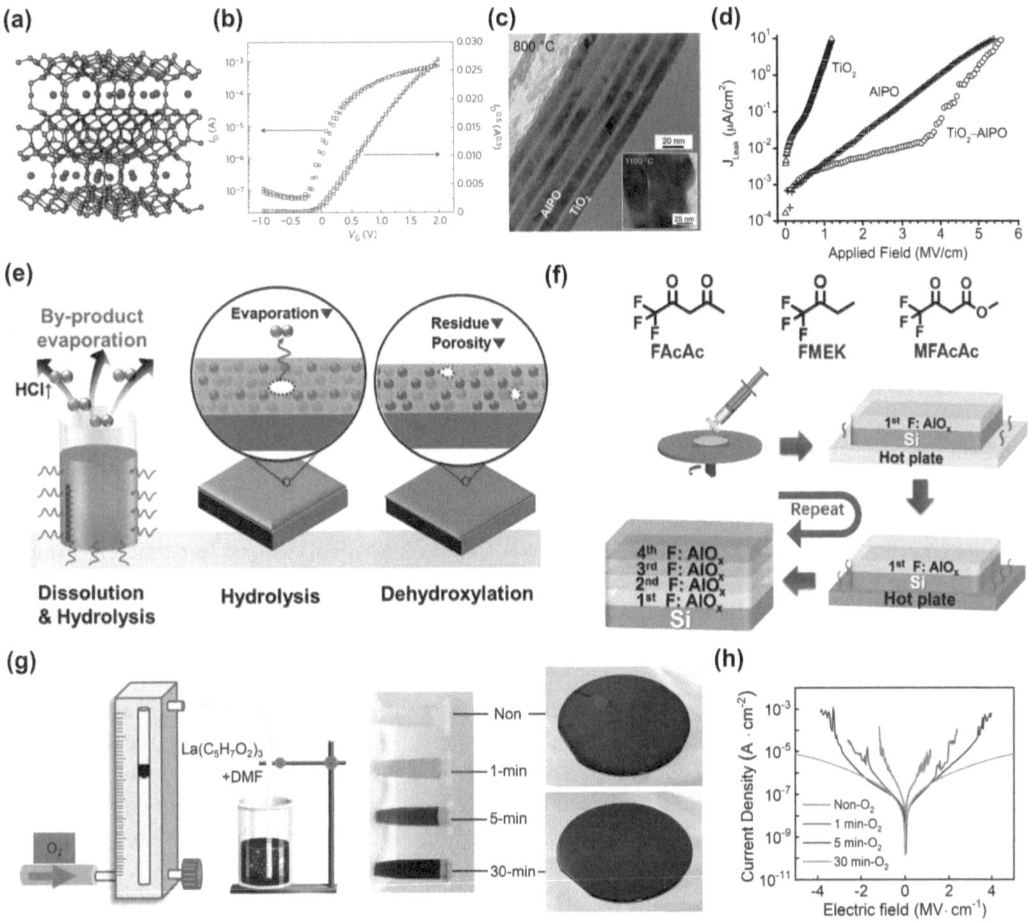

FIGURE 10.4 Sol-gel metal oxide dielectrics via various synthesis design engineering. (a) sodium ion-doping methods on AlO_x dielectric (SBA); crystal structure of SBA. The mobile sodium ions are depicted by blue dots, while oxygen and aluminum atoms are described in each red and yellow color. (b) Transfer characteristics for a ZTO transistor fabricated on SBA/ITO glass substrate with a W/L ratio ~10. (Reproduced with permission [33]. Copyright 2009, Nature Publishing Group.) (c) A cross-sectional TEM image of 6 nanolaminate bilayer oxide dielectrics of TiO_2 (~10 nm)/$AlPO_x$ (~5 nm) after annealing for 1 h at 800 °C in air (inset: diffusing interlayer at 1100 °C) (d) Representative characteristic of leakage current densities for TiO_2-$AlPO_x$ nanolaminate, single-layer TiO_2 and $AlPO_x$ by annealing at 350 °C. (Reproduced with permission [64]. Copyright 2012, American Chemical Society.) (e) Schematic illustration of the thermally purified solution process to efficiently generated pre-hydrolysis conditions and further reduce the processing temperature of the HfO_x films. (Reproduced with permission [68]. Copyright 2018, Royal Society of Chemistry.) (f) Chemical structure of fluoro-involving additive molecules and schematic of the fabrication of the F:AlO_x dielectric films. (Reproduced with permission [71]. Copyright 2020, American Chemical Society.) (g) Schematic of the experimental setup for oxygen gas infusion into the lanthanum oxide precursor solutions with different oxygen infusion times after 12 h stirring and photo images of as-deposited films on silicon wafers. (h) Leakage current density for insulating properties of La_2O_3 dielectric films with different oxygen infusion times. (Reproduced with permission [72]. Copyright 2020, Royal Society of Chemistry.)

combinations, including In_2O_3 or ZnO-based oxide semiconductors with yttrium oxide (YtO_x) [50, 51], gadolinium oxide (GdO_x) [52, 53], hafnium oxide (HfO_x) [41, 43, 54–56], zirconium oxide (ZrO_x) [41, 57–61], or gallium oxide (GaO_x) dielectrics [62]. These results highlight the potential applications of less conventional metal oxide dielectric materials for use in high-performance metal oxide electronics with low operation voltages.

10.3.2 Multi-component (More Than Two Metal Species) Metal Oxide Dielectrics

The seminal research conducted by Katz [31–33] and Keszler [34, 55, 63] and groups inspired several researchers by proposing that smart compositional and structural engineering can achieve desired dielectric properties. Therefore, considerable efforts have been devoted to the creation of hetero-metal oxide structures with multi-component characteristics. The significant tunable parameters include bulk optical transparency, interfacial state density, and effective K value. Note that the sol–gel-processed route can be beneficial for realizing well-defined multi-component or multi-layered dielectrics because achieving considerable controllability of the film surface morphology (in nanoscale) and various composition via vacuum-based deposition techniques is often challenging.

Keszleret al. successfully developed multi-layered TiO_2-$AlPO_x$ nanolaminates with an extremely low leakage current of 10^{-8} A/cm^2 at 1 MV/cm and a high breakdown electric field of approximately 5.4 MV/cm (Figure 10.4c, d) [64]. This sequentially film-stacked structure was fabricated with the repetition of two alternating spin-coating processes. The authors determined that an appropriate combination of nanolaminate materials could compensate for the shortcoming of each component. In the preceding study, for example, the high hydration temperature of $AlPO_x$ and crystallinity-related current paths of TiO_2 disappeared in the resultant oxide nanolaminate multilayer, which exhibited a lower leakage current density than each single-component film, as illustrated in Figure 10.4d. Furthermore, the extracted K value was eight, which is greater than that of $AlPO_x$ by a factor of 1.9. Further, amorphous indium gallium zinc oxide (IGZO) TFTs with this oxide nanolaminate dielectric exhibited a good electron mobility (3.2 cm^2/V s) and a high current on/off ratio (> 10^5).

Park et al. demonstrated a Zr-incorporated AlO_x dielectric layer [65]. In their study, the surface functionality was effectively adjusted to produce an activated solid surface with a smooth interface. Based on X-ray photoelectron spectroscopy, the authors revealed that the surface reactivity and amenability for the sol–gel semiconductor on top were more pronounced on $ZrAlO_x$ than on pure AlO_x. This Zr-incorporated AlO_x dielectric was fabricated via UV photochemical activation, resulting in a high-K value of 8.82 (at Zr~ 9 mol% of Zr:Al molar ratios) and a high breakdown electric field greater than 7 MV/cm.

The Baik [66] and Moon [67]groups also reported low-temperature processed high-K dielectric materials using ion-doped dielectrics, such as boron(B)-doped ZrO_x and zirconium(Zr)-doped AlO_x, respectively. To achieve reduced annealing temperatures as low as 250 °C, the authors added small amounts of boron (B ~ 9 mol%) or zirconium (Zr ~ 10 mol%) precursors to the dominant stock solutions (i.e., mother solutions) of MO_x dielectrics. Note that the chemical bonding between a metal dopant and oxygen is stronger than that between aluminum and oxygen (Al–O), resulting in excellent dielectric performance owing to fewer hydroxide functionalities and lattice defects. Therefore, all-solution-processed In_2O_3 and ZnO TFTs can be fabricated on a flexible polyimide substrate, in combination with oxide dielectrics exhibiting extremely low gate leakage currents and high areal capacitances.

10.3.3 Precursor Engineering for High-quality Dielectric Film Formation

Chung et al. demonstrated a thermally purified solution (TPS) process, in which applying thermal energy to solutions before film formation could effectively create a hafnium oxide network structure [68]. The detailed mechanism of this scheme is illustrated in Figure 10.4e. The TPS facilitated the hydrolysis reaction in the solution phase, referred to as pre-hydrolysis, whereas a general hydrolysis reaction occurred after film casting. Owing to the evaporation of byproducts during pre-hydrolysis, the reaction only required a specific low thermal energy to form the oxide framework. In this study, conventionally synthesized HfO_x sol–gel solutions were heated at different temperatures from room temperature to 125 °C on a hot plate before spin-casting. The TPS-assisted HfO_x films annealed at 200 °C demonstrated dielectric performance similar to that of the conventionally annealed films at 350 °C. The TPS-assisted HfO_x gate dielectric layer exhibited excellent electrical properties, including a low leakage current density (~10^{-9} A/cm^2 at 2 MV/cm) and high breakdown electric field (6 MV/cm), resulting in good TFT device performance with a mobility of 3.67 cm^2/V s and an on/off current ratio (> 10^6).

Notably, the synthetic route of precursor solutions is highly crucial because the quality and performance of the desired oxide films are known to be influenced by the sequential chemical reactions for the formation of the M–O–M network. For example, acidity control of the precursor solution can improve the solubility of precursors, as well asthe hydrolysis reaction [69, 70]. Moreover, anion doping of the precursor solution can improve the dielectric stability over a wide range of frequencies. Zhuang et al. reported fluoride (F) doping in AlO_x dielectrics by introducing an F-involving small molecule, 1,1,1-trifluoro-2,4-pentanedione, to overcome the frequency-dependent dielectric capacitance characteristics, which yielded reliable film formation and stable device performance [71]. As illustrated in Figure 10.4f, such F sources were added to each precursor solution, and spin-cast F:AlO_x films were repeatedly deposited to obtain desired thicknesses. The authors determined that the optimized AlO_x dielectrics with an F content as high as 3.7 at .% exhibited a stable capacitance of 166 nF/cm^2 in the frequency range of 10^{-1}–10^4 Hz. Typically, TFT devices with both inorganic and organic semiconductors exhibit negligible *I–V* hysteresis owing to a reduction in the H^+ content. However, the foregoing chemical approaches for solution synthesis fail to produce high-quality TFTs with low hysteresis.

Notably, gas-infusion treatment during solution synthesis reactions can also improve the dielectric performance of oxide dielectric films. Yan et al. demonstrated the incorporation of oxygen gas into a La_2O_3 sol–gel solution at different gas flow rates to enhance the poor wettability of the synthesized sol–gel solution on the target substrate (Figure 10.4g, h) [72]. Correspondingly, an increase in the oxygen gas flow rate induced a color change in the solution with a decreasing contact angle (e.g., surface energy) caused by the generated hydroxide group. The La_2O_3 films subjected to gas infusion treatment revealed superior film densification compared to the typical oxide films prepared via ALD with outstanding dielectric and insulating performance (Figure 10.4h). The authors demonstrated that the oxidative gas treatment on a sol–gel solution during the synthetic reactions was applicable to other oxide dielectrics.

10.4 LARGE-AREA PRINTING METHODS OF HIGH-K METAL OXIDE DIELECTRIC FILMS

10.4.1 Inkjet Printing

The inkjet printing method, a noncontact printing technology, has been extensively employed for ultrafine resolution, direct solution printing in both academia and the industry. However, several factors are known to influence the printing quality and resulting performance (e.g., customized inkjet-printable ink materials, undesired nozzle clogging, typical coffee-ring effects of the droplets, and specific surface conditions for the optimal substrate). Thus, an optimization process of the inkjet printing technology, particularly for a sol–gel oxide precursor solution, is a prerequisite for achieving the desired performance in electronic applications. Subramanian et al. introduced all-inkjet-printed transparent metal oxide devices [61]. Here, the ink formulation was carefully optimized for all inkjet-printable metal oxides inks, such as Sb-doped SnO_2 conductors, SnO_2 semiconductors, and ZrO_2 dielectrics. The authors also manipulated the surface wettability using an ultrathin poly(methyl methacrylate) (PMMA) polymer layer to control the surface energy and suppress the coffee-ring effect of the typical ink-jetted droplets during printing. The wetting controllable polymer layer acted as a sacrificial layer that allowedthe creation of a flat metal oxide film; however, it was burned off readily during the high-temperature post-annealing process (> 500 °C) without affecting device performance. Figure 10.5a and b illustrate the height profiles of the annealed inkjet-printed ZrO_x dielectric layer with inkjet-printed SnO_2/Sb-doped SnO_2 (ATO)-channel and gate electrode layers. The inkjet-printed ZrO_2 dielectric layers exhibited excellent dielectric properties, such as a high

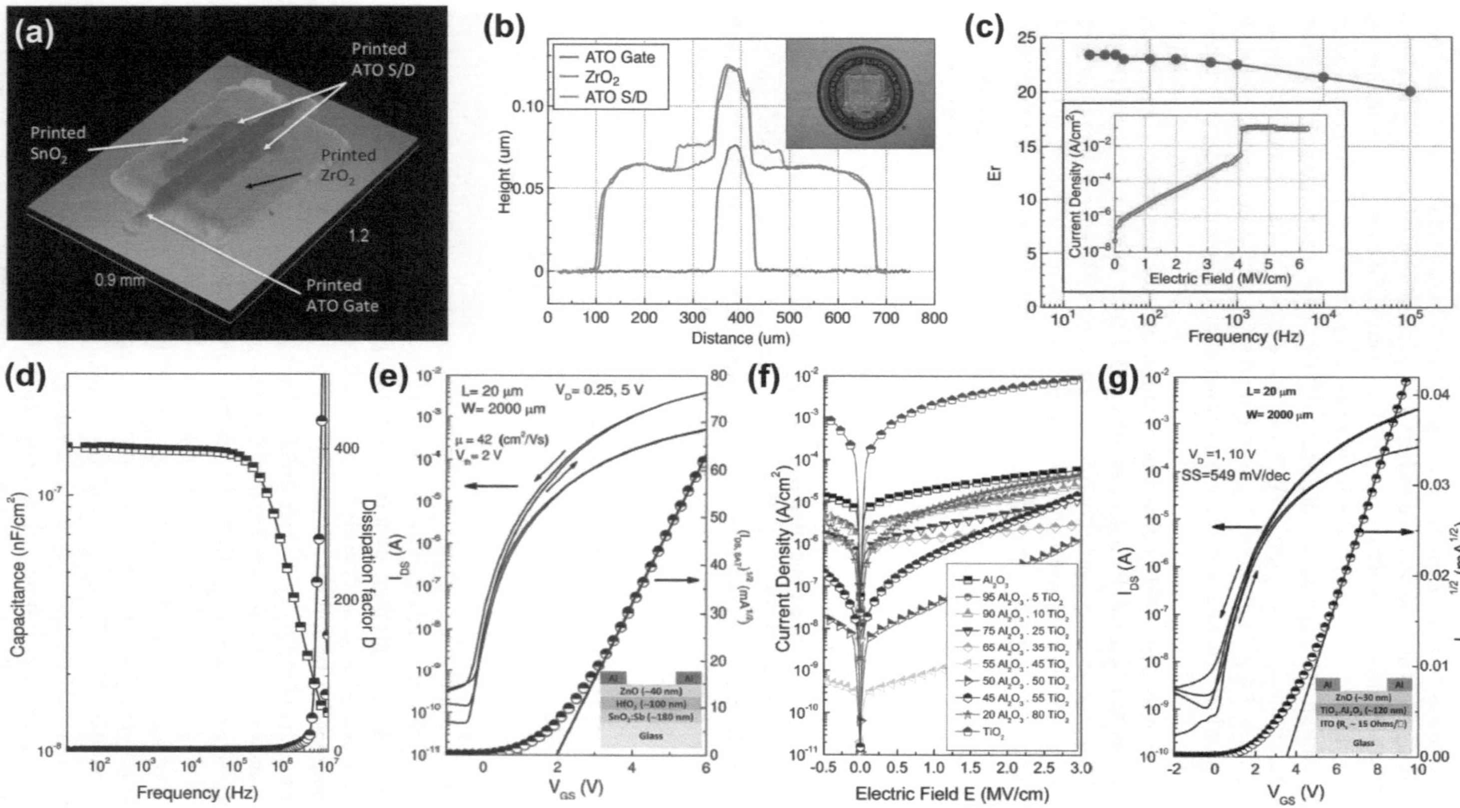

FIGURE 10.5 (a) 3D height profiles of all-oxide-ink-jet printed TFT device on a glass substrate. (b) Cross-sectional height profiles of the fabricated oxide layers in TFT structure. (Inset: the photo image of the ink-jet printed devices). (c) Electrical characteristics of ink-jet printed ZrO_2 dielectrics. (Reproduced with permission [61]. Copyright 2015, Royal Society of Chemistry.) (d) Areal capacitance and dissipation factor of HfO_2 dielectric (~104 nm) film deposited by spray pyrolysis (SP). (e) Linear and saturated (V_{DS} = 0.25 and 5 V) transfer characteristics of bottom gate, top-contact TFT devices (channel width and length, W/L = 2000/20 μm), fabricated with spray-coated ZnO films on a HfO_2 dielectric by SP (151 nF/cm²). (Reproduced with permission [54]. Copyright 2014, Wiley-VCH.) (f) Leakage current density-electric filed characteristics of $Ti_xAl_{2x-1}O_y$ solution-processed dielectrics as a function of the $[Ti^{4+}]/[Ti^{4+}+2Al^{3+}]$ atomic ratio. (g) Linear and saturated (V_{DS}=1 and 10 V) transfer characteristics of bottom-gate, top-contact TFTs with (channel width and length, W/L = 2000/20 μm), fabricated with SP-based ZnO films on a $Ti_xAl_{2x-1}O_y$ ($[Ti^{4+}]/[Ti^{4+}+2Al^{3+}]$~45%) dielectric by SP (90 nF/cm²). (Reproduced with permission [74]. Copyright 2015, American Chemical Society.)

dielectric constant (K ~ 23 at 20 Hz) and a breakdown electric field of 4.0 MV/cm(Figure 10.5c). The final all inkjet-printed TFT devices fabricated on a glass substrate presented excellent electrical characteristics with high electron mobilities (~11 cm^2/V s) and on/off current ratios (~10^6) under a low operation voltage of 5 V.

10.4.2 Spray Coating

Anthopoulos and Adamopoulos et al. pioneered the simple and large-area fabrication of metal oxide transistors onglass substrates with pre-patterned ITO electrodes using spray printing techniques. They successfully demonstrated the fabrication of sol–gel ZrO_2 [73] and HfO_2 [54] dielectric films with high-K values via ambient spray pyrolysis, an up-and-coming technique for large-scalable coating. The resultant ZrO_x and HfOx dielectric films exhibited decent film surface roughness values of ~3 and 0.8 nm, comparable with the surface roughness (~2.9 nm) of the ITO film on the glass substrate. Moreover, these films have excellent optical transparency with band gap energies of ~5.9 and ~5.7 eV and, particularly, high-K values of ~14 (126 nF/cm^2) and 18.5 (151 nF/cm^2), respectively (Figure 10.5d). The sequential spray-pyrolyzed TFT devices with multilayer structures of ITO/ZrO_2/Li-doped ZnO and SnO_2:Sb/HfO_2/ZnO at substrate temperatures of 400–450 °C exhibited impressive electron mobilities of up to ~85 and ~42 cm^2/V s, respectively, together with on/off current modulation ratios above 10^6, a negligible hysteresis window, and a low operation voltage (6 V) (Figure 10.5e). In particular, the multi-stack layer of SnO_2:Sb/HfO_2/ZnO obtained by the spray pyrolysis method exhibited remarkable optical device transparency with a high transmittance of ~80% in the visible range and could be deemed suitable for developing highly transparent metal oxide electronics.

In a follow-up study, Adamopoulos et al. presented the spray pyrolysis of AlO_x-incorporated ternary oxide component $Ti_xAl_{2x-1}O_y$ dielectric films with different compositions; these films exhibited much lower leakage currents than the binary oxide component TiO_2 and Al_2O_3 films (Figure 10.5f) [74]. Therefore, they determined the optimum composition ratio of TiO_x (45%) with AlO_x dielectric networks to be 45 $TiO_2 \cdot$ 55 Al_2O_3, thereby resulting in extremely low leakage currents of 5×10^{-9} A/cm^2 at 3.0 MV/cmwith high breakdown fields (> 5 MV/cm). Furthermore, in the frequency range from 10^2 to 10^5 Hz, the $Ti_{0.45}Al_{0.55}O_y$ films presented dielectric constants of 13, higher than that of an Al_2O_3 film (K ~ 9). Figure 10.5g illustrates a representative transfer characteristic obtained from a ZnO TFT based on the optimized 45 $TiO_2 \cdot$ 55 Al_2O_3 dielectric layer, exhibiting a high electron mobility (~10 cm^2/V s) and on/off current ratio exceeding 10^6 with a negligible hysteresis window. As another ternary configuration, the authors also reported a $La_xAl_{1-x}O_y$ dielectric layer by adjusting the composition ratio between Al_2O_3 and La_2O_3 (K ~ 27 with a large E_g of ~6 eV) [75]. They demonstrated that solution-processed $La_xAl_{1-x}O_y$ presents high stability against environmental moisture and good interface properties, unlike pristine-La_2O_3; it also presents a high dielectric constant (K ~ 16) and high optical band gap (~6.18 eV) with a very low leakage current density (3×10^{-9} A/cm^2 at 1.5 MV/cm). Following the above studies, the ambient spray pyrolysis technique presents the potential for the development of low-cost and large-area transparent oxide electronics.

10.4.3 Bar Coating

Lee et al. first demonstrated metal oxide TFT arrays over a large area using a modified wire-bar coating technique to fabricate ultrathin and dense dielectric oxide films, unlike conventional bar-coating that formed a microscale thick film [43]. For this, a minimal volume of a diluted $Al(NO_3)_3$ ink solution dissolved in 2-methoxyethanol was loaded onto a substrate and kept in contact with thin metal wire grooves from a wire-bar shearing system; consequently, the bar was horizontally moved at a constant rate. A wet-deposited film was subsequently formed behind the shearing bar; thereafter, the film was thermally annealed at 385 °C in ambient air. The bar pulling rates (10–90 mm/s),

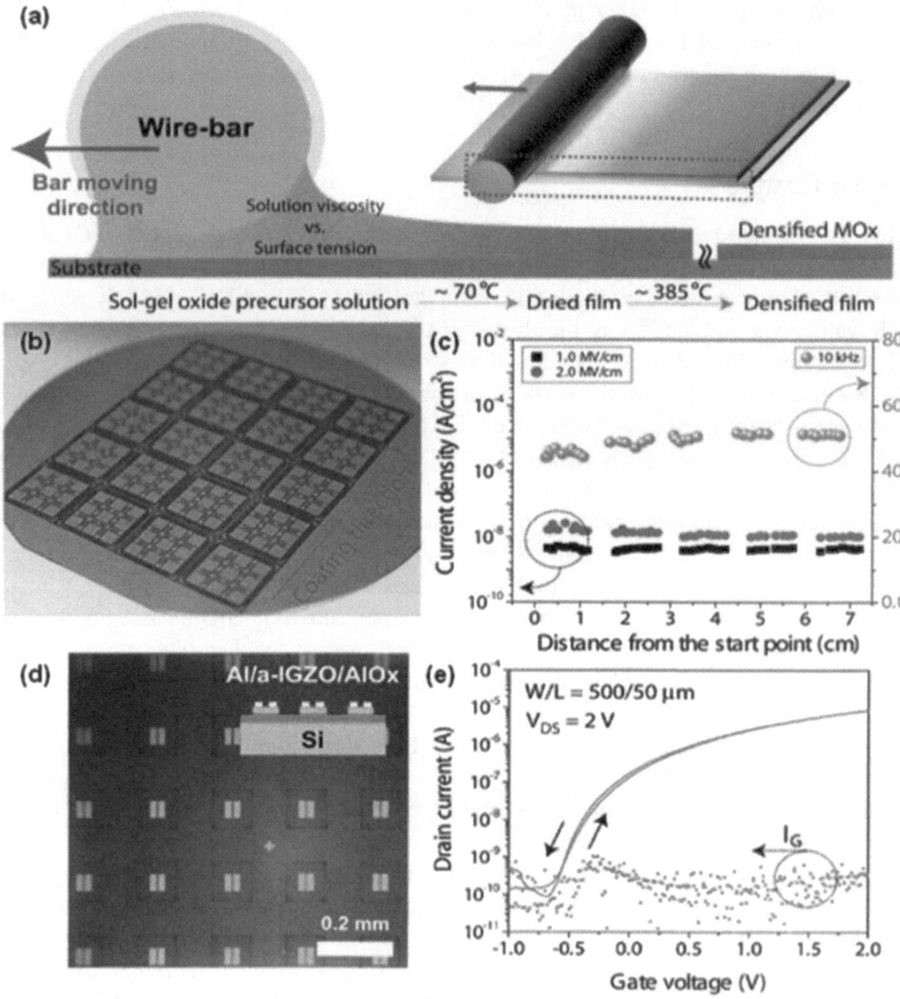

FIGURE 10.6 Large-area bar-coating and direct patterning of high-K metal oxide dielectrics. (a) Schematic illustration of wired-bar coating for depositing a very diluted sol-gel metal oxide precursor solution. (b) Photographic image of the uniformly bar-coated AlO_x film on the whole 4-inch silicon wafer finished with thermally deposited top gold electrodes. (c) Statistical distribution of leakage current density (at 1.0 or 2.0 MV/cm, left axis) and areal capacitance at 10 kHz (right axis) in the bar-coated sol-gel AlO_x dielectric device collected from different locations on the 4-inch silicon wafer. (d) Photo image and (e) electrical characteristic transfer curves of all bar-printed a-IGZO/AlO_x TFT arrays with top-contact aluminum source and drain electrodes. Reproduced with permission [43]. Copyright 2015 Wiley-VCH.

sol–gel precursor concentrations (0.2–0.4 M), and the number of coatings could adjust the AlO_x dielectric film thickness on the nanoscale, that is,10–40 nm(Figure 10.6a).

After thermal annealing, all the bar-coated thin films exhibited good film uniformity (~16 nm) over a large area and an extremely smooth surface (RMS < 0.1 nm), as well as excellent dielectric properties, including a low current density (~10^{-8} A/cm^2 at 2.0 MV/cm) and high areal capacitance (> 380 nF/cm^2 at 10 kHz, K ~ 7.9) (Figure 10.6b, c). Remarkably, this modified bar-coating method based on diluted sol–gel solutions allowed for the realization of fine controllability of the oxide film thickness in the nanoscale range, large-area uniformity in the film thickness and electrical performance, and minimal active solution usage, as indicated by the low consumption of 30 μL of the 0.1 M sol–gel solution for coating a four-inch silicon wafer. Further, different dielectric thin films with high-K values, HfO_x, were also fabricated in the same manner, resulting in a dielectric constant of K ~ 11.0, with similar good large-area uniformity. Furthermore, the authors combined this selective surface-wetting method via self-assembled monolayer patterning with the wire-bar printing method to achieve a direct patterned semiconducting IGZO film without photolithography, i.e., lithography-free direct printing [36] (Figure 10.6d). The resultant subsequently bar-coated IGZO/AlO_x TFT arrays exhibited reliable and improved electrical properties, such as an electron mobility of > 5 cm^2/V s and an on/off current ratio of 10^5, with a low-voltage operation (2 V), as illustrated in Figure 10.6e.

10.5 LOW-TEMPERATURE FABRICATION OF HIGH-K METAL OXIDE DIELECTRIC FILMS

Although the fabrication of solution-processed metal oxide films has demonstrated various advantageous features (e.g., simple process and broad properties design), high processing temperatures (> 400 °C) and long processing times remain indispensable for the realization of high-quality oxide films with high densities that are free of pinholes/cracks. In particular, these represent the initial conditions that must be satisfied for dielectric films to function as insulators, preventing leakage currents in TFTs. However, such high processing temperatures often limit the use of low-temperature plastic substrates in flexible electronics. Correspondingly, various approaches for lowering the processing temperatures have been adopted to compensate for the required thermal energy and reduce activation energy barriers during the formation of metal oxide dielectric layers using external energy resources, particularly photonic resources, from synthesized sol–gel solutions. As the most representative route, the photonic annealing process based on different variables (e.g., light emitting source, wavelength, intensity, irradiation time, and ambient gas) not only lowers the thermal energy barrier for the conversion and formation of metal–oxygen frameworks but also creates high-quality metal oxide thin films on flexible plastic substrates, with the maximum temperature not exceeding 300 °C.

10.5.1 Photonic Energy-derived (Post-)treatment of Solution-processed Metal Oxide Dielectric Films

Despite tremendous efforts devoted to the development of various low-processing temperature methods, the final processing temperatures are still too high for most plastic substrates (250–350 °C). Among these, the ultraviolet (UV) photonic process has significantly addressed this issue by further decreasing the overall processing temperature [76, 77]. Park et al. pioneered the development of a distinct photochemical route without intentional additives for low-temperature metal oxide photoactivation, which can be applied to several types of metal oxide materials in a generalizable manner [78]. They proposed an initial deep UV (DUV) photochemical conversion of metal nitrate-based sol–gel precursors to fabricate metal oxide TFT devices based on indium-based amorphous oxide semiconductors (e.g., IGZO, IZO, and In_2O_3) at temperatures close to the room temperature using low-pressure mercury lamps with radiation wavelengths of 254 (90%) and 185 nm (10%). They suggested that the DUV photochemical conversion eliminated organic impurities effectively from the solution-deposited films, resulted in photochemical cleavage, enhanced hydrolysis/condensation reactions, and efficiently developed M–O–M frameworks, even at low temperatures.

As comprehensive follow-up investigations, Park et al. conducted in-depth studies on the fundamental reaction mechanism underlying the DUV photoactivation of sol–gel metal oxide films, including various dielectric materials (e.g., AlO_x, ZrO_x, and HfO_x), based on spectroscopic techniques [41]. The experimental evidence proved the possibility of *in-situ* production of highly reactive radicals (e.g., NO_2 and OH radicals) from the photolysis of residual (metal) ligands (e.g., nitrate, isopropoxide, and acetylacetonate) inside porous films under DUV photoactivation. Notably, rapid radical-mediated reactions can effectively aid in eliminating organic impurities

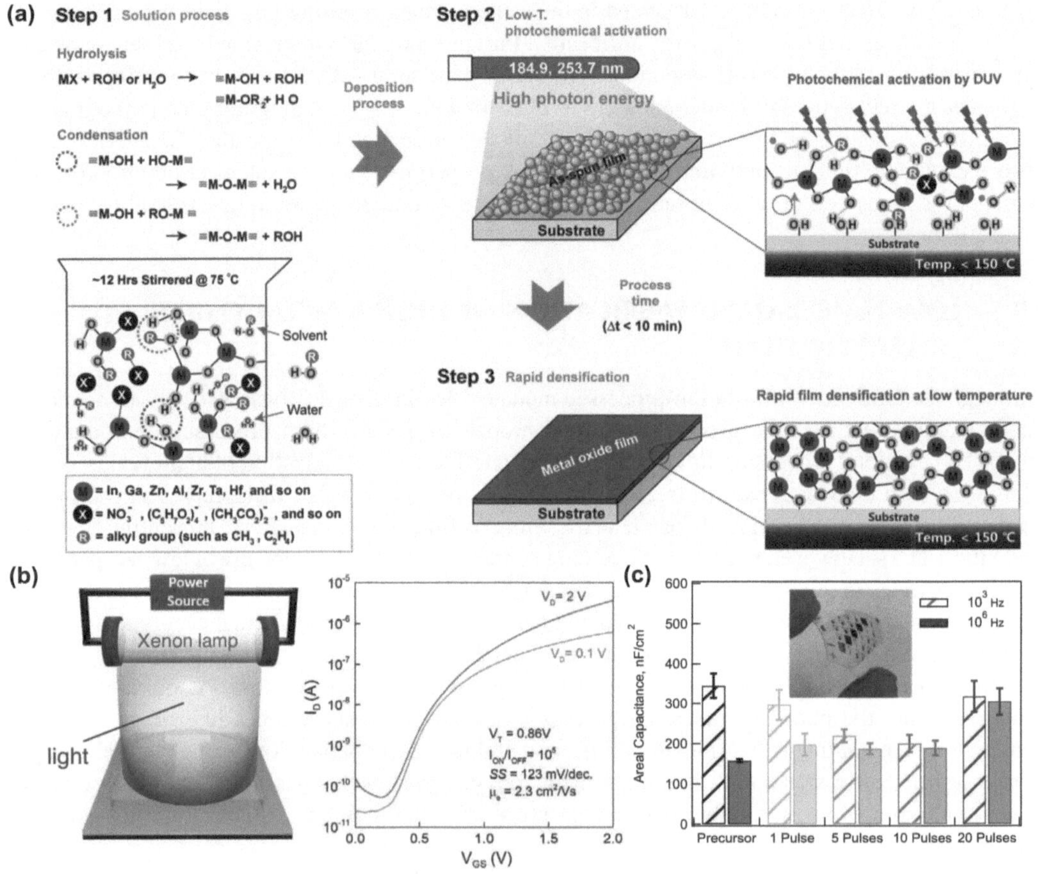

FIGURE 10.7 (a) Photochemical activation reaction pathways of sol-gel metal oxide materials and the proposed mechanism. Overall, schematic illustrations of the rapid low-temperature photoactivation of various sol-gel metal oxide films with three steps. (Reproduced with permission [41]. Copyright 2015 Wiley-VCH.) (b) Schematic of the photonic curing process using high-intensity light pulses generated by a xenon lamp. The deposited precursor films on top of the Al electrode converted to the densified oxide layer after the light pulse exposures onto the $In_2O_3/ZrO_2/Al_2O_3/Al$ electrode/glass substrate. (Right) Electrical characteristics of transfer curves of In_2O_3/ZnO heterojunction-structured TFT devices with Al_2O_3/ZrO_2 bilayer dielectrics. (Reproduced with permission [81]. Copyright 2020 Wiley-VCH.) (c) Bar-graph showing the low-frequency (10^2 Hz, hashed bars) and high-frequency (10^6 Hz, solid bars) areal capacitance for each condition of photonically cured films (6.32 J/cm^2 per pulse, 20 ms pulse length, 0.2 Hz repetition rate) using 1-20 pulses. (Inset) a photo image of a photonically fabricated device on PEN flexible substrate. (Reproduced with permission [82]. Copyright 2020 Nature Publishing Group.)

(direct photodecomposition) and promote the consequent radical-assisted local processes, such as polycondensation of metal–oxygen bond formation and rapid densification of the entire solid metal oxide film (Figure 10.7a). Further, the highly effective DUV photoactivation process at a small processing temperature (~150 °C for 30 min) was successfully implemented to fabricate a variety of high-performance dielectric and semiconducting sol–gel oxide films on glass and ultra-thin plastic substrates.

Jo et al. demonstrated bulky metal–oxo–hydroxyl nanoclusters with few organic impurities; these nanoclusters were subjected to a spatially controllable photoactivation process, resulting in highly uniform and dense AlO_x dielectric films at low temperatures (< 60 °C) [79]. The metal–oxo–hydroxy nanocluster-based AlO_x solutions could provide a preformed state of sol–gel metal–oxygen networks for not solely energy-efficient decomposition of organic residues and high condensation/densification at low temperatures, resulting in the direct fabrication on flexible polymeric and stretchable substrates with excellent dielectric properties, such as a low leakage current density (~10^{-8} A/cm^2 at 2 MV/cm), high breakdown electric field (> 6 MV/cm), and excellent electrical stability.

Lee et al. reported the rapid fabrication of a reliable sol–gel AlO_x dielectric bilayer on a silicon substrate via DUV photoactivation with a low light intensity [80]. When fabricating stable dielectric AlO_x thin films during DUV photoactivation on the silicon substrate, an aluminum silicate ($AlSiO_x$) thin film, as a stable oxide interfacial layer, was formed within an exceptionally short processing time (within 5 min), resulting in effective dielectric polarization and minimum current leakages. Based on various physicochemical characterizations, the authors also suggested a photoactivation mechanism of the $AlSiO_x$ layer that allowed the aluminum atoms in AlO_x to be ionically diffused into the silicon oxide layer. They also revealed a correlation between the maximum film thickness and solute concentration that could result in the maximum efficiency of the DUV photoactivation process via optical analysis. Finally, the rapidly fabricated sol–gel oxide TFTs were found to exhibit good electrical performance with extremely low operation voltages (~0.5 V).

Anthopoulos et al. reported the sequential sol–gel solution deposition of Al_2O_3/ZrO_2 and In_2O_3/ZnO bilayers, accompanied by the photonic annealing process using a pulsed xenon flash lamp, to overcome two dominant technical challenges: (1) high process temperatures and (2) prolonged annealing times (Figure 10.7b) [81]. Under an ambient atmosphere, pulsed-flash photonic processing provided a temporal and sharp increase in the temperature in the internal region of metal oxide precursor layers, while dramatically reducing the overall processing time (< 60 s), and this was achieved without compromising the film quality of the metal oxide layer. Notably, flash photonic annealing with a broad spectrum of 200–1,100 nm enables the use of different energy densities and simulated pulse lengths and numbers to minimize the degradation of the flexible plastic substrates vulnerable to high temperatures. The dielectric bilayers fabricated under photothermally induced precursor conversion exhibited good dielectric properties, such as leakage current densities of 10^{-6} A/cm^2 at 1 MV/cm and high areal capacitances of ~120 nF/cm^2. Though consecutive photonic conversion on a flexible polyethylene naphthalate substrate (PEN), the fabricated TFT devices with In_2O_3/ZnO heterojunction semiconductor layers on the fabricated dielectric layer exhibited high mobility (~2.3 cm^2/V s) under low operation voltages (~2 V).

Hsu et al. also presented the rapid photonic curingof sol–gel dielectric ZrO_2 layers with high-K values using a pulsed xenon flash system to address the two significant challenges (i.e., high processing temperature and long processing time during the (post-) annealing step) for low-cost manufacturing and high-throughput processing(Figure 10.7c) [82]. They concurrently employed the photothermal curing process and the self-aligned patterning process with a shadow mask for rapid and efficient precursor conversion of the oxide layer placed atop the patterned Al contact electrodes. Under varying pulsed photonic conditions, the fabricated ZrO_2 dielectric layers exhibited excellent dielectric properties with high areal capacitances (200 nF/cm^2), low leakage current densities (~10^{-7} A/cm^2 at 2 MV/cm), and high breakdown fields (~8 MV/cm from double ZrO_2 layers). They suggested that the photonic curing system under adjustable conditions could produce high-quality solution-based oxide dielectric films on flexible plastic substrates (e.g., PEN) based on a rapid condensation/densification reaction for a high throughput beyond 30 m/min.

10.5.2 Combustion-derived (Post-)treatment of Solution-processed Metal Oxide Dielectric Films

Solution-combustion synthesis (SCS) is an energy-efficient processing method that utilizes the self-sustained exothermic reactions between fuels (i.e., reducing agents) and oxidizers (i.e., oxidizing agents) in liquid metal–organic chemicals. Here, the liquid media, such as aqueous or sol–gel solutions, have fuels and oxidizers that can react in a self-propagating manner under the supply of localized thermal energy (ignition) to lower the processing temperature. This process produces

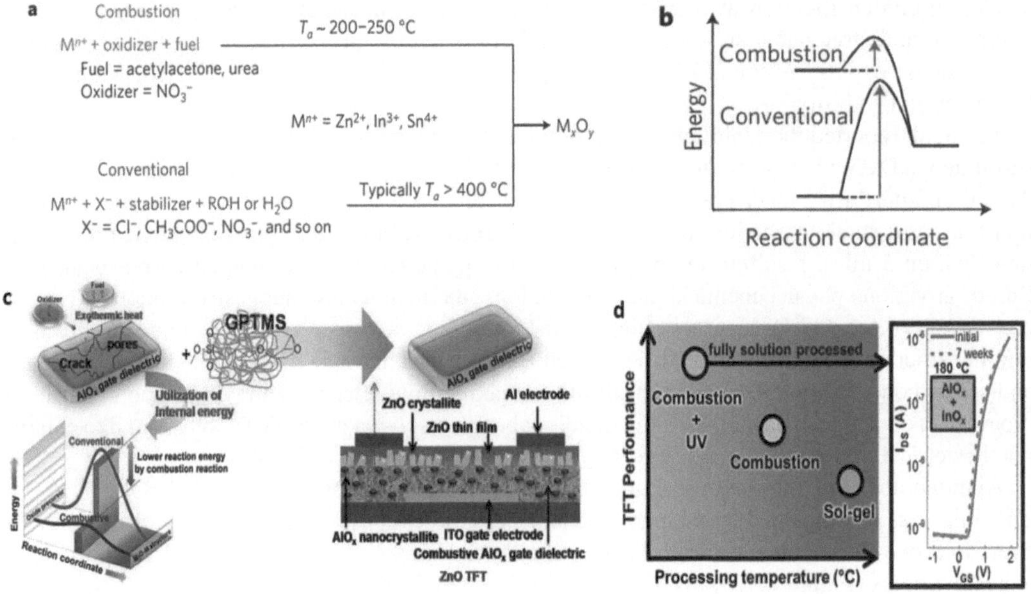

FIGURE 10.8 (a) Two different synthetic approaches of sol-gel processed metal oxide film formation: combustion and conventional routes, respectively. (b) Energetics and reaction coordinate diagram of combustion and conventional thermal processes. (Reproduced with permission [42]. Copyright 2011 Nature Publishing Group.) (c) Schematic illustration of the AlO_x gate dielectric films prepared by self-combustion synthesis of metal precursors mixed with organic compounds for the enhanced surface morphology. (Reproduced with permission [86]. Copyright 2014 Royal Society of Chemistry.) (d) Combustion synthesis process for lower processing temperatures by combining thermal annealing with far-UV irradiation for efficient organic residue degradation and promoted film densification of AlO_x/In_2O_3 thin films, resulting in negligible low hysteresis and very low leakage currents. (Reproduced with permission [87]. Copyright 2015, American Chemical Society.)

self-generated heat that converts the precursors into solid oxide products. Typically, SCS-based functional oxide materials can possess various structures and better qualities than those fabricated by conventional annealing processes [83]. In 2011, Marks et al. published the first study on SCS-based dense metal oxide thin film preparation at low processing temperatures (~200 °C) for TFT device fabrication (Figure 10.8a, b) [42]. Here, a highly exothermic combustion reaction created high localized temperatures inside the gel film by incorporating either acetylacetone or urea as fuel and nitrate as an oxidizer(Figure 10.8a). Figure 10.8b illustrates an energetic and reaction coordinated diagram between the combustion and conventional thermal pathways. Notably, for dielectric and insulating applications, minimizing the porosities and surface roughness of metal oxide films, which can induce significant leakage currents and interfacial trapping sites, is crucial. In particular, various combinations of fuels and oxidizers for functional combustible precursors have recently garnered significant attention, along with new fuels, such as representative conventional combustion fuels (e.g., urea, acetylacetone, and hydrazine) and various acidic fuels (e.g., ethylenediamine tetra-acetic acid, oxalic acid, oleic acid, citric acid, tartaric acid, and malic acid) [84, 85].

Cho et al. reported the use of two different types of aluminum precursor ligands, $Al(C_2H_5O_2)_3$ and $Al(NO_3)_3$, as the fuel (acetylacetone) and oxidizer (nitrate), respectively, as illustrated in Figure 10.8c [86]. They further utilized an additional polymer binder (3-glycidoxypropyl)trimethoxysilane (GPTMS) to enhance the resulting surface morphology and stable dielectric properties. The two different aluminum precursor ligands exhibited low-temperature self-combustive reactions with intense exothermic behavior at approximately 170 °C based on thermogravimetry-differential scanning calorimetry. However, when only the AlO_x self-combustible precursor without organic binders was deposited, the resultant dielectric films exhibited poor surface morphologies with numerous cracks and nano-pores during film formation, leading to high leakage currents and numerous charge-trapping sites. The authors determined that epoxy moieties of GPTMS (K ~ 16 at 100 Hz) could be polymerized with the aluminum precursor network during annealing to overcome the morphology issue. The final dielectric layer (~200 nm) with the optimal molar ratio between the aluminum self-combustive precursor and GPTMS binder (1:8) exhibited a smooth surface roughness with a very low leakage current density (3.3×10^{-9} A/cm^2 at 2 MV/cm) and a K value of ~ 8.7 at 1 Hz after annealing at 250 °C.

In 2014, Yang et al. proposed a facile route for fabricating patterned metal oxide films at low temperatures by combining combustion chemistry with direct light patterning [39]. They synthesized AlO_x precursor solutions with additional additives of acetylacetone (AcAcH) and ammonium hydroxide (NH_4OH), resulting in high photosensitivity to UV light. Using a photo-pattern shadow mask, the selectively photo-exposed sol–gel oxide films could realize insoluble patterned oxide formation induced by the radicals photogenerated from the photoreactive species and exothermic reaction inside the layer. The IGZO/AlO_x TFT devices fabricated on a flexible polyimide substrate exhibited impressive electrical properties and a high electron mobility of 84.4 cm^2/V s with a current on/off ratio of > 10^5.

Fortunato et al. demonstrated a combination of SCS and far UV (FUV) irradiation for the fabrication of high-quality aluminum oxide dielectric films, that is, higher densification of the solution-processed metal oxide thin films at low processing temperatures (< 200 °C) (Figure 10.8d) [87]. As mentioned, UV photochemical activation can promote the formation of metal–oxygen (M–O) bonds via the successive and homogeneous photo-cleavage of alkoxy groups and organic residues. The proposed integrated approach comprising two low-temperature processing techniques between external photonic and internal combustion routes produced high-quality aluminum oxide layers with improved dielectric properties for outstanding TFT applications. Moreover, the dielectric layers were successfully employed in solution-processed indium oxide TFTs, resulting in improved electrical performance under low operation voltages and good uniformity and reproducibility under stress and aging conditions.

10.6 COMBINED APPROACHES FOR HIGH-THROUGHPUT METAL OXIDE DIELECTRIC FILM FORMATION

Although ensuring the high quality and reliability of solution-based oxide dielectric films is essential, the incorporation of such films into continuous coating/printing processes of oxide dielectric films within a short processing time is crucial for high-throughput manufacturing. Accordingly, lowering the high processing temperatures and times with complete conversion from chemical precursor films to metal oxide networks is necessary to overcome the poor productivity encountered in fabricating condensedand uniform oxide dielectric layers on flexible plastic substrates. Several research groups have reported the rapid fabrication of functional metal oxide layers with good uniformity and high densification for entirely printed electronic devices on flexible plastic substrates. This has been accomplished by improving existing technologies and investigating unknown mechanisms or by combining various approaches among novel synthetic designs for improved oxide precursor solutions and facile routes to supply external energy (e.g., photonic irradiation) and precisely control the environmental conditions for the ink-deposited films.

10.6.1 Multi-nozzle Inkjet Printing with UV-photonic Annealing

Subramanian et al. presented integrated techniques comprising the high-throughput multi-nozzle inkjet printing and subsequent UV-photonic (post-)annealing process for the fabrication of low-temperature (250 °C) solution-printed InO_x TFT devices with UV-annealed high-K AlO_x gate dielectrics [88]. For this, first, they considered the design of AlO_x solution inks (e.g., varying solute concentrations for the ink viscosity and desired nanoscale thickness for printing uniformity), compatible with multi-nozzle inkjet printing methods to produce good film uniformity. They highlighted the difference between single-nozzle and multi-nozzle inkjet printing methods to eliminate intermittent drying and enhance film uniformity with consecutively printed lines. They also addressed the influence of UV-photonic annealing during the formation ofAlO_x dielectric layers based on thickness changes and device bias-stress tests. Finally, they successfully fabricated all-inkjet-printed oxide TFT devices (i.e., high mobility (~12 cm^2/V s) with negligible hysteresis and improved operational stability) without additional lithographic patterning and vacuum processing, which are known to impede high-throughput production otherwise.

10.6.1.1 Combustion Synthesis Using Scalable Coating Methods

Facchetti et al. demonstrated ascalable and rapid spray-combustion synthesis (spray-SCS), allowing the fabrication of high-quality metal oxide semiconducting and dielectric layers within minutes, rivaling that of the vacuum-based sputtered films (Figure 10.9a) [89]. Notably, conventional solution processes require inefficient multiple depositions and time-consuming annealing sequences to achieve the desired film thicknesses in the nanoscale range (> 5 nm), which inevitably produces bulk trap states at the semiconductor/dielectric interfaces. In the preceding study, the SCS process produced high-quality ZrO_x gate dielectric films at 300 °C, which exhibited a smooth surface morphology with low roughness, high capacitance (> 600 nF/cm^2), low leakage current density (10^{-7} A/cm^2 at 2 MV/cm), and high breakdown field (~9 MV/cm). After a year, the authors employed an automated spray-coating system with an ultrasonic nozzle to modulate mathematically defined droplet sizes [90]. The ultrasonic-enhanced SCS approach developed all-solution-processed metal oxide layers in TFTs without other vacuum-based electrode deposition processes that are incompatible with high-throughput roll-to-roll fabrication. The all-oxide SCS-derived TFT devices, including InGaZnO semiconductors, ZrO_x gate dielectrics, and ITO contacts on glass substrates, exhibited good electron mobilities up to 5.3 cm^2/V s with good dielectric–semiconductor interfacial qualities. Thus, spray-SCS-based printing is a promising solution-based oxide film deposition method, which is compatible with large-scale high-throughput fabrication (e.g., R2R) of future printed electronics.

The foregoing research group subsequently introduced a novel co-fuel-assisted solution-combustion method that effectively reducedthe general ignition temperature of the combustion process and promoted rapid completion of the combustion reactions and condensation of metal–oxygen bonds. In addition, the authors utilized several blade-coating techniques simultaneously to realize a highly efficient solution deposition process, combustion blade-coating (blade-SCS), on the wafer-scale substrate [91]. In this study, a fluorinated co-fuel (1,1,1-Trifuluoroacetylacetone, FAcAcH) was introduced in the redox reactions between the nitrate oxidizers and acetylacetone fuel for both semiconductors and dielectric films (Figure 10.9b). The film formation for the aluminum oxide dielectrics occurred within a short time (less than 1 min) via the ultrafast reactions and metal–oxygen lattice condensation in the initial pre-annealing state with a temperature of 120 °C. Therefore, the co-fuel-assisted blade-SCS IGZO/AlO_x TFTs fabricated in a minute presented high field-effect mobility (> 25 cm^2/V s) and stable operational performance under a bias stress of over 1 h. Further, wafer-scale TFT arrays (> 100 device units) presented high-quality device characteristics with average electron mobilities (~18 cm^2/Vs) and negligible hysteresis at low voltages (< 2 V) (Figure 10.9c–e). Thus, in conjunction with the SCS process, the co-fuel-assisted blade-SCS method is expected to advance efficient and continuous additive manufacturing, as well as the production of R2R-compatible large-scale printed electronic systems.

10.6.1.2 Photo-combustion Synthesis for High-throughput Metal Oxide Dielectric Film Formation

A few other research groups also successfully demonstrated advanced integrated combustion synthesis routes for low-temperature solution-processed oxide dielectrics, particularly for AlO_x dielectric layers, for high-throughput fabrication techniques. In particular, Fortunato et al. presented the fabrication of high-quality large-area AlO_x thin films via a novel SCS route photo-triggered by

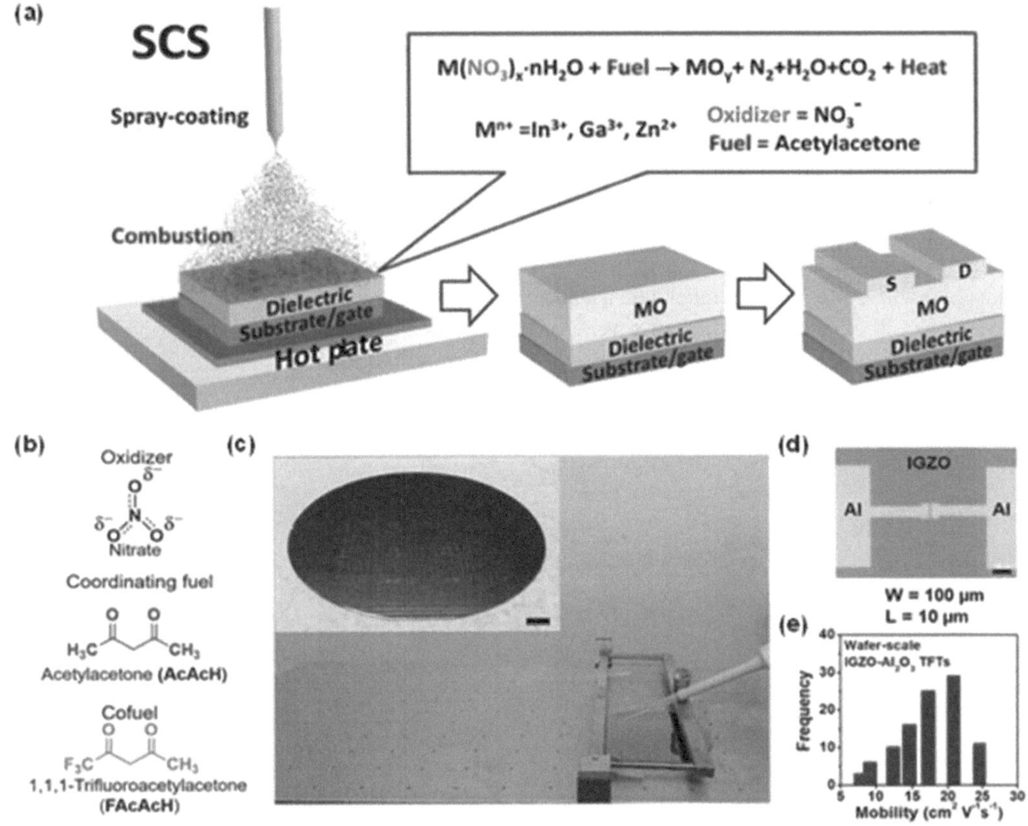

FIGURE 10.9 (a) Spray-based solution combustion synthesis (Spray-SCS) of metal oxide films. A schematic of the Spray-SCS process used for depositing metal oxide films under ambient conditions and the corresponding bottom gate top contact TFT structure. (Inset) The simple chemical formula of a combustion redox reaction where metal nitrate and acetylacetone act as an oxidizer and fuel, respectively. (Reproduced with permission [89]. Copyright 2015, Proceedings of the National Academy of Sciences.) (b) Chemical structures of nitrates (as an oxidizer), acetylacetone (AcAcH, as a coordinating fuel), and 1,1,-trifuloroacetylacetone (FAcAcH, as a cofuel) for efficient solution combustion synthesis reactions. (c) Photo of the scalable blade coating process with SCS reactions (so-called blade-SCS) (Inset: blade-coated device array on a 4-inch silicon wafer). (d) An optical image of a patterned low-voltage IGZO/AlO_x TFTs with aluminum source/drain electrodes. (e) Statistical distribution of electrical mobilities of the large-scale fabricated device arrays. (Reproduced with permission [91]. Copyright 2019 National Academy of Sciences.)

(KrF pulsed) excimer laser annealing (ELA), and they demonstrated a cost-efficient high-throughput manufacturing process (Figure 10.10a) [92]. By integrating the SCS and ELA approaches, the fabricated AlO_x dielectric thin films achieved rapid densification within a short processing time (less than 15 min). The optimized dielectric layers were well-formed through short drying cycles at a low temperature of 150 °C for 15 min, followed by the ELA process (fluence/number of pulses). They exhibited a low surface roughness (< 2.8 nm) and optimized dielectric properties compared to the reference samples (thermally annealed at 300°C for 1 h): low leakage current density (10^{-6} A/cm^2 at 1 MV/cm), high breakdown voltage (up to 4 MV/cm), and dielectric constant (K ~ 9 at 1 kHz). The fabricated TFT devices comprising ELA-SCS AlO_x dielectric layers with sputtered IGZO semiconducting layers exhibited improved saturation mobility (~20.4 cm^2/Vs) and turn-on voltage (~0 V) (Figure 10.10b). Because the ELA method has unique features, such as excellent surface/area selectivity and a faster processing time, this combination approach presents significant potential for further development in the high-throughput printed electronics industry.

Furthermore, first, the same group subsequently demonstrated the implementation of the combined technique in large-area AlO_x dielectric thin films on a flexible polyimide (PI) substrate, using flexographic printing, an R2R-compatible and industrially scalable printing method (Figure 10.10c) [93]. Accordingly, they reported the large-area printing of highly stable AlO_x dielectric thin films with high-K values on flexible substrates via a combined low-temperature (200 °C) annealing process involving SCS and UV lamp irradiation (with a high-throughput of 50 m/min). The fabricated AlOx dielectric exhibited a low leakage current density (10^{-6} A/cm^2 at 1 MV/cm) with a K value of ~ 8.2 (at 1 Hz). As illustrated in Figure 10.10d, the flexible TFT devices with the ink-jetted In_2O_3 layer fabricated on the flexographically-printed AlO_x layer demonstrated high operational stability with an insignificant voltage shift under electrical/mechanical stresses. Thus, based on the foregoing analysis, the effective combination of several techniques is expected to facilitate the production of large-area printed flexible electronic devices for high-throughput manufacturing.

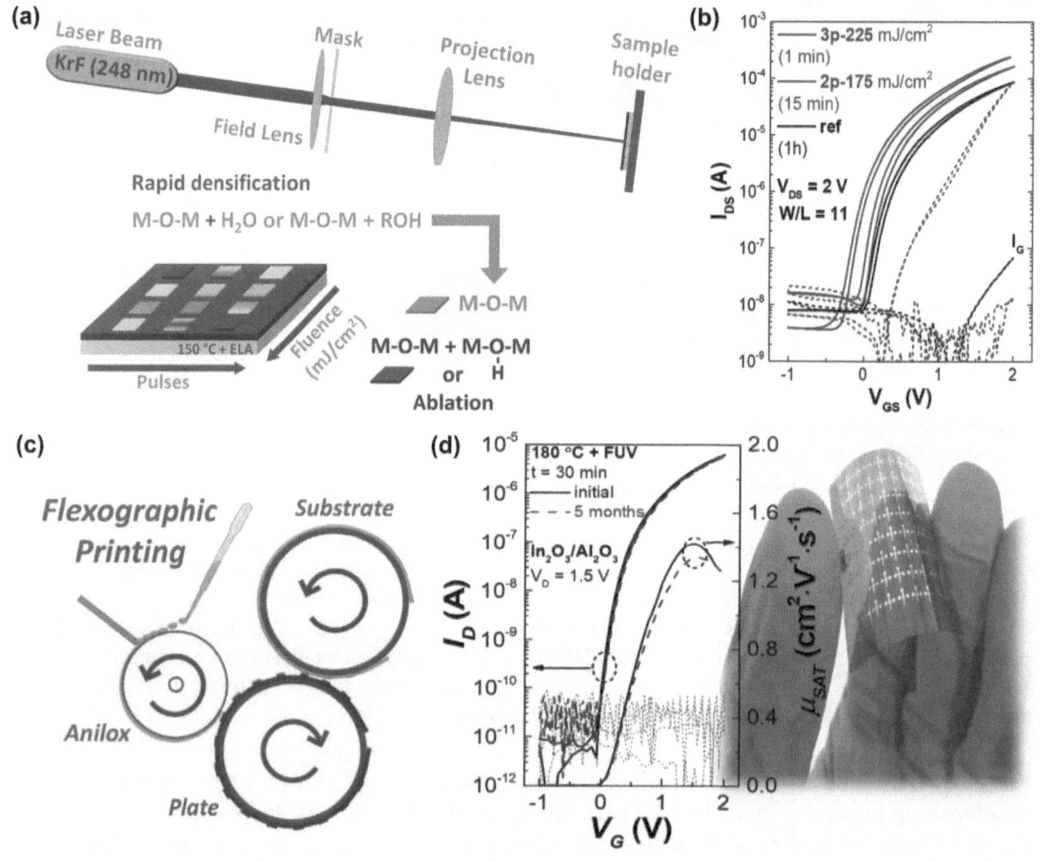

FIGURE 10.10 (a) Overall schematic illustration of the photonic annealing process with excimer laser for SCS-based AlOx thin films with varying light fluences and number of pulses, right after soft-baking for a gentle solution-film drying at 150 °C with different times (less than 15 min). (b) Electrical characteristic transfer curves of IGZO TFT devices with the excimer laser-assisted AlO_x dielectric layers prepared with different light fluence, pulse number, and drying times. (Reproduced with permission [92]. Copyright 2020 Royal Society of Chemistry.) (c) Schematic of the flexographic printing process of the SCS-based AlO_x dielectric ink on top of the thermally deposited bottom electrodes. (d) Electrical characteristic transfer curves of initial and aged (over five months) In_2O_3 TFT devices with the AlO_x dielectric layer through the subsequent printing processes (1: flexographic printing, 2: UV photo-assisted annealing with SCS reactions, 3: inkjet printing, and 4: S/D electrode deposition by thermal evaporation). (Reproduced with permission [93]. Copyright 2020 John Wiley and Sons, Inc.)

10.7 CONCLUSIONS

Over the years, the technological advancements have been dramatic, primarily focused on the realization of high-performance, low-power-operative, easily portable and light electronics and the implementation of scalable processing conditions in silicon-based electronics. However, limitations on the scaling of SiO_2 pioneered the extensive research on high-K oxide dielectrics that could be used as alternative dielectric materials in low-power large-scale integrated circuits. In parallel, several researchers attempted to replace conventional vacuum deposition techniques with advanced solution-processing technologies, particularly aimed at realizing industrial applications of flexible printed electronics in a cost-effective manner. Owing to the aforementioned potential attributes, sol–gel metal oxide dielectric materials can play a vital role in creating and leading a new generation of high-performance printed electronic systems, including disposable/transient/environmentally friendly electronics, radio-frequency identification (RFID) tags and intelligent display labels.

This chapter extensively reviewed various types of solution-processed metal oxide dielectric materials and their synthetic routes, large-scale and rapid oxide film fabrication methods with low-temperature techniques to overcome current technological limits. Despite unprecedented progress and developments in the research field of sol–gel metal oxide dielectrics, several challenges with regard to the materials, processing, and device architectures/performance still persist. Furthermore, combined withthe surface chemical modification and the expeditious and low-temperature oxide activation processes, cost-effective fabrication methods for large-area sol–gel ink printing, such as gravure and flexography, are desired for the industry-scale fabrication of metal oxide electronic devices and complex circuits. We envision that the advancements in sol–gel metal oxide electronics can be accelerated with a more comprehensive understanding of fundamental material perspectives and the innovative development of device engineering approaches. We expect that significant research efforts would be devoted to overcoming the remaining technical and technological hurdles, further paving the way to numerous functional oxide applications, such as advanced low/high-power electronics, photovoltaic cells, and energy storage and conversion systems.

REFERENCES

[1] Wang C, Hwang D, Yu Z, et al. User-interactive electronic skin for instantaneous pressure visualization. *Nat Mater.* 2013;12:899–904.

[2] Son D, Lee J, Qiao S, et al. Multifunctional wearable devices for diagnosis and therapy of movement disorders. *Nat Nanotechnol.* 2014;9:397–404.

[3] Schwartz G, Tee BC-K, Mei J, et al. Flexible polymer transistors with high pressure sensitivity for application in electronic skin and health monitoring. *Nat Commun.* 2013;4:1859.

[4] Hwang S-W, Tao H, Kim D-H, et al. A Physically Transient Form of Silicon Electronics. *Science.* 2012;337:1640–1644.

[5] Aleeva Y, Pignataro B. Recent advances in upscalable wet methods and ink formulations for printed electronics. *J Mater Chem C.* 2014;2:6436–6453.

[6] Pasquarelli RM, Ginley DS, O'Hayre R. Solution processing of transparent conductors: from flask to film. *Chem Soc Rev.* 2011;40:5406–5441.

[7] Fukuda K, Someya T. Recent Progress in the Development of Printed Thin-Film Transistors and Circuits with High-Resolution Printing Technology. *Adv Mater.* 2017;29:1602736.

[8] Garlapati SK, Divya M, Breitung B, et al. Printed Electronics Based on Inorganic Semiconductors: From Processes and Materials to Devices. *Adv Mater.* 2018;30:1707600.

[9] Carlos E, Branquinho R, Martins R, et al. Recent Progress in Solution-Based Metal Oxide Resistive Switching Devices. *Adv Mater.* 2021;33:2004328.

[10] Park S, Kim C-H, Lee W-J, et al. Sol-gel metal oxide dielectrics for all-solution-processed electronics. *Mater Sci Eng, R.* 2017;114:1–22.

[11] Lee W-J, Kwak T, Choi J-G, et al. Solution-processed metal oxide dielectric films: Progress and outlook. *APL Mater.* 2021;9:120701.

[12] Liu A, Zhu H, Sun H, et al. Solution Processed Metal Oxide High-κ Dielectrics for Emerging Transistors and Circuits. *Adv Mater.* 2018;30:1706364.

[13] Wang B, Huang W, Chi L, et al. High-K Gate Dielectrics for Emerging Flexible and Stretchable Electronics. *Chem Rev*. 2018;118:5690–5754.
[14] Glynn C, O'Dwyer C. Solution Processable Metal Oxide Thin Film Deposition and Material Growth for Electronic and Photonic Devices. *Adv Mater Interfaces*. 2017;4:1600610.
[15] Bécu S, Crémer S, Autran JL. Capacitance non-linearity study in Al_2O_3 MIM capacitors using an ionic polarization model. *Microelectron Eng*. 2006;83:2422–2426.
[16] Smith WF, Hashemi J, Presuel-Moreno F. *Foundations of materials science and engineering*. McGraw-hill New York; 2006.
[17] Ortiz RP, Facchetti A, Marks TJ. High-K Organic, Inorganic, and Hybrid Dielectrics for Low-Voltage Organic Field-Effect Transistors. *Chem Rev*. 2010;110:205–239.
[18] Barquinha P, Martins R, Pereira L, et al. *Transparent oxide electronics: from materials to devices*. John Wiley & Sons; 2012.
[19] Houssa M. *High k gate dielectrics*. CRC Press; 2003.
[20] Robertson J. High dielectric constant gate oxides for metal oxide Si transistors. *Rep Prog Phys*. 2005;69:327.
[21] Barquinha P, Martins R, Fortunato E. *Gate Dielectrics in Oxide Electronics*. Transparent Oxide Electronics [Internet]. John Wiley & Sons, Ltd; 2012 [cited 2022 Dec 16]. p. 101–153. Available from: DOI: 10.1002/9781119966999.ch4
[22] Robertson J, Falabretti B. Band offsets of high K gate oxides on high mobility semiconductors. *Mater Sci Eng, B*. 2006;135:267–271.
[23] Leskelä M, Ritala M. Atomic Layer Deposition Chemistry: Recent Developments and Future Challenges. *Angew Chem, Int Ed*. 2003;42:5548–5554.
[24] Niinistö L, Nieminen M, Päiväsaari J, et al. Advanced electronic and optoelectronic materials by Atomic Layer Deposition: An overview with special emphasis on recent progress in processing of high-K dielectrics and other oxide materials. *Phys Status Solidi A*. 2004;201:1443–1452.
[25] Johnson RW, Hultqvist A, Bent SF. A brief review of atomic layer deposition: from fundamentals to applications. *Mater Today*. 2014;17:236–246.
[26] Palneedi H, Park JH, Maurya D, et al. Laser Irradiation of Metal Oxide Films and Nanostructures: Applications and Advances. *Adv Mater*. 2018;30:1705148.
[27] Kim HJ, Park K, Kim HJ. High-performance vacuum-processed metal oxide thin-film transistors: A review of recent developments. *J Soc Inf Disp*. 2020;28:591–622.
[28] Hench LL, West JK. The sol-gel process. Chem Rev. 1990;90:33–72.
[29] Brinker CJ, Scherer GW. *Sol-gel science: the physics and chemistry of sol-gel processing*. Academic Press; 2013.
[30] Song K, Yang W, Jung Y, et al. A solution-processed yttrium oxide gate insulator for high-performance all-solution-processed fully transparent thin film transistors. *J Mater Chem*. 2012;22:21265–21271.
[31] Liu Y, Guan P, Zhang B, et al. Ion Dependence of Gate Dielectric Behavior of Alkali Metal Ion-Incorporated Aluminas in Oxide Field-Effect Transistors. *Chem Mater*. 2013;25:3788–3796.
[32] Zhang B, Liu Y, Agarwal S, et al. Structure, Sodium Ion Role, and Practical Issues for β-alumina as a High-K Solution-Processed Gate Layer for Transparent and Low-Voltage Electronics. *ACS Appl Mater Interfaces*. 2011;3:4254–4261.
[33] Pal BN, Dhar BM, See KC, et al. Solution-deposited sodium beta-alumina gate dielectrics for low-voltage and transparent field-effect transistors. *Nat Mater*. 2009;8:898–903.
[34] Meyers ST, Anderson JT, Hong D, et al. Solution-Processed Aluminum Oxide Phosphate Thin-Film Dielectrics. *Chem Mater*. 2007;19:4023–4029.
[35] Avis C, Jang J. High-performance solution processed oxide TFT with aluminum oxide gate dielectric fabricated by a sol–gel method. *J Mater Chem*. 2011;21:10649–10652.
[36] Sung S, Park S, Cha S, et al. Direct patterning of sol–gel metal oxide semiconductor and dielectric films via selective surface wetting. *RSC Adv*. 2015;5:38125–38129.
[37] Xu W, Wang H, Xie F, et al. Facile and Environmentally Friendly Solution-Processed Aluminum Oxide Dielectric for Low-Temperature, High-Performance Oxide Thin-Film Transistors. *ACS Appl Mater Interfaces*. 2015;7:5803–5810.
[38] Branquinho R, Salgueiro D, Santos L, et al. Aqueous Combustion Synthesis of Aluminum Oxide Thin Films and Application as Gate Dielectric in GZTO Solution-Based TFTs. *ACS Appl Mater Interfaces*. 2014;6:19592–19599.

[39] Rim YS, Chen H, Liu Y, et al. Direct Light Pattern Integration of Low-Temperature Solution-Processed All-Oxide Flexible Electronics. *ACS Nano*. 2014;8:9680–9686.

[40] Park JH, Kim K, Yoo YB, et al. Water adsorption effects of nitrate ion coordinated Al_2O_3 dielectric for high performance metal-oxide thin-film transistor. *J Mater Chem C*. 2013;1:7166–7174.

[41] Park S, Kim K-H, Jo J-W, et al. In-Depth Studies on Rapid Photochemical Activation of Various Sol–Gel Metal Oxide Films for Flexible Transparent Electronics. *Adv Funct Mater*. 2015;25:2807–2815.

[42] Kim M-G, Kanatzidis MG, Facchetti A, et al. Low-temperature fabrication of high-performance metal oxide thin-film electronics via combustion processing. *Nat Mater*. 2011;10:382–388.

[43] Lee W-J, Park W-T, Park S, et al. Large-Scale Precise Printing of Ultrathin Sol–Gel Oxide Dielectrics for Directly Patterned Solution-Processed Metal Oxide Transistor Arrays. *Adv Mater*. 2015;27:5043–5048.

[44] Islam MM, Saha JK, Hasan MdM, et al. Spray-Pyrolyzed High-K Zirconium-Aluminum-Oxide Dielectric for High Performance Metal-Oxide Thin-Film Transistors for Low Power Displays. *Adv Mater Interfaces*. 2021;8:2100600.

[45] Shamala KS, Murthy LCS, Narasimha Rao K. Studies on optical and dielectric properties of Al_2O_3 thin films prepared by electron beam evaporation and spray pyrolysis method. *Mater Sci Eng, B*. 2004;106:269–274.

[46] Tan H, Liu G, Liu A, et al. The annealing effects on the properties of solution-processed alumina thin film and its application in TFTs. *Ceramics International*. 2015;41:S349–S355.

[47] Koslowski N, Sanctis S, Hoffmann RC, et al. Synthesis, dielectric properties and application in a thin film transistor device of amorphous aluminum oxide AlxOy using a molecular based precursor route. *J Mater Chem C*. 2019;7:1048–1056.

[48] Cochran EA, Woods KN, Johnson DW, et al. Unique chemistries of metal-nitrate precursors to form metal-oxide thin films from solution: materials for electronic and energy applications. *J Mater Chem A*. 2019;7:24124–24149.

[49] Li J, Pan Y, Xiang C, et al. Low temperature synthesis of ultrafine α-Al_2O_3 powder by a simple aqueous sol–gel process. *Ceramics International*. 2006;32:587–591.

[50] Xu W, Wang H, Ye L, et al. The role of solution-processed high-κ gate dielectrics in electrical performance of oxide thin-film transistors. *J Mater Chem C*. 2014;2:5389–5396.

[51] Liu G, Liu A, Zhu H, et al. Low-Temperature, Nontoxic Water-Induced Metal-Oxide Thin Films and Their Application in Thin-Film Transistors. *Adv Funct Mater*. 2015;25:2564–2572.

[52] Choi S, Park B-Y, Jeong S, et al. Low voltage operated, sol–gel derived oxide thin film transistor based on high-K Gd_2O_3 gate dielectric. *Mater Chem Phys*. 2013;138:1–4.

[53] Choi S, Park B-Y, Jung H-K. Growth and characterization of sol–gel prepared Gd_2O_3 films as gate insulators for Zn–Sn–O thin film transistors. *Thin Solid Films*. 2013;534:291–295.

[54] Esro M, Vourlias G, Somerton C, et al. High-Mobility ZnO Thin Film Transistors Based on Solution-processed Hafnium Oxide Gate Dielectrics. *Adv Funct Mater*. 2015;25:134–141.

[55] Jiang K, Anderson JT, Hoshino K, et al. Low-Energy Path to Dense HfO_2 Thin Films with Aqueous Precursor. *Chem Mater*. 2011;23:945–952.

[56] Avis C, Kim YG, Jang J. Solution processed hafnium oxide as a gate insulator for low-voltage oxide thin-film transistors. *J Mater Chem*. 2012;22:17415–17420.

[57] He W, Xu W, Peng Q, et al. Surface Modification on Solution Processable ZrO_2 High-K Dielectrics for Low Voltage Operations of Organic Thin Film Transistors. *J Phys Chem C*. 2016;120:9949–9957.

[58] Oluwabi A, Katerski A, Carlos E, et al. Application of ultrasonic sprayed zirconium oxide dielectric in zinc tin oxide-based thin film transistor. *J Mater Chem C*. 2020;8:3730–3739.

[59] Liu A, Liu GX, Zhu HH, et al. Fully Solution-Processed Low-Voltage Aqueous In_2O_3 Thin-Film Transistors Using an Ultrathin ZrOx Dielectric. *ACS Appl Mater Interfaces*. 2014;6:17364–17369.

[60] Park JH, Yoo YB, Lee KH, et al. Low-Temperature, High-Performance Solution-Processed Thin-Film Transistors with Peroxo-Zirconium Oxide Dielectric. *ACS Appl Mater Interfaces*. 2013;5:410–417.

[61] Jang J, Kang H, Chakravarthula HCN, et al. Fully Inkjet-Printed Transparent Oxide Thin Film Transistors Using a Fugitive Wettability Switch. *Adv Electron Mater*. 2015;1:1500086.

[62] Xu W, Cao H, Liang L, et al. Aqueous Solution-Deposited Gallium Oxide Dielectric for Low-Temperature, Low-Operating-Voltage Indium Oxide Thin-Film Transistors: A Facile Route to Green Oxide Electronics. *ACS Appl Mater Interfaces*. 2015;7:14720–14725.

[63] Anderson JT, Munsee CL, Hung CM, et al. Solution-Processed HafSOx and ZircSOx Inorganic Thin-Film Dielectrics and Nanolaminates. *Adv Funct Mater*. 2007;17:2117–2124.

[64] Jiang K, Meyers ST, Anderson MD, et al. Functional Ultrathin Films and Nanolaminates from Aqueous Solutions. *Chem Mater*. 2013;25:210–214.

[65] Jo J-W, Kim J, Kim K-T, et al. Highly Stable and Imperceptible Electronics Utilizing Photoactivated Heterogeneous Sol-Gel Metal–Oxide Dielectrics and Semiconductors. *Adv Mater*. 2015;27:1182–1188.

[66] Park JH, Yoo YB, Lee KH, et al. Boron-Doped Peroxo-Zirconium Oxide Dielectric for High-Performance, Low-Temperature, Solution-Processed Indium Oxide Thin-Film Transistor. *ACS Appl Mater Interfaces*. 2013;5:8067–8075.

[67] Yang W, Song K, Jung Y, et al. Solution-deposited Zr-doped AlOx gate dielectrics enabling high-performance flexible transparent thin film transistors. *J Mater Chem C*. 2013;1:4275–4282.

[68] Chung J, Tak YJ, Kim W-G, et al. Low-temperature fabrication of solution-processed hafnium oxide gate insulator films using a thermally purified solution process. *J Mater Chem C*. 2018;6:4928–4935.

[69] Park G, Yang H, Lee JH, et al. Polymer-Assisted Deposition of Al-Doped HfO_2 Thin Film with Excellent Dielectric Properties. *Adv Mater Interfaces*. 2019;6:1900588.

[70] Lee S, Lee SH, On N, et al. A solution-processed La–Zr–O dielectric at a low temperature for high-performance In-Ga-O transistors: Engineering a precursor solution. *Ceram Int*. 2021;47:6918–6927.

[71] Zhuang X, Patel S, Zhang C, et al. Frequency-Agile Low-Temperature Solution-Processed Alumina Dielectrics for Inorganic and Organic Electronics Enhanced by Fluoride Doping. *J Am Chem Soc*. 2020;142:12440–12452.

[72] Yan L, He W, Liang X, et al. Oxygen incorporated solution-processed high-κ La_2O_3 dielectrics with large-area uniformity, low leakage and high breakdown field comparable with ALD deposited films. *J Mater Chem C*. 2020;8:5163–5173.

[73] Adamopoulos G, Thomas S, Wöbkenberg PH, et al. High-Mobility Low-Voltage ZnO and Li-Doped ZnO Transistors Based on ZrO_2 High-K Dielectric Grown by Spray Pyrolysis in Ambient Air. *Adv Mater*. 2011;23:1894–1898.

[74] Afouxenidis D, Mazzocco R, Vourlias G, et al. ZnO-based Thin Film Transistors Employing Aluminum Titanate Gate Dielectrics Deposited by Spray Pyrolysis at Ambient Air. *ACS Appl Mater Interfaces*. 2015;7:7334–7341.

[75] Esro M, Mazzocco R, Vourlias G, et al. Solution processed lanthanum aluminate gate dielectrics for use in metal oxide-based thin film transistors. *Appl Phys Lett*. 2015;106:203507.

[76] Park JW, Kang BH, Kim HJ. A Review of Low-Temperature Solution-Processed Metal Oxide Thin-Film Transistors for Flexible Electronics. *Adv Funct Mater*. 2020;30:1904632.

[77] Yarali E, Koutsiaki C, Faber H, et al. Recent Progress in Photonic Processing of Metal-Oxide Transistors. *Adv Funct Mater*. 2020;30:1906022.

[78] Kim Y-H, Heo J-S, Kim T-H, et al. Flexible metal-oxide devices made by room-temperature photochemical activation of sol–gel films. *Nature*. 2012;489:128–132.

[79] Jo J-W, Kim Y-H, Park J, et al. Ultralow-Temperature Solution-Processed Aluminum Oxide Dielectrics via Local Structure Control of Nanoclusters. *ACS Appl Mater Interfaces*. 2017;9:35114–35124.

[80] Lee W-J, Choi J-G, Sung S, et al. Rapid and Reliable Formation of Highly Densified Bilayer Oxide Dielectrics on Silicon Substrates via DUV Photoactivation for Low-Voltage Solution-Processed Oxide Thin-Film Transistors. *ACS Appl Mater Interfaces*. 2021;13:2820–2828.

[81] Yarali E, Faber H, Yengel E, et al. Low-Voltage Heterojunction Metal Oxide Transistors via Rapid Photonic Processing. *Adv Electron Mater*. 2020;6:2000028.

[82] Daunis TB, Schroder KA, Hsu JWP. Photonic curing of solution-deposited ZrO_2 dielectric on PEN: a path towards high-throughput processing of oxide electronics. *npj Flexible Electron*. 2020;4:1–7.

[83] Varma A, Mukasyan AS, Rogachev AS, et al. Solution Combustion Synthesis of Nanoscale Materials. *Chem Rev*. 2016;116:14493–14586.

[84] Carlos E, Martins R, Fortunato E, et al. Solution Combustion Synthesis: Towards a Sustainable Approach for Metal Oxides. *Chem - Eur J*. 2020;26:9099–9125.

[85] Pujar P, Gandla S, Gupta D, et al. Trends in Low-Temperature Combustion Derived Thin Films for Solution-Processed Electronics. *Adv Electron Mater*. 2020;6:2000464.

[86] Bae EJ, Kang YH, Han M, et al. Soluble oxide gate dielectrics prepared using the self-combustion reaction for high-performance thin-film transistors. *J Mater Chem C*. 2014;2:5695–5703.

[87] Carlos E, Branquinho R, Kiazadeh A, et al. UV-Mediated Photochemical Treatment for Low-Temperature Oxide-Based Thin-Film Transistors. *ACS Appl Mater Interfaces*. 2016;8:31100–31108.

[88] Scheideler WJ, McPhail MW, Kumar R, et al. Scalable, High-Performance Printed InOx Transistors Enabled by Ultraviolet-Annealed Printed High-K AlOx Gate Dielectrics. *ACS Appl Mater Interfaces.* 2018;10:37277–37286.
[89] Yu X, Smith J, Zhou N, et al. Spray-combustion synthesis: Efficient solution route to high-performance oxide transistors. *Proc Natl Acad Sci.* 2015;112:3217–3222.
[90] Wang B, Yu X, Guo P, et al. Solution-Processed All-Oxide Transparent High-Performance Transistors Fabricated by Spray-Combustion Synthesis. *Adv Electron Mater.* 2016;2:1500427.
[91] Wang B, Guo P, Zeng L, et al. Expeditious, scalable solution growth of metal oxide films by combustion blade coating for flexible electronics. *Proc Natl Acad Sci.* 2019;116:9230–9238.
[92] Carlos E, Dellis S, Kalfagiannis N, et al. Laser induced ultrafast combustion synthesis of solution-basedAlOx for thin film transistors. *J Mater Chem C.* 2020;8:6176–6184.
[93] Carlos E, Leppäniemi J, Sneck A, et al. Printed, Highly Stable Metal Oxide Thin-Film Transistors with Ultra-Thin High-κ Oxide Dielectric. *Adv Electron Mater.* 2020;6:1901071.

11 High-K Pocket in Tunneling FET Devices

Aashebir Taresa, Avtar Singh, Dereje Tekilu, and Chandan Kumar Pandey

11.1 INTRODUCTION

In classical mechanics, particles are considered as localized objects which have exact limits of existence. Every single one of the particles has both properties (i.e., mass and volume). Waves are the disturbances in the medium that transfer the energy without a net displacement of the particle. In 1929, Louis de Broglie was awarded with the Nobel Prize for his hypothetical statement that matter exhibits both the particle nature as well as wave nature. This statement is also termed as one of the pillars of Quantum physics. Further, the postulate was supported by the double slit experiment performed by Thomas Young. This experiment reported the conventional fundamental nature (i.e., both particle and wave) of electrons. Waves are of many types: like mechanical wave, electromagnetic wave, sound wave, etc. After that, extensive research was performed by researchers all over the globe to find the type of wave that resembled the electron. Taking the evidence from the existing quantum equation, researchers gave a very similar wave function to the quantum particle. Generally, wave functions are used to express the operation of the wave at all the points in a specific domain. Using these quantum wave functions, the time-independent form of Schrodinger's equation is formulated, which will be the basic method to solve the basic and complex quantum problems. By using these equations, we are able to find the relations between the electron current probability and the quantum mechanical behavior of charge carriers.

Basically, tunneling is a quantum happening, in which a charge carrier can transfer a potential barrier, despite the fact that it lacks the energy to do so. In classical physics, this type of theory is not found. It is better to imagine the quantum particle as a wave and link it with the charge carriers' tunneling phenomenon. It has been found that when an electromagnetic wave passes through a metal (conductor), the amplitude of the wave decreases and is dependent on the thickness of the conductor region, which means the damping ratio of the wave oscillations decreases when a wave passes from the thicker region of the conductor medium. If we imagine the EM waves as a quantum particle and metal as a potential barrier, the same situation again arises. If the quantum particle does not have sufficient energy to cross a potential barrier, but if the quantum particle is found on the other side of the potential barrier, anyway, then the process of transferring the charge carriers is termed as quantum mechanical tunneling.

Tunnel field-effect transistors are unipolar transistors that work based on the voltage-controlled current device concept [1]. They are devices in which the action of voltage (an electric field) at a third terminal (called the electrode gate), which has a thin metal oxide layer between the semiconductor and the metal, and current control between two electrodes drain and the source is controlled by the third terminal (gate). The positive voltage at the gate region for n-TFETs causes a negative charge to accumulate at the oxide-silicon interface, forming a channel for current conduction. The channel connects the path between both the source and the drain regions. Furthermore, the third terminal electrode is usually a heavily doped metal work function and is isolated by a thin oxide gate from the silicon channel. The device structure of TFET is shown in Figure 11.1. MOSFETs have N-P-N diode and doped regions as the source, channel, and drain. Additionally, in traditional FETs, the carrier injection method is based on thermal injection [2], but Tunneling FETs utilize band to

DOI: 10.1201/9781003352365-11

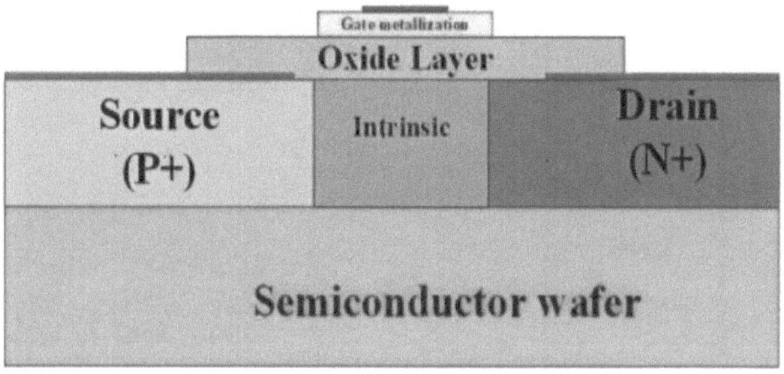

FIGURE 11.1 Device structure of Tunnel FET.

band tunneling (BTBT) as a source carrier injection mechanism. Also, TFETs differ from MOSFETs structure in that, in TFETs, the drain region and the source region are doped with different types of dopants (i.e., the source region doped in a MOSFET is N-type concentration, whereas it is P-type concentration in the TFET) and the steep slope current-voltage characteristics are achieved using a band-to-band tunneling mechanism through a reverse-biased junction [3]. This is the significant difference between MOSFETs and TFETs.

11.2 OFF STATE AND ON STATE

In the OFF-state, when gate to source voltage is zero and drain to source voltage is greater than zero, than charge carriers that are present inside the conduction band of the channel will tend to shift to the drain and thus forms a current [4]. However, because the source is p-type, the conduction band contains relatively few electrons, which can inject very few electrons into the channel. This leads to an insignificant OFF-state current [5]. The MOSFET current generation principle depends on thermionic emission (a few electrons would be penetrated into the channel through the potential barrier at source-channel interface). When MOSFET is compared with TFET, the OFF-state leakage current in MOSFET is high, limiting the device's performance. In the ON-state current, when gate to source voltage and drain to source voltage both are positive, current will be flowing in the device and charge carriers (electrons) are transferring into the conduction band of the channel from the valance band of source[6]. In TFETs, the valance band of the source area serves as the source of the carriers, which are slightly free electrons that are present in the conduction band of the source region. With a rise in gate voltage (VGS), the energy bands in the channel area change. The valance bands inside the source region, as well as the conduction band in the channel, meet when the gate is at a positive voltage. The electrons in the valence band of said source had no attainable free energy inside the channel toward the tunnel once the device was turned OFF. After determining the valence band of the source and the conduction band of a channel, the electron could tunnel through the potential barrier imposed by the band gap. The gate voltage that causes this configuration of a source valance band with channel conduction band emerging indicates the start of the TFET's ON-state [7]. In Figure 11.2, the charge carriers from the gate tunnel into the channel through the potential barrier in the ON-state of TFETs, use the band to band tunneling operating principle (BTBT). The channel barrier is high in the OFF-state of a TFET, so there is no net current flow. But, due to the thermal distribution of carriers, ideally it is not zero, only the leakage current flows between the sources and drain region.

11.3 AMBIPOLAR BEHAVIOR

A positive gate voltage is applied in the ON-state with the drain voltage $V_{DS}>0$ [8]. Let's discuss the behavior of a TFET when a negative gate voltage ($V_{GS}<0$) is applied [1]. As $V_{GS}=0$, no electron can tunnel from the source valance band to the channel conduction band; as the gate voltage is

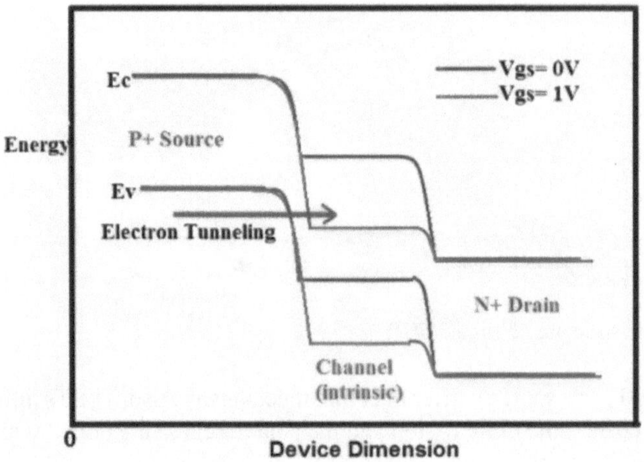

FIGURE 11.2 Energy band diagram of TFETs where the blue line indicates ON-condition and red lines OFF-condition [1].

reduced below 0 V (V_{GS}<0), the energy band of the channel travels upward through to the source region. When the channel valence band aligns with the drain conduction band, an electron from the channel valence band can tunnel into the drain conduction band, resulting in opposite direction current flow [9]. Since the negative gate bias is risen, the drain current rises significantly. Primarily, this action is due to two factors: (i) a shorter tunneling distance at the channel-drain junction, and (ii) an expansion in the number of states in the channel valence band in which an electron could tunnel to the drain's conduction band [7]. Ambipolar current is the name given to the current generated by the above-mentioned mechanism.

11.4 BAND DIAGRAM

The current conduction mechanism in MOSFETs is thermal carrier injection (over the barrier), while the current conduction mechanism in TFETs is band-to-band tunneling (through the barrier). In an N-TFET, electrons penetrate from the source valence band toward the drain conduction band, changing the gate potential controls, the BTBT probability (consequently, the current) [10]. The conduction and valence band energies of MOSFETs and n-TFETs are shown in Figure 11.3. The source-side p-n junction barrier limits conduction in MOSFETs in the OFF state (V_{GS}=0 V, V_{DS}>1 V), limiting thermionic carrier emission, but in the ON state current (V_{GS}>V_{DS}>0 V), the source barrier is insignificant, allowing thermionic carrier emission. Because of the large source-to-channel tunnel junction barrier (Low Electric Field), the propagation efficiency of TFETs during off state is low, resulting in very low OFF currents. The tunnel barrier thickness permits carriers that tunnel into the channel while it is in the ON state (Koswatta et al., 2009). The BTBT producing mechanism in TFETs improves subthreshold swing by less than 60 mV/dec (Park et al., 2018), which means by reducing the device's size, the circuit speeds are also improved by a subthreshold swing. In addition, the most attractive result of scaling is that the subthreshold swing reduces the power dissipation or increases power consumption (energy efficiency).

11.5 RECENT WORKS IN TUNNEL FET

The main challenges in TFETs are the low ON-state current (I_{ON}), leakage current (I_{OFF}), and ambipolar conduction, which reduce the performance of devices because when TFETs have a low ON current and ambipolar behavior, there is more delay in the circuits [12]. Due to this issue, many researchers have found that the low sub-threshold current leads to low leakage per device, and the TFET's high I_{ON}/I_{OFF}

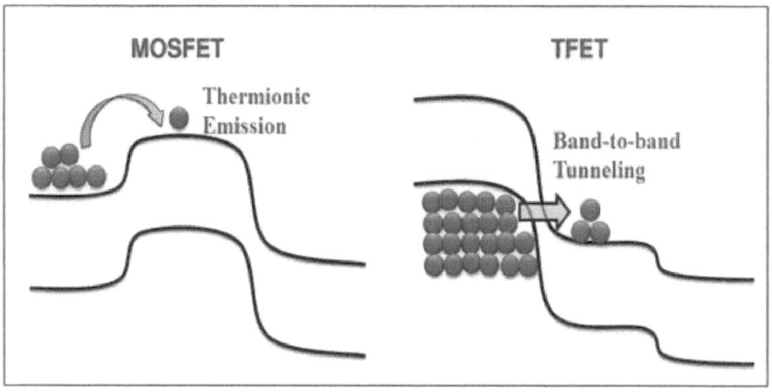

FIGURE 11.3 Device principle of the MOSFETs and TFETs [11].

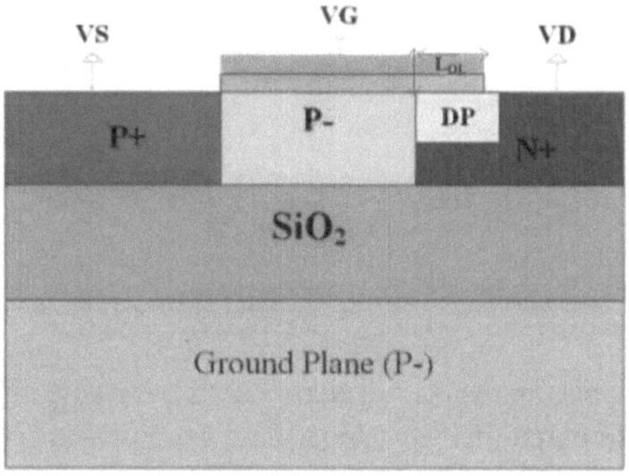

FIGURE 11.4 Cross-sectional view of DP-SG-TFETs [17].

ratio makes it suited for analog/RF applications, although the majority of researchers have focused on TFETs for low-power and high speed switching applications. Therefore, many efforts are being made around the globe by the device scientists to suppress ambipolar conduction and improve I_{ON}, which includes the hetero-junction source [13], a high-k gate dielectric [14], a double-gate architecture [6], Ge source TFETs [6], a low-bandgap material in the tunneling region [2], gate overlapped TFETs (Vertical tunneling TFETs) [15], pocket doped double gate structure [16], extended source TFETs [9], and L-shaped TFETs [10] that have all been used to improve ON-state current. All of the approaches discussed above improve considerably the drive capability of TFETS; nevertheless, the ambipolar current and leakage current are not improved; this is a constraint of TFETs that degrades the analog/RF efficiency of the TFETs. For low-power switching and analog/RF circuits, low subthreshold swing (SS), low ambipolar conduction (leakage current), and a high ON-state current (I_{ON}) to OFF current (I_{OFF}) ratio are all desirable performance characteristics, including a gain-bandwidth product (GBP), transconductance (g_m) and high cut-off frequency parameters (f_T) [10]. Limiting ambipolar conduction in a Tunnel FET by means of a dielectric pocket has been investigated [17]. A dielectric pocket single gate tunnel field-effect transistor (DP-SG-TFET) structure is analyzed to minimize ambipolar conduction in tunnel field-effect transistor devices, which is also shown in Figure 11.4. The dielectric pocket structure of single gate tunnel field-effect transistors improves system performance and prevents ambipolar conduction. The researcher focuses on the DP SG-TFET and L_{OL} structures, which use a 50 nm channel length and a high-K at dielectric pocket (DP) to reduce ambipolar conduction at the channel-drain junction without changing the ON-state current. It is observed that when using gate-drain overlap (L_{OL}), ambipolar current (I_{AMB}) in TFET devices decreases. However, the additional gate-drain overlap raises the needed distance, wasting silicon area and increasing manufacturing costs [17].

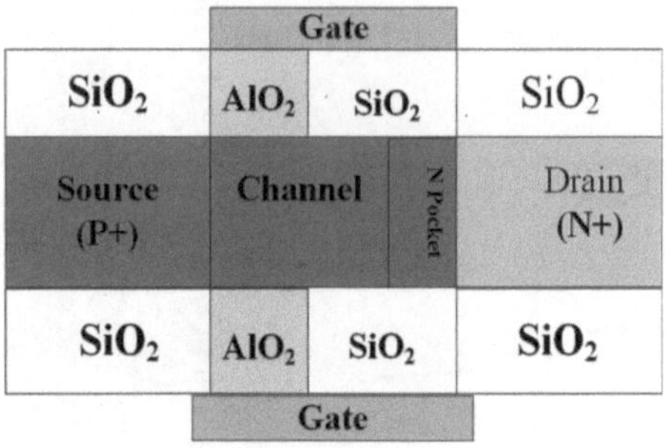

FIGURE 11.5 Schematic view of DP-HD-DGTFET [18].

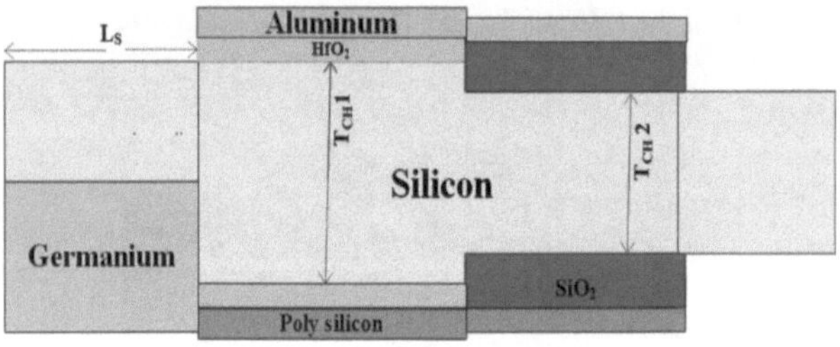

FIGURE 11.6 Schematic view of nonuniform silicon TFET [19].

In a paper entitled "Performance analysis of drain pocket hetero gate dielectric DG-TFET," authored by Goyal et al., the group documented their investigation for ambipolar conduction and enhanced drive current. The 2D structure is shown in Figure 11.5 [18]. The impact of drain pocket and heterogeneous gate dielectric (gate stack) were analyzed on the performance of double gate tunnel field-effect. By suppressing BTBT at the channel-Drain junction, the authors noticed that an area of concentration optimized a drain pocket near the drain region which lowered ambipolar conduction by 7.2×10^{-15} A/µm. Furthermore, the existence of Al_2O_3 as a heterogeneous dielectric (HD) on the source region (side) causes charge production, resulting in a steeper source-channel junction. Compared to the conventional counterpart, using Al_2O_3 as HD improved I_{ON} to an order of 10^{-5} A/m while avoiding a rise in I_{OFF}, increasing the I_{ON}/I_{OFF} ratio to 10^{11}[18]. The ambipolar current and I_{ON} are enhanced by utilizing the above method. However, the drain-gate overlap (DG_{OL}) increases capacitances, limiting the TFET performances at high frequencies. Similarly, a multi-gate structure or reduced channel thickness (lesser channel thickness leads to a larger BTBT rate) does not affect the performance of the hetero dielectric structure.

A non-uniform silicon TFET design with dual material source and compressed drain has been studied in Figure 11.6[19]. This study introduced a non-uniform channel TFET structure with a double-material source and compressed drain. This device results show that the suggested structure has a lesser value of SS (subthreshold slope) for gate voltages.

Furthermore, the ambipolar current is improved by 10^{-14} A/µm, I_{ON} current is 10^{-5} A/µm, I_{OFF} values 10^{-16} A/µm, and an I_{ON}/I_{OFF} ratio around 10^{11}. In this study, the impact of trap charge at the oxide/semiconductor interface has been studied for various trap charge distributions[19]. However, gate-to-drain capacitance increases because of the trap charge dispersion at the oxide-semiconductor junction, leading to poor analog/RF performance (i.e., limiting the cut-off frequency).

Suppression of ambipolar current in tunnel FETs using a drain-pocket has been studied in [20]. The effect of a drain pocket (DP) close to the drain region in TFETs on suppressing the ambipolar current is investigated and shown in Figure 11.7. Even if the TFET is biased at large negative bias voltage and also the drain doping is maintained at the same level as the source doping, the addition of DP to a TFET is able to suppress the ambipolar current entirely. Furthermore, because both require similar doping types and concentrations, adding DP to a TFET devices is complementary to

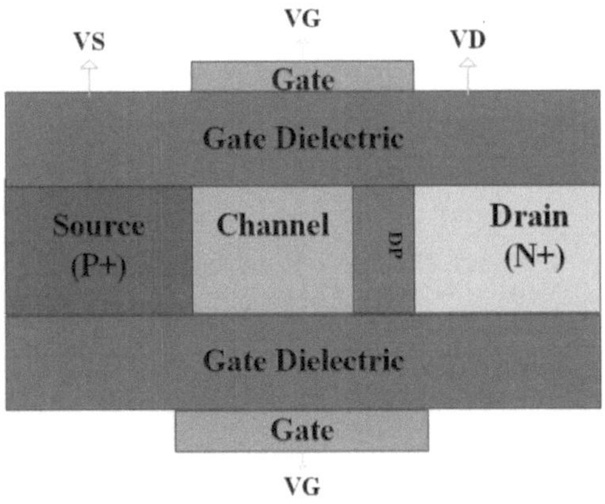

FIGURE 11.7 Cross-sectional view of DP-DGTFET.

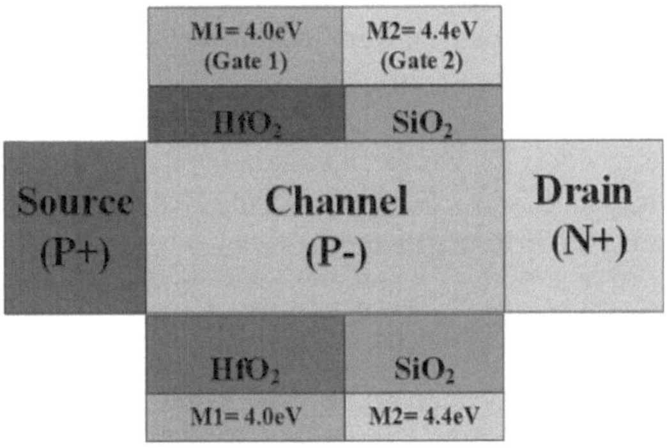

FIGURE 11.8 Two-dimensional structure of dual metal gate – double gate TFET.

the well-known source-pocket technique in a TFET (Garg & Saurabh, 2018). However, previously, lowering TFET ambipolar current with a high drain doping concentration was effective. High drain region doping concentration, on the other hand, reduces the tunneling barrier at the drain-channel interface. It increases the OFF-state leakage current.

Ge-based double gate dual metal tunnel FET new architecture investigation employing different hetero dielectric materials has been studied [21]. In order to address the limitations of traditional silicon (Si)-based TFETs, a germanium-based dual metal double gate tunnel field-effect transistor device was investigated in this study. The performance investigation is carried out for various doping concentrations and hetero dielectric materials. The structure is shown in Figure 11.8. It has a leakage current of 9.27×10^{-13} A/m and an I_{ON} of 1.3×10^{-4} A/m. This indicates that the device is more suited to low-power applications [21]. Furthermore, Garcia presented a dual metal gate TFET, in which the ambipolar current is lowered, but the drain-gate overlap capacitance is increased, degrading the performance of the TFET devices at high frequencies.

Figure 11.9 shows a redrawn structure of a tunnel FET with the drain side tunneling thickness controlled to decrease the ambipolar current. Employing 2-D simulations, this work showed that placing a hetero dielectric BOX (HDB) over a heavily doped ground plane may regulate the tunneling width at the channel-drain junction and reduce the ambipolar current in tunnel FETs (TFETs).

The buried oxide is made up of SiO_2 for source and channel regions and HfO_2 for the drain. Note that the drain region at the channel–drain interface is fully depleted when the HDB thickness is 25 nm and the ground plane is heavily doped. As a result, the tunneling thickness at the channel–drain interface increases, suppressing ambipolar conduction in a TFET [7]. Various "L-shaped architectures based on gate-source overlap structures and vertical channels" have been studied to increase

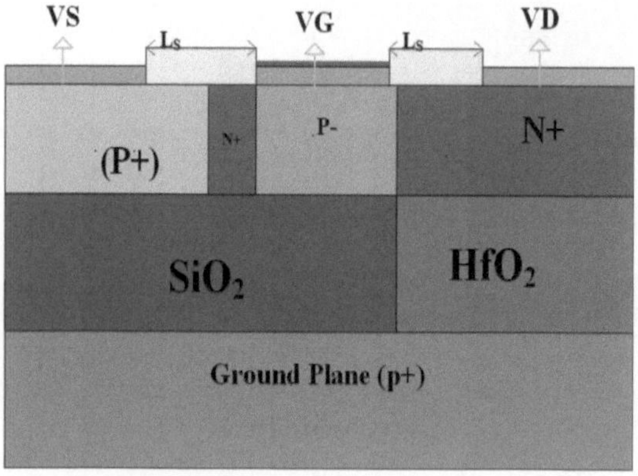

FIGURE 11.9 Schematic of HDB TFET.

the current drivability of the devices and minimize the TFET footprints. On the other hand, L shaped TFETs structures have been shown to cause a significant increase in ambipolar behavior (high ambipolar current), which needs further tuning [10]. Recently, "T-shaped TFETs with a gate-drain overlap (GDO) structure" was studied to enhance the driving speed simultaneously and suppress the ambipolar current. This T-shaped TFET with a gate-drain overlap structure, however, can only suppress ambipolar current when VDS is low. Furthermore, the GDO topology raises gate-drain capacitance, leading in poor analog/RF performance metrics [15].

11.6 DESIGN OF DIELECTRIC POCKET IN TFET

The design device structure is a silicon-source-based SOI-TFET with a dielectric pocket (DP) at the top of the drain region. A study of the device structure (DP-TFET) has been compared to a conventional TFET (without DP) and, as shown in Figure 11.10, the optimum thickness and length of DP (dielectric pocket) at the drain region were found to be 6 nm and 5nm, respectively. DP is made up of dielectric material to achieve low leakage current, which improves the device performance. Additionally, the device structure design parameters, such as gate oxide, are made up of high-k dielectric (HfO_2), and the dielectric thickness is 5nm. The Insulator region of SOI made up of SiO_2 is a place to provide isolation between drain and source. The gate length is 45 nm and is made up of metal work functions that are kept constant at 4.9eV. The length of the source and drain regions is fixed at 100 nm each, respectively. A doping concentration of 1×10^{20}, 1×10^{16}, and 5×10^{18} cm^{-3} respectively, is taken for source, channel, and drain regions. A lightly p-type doped layer at a substrate is added in the BOX (buried oxide) at a certain depth from the back silicon dioxide (SiO_2) interface. Adding a BOX in the device structure depletes the drain region at the channel-drain junction and reduces the leakage current. The BOX also controls the short channel effect (SCE), which helps with device reliability [7].

Table 11.1 lists the device parameters that have been used in the design of the TFET simulation structure. The parameter numbers used here are taken from the International Technology roadmap of Semiconductors (ITRS) roadmap.

In this TFET, the structure source region is heavily P-type doped. A drain region is also heavily doped with n+ type, while the channel is lightly doped with P-type. The gate terminal controls the electrostatic potential and the charge concentration, thus increasing the junction between the semiconductor and the extrinsic region [22].

11.6.1 Dielectric Materials Used and Their Properties

The two-dimensional device structure of the conventional TFETs and the proposed DP-TFETs structures are shown in Figures 11.3(a) and (b). However, Dielectric materials parameters are as shown in the Table 11.2.

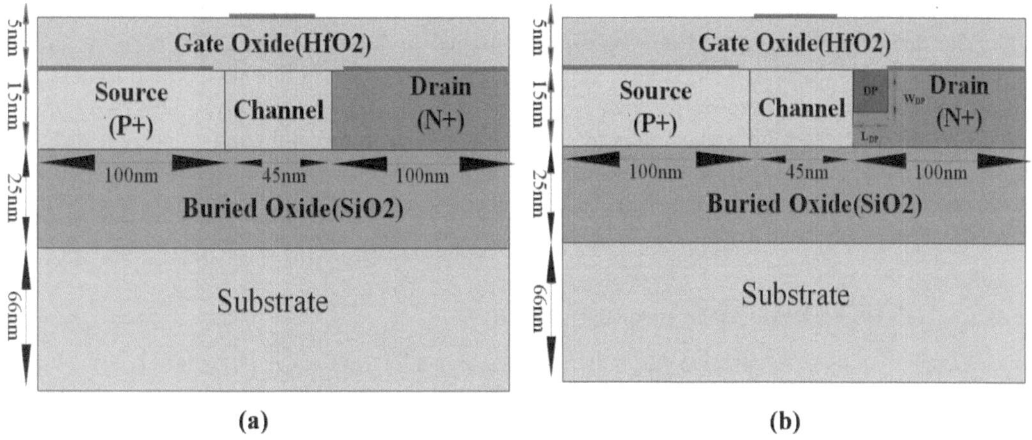

FIGURE 11.10 Structure of (a) TFETs (without DP) and (b) structure of TFETs (DP-TFETs).

TABLE 11.1
List of Parameters in TFETs Device Simulations

Parameters	Values
Channel length and thickness	45 and 15 nm
Gate work function,	4.9 eV
Source length and thickness	100 and 15 nm
Drain lengths and thickness,	100 and 15 nm
Source doping	1e 20 cm^{-3}
Channel doping	1e 16 cm^{-3}
Drain doping	5e 18 cm^{-3}
Gate oxide (HfO_2) thickness	5 nm
Silicon-On-Insulator (SOI) thickness (tsi)	25 nm
Silicon-On-Insulator (SOI) length	100 nm
Substrate doping (NA)	1e 16 cm^{-3}
Drain pocket (material HfO_2) length	2 nm to10 nm
Drain pocket (material HfO_2) thickness	2 nm to 6 nm

TABLE 11.2
Dielectric Materials Parameter

Dielectric Materials	Properties	
	Energy Band-Gap (eV)	Permittivity (K)
SiO_2	9	3.9
Si_3N_4	5	7.5
HfO_2	5.7	25

Figure 11.10(a) shows the TFET and Figure 11.10(b) depicts the DPTFET with dielectric Pocket (DP). The device dimension used are same as mentioned in Table 11.1 In DP-TFETs, the drain region of the device's structure has been scaled down to its optimal length of dielectric pocket (L_{DP}) and pocket thickness of dielectric pocket (T_{DP}). Furthermore, a dielectric pocket (DP) or a high-k dielectric (HfO_2) at the length and thickness drain side has been mounted above the scaled drain regions. Similarly, adding a dielectric pocket (DP) in the portion of the drain is shown to increase the tunneling barrier thickness at the channel-drain junction, resulting in the ambipolar current being reduced.

11.6.2 DP-TFETs Parameters Variation Study

This section contains an exhaustive parameters variation in a DP (dielectric pocket) of length (L_{DP}), and thickness of the DP (T_{DP}), and dielectric materials of DP on the TFETs device is investigated to select optimum parameter values.

11.6.2.1 Impact of DP Thickness (T_{DP}) Variation on Device Performance

Figure 11.11(a) compares the transfer characteristics of the device for various dielectric pocket (DP) thickness values. The rate at which charge carriers travel through the junction determines ambipolar conduction at the drain-channel interface. When a dielectric pocket is present above a scaled drain terminal at the drain-channel junction, the underlying drain region is significantly depleted. During simulations, the DP length is kept fixed at 5 nm, and HfO_2 added in the dielectric pocket area is fixed, because a dielectric with a high-K (dielectric constant) value has a more significant effect on the electrostatic behavior in the drain region. As observed in the figure, an ambipolar current is reduced from 10^{-9} to 10^{-17} A/μm when T_{DP} has increased from 2 nm to 6 nm, respectively, and ambipolarity is nearly removed up to 6 nm thickness of DP (T_{DP}). Figure 11.11(b) illustrates the electric field along the vertical direction of the drain-channel junction with the dielectric pocket thickness varying. The figure shows that the electrical field can be reduced at the drain-channel junction as the thickness of DP increases due to the larger depletion of high-K in the T_{DP}, which decreases I_{OFF} and ambipolar current. Figure 11.11(c) shows the result of increasing the dielectric pocket thickness (T_{DP}) on the horizontal electric field at a channel-drain (C-D) junction. The horizontal electric field at a channel to drain interface significantly decreases when the thickness of DP increases.

Since the fringing field comes from the dielectric pocket, the output tunneling junction is widely spread over the depleted drain region below DP; due to these causes, the electric field (flow of charge carriers tunneling) is reduced at a C-D junction. The electron concentration of the DP-TFET along the C-D junction is illustrated in Figure 11.11(d). Due to a dielectric pocket with a thickness of 6 nm having a broader or wider electron concentration, the tunneling width at the output tunneling interface increases, which decreases the charge carrier tunneling rate through the drain-to-channel junction. The effect of T_{DP} on the BTBT electron in TFET is shown in Figure 11.11(e), in which the electron BTBT (band to band tunnel) generation rate of charge carriers is decreased when the T_{DP} is increased. When TDP is increased, the electron BTBT is lowered; due to the reduced BTBT generation rate of electron charge carriers, the leakage current and ambipolar current in DP-TFET are reduced.

11.6.2.2 Impact of DP Length (L_{DP}) Variation on Device Performance

Figure 11.12(a) shows the effect of varying the dielectric pocket (L_{DP}) length at the drain region on the I_D-V_G curve of DP-TFETs.

Throughout simulations, the thickness of the DP at the drain region is kept at its optimum of 6 nm, and the dielectric material is constant with a high-K (HfO_2) value. The I_{AMB} is about 10^{-17} A/μm when the length of the DP is 5 nm (optimum). Furthermore, for dielectric pocket lengths greater than 5 nm, I_{AMB} reduction stopped rapidly as DP length increased. Therefore, I_{AMB} is more suppressed when the length of DP is 5 nm. In Figure 11.12(b), the electric field in the

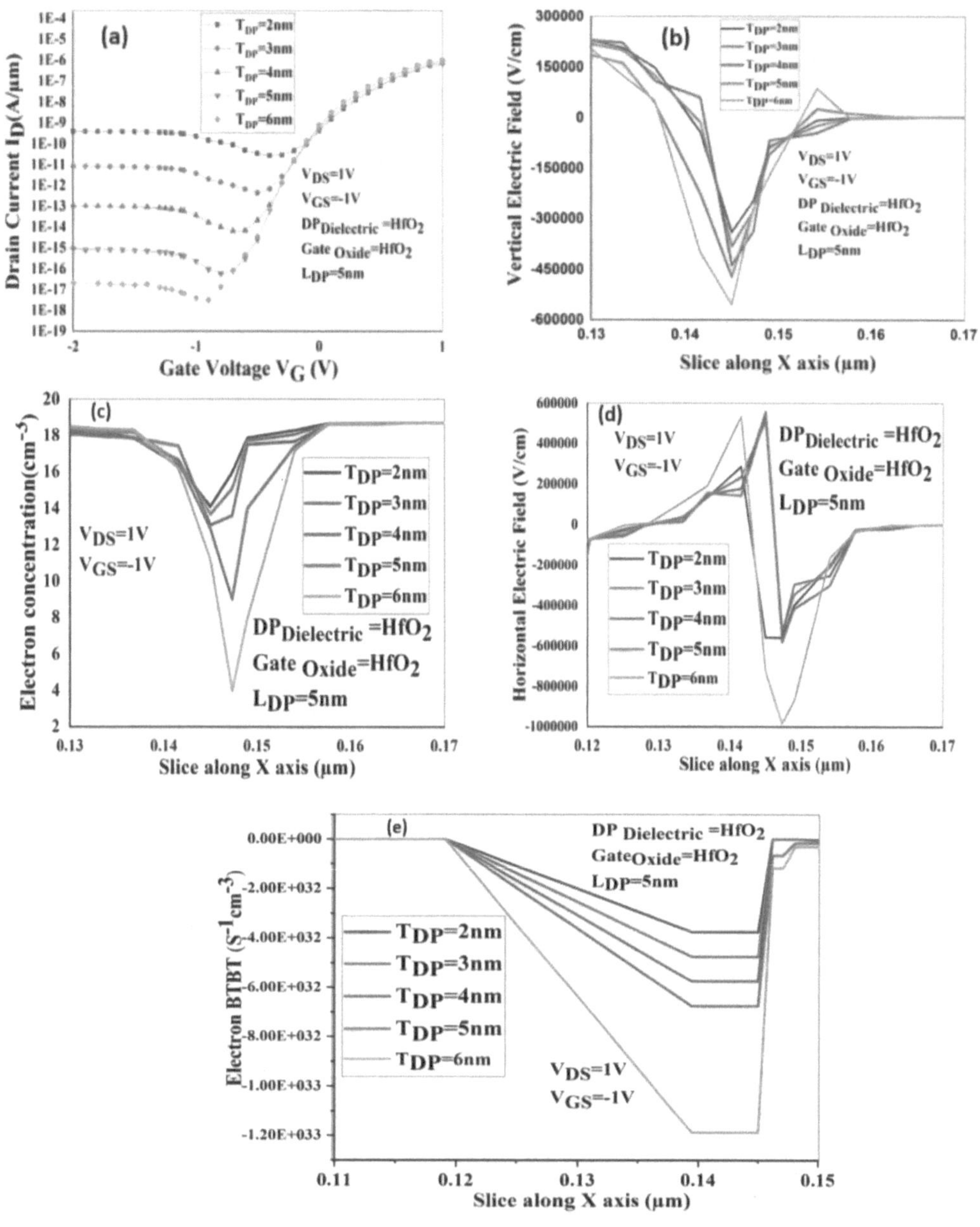

FIGURES 11.11 DC characteristics of the DP-TFETs (a) transfer characteristics, (b) vertical electric field, (c) horizontal concentration, (d) electron concentration, (e) electron-BTBT generation rate, on variation of T_{DP}.

horizontal direction is plotted for the L_{DP} varying from 2 nm to 10 nm at the ambipolar state. When DP is 5 nm, an ambipolar current and leakage current are reduced. In other words, this further causes a significant increase in the tunneling barrier thickness of charge carriers at a channel to drain junction. Figure 11.12(c) shows that the electric field in a vertical direction is designed to change the electric field when the DP-TFETs device is biased at the ambipolar state. With L_{DP}=5 nm (optimum), the vertical electrical field induces rapid charge carrier recombination in the drain region below DP, and this causes the depletion area to appear at the channel drain interface of the DP-TFETs, causes a greater expansion in the output tunneling barrier thickness at a C-D junction when L_{DP} = 5 nm.

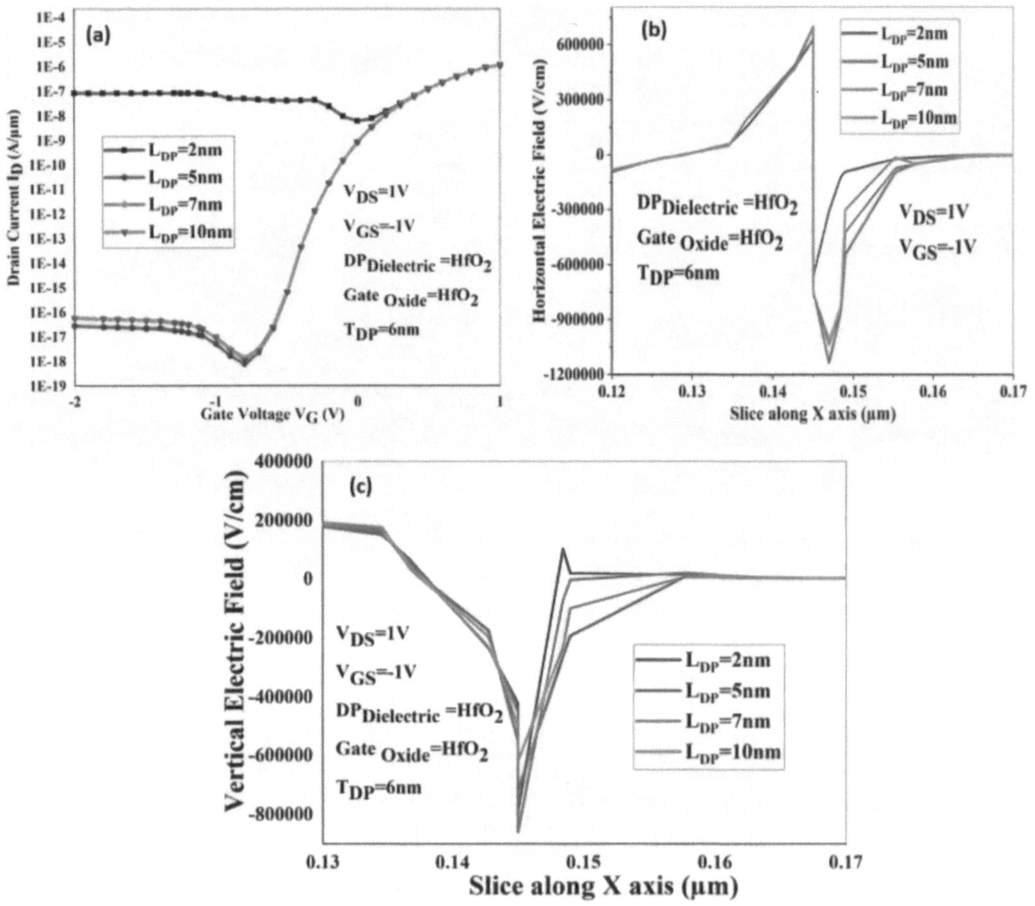

FIGURES 11.12 DC characteristics of the DP-TFETs (a) transfer characteristics, (b) horizontal electric field, (c) vertical electrical field, on a variation of L_{DP}.

11.6.2.3 Effects of Dielectric Materials

In this section, the impact of dielectric materials variation at the DP has been analyzed appropriately to achieve the low ambipolar current and low OFF-state leakage current. The physical DP dimension thickness is 6 nm and length of 5 nm, have been considered (optimum) with various dielectric materials, such as SiO_2, Si_3N_4, and HfO_2, which have different properties. The device's I_D-V_G is exposed in Figure 11.13(a) by varying the dielectric material at the DP. The graph result shows that high dielectric material HfO_2 shows higher performance (preferred) over two materials, Si_3N_4 and SiO_2. So high-K-values have a more significant impact on reducing ambipolarity conduction and leakage current. Moreover, the fundamental goal of this study is to find a device structure that reduces I_{AMB} significantly without decreasing (affecting) I_{ON}. Because the existence of a dielectric pocket on the drain region has no effect on charge carriers' tunneling from the source-to-channel junction. Figure 11.13(b), in which electron-BTBT generation rate of charge transfer by varying dielectric material (SiO_2=3.9, Si_3N_4=7.5, and HfO_2=25) device is biased at ambipolar state. The electron BTBT is reduced when DP is incorporated with TFETs; due to the reduction of BTBT generation rate at the C-D junction, an OFF-state leakage current and I_{AMB} are reduced in DP-TFET, and the order of this decrease is when a high-K (HfO_2) material is utilized in the DP, the result is found to be high.

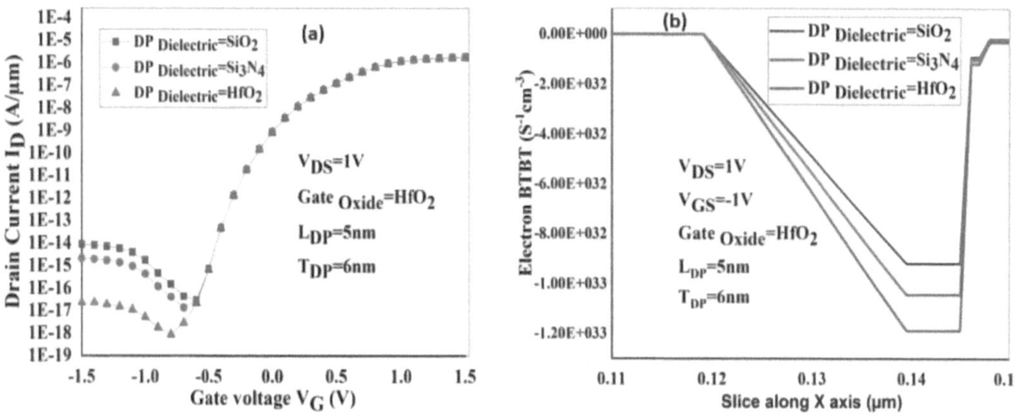

FIGURES 11.13 DC characteristics of the DP-TFETs (a) transfer characteristics or $I_{D-}V_G$ curve (b) electron-BTBT generation rate for different dielectric materials at DP.

11.6.3 Device Optimization

This section analyses the impact of dielectric pocket with (proposed study) and without DP (conventional TFETs). To suppress the ambipolar current, reduce OFF-state current, and improve electrical characteristics of TFETs the TFET structure is modified and a DP is introduced at the drain region.

11.6.3.1 Performance Comparison of Conventional TFET and Proposed TFET (DP-TFET)

Figure 11.14(a) compares the $I_{D-}V_G$ curve of the conventional TFETs (without dielectric-pocket) and with dielectric pocket or DP-TFETs. Primarily, it is noted that the transfer characteristics for conventional TFETs and DP-TFETs are similar in ON-state current (in positive gate voltages). This means the insertion of DP at the drain region does not affect (change) the I_{ON} performance. However, as shown in Table 11.3, different electrical parameter characteristics, such as I_{OFF}, I_{ON}/I_{OFF}, and SS, have improved, while for the ambipolar state, it is noticed that I_{AMB} is suppressed from 9.38×10^{-7} A/µmin the conventional TFETs to 1.22×10^{-17} A/µm in the DP-TFET, thus decreasing the ambipolar conduction (I_{AMB}) by ten (10) orders of magnitude. Furthermore, it is demonstrated in Figure 11.13(b) that the studied structure with DP offers a lesser value of SS, ambipolar current, I_{OFF}, and a higher value of I_{ON}/I_{OFF} rate when compared with conventional TFET, thus improving the device performance special for low power operation and high-speed applications, as well as digital and analog and radio frequency (RF) amplification. So, with this small trade-off between transfer characteristics, the structure presented as better structure TFETs include those having a DP on the drain region.

Figure 11.14(b) shows the comparison electric field of conventional TFETs and DP-TFETs in the horizontal direction at the C-D interface in the TFETs device. As observed from the figure, due to the addition of a dielectric pocket (DP), the electric field is decreased in the channel-drain interface, which means when an electric field is decreased at a C-D junction, the ambipolar current is significantly reduced in DP-TFET because the output tunnel barrier thickness is increased.

Yet again, as observed from Figure 11.14(c), electron concentration in the drain side of the conventional and proposed study is plotted at ambipolar current. As observed from the figure when comparing conventional TFETs with DP-TFETs, the electron charge carrier's concentration is reduced in DP-TFET because the insertion or addition of a dielectric pocket (DP) in the drain region decreases the electrostatic behavior of DP-TFETs devices. Figure 11.14(d) shows the effects of DP in the ambipolar state, in which the electron BTBT generation rate is compared with conventional TFETs and DP-TFETs devices. The electron BTBT is reduced when DP is incorporated with TFETs. Due to this reduction in the BTBT generation rate of electron charge carriers, the ambipolar current

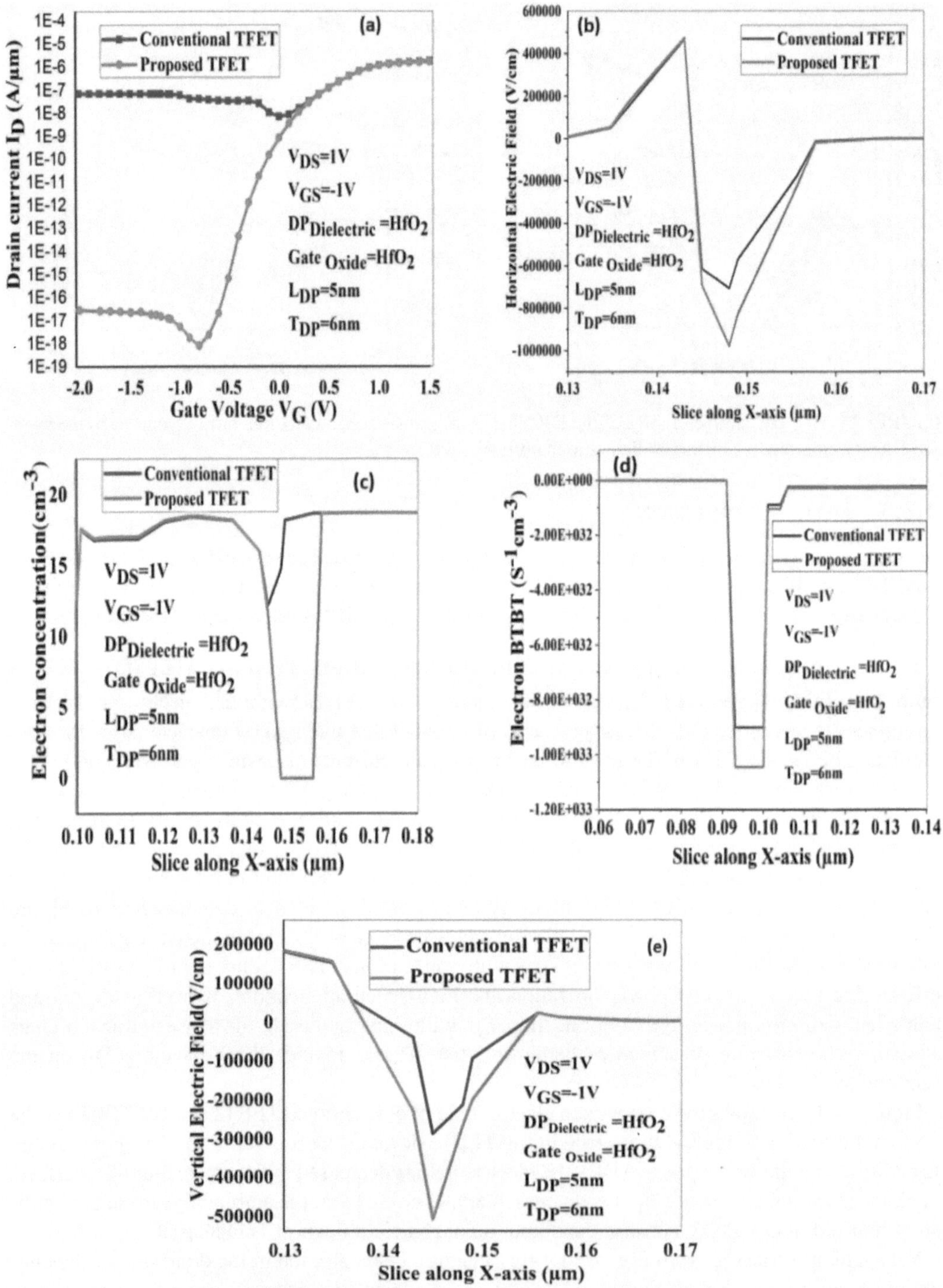

FIGURE 11.14 DC characteristics of (a) transfer characteristics (b) horizontal electric field, (c) electron concentration, (d) electron-BTBT generation rate, (e) vertical electrical field, between conventional TFETs and DP-TFETs.

TABLE 11.3
Comparison of Performance Parameters between the Conventional TFET and Proposed TFET (DP-TFET)

Performance Parameters	Conventional TFET	Proposed TFET (DP-TFET)
I_{ON} (A/μm)	2.09×10^{-6}	2.09×10^{-6}
I_{OFF} (A/μm)	4.13×10^{-8}	7.88×10^{-19}
I_{AMB} (A/μm)	9.38×10^{-7}	1.22×10^{-17}
SS (mV/dec)	~78	~58

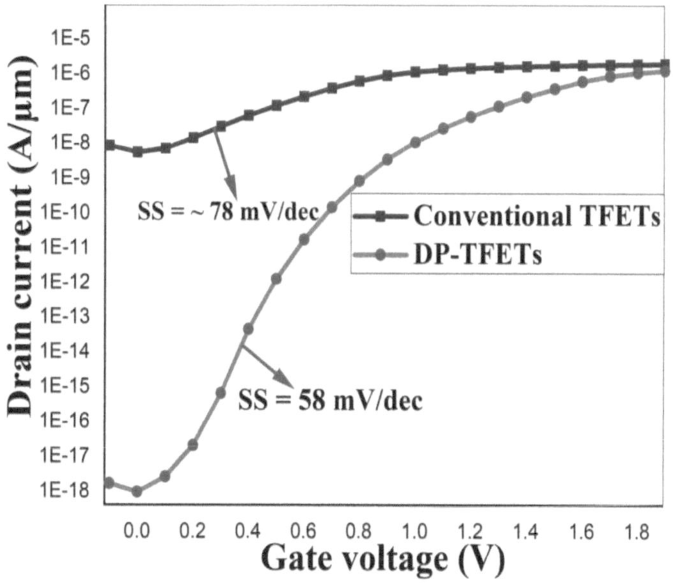

FIGURE 11.15 Comparison of transfer characteristics of conventional and DP-TFETs.

is suppressed in DP-TFET, and the order of this decrease is found to be high when HfO_2 dielectric material is used at DP.

Figure 11.14(e) shows the electric field in the vertical direction at the channel-drain junctions without DP (conventional) and the proposed study based on the ambipolar state. The decreased value of the vertical electric field at the channel-drain junction using a dielectric pocket is better than that of conventional (without DP). This ultimately causes a wider reduction in output tunneling thickness in the channel-drain junction of DP-TFET, which leads to suppressed or reduced ambipolarity and OFF-state leakage current. Depending on the results of the study, an optimal value for all three design parameters of the DP dielectric, LDP, and TDP is determined, culminating with in proposed device having the lowest OFF-state leakage current and ambipolar current when compared to the traditional TFET. Additionally, a subthreshold slope (SS) comparison between conventional and DP-TFETs is shown in Figure 11.15. A subthreshold parameter slope is defined as the change in gate voltage required producing a one-decade increase in output current. The figure shows that the subthreshold slope for thesis study (DP-TFETs) is better than conventional TFETs. This indicates that the addition of DP at the drain region significantly impacts the reduction of subthreshold slope and the improvement of device performance. It is expressed as equation (4.1).

$$SS = \frac{\partial V_{GS}}{\partial \log I_{DS}}$$

In this chapter we have investigated the effect of dielectric pocket on the tunnel FET structure, although we have taken the silicon on insulator-based Tunnel FET. When the dielectric pocket (DP) is placed at the channel drain interface junction, ambipolar current is reduced. The effect of parameter variation such as TDP, L_{DP}, and dielectric DP on the device's performance tunnel FETs (TFET) we also investigated in this chapter. It is observed that the dielectric pocket (DP) is an important parameter that affects the TFET device's ambipolar behavior. Further, the ambipolar current is also reduced when the low K (SiO_2) material is replaced by the high-K material (HfO_2). Furthermore, it has been added (insertion) that DP at the drain side does not degrade the ON-state performance of TFET, thus improving the current drive capacity of the device. The reduced ambipolar current (I_{AMB}) is 1.22×10^{-17}A/μm in this improved structure with a drain side dielectric pocket, while I_{OFF} (leakage current) is around 7.88×10^{-19}A/μm, which is lower by approximately ten (10) magnitude than conventional TFETs. Furthermore, those DP-TFETs devices are suitable for low-power operation, which can be working at high frequencies, and different applications of tunnel FETs consist of low-power circuits. As compared to the other two dielectric materials (SiO_2 and Si_3N_4), a high-K dielectric material (HfO_2) delivers a more significant reduction in OFF-state current (leakage current) and ambipolar current. However, the analog/RF characteristic of DP-TFETs with low k-dielectric material (SiO_2) is significantly better than that of high-K dielectric (HfO_2) and medium K dielectric material (Si_3N_4), and has a g_m of 2.7×10^6 S/μm, a gain-bandwidth product of 0.3GHz, and cut-off frequency of 0.23GHz.

REFERENCES

[1] S. M. Turkane and A. K. Kureshi, "Review of tunnel field effect transistor (TFET)," *Int. J. Appl. Eng. Res.*, vol. 11, no. 7, pp. 4922–4929, 2016.

[2] B. V. V. Satyanarayana and M. D. Prakash, "Lower subthreshold swing and improved miller capacitance heterojunction tunneling transistor with overlapping gate," *Mater. Today Proc.*, vol. 45, no. xxxx, pp. 1997–2001, 2021, doi: 10.1016/j.matpr.2020.09.420

[3] S. O. Koswatta, M. S. Lundstrom, and D. E. Nikonov, "Performance comparison between p-i-n tunneling transistors and conventional MOSFETs," *IEEE Trans. Electron Devices*, vol. 56, no. 3, pp. 456–465, 2009, doi: 10.1109/TED.2008.2011934

[4] R. Dutta, N. Paitya, and A. Majumdar, "Ambipolar reduction methodology for SOI tunnel FETs in low power applications: A performance report," *Int. J. Recent Technol. Eng.*, vol. 8, no. 5, pp. 1894–1897, 2020, doi: 10.35940/ijrte.e6271.018520

[5] A. M. Jossy and T. Vigneswaran, "A perspective review of tunnel field effect transistor with steeper switching behavior and low off current (I OFF) for ultra low power applications," *Int. J. Eng. Technol.*, vol. 6, no. 5, pp. 2092–2104, 2014.

[6] S. Wang et al., "Device simulation of GeSn/GeSiSn pocket n-type tunnel field-effect transistor for analog and RF applications," *Superlattices Microstruct.*, vol. 111, pp. 286–292, Nov. 2017, doi: 10.1016/j.spmi.2017.06.037

[7] S. Sahay and M. J. Kumar, "Controlling the drain side tunneling width to reduce ambipolar current in tunnel FETs using heterodielectric BOX," *IEEE Trans. Electron Devices*, vol. 62, no. 11, pp. 3882–3886, 2015, doi: 10.1109/TED.2015.2478955

[8] R. Dutta, M. Rahaman, A. Guha, and N. Paitya, "Study of gate source-drain overlap/gate-channel underlap in Heteojunction (50nm Ge channel) n-Double Gate TFET for different?-spacer," *Proc. 2nd Int. Conf. Smart Syst. Inven. Technol. ICSSIT 2019*, vol. 10, no. Icssit, pp. 672–676, 2019, doi: 10.1109/ICSSIT46314.2019.8987903

[9] R. Dutta, T. D. Subash, and N. Paitya, "Improved DC performance analysis of a novel asymmetric extended source tunnel FET (AES-TFET) for fast switching application," *Silicon*, 2021, doi: 10.1007/s12633-021-01147-8
[10] C. Li, X. Zhao, Y. Zhuang, Z. Yan, J. Guo, and R. Han, "Optimization of L-shaped tunneling field-effect transistor for ambipolar current suppression and Analog/RF performance enhancement," *Superlattices Microstruct.*, vol. 115, pp. 154–167, Mar. 2018, doi: 10.1016/j.spmi.2018.01.025
[11] N. Mohankumar, B. Syamal, and C. K. Sarkar, "Investigation of novel attributes of single halo dual-material double gate MOSFETs for analog/RF applications," *Microelectron. Reliab.*, vol. 49, no. 12, pp. 1491–1497, Dec. 2009, doi: 10.1016/j.microrel.2009.06.006
[12] D. B. Abdi and M. J. Kumar, "Controlling ambipolar current in tunneling FETs using overlapping gate-on-drain," *IEEE J. Electron Devices Soc.*, vol. 2, no. 6, pp. 187–190, 2014, doi: 10.1109/JEDS.2014.2327626
[13] I. A. Pindoo, S. K. Sinha, and S. Chander, "Improvement of electrical characteristics of SiGe source based tunnel FET device," *Silicon*, vol. 13, no. 9, pp. 3209–3215, 2021, doi: 10.1007/s12633-020-00674-0
[14] S. B. Rahi, P. Asthana, and S. Gupta, "Heterogate junctionless tunnel field-effect transistor: future of low-power devices," *J. Comput. Electron.*, vol. 16, no. 1, pp. 30–38, 2017, doi: 10.1007/s10825-016-0936-9
[15] S. Chen, H. Liu, S. Wang, W. Li, X. Wang, and L. Zhao, "Analog/RF performance of T-shape gate dual-source tunnel field-effect transistor," *Nanoscale Res. Lett.*, vol. 13, 2018, doi: 10.1186/s11671-018-2723-y
[16] B. R. Raad, D. Sharma, P. Kondekar, K. Nigam, and S. Baronia, "DC and analog/RF performance optimisation of source pocket dual work function TFET," *Int. J. Electron.*, vol. 104, no. 12, pp. 1992–2006, Dec. 2017, doi: 10.1080/00207217.2017.1335788
[17] C. K. Pandey, D. Dash, and S. Chaudhury, "Approach to suppress ambipolar conduction in Tunnel FET using dielectric pocket," *Micro Nano Lett.*, vol. 14, no. 1, pp. 86–90, 2019, doi: 10.1049/mnl.2018.5276
[18] P. Goyal, J. Madan, G. Srivastava, R. Pandey, and R. S. Gupta, "Performance analysis of drain pocket hetero gate dielectric DG-TFET: Solution for ambipolar conduction and enhanced drive current," *Silicon*, 2022, doi: 10.1007/s12633-021-01564-9
[19] J. Talukdar and K. Mummaneni, "A non-uniform silicon TFET design with dual-material source and compressed drain," *Appl. Phys. A Mater. Sci. Process.*, vol. 126, no. 1, pp. 1–9, 2020, doi: 10.1007/s00339-019-3266-5
[20] S. Garg and S. Saurabh, "Suppression of ambipolar current in tunnel FETs using drain-pocket: Proposal and analysis," *Superlattices Microstruct.*, vol. 113, pp. 261–270, 2018, doi: 10.1016/j.spmi.2017.11.002
[21] D. Gracia, D. Nirmal, and A. Nisha Justeena, "Investigation of Ge based double gate dual metal tunnel FET novel architecture using various hetero dielectric materials," *Superlattices Microstruct.*, vol. 109, pp. 154–160, 2017, doi: 10.1016/j.spmi.2017.04.045
[22] Pasupathy, K. R., Manivannan, T. S., & Lakshminarayanan, G., "A review of engineering techniques to suppress ambipolarity in tunnel FET," *Silicon*, vol. 14, no. 5, pp. 1887–1894, 2022, doi: 10.1007/s12633-021-01018-2

12 Metal Oxide-based High-K Dielectrics for High Density MIM Capacitor Applications

Ai-Dong Li and Song Sun

12.1 INTRODUCTION

With the integrated circuits (ICs) booming up, the importance of System-on-Chip (SoC) designs increases exponentially in microelectronics, and traditional structure capacitors, such as polysilicon-insulator-polysilicon (PIP) or metal-oxide-silicon (MOS), cannot satisfy the stringent requirements of next-generation analog/mixed-signal and radio frequency (RF) ICs because of the problems of parasitic capacitance, inferior figure of merit (FOM), and voltage linearity [1, 2]. New MIM capacitors with metal electrodes have been the first choice for SoC design so as to avoid the parasitic capacitance, contact resistance, and depletion effect [3]. With the continuous scaling down of ICs, conventional SiO_2 and Si_3N_4have to be substituted with high-K dielectrics so as to keep the high capacitance value. However, single high-K dielectrics, such as ZrO_2 [4], Al_2O_3 [5], TiO_2 [6], also generate new drawbacks, such as high leakage current density, large voltage linearity, and poor thermal stability. It is urgent to develop high-performance composite high-K dielectric MIM capacitors, e.g., stacking structures ($Al_2O_3/HfO_2/Al_2$ [7], $Al_2O_3/ZrO_2/SiO_2$ [8], $HfO_2/La_2O_3/HfO_2$ [9], and nanolaminate structures $(HfO_2/SiO_2)_m$) [10]. The improvement of electrical properties of high-density MIM capacitors by ALD is still needed, although some advances have been achieved.

In addition, the tremendous demand for wearable devices and flexible electronics is rising rapidly [11–13]. Thus, the preparation of flexible electronic circuits is increasingly urgent. A passive basic capacitor must also meet the flexible device requirements, including bendability, lightweight, and robustness. Polymer is one of the attractive substrate candidates for flexible MIM capacitors because of great bendability, low cost, and large-scale producibility. Nonetheless, its poor thermal stability impedes the development of flexible MIM capacitors.

At present, the fabrication technology of high-K ultrathin film compatible with the back-end semiconductor processing (<450°C) or on flexible polymer substrates is lacking, which has severely impeded these high-K materials' practical applications. Since 2001 ALD has been regarded as one of the main candidate technologies for the IC industry, with plasma-enhanced chemical vapor deposition (PECVD) and metal-organic CVD by the international technology roadmap for semiconductors (ITRS) [14], the rapid development for ALD in the film deposition field has been achieved, owing to a unique self-limiting reaction mechanism, enabling some advantages such as large-area uniformity, excellent three-dimensional (3D) conformality, accurate and simple thickness control, and low deposition temperature (<400°C) [15–18]. ALD-derived high-K dielectric layers have been successfully applied in commercial metal-oxide-semiconductor field effect transistor (MOSFET) devices. Therefore, it is tempting to prepare high-density MIM capacitors on Si or flexible MIM capacitors on polymer by means of ALD. Moreover, ALD is particularly suitable for rational design and controllable fabrication of stacking structures, nanocomposite, and nano-laminate high-K thin films [19]. Compared to thermally-driven ALD (TALD), plasma-enhanced ALD (PEALD) is an energy-assisted ALD method, where plasma species as co-reactants are exploited for thin film deposition [18–20]. PEALD

DOI: 10.1201/9781003352365-12

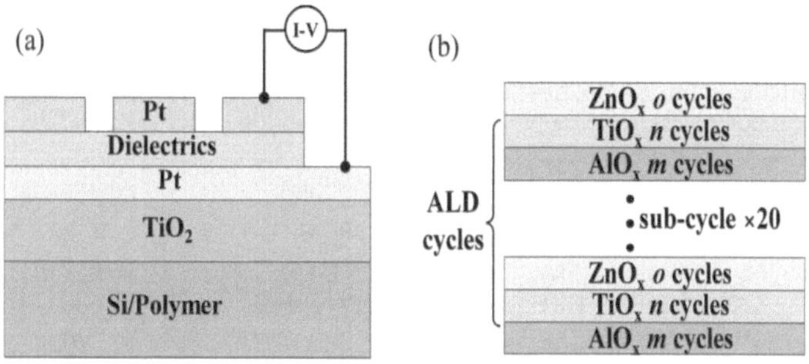

FIGURE 12.1 Schematics of (a) Si-based and (b) polymer-based MIM capacitor. Schematics by the author.

can provide some merits over TALD, such as low growth temperature and impurity residues, increased deposition rate, improved film properties, and more precursor/materials/substrate choice.

This chapter focuses on the exploration of the new metal oxide-based high-K dielectric system on rigid Si and flexible polymer substrates by the ALD method. Figure 12.1 (a) shows the schematics of Si-based and polymer-based MIM capacitors [21]. A series of high-K nanolaminate systems, including Zr-Ti-O, Zr-Sn-Ti-O, Hf-Ti-O, Hf-Sn-Ti-O, Al-Ti-O, Al-Ti-Sn-O, and Al-Ti-Zn-O, have been designed and constructed for optimal processing/composition and electrical properties, such as capacitor density, leakage current density, and voltage linearity. Figure 12.1(b) illustrates the ALD layer sequence of Al-Ti-Zn-O thin film, including the sub-cycle of $AlO_x/TiO_x/ZnO_x$ of *m*: *n*: *o*. Moreover, the effect of different bending radii on the electrical properties of polymer-based MIM capacitors has been evaluated, and the dynamic bending tests have been performed in the optimal polymer-based MIM capacitor system. All these indicate that ALD-derived metal oxide high-K nanolaminate/nanocomposite dielectrics are potential candidates for next-generation high-density MIM capacitors in Si-based ICs and flexible electronics.

12.2 HIGH-K DIELECTRICS FOR SI-BASED MIM CAPACITORS

In this section, thermal ALD (TALD) and plasma-enhanced ALD (PEALD) were used to design and fabricate a series of high-K dielectric thin films, including single, binary, and ternary metal oxides, for Si-based MIM capacitors. Commercial $Si/SiO_2/TiO_2/Pt$ was chosen as grown substrate and RF-sputtered Pt as a top electrode. After the deposition of MIM capacitors, 400°C post-annealing was conducted for 30 min. The C-V and I-V curves were examined by a Keithley 4200-SCS semiconductor characterization system on a probe station (CasCade Summit 12000B-M). These MIM capacitors were compared and optimized based on several key electrical characteristics, such as high capacitance density (10-15fF/μm^2), low leakage current density (~10^{-8} A/cm^2), and low voltage linearity (~100 pm/V^2) [14, 22]. It is found that PEALD-derived Zr-Sn-Ti-O dielectrics ($Zr_{0.45}Sn_{0.06}Ti_{0.48}O_2$) exhibit the best electrical performance to meet the requirements of SoC [23]. Next, we'll introduce the relevant work on ALD-derived high-K dielectrics of single, binary, and ternary metal oxides for Si-based MIM capacitors, in turn [24, 25].

12.2.1 SINGLE METAL OXIDE

The dielectric constant (k) and band gap (E_g) of several metal oxides are 9/8.7 eV forAl_2O_3, 25/5.7 eV for HfO_2, and 25/7.8 eV for ZrO_2 [26]. Considering the trade-off between the dielectric constant and band gap of metal oxides, Al_2O_3, HfO_2 and ZrO_2 were chosen and prepared on Pt-coated Si substrates as high-K dielectric layers by TALD for MIM capacitor applications. We also tried to grow AlN (k=8.9, E_g=6.2 eV) as a dielectric layer by PEALD. Table 12.1 summarizes the ALD processing and electrical properties of several single metal oxide/nitride Si-based MIM capacitors.

TABLE 12.1
ALD Processing and Electrical Properties of Several Single Metal Oxide/Nitride Si-based MIM Capacitors

High-K Materials	Al_2O_3	AlN	HfO_2	ZrO_2
Deposition method	TALD	PEALD	TALD	TALD
Precursor	TMA[a]	TMA[a]	TEMAH[a]	TEMAZ[a]
Co-reactant	H_2O	NH_3 plasma	H_2O	H_2O
Deposition temperature (°C)	300	250	200	250
Cycles	200	300	500	500
Thickness (nm)	20	17	26.5	26.5
Electrode	Pt	Pt	Pt	Pt
C_d (fF/μm²)	9.5	5.3	3.4	2.6
VCC (ppm/V²)	375	1600	165	116
J_A (A/cm²)@ ±3 V	10^{-8}	10^{-7}	2.77×10^{-2}	3.3×10^{-8}

[a] TMA:$Al(CH_3)_3$; TEMAH:$Hf[N(CH_3)(C_2H_5)]_4$; TEMAZ:$Zr[N(CH_3)(C_2H_5)]_4$.

The processing parameters mainly contain used precursor and co-reactant, deposition temperature, and ALD cycles, while the electrical properties involve three key values of capacitance density (C_d), leakage current density (J_A), and voltage linearity (VCC: quadratic voltage coefficient of the capacitance, α). It can be seen that all single high-K metal oxide MIM capacitors cannot satisfy the strict requirements of C_d, J_A, and VCC. Among them, the ZrO_2 sample shows a lower VCC value of 116 pm/V², close to the target value of 100 pm/V².

12.2.2 Binary Metal Oxide

Table 12.2 summarizes the processing conditions and electrical performances of several binary metal oxide Si-based MIM capacitors by TALD, including stacking structures of $HfO_2/Al_2O_3/HfO_2$ (HAH) and $TiO_2/Al_2O_3/TiO_2$ (TAT), nanocomposites of Zr-Al-O and Zr-Ti-O. $HfO_2/Al_2O_3/HfO_2$ and $TiO_2/Al_2O_3/TiO_2$ samples have a larger capacitance density than 10 fF/μm², however their voltage linearity is extremely high (> 1800 pm/V²). The Zr-Al-O sample also possesses a relatively high VCC value of 860 pm/V². Compared to the above binary metal oxide samples, the TALD-derived $ZrTiO_x$ MIM capacitor, using inorganic Zr resource, shows low voltage linearity of 295 pm/V² when the Zr/TI cycle ratio is 5. As a result, we'll focus on Zr-Ti-O high-K composite system to tune its electrical properties by optimizing the ALD processing, such as the deposition temperature and composition ratio of Zr/Ti, using various Zr sources of organic TEMAZ, inorganic $ZrCl_4$, and oxygen sources of H_2O and O_2 Plasma.

12.2.3 Zr-Ti-O MIM Capacitors

At first, we optimized the deposition temperature of Zr-Ti-O dielectric films on Pt-coated Si at 200, 250 and 300°C by TALD, using two Zr sources of TEMAZ and $ZrCl_4$. Zr-Ti-O samples grown at 300°C had the lowest C or Cl residue, with a smaller VCC value, hence the ALD growth temperature was set to 300°C. Then Zr-Ti-O ones with various Zr/Ti cycle ratios from 5:1 to 1:1 were prepared. When the Zr/Ti cycle ratio was 1:1, the obtained Zr-Ti-O films had the Zr/Ti atomic ratio close to 1:1, exhibiting larger capacitance density with higher k of ~40. Accordingly, the Zr/Ti cycle ratio was fixed to 1:1. Subsequently, PEALD was exploited to deposit Zr-Ti-O samples using $ZrCl_4$ and O_2 plasma. Two TALD control samples from $ZrCl_4$/TEMAZ and H_2O under similar conditions were also obtained. Table 12.3 lists the TALD/PEALD processing and electrical properties of three

TABLE 12.2
ALD Processing and Electrical Properties of Several Binary Metal Oxide Si-based MIM Capacitors

High-K Materials	$HfO_2/Al_2O_3/HfO_2$	$TiO_2/Al_2O_3/TiO_2$	$ZrAlO_x$	$ZrTiO_x$	$ZrTiO_x$
Deposition method	TALD	TALD	TALD	TALD	TALD
Precursor Co-reactant	TEMAH TMA H_2O	$TiCl_4$ TMA H_2O	$ZrCl_4$ TMA H_2O	$ZrCl_4$ $TiCl_4$ H_2O	$ZrCl_4$ $TiCl_4$ H_2O
Deposition temperature (°C)	300	300	300	300	300
Cycles	20/110/20	150/3/150	(1+1)×100	(5+1)×100	(7+6)×45
Thickness (nm)	15	10.8	14.6	26.4	26.4
Electrode	Pt	Pt	Pt	Pt	Pt
C_d (fF/μm²)	13.8	27	8.4	4.8	10.8
VCC (ppm/V²)	1875	30000	860	295	1637
J_A (A/cm²) @ ±3 V	10^{-7}	1.5×10^{-5}	10^{-7}	10^{-7}	10^{-8}

TABLE 12.3
TALD/PEALD Processing and Electrical Properties of Three Zr-Ti-O Samples

Deposition Method	TALD	TALD	PEALD
Precursor Co-reactant	$ZrCl_4$, $TiCl_4$, H_2O	TEMAZ, $TiCl_4$, H_2O	$ZrCl_4$, $TiCl_4$, O_2 Plasma
Cycles	(1+1) ×300	(1+1) ×300	(1+1) ×300
Deposition T (°C)	300	300	300
Thickness (nm)	25.5	31.5	19.5
C_d(fF/μm²)	8.8	8	7.4
VCC(ppm/V²)	610	536	-1775
J_A(A/cm²)@ ±3 V	1.1×10^{-7}	6.2×10^{-5}	6.7×10^{-7}

Zr-Ti-O samples. It is found that Pt/Zr-Ti-O/Pt capacitors by TALD show relatively larger positive VCC of 536~610 ppm/V², whereas Pt/Zr-Ti-O/Pt capacitors by PEALD have larger negative VCC of -1775 ppm/V². Figures 12.2 and 12.3 show the electrical curves of TALD and PEALD-derived Zr-Ti-O thin film capacitors using $ZrCl_4$, respectively. The normalized C/C_0–V curve can be proximately simulated as $C/C_0\ (V) = \alpha V^2 + \beta V + 1$ [27], where $C_0$is the capacitance at zero bias, and α and β of the quadratic and linear VCCs, respectively.

So Zr-Ti-O dielectrics were fabricated by a combination method of TALD and PEALD using TEMAZ and $ZrCl_4$, in order to acquire a relatively smaller VCC value by means of the offset effect, as indicated in Table 12.4. It can be seen in Figure 12.4 that its C_d and J_A of Zr-Ti-O capacitor is 9.4 fF/ μm² and 6.8×10^{-8} A/ cm² using inorganic Zr source, which met the requirements, but its voltage linearity is the rather larger negative value of –3200 ppm/ V², which might be related to the larger Cl impurity in the dielectric layer. As a consequence, the organic Zr source of TEMAZ was utilized to prepare the Zr-Ti-O capacitor on Pt-coated Si by TALD and PEALD. Its capacitance density is 4.1 fF/ μm², with leakage current density of 2.5×10^{-7} A/cm², voltage linearity is –102 ppm/ V², as shown in Figure 12.5. It is worth noticing that the organic zirconium source satisfies the requirement (VCC= –102 ppm/V²), and the leakage current is also small, but the capacitance density is lower,

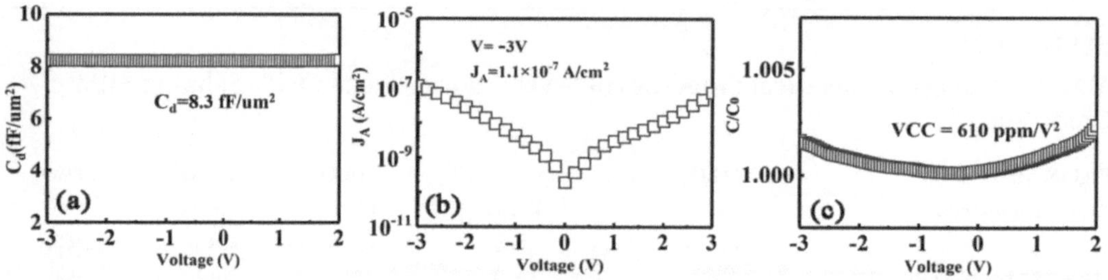

FIGURE 12.2 (a) C_d–V, (b) J_{A}-V and (c) C/C_0-V curves of Zr-Ti-OMIM capacitors by TALD with $ZrCl_4$ as Zr source. (Figure by the authors.)

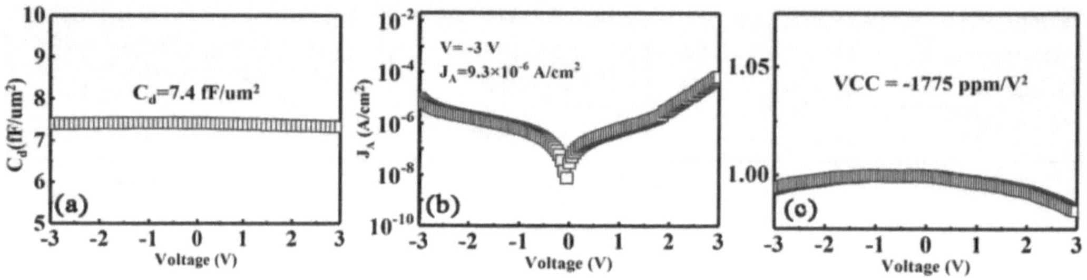

FIGURE 12.3 (a) C_d–V, (b) J_{A}-V and (c) C/C_0-V curves of Zr-Ti-OMIM capacitors by PEALD with $ZrCl_4$ as Zr source. (Figure by the authors.)

TABLE 12.4
Processing Parameters and Electrical Properties of Two Zr-Ti-O Samples by Combination of TALD and PEALD

Deposition Method	TALD+PEALD	TALD+PEALD
Precursor	a-$ZrCl_4$, $TiCl_4$, H_2O	a-TEMAZ, $TiCl_4$, H_2O
Co-reactant	b- $ZrCl_4$, $TiCl_4$, O_2 Plasma	b-TEMAZ, $TiCl_4$, O_2 Plasma
Cycles	150a + 150b	150a + 150b
Deposition T (°C)	300	300
Thickness (nm)	22.5	26.3
C_d(fF/μm^2)	9.4	4.1
VCC(ppm/V^2)	-3200	-102
J_A(A/cm^2)@ ±3 V	9.0×10^{-8}	4.7×10^{-8}

only 4.1 fF/um^2, which is related to the C (2%) and Cl (5%) impurity residue with decreased k [28, 29]. However, as though the combination method of TALD and PEALD may reduce the VCC value with elaborated and complicated control, it is hard to modulate the TALD/ PEALD processing parameters to get a lower VCC yet. Therefore, the process and performance of Zr-Ti-O MIM capacitors still need further optimization by the introduction of new metal elements, such as Sn [23].

12.3 TERNARY METAL OXIDE

After a careful literature survey, it is found that the Zr-Sn-Ti-O system exhibited superb microwave properties, such as a high dielectric constant of ~38 and FOM and a nearly zero temperature

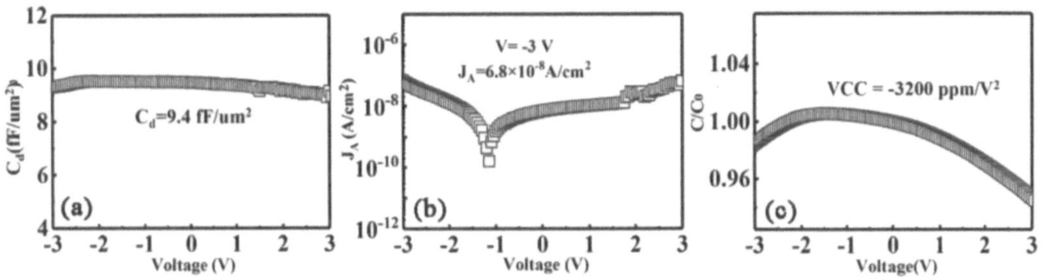

FIGURE 12.4 (a) C_d–V, (b) J_{A}-V, and (c) C/C_0-V curves of Zr-Ti-O MIM capacitors by combination of TALD and PEALD, with $ZrCl_4$ as Zr source. (Figure by the authors.)

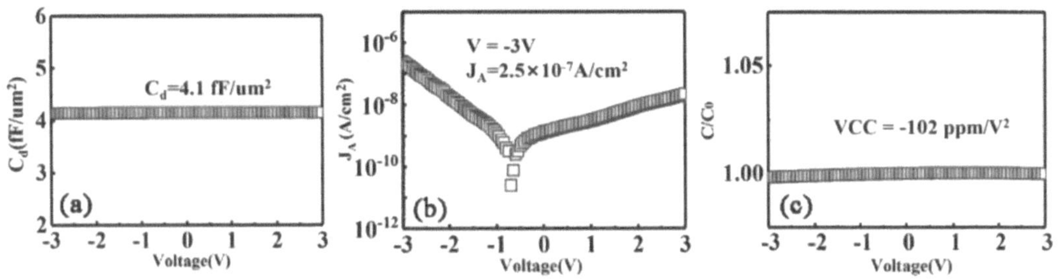

FIGURE 12.5 (a) C_d–V, (b) J_{A}-V, and (c) C/C_0-V curves of Zr-Ti-O MIM capacitors by combination of TALD and PEALD, with TEMAZ as Zr source. (Figure by the authors.)

co-efficient at resonant frequency [30] for microwave RF circuit applications. Furthermore, amorphous Zr-Sn-Ti-O thin films with $Zr_{0.15}Sn_{0.3}Ti_{0.55}O_{2-\delta}$ had the largest k value of 62.5 and superior FOM of 24.3 μC/cm² [31]. Thereby based on Zr-Ti-O MIM capacitors, we explored the effect of the Sn element incorporation into TALD and PEALD Zr-Ti-O dielectric films on the electrical characteristics of MIM capacitors, especially for the VCC [23].

12.3.1 Zr-Sn-Ti-O MIM Capacitors

Zr-Sn-Ti-O thin films were prepared on $Pt/TiO_2/SiO_2/Si$ at 300°C by TALD and remote PEALD using organic TEMAZ and inorganic $SnCl_4/TiCl_4$ as metal sources, and H_2O or O_2 plasma as the oxygen source, respectively. The m (m = 2~4) cycles of ZrO_2,n (n = 1~14) cycles of SnO_2, and k (k = 2~8) cycles of TiO_2 were in turn deposited on Pt-coated Si as sub-cycles. By tuning the m, n, and k values and sub-cycle numbers from 30 to 60, the composition and thickness impact on the electrical characteristics of Zr-Sn-Ti-O capacitors was investigated. Thanks to the evident inter diffusions after 400 °C post-anneal, Zr-Sn-Ti-O dielectric films by TALD and PEALD were nanocomposite oxides, as proved by X-ray photoelectron spectroscopy (XPS) results.

At ZrO_2: SnO_2: TiO_2 cycle ratio of 4: 14: 8 with a similar total cycle number of ~1000, the electrical curves of TALD and PEALD Zr-Sn-Ti-O thin film capacitors are shown in Figure 12.6(a)–(c), respectively. The capacitance density of 8.7 fF/μm² for the PEALD-derived sample is slightly higher than that of 7.7 fF/μm² for TALD one. Furthermore, the quadratic VCC value (-173 ppm/V²) of the PEALD sample is much smaller than that of the TALD one, with a positive VCC of 525 ppm/V². But the PEALD sample has a relatively higher J_A (1.6×10^{-6} A/cm²), compared to the TALD sample, with 1.5×10^{-7} A/cm. XPS measurements reveal that the PEALD and TALD samples show a Zr, Sn, Ti and O atomic ratio of 0.35: 0.03: 0.62: 2.00 and 0.27: 0.10: 0.65: 2.00, respectively. The corresponding atomic percentages of Zr, Sn, Ti and O are 11.9%, 1.1%, 20.7% and 66.3% in the PEALD one, and 9.0%, 3.4%, 21.6% and 66.0% in the TALD one. At the nominal same cycle ratio of ZrO_2: SnO_2:

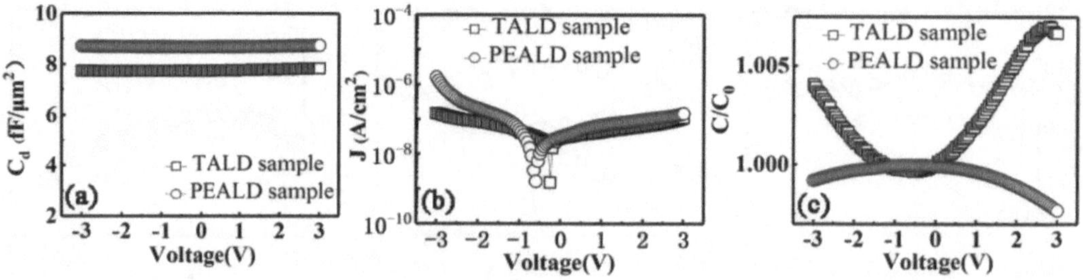

FIGURE 12.6 (a) C_d–V, (b) J_A-V and (c) C/C_0-V curves of TALD and PEALD-derived Zr-Sn-Ti-O_2 MIM capacitors [23]. (Copyright 2019. Reproduced with permission from IEEE.)

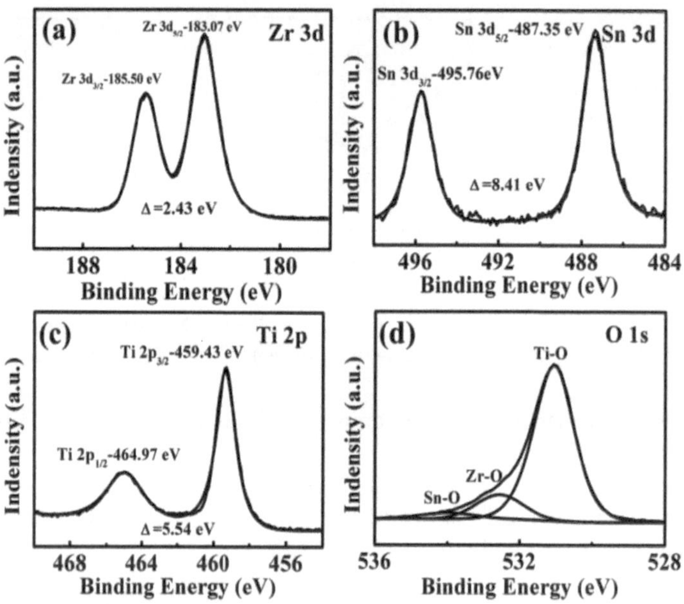

FIGURE 12.7 XPS spectra of $Zr_{0.45}Sn_{0.06}Ti_{0.48}O_2$ thin films by PEALD: (a) Zr 3d, (b) Sn 3d, (c) Ti 2p, (d) O 1s [23]. (Copyright 2019. Reproduced with permission from IEEE.)

TiO_2, the actual PEALD and TALD Zr-Sn-Ti-O film compositions are not the same. The PELAD sample owns a higher Zr amount and lower Sn amount, with a deviated Zr: Ti ratio of 1: 1. Among high-density MIM capacitors, there is a close and subtle relationship between the C_d, J_A, and VCC. The most effective way to reduce the leakage current density is to increase the film thickness. However, the enhanced film thickness implies the reduced capacitance density. Therefore, a compromise between the two factors has to be made. The influence factors of voltage linearity are more complex, and are related to the intrinsic characteristics of the selected material system, as well as the preparation process, film thickness, and quality.

In view of the PEALD Zr-Sn-Ti-O sample with higher C_d and lower VCC, the Sn content influence on the electrical characteristics of PEALD Zr-Sn-Ti-O MIM capacitors was studied and optimized. It is found that the Sn element's role in Zr-Ti-O thin films is to reduce the voltage linearity. For ~0.2% doped Sn with trace amounts, the VCC decreases to -400~-500 ppm/V^2. When the Sn concentration rises to ~1%, the VCC further declines to -130~-200 ppm/V^2. The optimal Sn content is close to 1.9%, with a VCC value of -83 ppm/V^2. Where the Sn concentration rises to 3%, the VCC increases to-150~-200 ppm/V^2. The Sn content in Zr-Ti-O thin films exceeds 5%, which reduces the

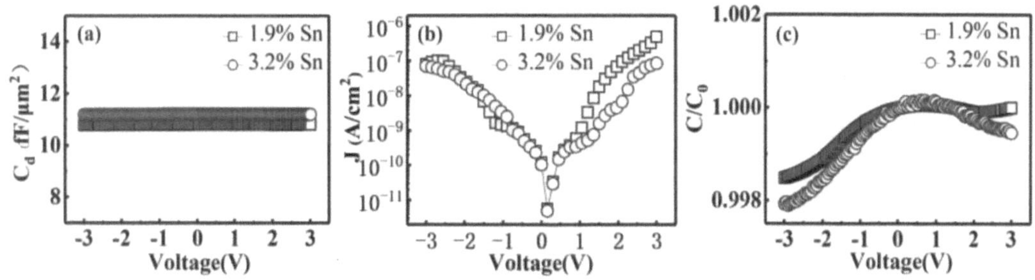

FIGURE 12.8 (a) C_d–V, (b) J-V and (c) C/C_0-V curves of $Zr_{0.45}Sn_{0.06}Ti_{0.48}O_2$ and $Zr_{0.43}Sn_{0.096}Ti_{0.48}O_2$ MIM capacitors with a Sn content of 1.9% and 3.2% [23]. (Copyright 2019. Reproduced with permission from IEEE.)

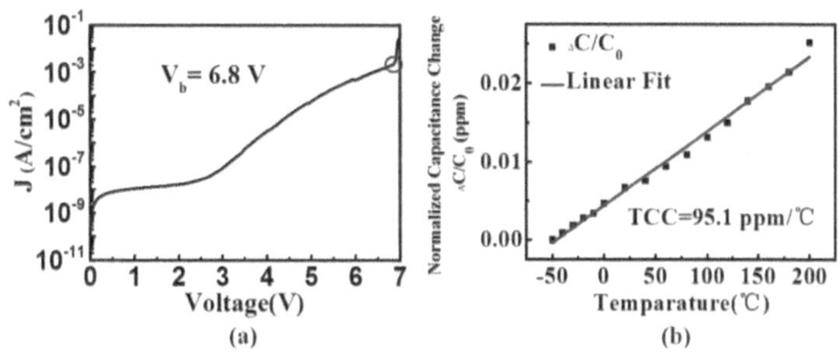

FIGURE 12.9 (a) Breakdown voltage and (b) temperature linearity curves of $Zr_{0.45}Sn_{0.06}Ti_{0.48}O_2$ MIM capacitors [23]. (Copyright 2019. Reproduced with permission from IEEE.)

TABLE 12.5
Effect of Various Electrode Systems on PEALD $Zr_{0.45}Sn_{0.06}Ti_{0.48}O_2$ MIM Capacitors

BE	Pt	Pt	Pt	Pt	TiN	Au	ITO	Ir	ITO	Au
W[a] (eV)	5.65	5.65	5.65	5.65	4.5	5.1	3.5–4.3	5.27	3.5-4.3	5.1
TE	Pt	ITO	Al	Au	Pt	Pt	Pt	Pt	ITO	Au
W[a] (eV)	5.65	3.5-4.3	4.28	5.1	5.65	5.65	5.65	5.65	3.5-4.3	5.1
C_d (fF/μm²)	10.8	1.2	1.5	5.4	2.4	14.9	12.5	4.8	6.03	11.8
VCC (ppm/V²)	-83	10000	332	-81	-976	2940	5800	9200	2900	1190
J_A (A/cm²)	5×10^{-7}	1×10^{-4}	7×10^{-8}	1×10^{-5}	4×10^{-2}	5×10^{-2}	4×10^{-2}	1×10^{-4}	1×10^{-4}	1×10^{-2}

[a] W:metal work function

bandgap and increases the leakage current density. For the optimal Zr-Sn-Ti-O sample, its formula is $Zr_{0.45}Sn_{0.06}Ti_{0.48}O_2$, with a Zr: Ti ratio of 1.00:1.07, close to 1:1, when the ZrO_2: SnO_2: TiO_2 cycle ratio is 2: 10: 2, along with total 700 cycles. The corresponding XPS spectra are illustrated in Figure 12.7, manifesting that the Sn atomic percentage is 1.9%.

Figure 12.8 show best electrical characteristics of the PEALD Zr-Sn-Ti-O sample, with larger C_d of 10.8 fF/μm², much lower VCC of -83 ppm/V², and J_A of 1.2×10^{-9} A/cm² at 1 V and 5×10^{-7} A/cm² at 3 V. Meanwhile, the $Zr_{0.43}Sn_{0.096}Ti_{0.48}O_2$MIM capacitor (Sn content of ~3.2%) has a C_d of 11.2fF/μm², VCC of -158 ppm/V², and J_A of 8.6×10^{-8} A/cm² at 3 V, as indicated in Figure 12.8.

TABLE 12.6
Comparison of the Pt/$Zr_{0.45}Sn_{0.06}Ti_{0.48}O_2$/Pt Capacitor with Other MIM Capacitors [23].

High-K Materials [Ref.]	Process	C_d (fF/μm^2)	J_A (A/cm^2) @RT	VCC (ppm/V^2)
ZrO_2 [33]	PEALD	6.8	1.7×10^{-9} @5 V	—
TiO_2 [34]	PECVD	11.2	—	30
$Al_2O_3/HfO_2/Al_2O_3$ [35]	ALD	8	$<10^{-8}$ @1 V	<100
Si_3N_4/LAHO[a] [36]	ALD	4.2	—	106
La doped HfO_2 [9]	ALD	16	$<1\times10^{-7}$ @-3.3 V	—
HfLaO/$LaAlO_3$/HfLaO [9]		9	—	<1000
$TiO_2/ZrO_2/TiO_2$ [37]	ALD	46.6	4.9×10^{-8} @-1 V	—
Ti-Ta-O [38]	AVD	11	~10^{-7} @2 V	<100
AZSZA[b] [39]	ALD	7.4	3.08×10^{-8} @5 V	-121
Zr-Sn-Ti-O [23]	PEALD	10.8	1.2×10^{-9} @1 V 5×10^{-7} @3 V	-83

[a] LAHO-laminated Al_2O_3/HfO_2.

[b] AZSZA: $Al_2O_3/ZrO_2/SiO_2/Al_2O_3/ZrO_2$.

The related results of $Zr_{0.45}Sn_{0.06}Ti_{0.48}O_2$ MIM capacitors on the breakdown voltage and the temperature linearity are displayed in Figure 12.9(a) and (b). The breakdown voltage and field are 6.8 V and 3.32 MV/cm, respectively. For the $Zr_{0.45}Sn_{0.06}Ti_{0.48}O_2$ MIM capacitor, the temperature linearity coefficient (TCC) is 95.1 ppm/°C, lower than 100 ppm/°C between -55~200°C. Thus, the $Zr_{0.45}Sn_{0.06}Ti_{0.48}O_2$ MIM capacitors have better temperature stability and higher breakdown resistance.

In addition, the impact of different electrode systems on PEALD Zr-Sn-Ti-O MIM capacitors with optimal composition was compared [32], as indicated in Table 12.5. The bottom/top electrode (BE/TE) pairs contain Pt/Pt, Pt/ITO, Pt/Al, Pt/Au, TiN/Pt, Au/Pt, ITO/Put, Ir/Pt, ITO/ITO, and Au/Au. It can be observed that a Zr-Sn-Ti-O MIM with bottom and top electrodes of Pt exhibits the best electrical performance.

Finally, Zr-Sn-Ti-O capacitors, such as $Zr_{0.45}Sn_{0.06}Ti_{0.48}O_2$, are compared with other MIM capacitors fabricated by ALD and CVD [33–39], as summarized in Table 12.6. Pt/$Zr_{0.45}Sn_{0.06}Ti_{0.48}O_2$/Pt capacitors possess optimal electrical properties comprehensively, such as a larger C_d, lower J_A, and superior VCC value. Our study elucidates that PEALD-derived Si-based Zr-Sn-Ti-O dielectrics are competitive candidates as high-density MIM capacitors applied in RF and analog/mixed signal ICs.

12.4 HIGH-K DIELECTRICS FOR FLEXIBLE POLYMER-BASED MIM CAPACITORS

With the flexible electronics booming up, the fabrication of high-K dielectric MIM capacitors on flexible polymer substrates becomes increasingly important, owing to the low cost, good flexibility, and easy manufacturing processing. However, the low-temperature deposition processing for high-K dielectric thin films on polymer substrates is still rather lacking [40, 41]. As described above, ALD is a kind of commercial deposition technology compatible with microelectronics, for large-scale application in MOSFETs. One intriguing advantage is its low deposition temperature. Even low defect-density thin films can be fabricated by PEALD at room temperature (RT) [17, 42–44]. Obviously, low/room temperature ALD can broaden the flexible substrate range and overcome the thermal obstacle. So, in this section, we pioneered to fabricate a series of high-K nanolaminates MIM capacitors on polyethylene terephthalate (PET), polyimide (PI) and epoxy resin (epoxy) at low temperature (80°C) or RT by TALD/PEALD, including Hf-Ti-O, Hf-Sn-Ti-O, Al-Ti-O, Al-Ti-Sn-O, and Al-Ti-Zn-O [21, 25, 45]. The bottom/top electrodes were Pt sputtered at

RT. The effect of composition, Ti precursors, and oxygen sources on the electrical properties of polymer-based MIM capacitors has been investigated and optimized. The device bendability has been evaluated.

12.5 FLEXIBLE HIGH-K NANOLAMINATE CAPACITORS FABRICATED AT LOW TEMPERATURE

Based on the previous progress on Si-based MIM capacitors using the high-K dielectric layer of Zr-Ti-O and Zr-Sn-Ti-O at 300°C by ALD [23–25], herein we focused on similar high-K systems of Hf-Ti-O and Hf-Sn-Ti-O nanolaminate thin films, which were prepared on PET, PI, and epoxy substrates at low temperature of 80°C, with proper RT/low temperature metal precursors by PEALD. The influence of inorganic $TiCl_4$ and organic amino titanium source ($Ti[N(CH_3)_2]_4$, TDMAT) on electrical properties were investigated thoroughly.

12.5.1 Hf-Ti-O and Hf-Sn-Ti-O MIM Capacitors from Inorganic $TiCl_4$ Precursor

Table 12.7 lists the processing parameters of Hf-Ti-O and Hf-Sn-Ti-O thin films at 80°C, using $TiCl_4$ by PEALD. Due to the low temperature deposition, the surface desorption of the precursor became slow, so the purge time was extended over 15 s so as to maintain the normal reaction growth. The electrical properties of polymer and Si-based Hf-Ti-O and Hf-Sn-Ti-O MIM capacitors are summarized in Tables 12.8 and 12.9, respectively. It can be seen that all Hf-Ti-O and Hf-Sn-Ti-O samples deposited at 80° C on PET, PI, epoxy, and Si show similar performances, such as relatively low C_d of ~5fF/μm², high J_A of 10^{-6}~10^{-5} A/cm², and very large VCC of >1500 ppm/V². Based on the XPS analyses, for Hf-Ti-O sample with a sub-cycle ratio of 2:8, the atomic ratio of Hf: Ti: O is 0.82: 0.43: 2.00; for the Hf-Sn-Ti-O sample with a sub-cycle ratio of 6:5:8, the atomic ratio of Hf: Sn: Ti: O is 0.73: 0.34: 0.28: 2.00. Nevertheless, the 80°C growth leads to high carbon (6.4%/6.8%) and Cl residue (8.7%/5.5%) in Hf-Ti-O/Hf-Sn-Ti-O thin films, respectively, which evidently degrades the electrical properties of MIM capacitors. In addition, the Sn element introduction into Hf-Ti-O samples reduce the VCC value by 300-500 ppm/V², which is consistent with the result of the Zr-Sn-Ti-O ones [23]. In order to lessen the influence of Cl impurity, next, the inorganic $TiCl_4$ was replaced with an organic TDMAT precursor for the fabrication of Hf-Ti-O and Hf-Sn-Ti-O by PEALD.

TABLE 12.7
Processing Parameters of Hf-Ti-O and Hf-Sn-Ti-O Thin Films at 80°C Using $TiCl_4$ by PEALD

Thin film	Total cycles	Sub-cycle	Precursor(Work T)	Pulse time	Purge time
Hf-Ti-O	350	2	TDMAH (75°C)	0.1	15
			O_2 plasma	30	25
		8	$TiCl_4$ (RT)	0.1	15
			O_2 plasma-	30	25
Hf-Sn-Ti-O	380	6	TDMAH (75°C)	0.1	15
			O_2 plasma	30	25
		5	$SnCl_4$ (RT)	0.3	15
			O_2 plasma	30	25
		8	$TiCl_4$ (RT)	0.1	15
			O_2 plasma	30	25

TABLE 12.8
Electrical Properties of Polymer and Si-based Hf-Ti-O MIM Capacitors

Substrate	Si	PET	PI	Epoxy
BE/TE		Pt/Pt		
Sub-cycle ratio		Hf: Ti = 2: 8		
Thickness (nm)		19.92		
C_d (fF/μm²)	5.6	5.2	5.8	5.5
J_A (A/cm²)	6.2×10^{-6}	4.5×10^{-5}	1.1×10^{-6}	5.8×10^{-5}
VCC (ppm/V²)	2120	2360	2250	2430

TABLE 12.9
Electrical Properties of Polymer and Si-based Hf-Sn-Ti-O MIM Capacitors

Substrate	Si	PET	PI	Epoxy
BE/TE		Pt/Pt		
Sub-cycle ratio		Hf: Sn: Ti= 6: 5: 8		
Thickness (nm)		28.36		
C_d (fF/μm²)	5.0	4.4	5.3	5.1
J_A (A/cm²)	4.3×10^{-6}	5.3×10^{-6}	3.2×10^{-6}	5.7×10^{-5}
VCC (ppm/V²)	1870	1590	2470	1860

TABLE 12.10
Electrical Properties of Hf-Ti-O and Hf-Sn-Ti-O MIM Capacitors with Various Sub-cycle Ratios

Materials	Sub-cycle	Sub-cycle ratio	C_d (fF/μm²)	J_A (A/cm²)	VCC (ppm/V²)
Hf-Ti-O	35	2: 2	9.5	8.2×10^{-7}	703
		2: 4	9.3	9.2×10^{-7}	856
		2: 6	8.6	7.1×10^{-7}	878
		2: 8	8.2	5.6×10^{-7}	1016
Hf-Sn-Ti-O	20	6: 5:2	4.2	3.3×10^{-5}	450
		6: 5:4	8.0	1.9×10^{-5}	548
		6: 5:6	5.4	2.8×10^{-5}	936
		6: 5:8	5.2	4.2×10^{-5}	1140

12.5.2 Hf- Ti-O and Hf-Sn-Ti-O MIM Capacitors from the Organic TDMAT Precursor

The processing parameters of Hf-Ti-O and Hf-Sn-Ti-O thin films using the organic TDMAT precursor are similar to those using inorganic $TiCl_4$, as shown in Table 12.7, except the work temperature of TDMAT (75°C) and cycle ratios. At first, the effect of cycle ratios of Hf: Ti and Hf: Sn: Ti on the electrical properties of Hf-Ti-O and Hf-Sn-Ti-O capacitors on Si, using TDMAT as the Ti source, has been explored and optimized, as seen in Table 12.10.

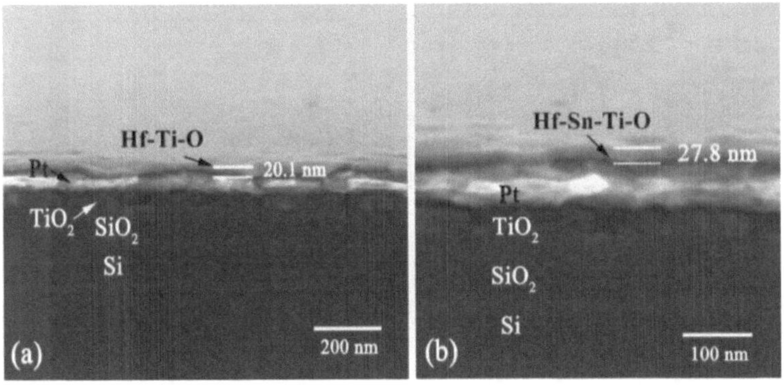

FIGURE 12.10 SEM cross-section images of (a) Hf-Ti-O and (b) Hf-Sn-Ti-O MIM capacitors. Figure by the authors.

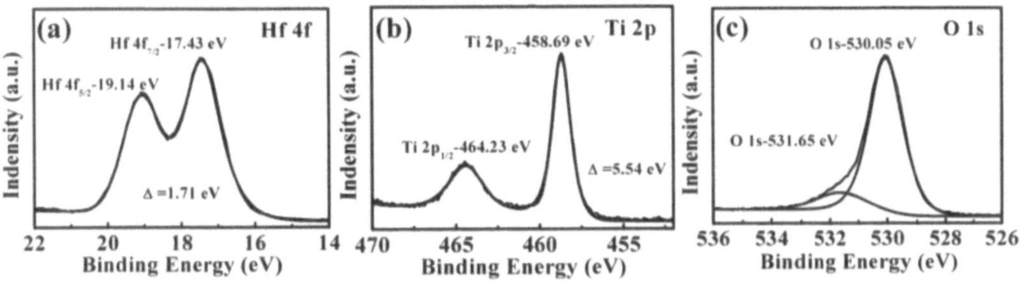

FIGURE 12.11 XPS spectra of Hf-Ti-O thin films with sub-cycle ratio of 2: 2 by PEALD: (a) Hf 4f, (b) Ti 2p, (c) O 1s. (Figure by the authors.)

For the Hf-Ti-O system, with the sub-cycle ratio of Hf: Ti varies from 2:2 to 2:8, the capacitance density and leakage current density reduce slightly from 9.5fF/μm^{2}and 8.2×10^{-7} A/cm^2 to 8.2fF/μm^2 and 5.6×10^{-7} A/cm^2, along with the increased VCC from 703ppm/V^2 and 1016ppm/V^2. As a result, Hf-Ti-O MIM capacitor with a sub-cycle ratio of 2:2 has relatively better comprehensive performances. Its SEM cross-section is shown in Figure 12.10(a) with an Hf-Ti-O thickness (t) of 20.1 nm. The calculated K value of Hf-Ti-O thin film is 21.6 based on $C_d = k\varepsilon_0/t$. The corresponding narrow-scan XPS spectra of Hf 4f, Ti 2p and O 1s are recorded in Figure 12.11(a)–(c), respectively. The doublet at 19.14 and 17.43 eV are assigned to Hf 4f$_{5/2}$ and Hf 4f$_{7/2}$ from Hf-O bonding (Hf^{4+}). The two peaks at 464.23 and 458.69 eV with spin-orbit splitting energy of 5.54 eV originate from Ti^{4+} 2p$_{1/2}$ and Ti^{4+} 2p$_{3/2}$, basically consistent with the values of TiO_2. The O 1s spectra can be deconvoluted into two peaks. The peaks at around 531.65 and 530.05 eV correspond to Hf-O and Ti-O bonding in Hf-Ti-O thin film, respectively. At the sub-cycle ratio of Hf: Ti of 2:2, the atomic ratio of Hf:Ti:O is 0.37: 0.75: 2.00, with carbon residue of 6.26% and Cl impurity of 2.27%. Compared to that sample using inorganic $TiCl_4$ source, the Cl residue in Hf-Ti-O film from organic TDMAT declines, evidently.

For Hf-Sn-Ti-O, when the sub-cycle ratio of Hf: Sn:Ti changes from 6:5:2 to 6:5:8 with the enhanced Ti content, the capacitance density first increases and then decreases. At the sub-cycle ratio of 6: 5:4, the capacitance density attains the maximum value of 8.0 fF/μm^2. The leakage current density of four Hf-Sn-Ti-O samples is about 10^{-5} A/cm^2, higher than the Hf-Ti-O ones, which may be related to the Cl residue of $SnCl_4$ source. The VCC value becomes large, with the Ti content increasing, but Hf-Sn-Ti-O samples with sub-cycle ratios of 6: 5:2and 6: 5:4 have a relatively smaller VCC of 450ppm/V^{2}and 548ppm/V^2 because of the Sn doping of Hf- Ti-O, respectively.

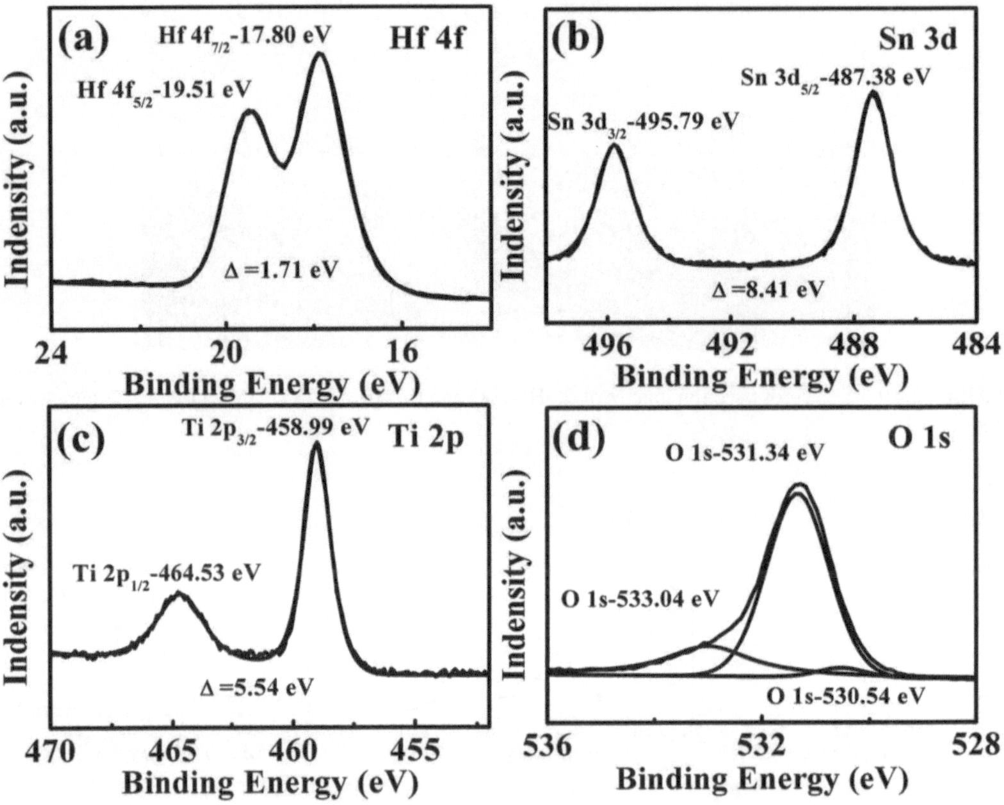

FIGURE 12.12 XPS spectra of Hf-Sn-Ti-O thin films with sub-cycle ratio of 6: 5: 4 by PEALD: (a) Hf 4f, (b) Sn 3d, (c) Ti 2p, (d) O 1s. (Figure by the authors.)

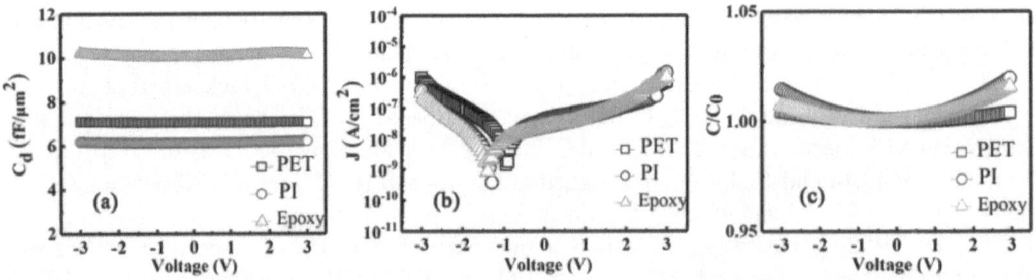

FIGURE 12.13 (a) C_d–V, (b) J_A-V, and (c) C/C_0-V curves of Hf-Ti-O MIM capacitors on PET, PI, and epoxy, with the sub-cycle ratio of 2: 2. (Figure by the authors.)

Therefore, a Hf-Sn-Ti-O MIM capacitor with a sub-cycle ratio of 6: 5:4 exhibits better electrical properties with the K Value of 25.1, whose thickness is ~27.8 nm, as shown in Figure 12.10(b). The corresponding XPS spectra of Hf 4f, Sn 3d, Ti 2p, and O 1s are shown in Figure 12.12(a)–(d), respectively. The double peaks at 19.51/17.80 eV, 495.79/487.38 eV, and 464.53/458.99 eV come from the Hf-O, Sn-O, and Ti-O bonds, respectively. The O 1s spectra can be deconvoluted into three peaks at 533.04, 531.34 and 530.54 eV, which belong to Sn-O (Sn^{4+}), Hf-O, and Ti-O bonds, respectively. The Hf-Sn-Ti-O sample with a sub-cycle ratio of 6: 5:4 has the atomic ratio of Hf:

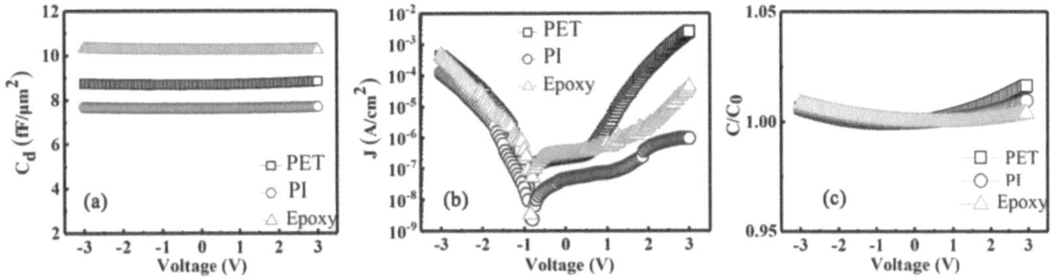

FIGURE 12.14 (a) C_d–V, (b) J_{A}-V, and (c) C/C_0-V curves of Hf-Sn-Ti-O MIM capacitors on PET, PI, and epoxy, with sub-cycle ratio of 6: 5: 4. (Figure by the authors.)

TABLE 12.11
Electrical Properties of Hf-Ti-O and Hf-Sn-Ti-O MIM Capacitors on Polymer Substrates Using Organic TDMAT Sources

Materials	Sub-cycle ratio	Substrate	C_d (fF/μm²)	J_A (A/cm²)	VCC (ppm/V²)
Hf-Ti-O	2 : 2	PET	7.1	9.7×10^{-7}	483
		PI	6.2	3.8×10^{-7}	768
		Epoxy	10.0	3.1×10^{-7}	616
		Si	9.5	8.2×10^{-7}	703
Hf-Sn-Ti-O	6: 5: 4	PET	8.7	4.1×10^{-4}	1290
		PI	7.6	1.2×10^{-4}	917
		Epoxy	10.2	4.5×10^{-4}	678
		Si	8.0	1.9×10^{-5}	548

Sn: Ti:O of 0.59:0.27: 0.34: 2.00, along with carbon residue of 7.5% and Cl impurity of 2.35%. Simultaneously, the Cl impurity content is reduced markedly in Hf-Sn-Ti-O thin film, due to using organic Ti precursor.

After obtaining the optimal processing of Hf-Ti-O and Hf-Sn-Ti-O MIM capacitors, three kinds of polymer-based samples, including PET, PI, and epoxy, were prepared at 80 °C, using organic TDMAT by PEALD. Figures 12.13 and 12.14 display the C_d–V, J_{A}-V and C/C_0-V curves of Hf-Ti-O with a sub-cycle ratio of 2: 2 and Hf-Sn-Ti-O with a sub-cycle ratio of 6: 5: 4 MIM capacitors on PET, PI, and epoxy, respectively.

Table 12.11 lists their electrical properties of Hf-Ti-O and Hf-Sn-Ti-O MIM capacitors on polymers and Si. It can be observed that Hf-Ti-O MIM capacitors on polymer substrates and Si have similar electrical properties of C_d, J_A, and VCC. In comparison with that from inorganic $TiCl_4$ source (Table 12.8), the Hf-Ti-O sample from organic TDMAT shows better electrical performances, such as enhanced C_d by 20%, reduced J_A by at least one order of magnitude, and declined VCC by ~70%. It indicates that the organic Ti precursor decreases the Cl impurity content and improves the electrical properties of Hf-Ti-O MIM capacitors.

However, compared with that using inorganic $TiCl_4$, the polymer-based Hf-Sn-Ti-O capacitors prepared by organic Ti source has significantly increased capacitance density (8.7–10.2 fF/μm²) and reduced voltage linearity (680-1290 ppm/V²), but the leakage current density increases by 1-2 orders of magnitude, which is related to the Cl impurity introduced by the $SnCl_4$ source. Further improvement of Hf-Sn-Ti-O MIM capacitors needs utilization of organic Sn precursor, such as tetrakis(dimethylamino) tin (TDMASn) [46].

TABLE 12.12
Processing Parameters of Al-Ti-O, Al-Ti-Sn-O and Al-Ti-Zn-O MIM Capacitors at RT by a Combination Method of TALD and PEALD

High-K	Precursor	Source T	Pulse time	Purge time
Al-Ti-O	TMA	RT	0.2s	10s
	O_2 plasma	-	30s	25s
	$TiCl_4$	RT	0.3s	15s
	H_2O	RT	2s	8s
Al-Ti-Sn-O	TMA	RT	0.2s	10s
	O_2 plasma	-	30s	25s
	$TiCl_4$	RT	0.3s	15s
	H_2O	RT	2s	8s
	$SnCl_4$	RT	0.3s	10s
	H_2O	RT	2s	8s
Al-Ti-Zn-O	TMA	RT	0.2s	10s
	O_2 plasma	-	30s	25s
	$TiCl_4$	RT	0.3s	15s
	H_2O	RT	2s	8s
	DEZ[a]	RT	0.2s	8s
	O_2 plasma	-	30s	25s

[a] DEZ: $(C_2H_5)_2Zn$.

12.6 FLEXIBLE HIGH-K NANOLAMINATE CAPACITORS FABRICATED AT ROOM TEMPERATURE

In the above section, we prepared and studied several polymer-based Hf- Ti-O and Hf-Sn-Ti-O MIM capacitors at 80°C by PEALD. In this section, we deposited Al-Ti-O, Al-Ti-Sn-O and Al-Ti-Zn-O MIM capacitors on PET, PI, epoxy, and Si at room temperature (RT), by a combination method of TALD and PEALD. The bottom/top electrodes were still metal Pt. The effect of various Al, Sn, and Zn contents on electrical properties of MIM capacitors was investigated and optimized. Table 12.12 summarizes the related processing parameters. We first characterized the optimal Al content in Al-Ti-O MIM capacitors. It is found that when the Al: Ti sub-cycle ratio is 4: 8 with total 300 cycles, the Si-based Al-Ti-O MIM capacitors show excellent electrical properties with the C_d of 5.5 fF/μm^2, the J_A of 2.8×10^{-6} A/cm^2 and the VCC of 172 ppm/V^2. Based on optimal Al-Ti-O samples, next we focused on Al-Ti-Sn-O and Al-Ti-Zn-O MIM capacitors on PET, PI, epoxy, and Si at RT by TALD and PEALD [21].

12.6.1 Al-Ti-Sn-O MIM Capacitors

At first, Al-Ti-Sn-O MIM capacitors with various Sn contents on Si were prepared and studied, as listed in Tables 12.13 and 12.14. When the Al: Ti: Sn sub-cycle ratio is 4: 8: 1 with a total of 325 cycles, Al-Ti-Sn-O sample on Si shows better electrical characteristics with the VCC of only 15 ppm/V^2, confirming the Sn element role in improving the voltage linearity [30]. In addition, Al-Ti-Sn-O thin films grown at RT have relatively Cl and carbon residues, which produces a larger effect on current leakage density. For the optimal Al-Ti-Sn-O sample, the atomic ratio of Al:Ti:Sn:O is 1.91: 0.24: 0.05:3.00, along with Cl content of 8.22 % and C content of 4.62 % from XPS analyses. The corresponding Al-Ti-Sn-O dielectric thin film is 46.8 nm with K value of 21.1.

TABLE 12.13
Electrical Properties of Al-Ti-Sn-O MIM Capacitors with Various Sn Contents

Dielectric layer	Total cycles	Sub-cycle ratio	C_d (fF/μm²)	J_A (A/cm²)	VCC (ppm/V²)
AlO_x:TiO_x:SnO_x	300	4: 8:0	5.5	2.8×10^{-6}	172
	325	4: 8:1	4.0	5.8×10^{-5}	15
	350	4: 8:2	3.7	4.0×10^{-5}	211
	375	4: 8:3	3.3	2.7×10^{-5}	190
	400	4: 8:4	3.1	1.3×10^{-4}	230

TABLE 12.14
Atomic Percentages of Al-Ti-Sn-O Thin Films with Various Sub-cycle Ratios

Sub-cycle ratio	Al (%)	Ti (%)	Sn (%)	O (%)	C (%)	Cl (%)
4: 8: 0	27.23	8.50	-	57.20	3.60	3.47
4: 8: 1	31.92	4.07	0.88	50.30	4.62	8.22
4: 8: 2	34.96	4.58	0.98	40.91	11.22	7.53
4: 8: 3	28.26	3.71	1.05	52.01	4.56	10.41
4: 8: 4	26.58	5.83	1.16	49.70	7.50	8.96

TABLE 12.15
Electrical Properties of Al-Ti-Sn-O MIM Capacitors on Various Substrates under Optimal Conditions

Al-Ti-Sn-O	C_d (fF/μm²)	J_A (A/cm²)	VCC (ppm/V²)
PET	4.2	8.1×10^{-5}	21
PI	5.1	7.1×10^{-6}	76
Epoxy	5.6	8.7×10^{-5}	30
Si	4.0	5.8×10^{-5}	15

Under optimal conditions, Al-Ti-Sn-OMIM capacitors with a sub-cycle ratio of 4: 8: 1 and a total 325 cycles were fabricated on PET, PI, and epoxy at RT by a combination of TALD and PEALD. The electrical properties are summarized in Table 12.15 and the corresponding C_d–V, J_{A-}V, and C/C_0-V curves are plotted in Figure 12.15. The polymer-based Al-Ti-Sn-O MIM capacitors have better electrical properties, such as the comparable capacitance density of 4.2-5.6 fF/μm² and the best voltage linearity of 21, 76, and 30 ppm/V²for PET, PI, and epoxy resin, respectively, all of which are less than 100 ppm/V². The corresponding leakage current density rises by one order of magnitude due to the addition of the Sn element.

12.6.2 Al-Ti-Zn-O MIM Capacitors

Al-Ti-Sn-OMIM capacitors have a disadvantage of higher leakage current density of around 10^{-5} A/cm², owing to the addition of the Sn element. In order to solve this problem, we attempted to replace

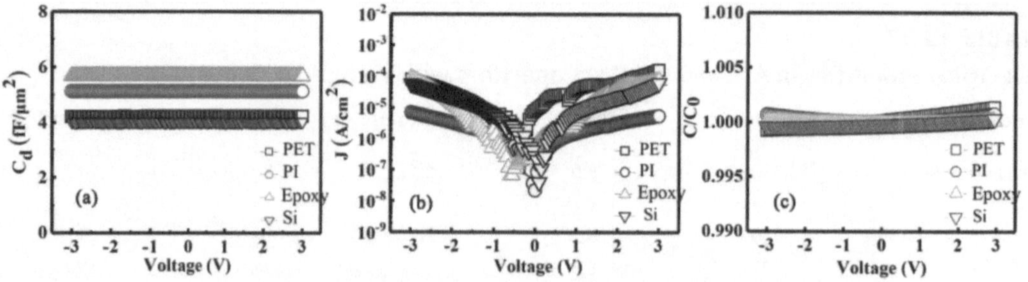

FIGURE 12.15 (a) C_d–V, (b) J_A-V, and (c) C/C_0-V curves of Al-Ti-Sn-O MIM capacitors on PET, PI, epoxy, and Si under optimal conditions. (Figure by the authors.)

TABLE 12.16
Electrical Characteristics of Al-Ti-Zn-O MIM Capacitors on Si with Different Zn Concentrations [21]

Dielectric layer	Total cycles	Sub-cycle ratio	C_d (fF/μm²)	J_A (A/cm²)	VCC (ppm/V²)
AlO_x:TiO_x:ZnO_x	300	4: 8: 0	5.5	2.8×10^{-6}	172
	260	4: 8: 1	5.1	7.9×10^{-6}	187
	280	4: 8: 2	6.0	1.3×10^{-7}	322
	300	4: 8: 3	7.2	9.5×10^{-8}	457
	320	4: 8: 4	8.6	8.9×10^{-8}	663

TABLE 12.17
Atomic Percentages of Al-Ti-Zn-O Thin Films with Different Sub-cycle Ratios [21]

Sub-cycle ratio	Al (%)	Ti (%)	Zn (%)	O (%)	C (%)	Cl (%)
4: 8: 0	27.23	8.50	-	57.20	3.60	3.47
4: 8: 1	18.78	8.02	0.03	64.61	5.03	3.52
4: 8: 2	19.17	7.43	1.02	66.71	3.98	2.65
4: 8: 3	27.83	8.32	1.10	53.59	6.38	2.77
4: 8: 4	25.43	7.90	1.38	48.42	8.54	8.32

the Sn element with the Zn one in Al-Ti-O, using a room temperature source of DEZ for the fabrication of Al-Ti-Zn-OMIM capacitors [21]. Tables 12.16 and 12.17 list the electrical properties and atomic percentages of Si-based Al-Ti-Zn-O MIM capacitors with different sub-cycle ratios, respectively [21]. With the Zn content increasing, the C_d and J_A of Al-Ti-Zn-O MIM capacitors are improved, evidently, although the VCC value increases in the same time. When the $AlO_x/TiO_x/ZnO_x$ cycle ratio is 4: 8: 3 with a total 300 ALD cycles, the Al-Ti-Zn-O sample shows optimal comprehensive performances. Compared to optimal Al-Ti-O MIM capacitors, the J_A(9.5×10^{-8} A/cm²) of Al-Ti-Zn-O on Pt-coated Si reduces by 30 times, with increased C_d of 7.2 fF/μm², despite that the VCC value slightly rises to457 ppm/V².

Figure 12.16(a) and (b) shows the thickness of optimal Al-Ti-Zn-O and Al-Ti-O thin films, respectively. The former is 50.2 nm-thick and the latter is 44.4 nm-thick. The calculated K value of Al-Ti-Zn-O and Al-Ti-O films is about 40.8 and 27.3, respectively. The former's K value is much

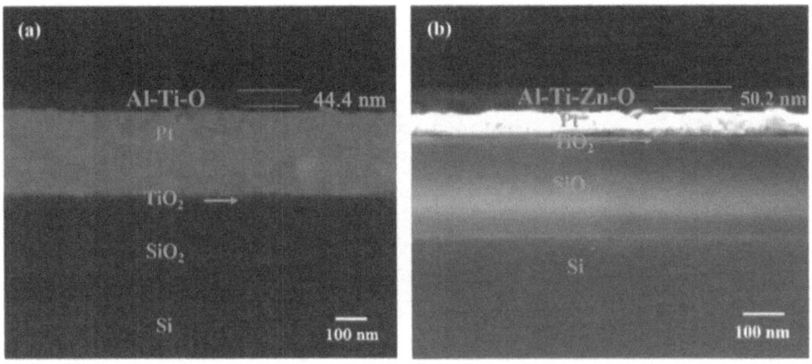

FIGURE 12.16 Cross-sectional SEM images of optimal Si-based MIM capacitors of (a) Al-Ti-Owith an AlO_x/TiO_x ratio of 4:8 and (b) Al-Ti-Zn-Owith an $AlO_x/TiO_x/ZnO_x$ ratio of 4:8:3 in total 300cycles [21]. (Copyright 2021. Reproduced with permission from Elsevier.)

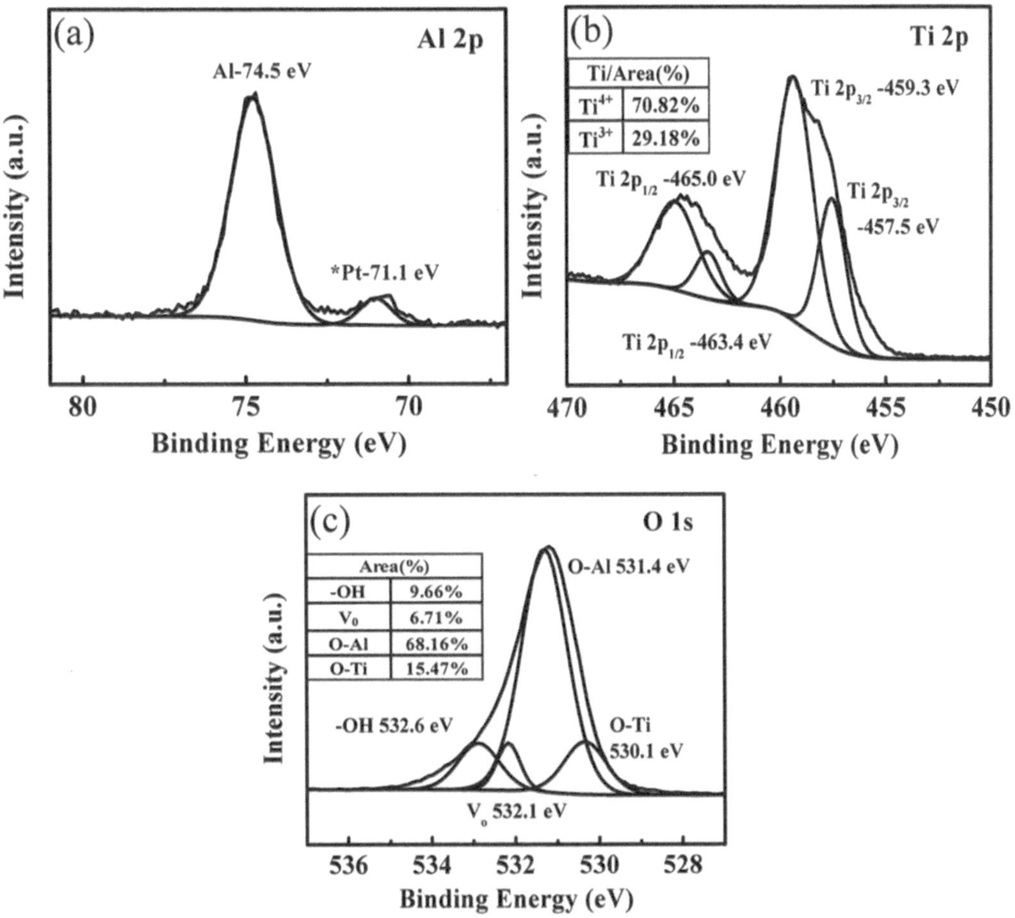

FIGURE 12.17 XPS spectra of optimal Al-Ti-O dielectrics on Si. (a) Al 2p, (b) Ti 2p, (c) O 1s. Herein, * peak at 71.1 eV represents Pt 4f signal from top electrode [21]. (Copyright 2021. Reproduced with permission from Elsevier.)

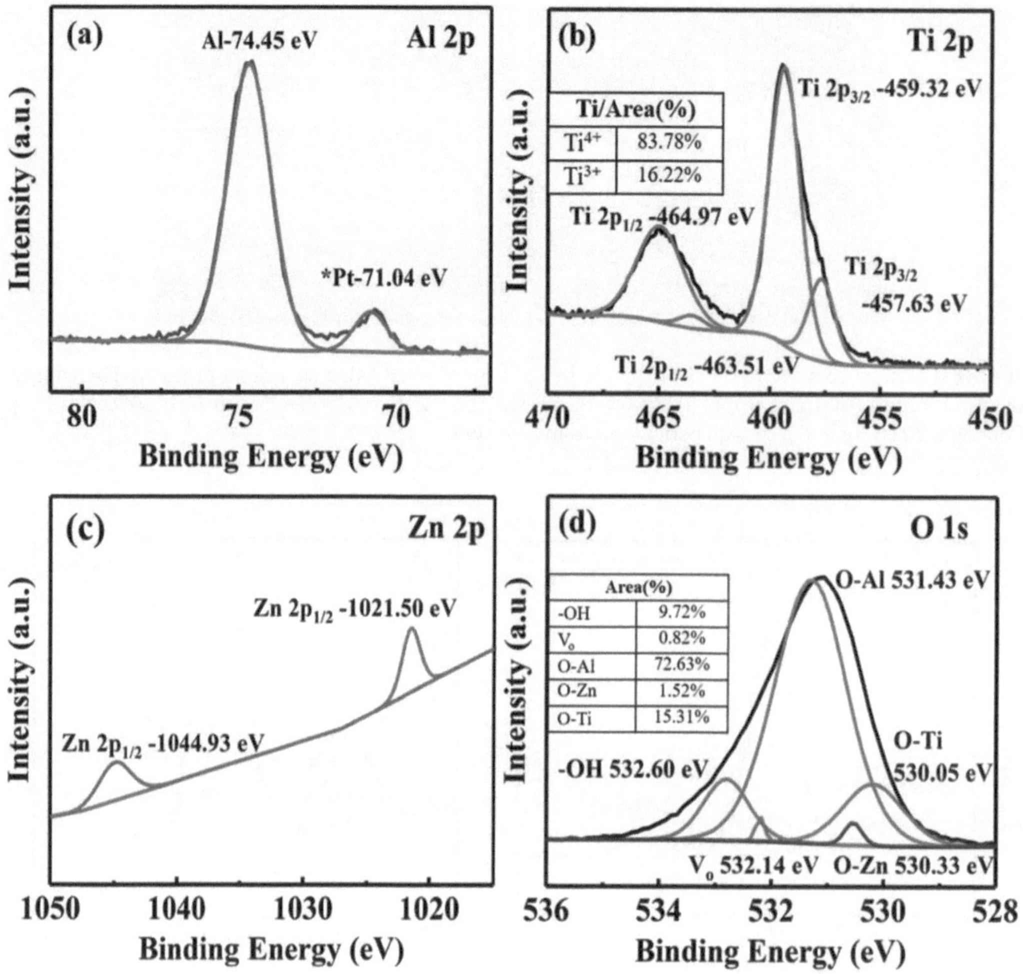

FIGURE 12.18 XPS spectra of optimal Al-Ti-Zn-O dielectrics on Si. (a) Al 2p, (b) Ti 2p, (c) Zn 2p, (d) O 1s. Herein, * peak at 71.0 eV represents Pt 4f signal from top electrode [21]. (Copyright 2021. Reproduced with permission from Elsevier.)

larger than that of the latter. XPS was exploited to examine their chemical states and composites, as shown in Figures 12.17 and 12.18, respectively. There are some C and Cl residues in two kinds of films because of the used TMA, $TiCl_4$, or DEZ precursors at RT. At an optimal cycle ratio of AlO_x/TiO_x with 4:8, Al-Ti-O film shows the Al/Ti/O atomic ratio of 1.43:0.44:3.00. At the optimal cycle ratio of AlO_x/TiO_x/$ZnO_x$with 4:8:3, Al-Ti-Zn-O film is close to $Al_{1.56}Ti_{0.47}Zn_{0.06}O_3$. As a result, Al-Ti-Zn-O film also refers to Zn-doped Al-Ti-O film (Ti: Al≈0.3). Due to the RT growth, the surface reactions cannot be finished well, producing C, Cl, and –OH residues in films [47].

There are nearly the same Al 2*p*XPS spectra in Al-Ti-O and Al-Ti-Zn-O films. In comparison with pure Al_2O_3 (74.7 eV), the Al 2*p* peak in two samples shifts to somewhat smaller 74.5 eV (Figures 12.17(a) and 12.18(a)), attributed to the Ti–O–Al bonding formation [17]. The double peaks of Ti 2*p* in two samples are deconvoluted into two group peaks with different peak intensity and shape of Ti 2*p*, as indicated in Figures 12.17(b) and 12.18(b). A stronger Ti2*p* doublet at 459.3 and 465.0 eV in Al-Ti-O films belongs to the Ti^{4+}–O bond of TiO_2. A weaker doublet at 457.5 and 463.4 eV comes from the Ti^{3+}–O bond with ~ 29% Ti^{3+}ions in all Ti ions, implies more oxygen vacancies in Al-Ti-O films. Whereas the Ti^{3+}peak in Al-Ti-Zn-O film declines markedly (~16% Ti^{3+} ions), suggesting the fewer

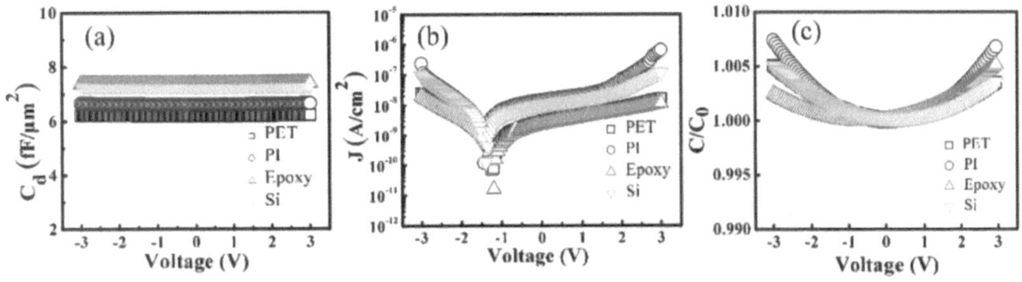

FIGURE 12.19 (a) C_d–V, (b) J_A-V, and (c) C/C_0-V curves of Al-Ti-Zn-O MIM capacitors on PET, PI, epoxy, and Si under optimal conditions [21]. (Copyright 2021. Reproduced with permission from Elsevier.)

TABLE 12.18
Electrical Performances of $Al_{1.56}Ti_{0.45}Zn_{0.06}O_3$ MIM Capacitors on PET, PI, Epoxy, and Si

Al-Ti-Zn-O	C_d (fF/µm²)	J_A (A/cm²)	VCC (ppm/V²)
PET	6.2	2.4×10^{-8}	456
PI	6.6	2.4×10^{-7}	810
Epoxy	7.4	2.0×10^{-8}	422
Si	7.2	9.5×10^{-8}	457

oxygen vacancies. Additionally, the weak Zn *2p* signals are observed in the Al-Ti-Zn-O sample with the enhanced splitting energy of 23.4 eV (Figure 12.18(c)), showing certain deviations from pure ZnO. This reveals the key role of the Zn element in Al-Ti-O dielectric films. Figures 12.17(c) and 12.18(d) show the deconvoluted O 1s spectra in two thin films, respectively. Except the stronger O-Ti and O-Al peaks at~530.1 and 531.4 eV, the weak O-Zn peak at ~530.3 eV and the highest 532.6 eV peak related to the –OH groups can be seen. The ~532.1 eV peak is thought from the oxygen vacancies in two thin films, respectively [48, 49]. Obviously, the V_O content of ~6.7% in Al-Ti-O is much higher than that of ~0.8% in Al-Ti-Zn-O, in good agreement with the Ti^{3+} ion concentration. XPS results elucidate that the addition of a little amount Zn into Al-Ti-O reduces the Ti^{3+} ion and oxygen vacancy concentrations and improves the electrical performance significantly.

Flexible Al-Ti-Zn-O MIM capacitors were fabricated on PET, PI, and epoxy by the combination method of TALD and PEALD at RT, based on optimal Si-based MIM capacitors. When the cycle ratio of $AlO_x/TiO_x/ZnO_x$ is 4:8:3 with total 300 cycles, the related electrical curves of optimal Al-Ti-Zn-O capacitors are plotted in Figure 12.19 (a)-(c). Three kinds of flexible capacitors on polymer substrate show similar electrical characteristics, as on Si (Table 12.18), with C_d of 6.2~7.4 fF/µm², J_A of 10^{-8} A/cm², and VCC value at around and 450 ppm/V². Flexible $Al_{1.56}Ti_{0.47}Zn_{0.06}O_3$ capacitors show a slight fluctuation of C_d between 6.2 and 7.4 fF/µm². The J_A and VCC value maintains around 10^{-8} A/cm² and 450 ppm/V², except the PI-based device. The PI rougher surface may be responsible for the poor VCC and leakage characteristics.

12.6.3 Experimental Results of Static and Dynamic Bending Tests

For the above-mentioned optimal polymer-based high-K nanolaminates MIM capacitors, the influence of different bending radii on electrical properties was studied by static and dynamic bending tests, as illustrated in Figure 12.20.

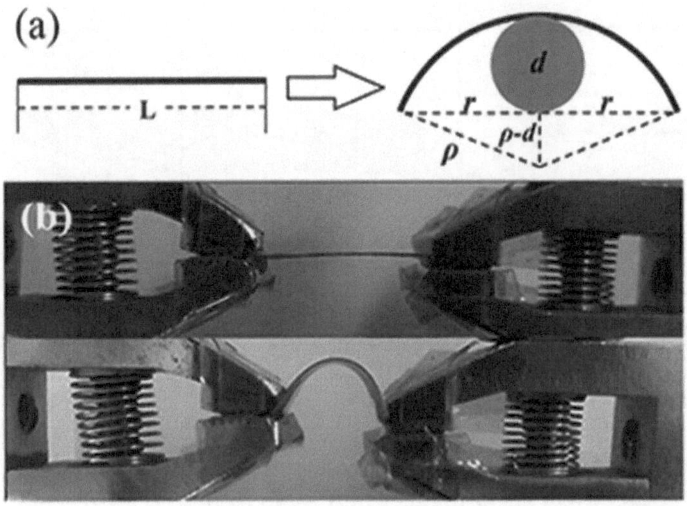

FIGURE 12.20 (a) Static bending with curvature radius ρ of 12.2 mm, 10.3 mm, 8.2 mm, and 6.4 mm by changing cylinder d value of 1.8 mm, 2.2 mm, 3.0 mm and 5.4 mm, respectively. ρ could be obtained by calculation. (b) Pictures of dynamic bending test [21]. (Copyright 2021. Reproduced with permission from Elsevier.)

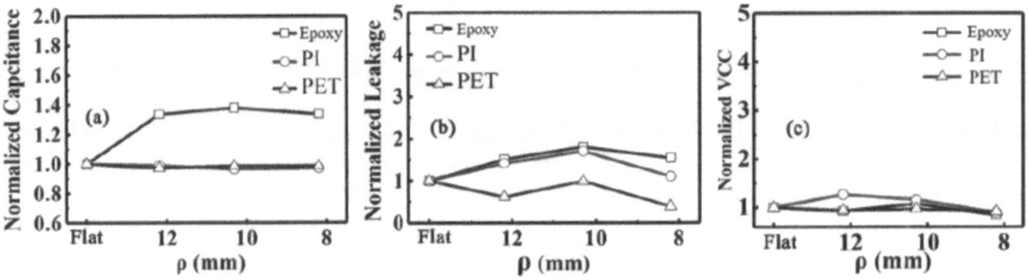

FIGURE 12.21 Static electrical properties of flexible Hf-Ti-O capacitors on three polymers from TDMAT source at bending radius ρ from 12 mm to 8.2 mm: normalized (a) capacitance, (b) leakage, and (c) VCC [21]. (Copyright 2021. Reproduced with permission from Elsevier.)

12.6.4 High-K MIM Capacitors by Low-Temperature PEALD

For the Hf-Ti-O and Hf-Sn-Ti-O polymer-based MIM capacitors prepared at 80°C by PEALD, when the bending radius ≥ 8.2mm, the electrical performances keep basically stable. The relative change of capacitance density is less than 10%, the leakage current density is relatively sensitive to the bending state, with the increase by no more than one time, and the voltage linearity is less sensitive to the bending state, with the relative change less than 25%. For example, static electrical properties of flexible Hf-Ti-O MIM capacitors on three polymers from the TDMAT source at several bending radii are shown in Figure 12.21(a)–c). Under the dynamic bending test, when the bending cycles are less than 40,000 to 50,000 times, the electrical performances of polymer-based MIM capacitors usually change little; when bending cycles reaches 70,000 times, the C_d drops suddenly, and the J_A increases rapidly, leading to the broken devices, as seen in Figure 12.22(a)–(c).

12.6.5 High-K MIM Capacitors by RTALD

For Al-Ti-Sn-O and Al-Ti-Zn-O MIM capacitors prepared at room temperature by the combination of PEALD and thermal ALD, the electrical properties of PET and epoxy-based MIM capacitors are

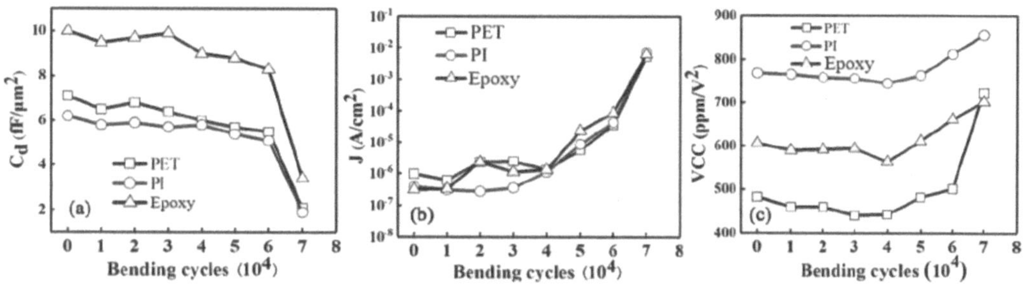

FIGURE 12.22 Electrical properties of flexible Hf-Ti-O capacitors on three polymers from the TDMAT source at various dynamic bending cycles: (a) C_d, (b) J_A, and (c) VCC. (Figure by the authors.)

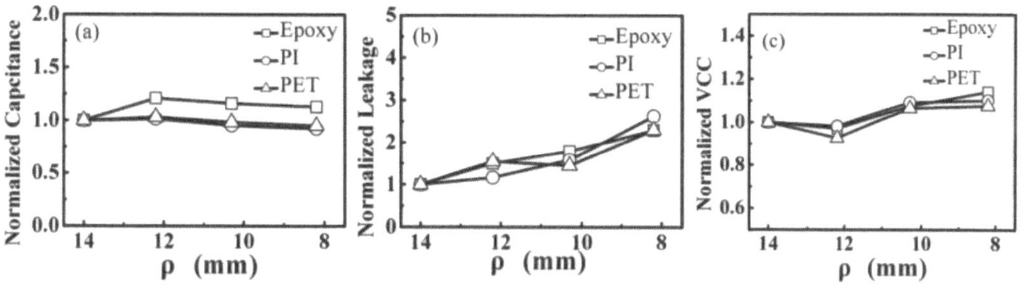

FIGURE 12.23 Static electrical characteristics of flexible $Al_{1.56}Ti_{0.47}Zn_{0.06}O_3$ capacitors on three polymers at various ρ:normalized (a) capacitance, (b) leakage, and (c) VCC [21]. (Copyright 2021. Reproduced with permission from Elsevier.)

basically stable when the bending radius is ≥ 8.2mm. The relative change in capacitance density is less than 15%. The leakage current density and voltage linearity are more sensitive to bending radii. The former usually increases by no more than 1.5 times, and the latter has a relative change of less than 1-5 times. For instance, when the ρ varies from flat state to 8.2 mm, the C_d of $Al_{1.56}Ti_{0.47}Zn_{0.06}O_3$ capacitors only decline within10% (Figure 12.23(a)). The J_A of capacitors gradually augments by 1.3 times on PET and epoxy and 1.6 times on PI (Figure 12.23(b)). The VCC values barely increase by 10~15% (Figure 12.23(c)). Despite a number of changes, flexible $Al_{1.56}Ti_{0.47}Zn_{0.06}O_3$ capacitors show better electrical properties of larger C_d of ~7fF/μm², lower J_A of 10^{-8} A/cm² and moderate VCC value less than 500 ppm/V².

Under a dynamic bending test, when the bending cycles are less than 4,000 times, the electrical performances show less change. When the bending cycles reaches 8000 times, the C_d reduces sharply, the J_A and voltage linearity rise dramatically, and the device fails. Compared with the result of 40,000 times bending resistance of polymer-based MIM capacitors prepared at 80°C by PEALD, the samples prepared at room temperature can only withstand about 4,000 fatigue tests. The difference of nearly 10 times may be related to the worse mechanical properties of the dielectric sample deposited at RT. The film strength and adhesion on polymer substrates fabricated at RT still need further improving. For example, the dynamic fatigue curves of $Al_{1.56}Ti_{0.47}Zn_{0.06}O_3$ MIM capacitors were acquired at the bending radius of 8.2mm for the 8,000 bending cycle, as shown in Figure 12.24(a)–(c). All three polymer-based Al-Ti-Zn-O samples could stand 4,000-cycle bending with stable electrical properties. After exceeding 4,000 bending, the flexible MIM capacitors' electrical properties start to decay significantly, attributed to the cracked Pt bottom electrode, to destroy MIM structure.

Recent advances of flexible polymer-based MIM capacitors prepared at low T/RT are summarized in Table 12.19. In comparison with the literature results, especially at room temperature, our optimal Al-Ti-Zn-O and Al-Ti-Sn-O flexible MIM capacitors by RTALD exhibit better

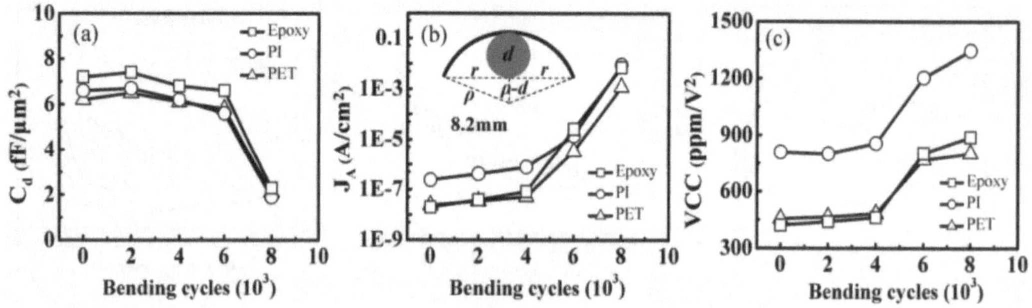

FIGURE 12.24 Electrical properties of $Al_{1.56}Ti_{0.47}Zn_{0.06}O_3$ flexible capacitors on three polymers under dynamic bending tests: (a) C_d, (b) J_A and the inset of 8.2-mm bending radius, (c) VCC [21]. (Copyright 2021. Reproduced with permission from Elsevier.)

TABLE 12.19
Comparison between This Work and Other Flexible MIM Capacitors by Low T/RT Processing [21]

Dielectric [Ref.]	Substrate	Process Method	Max T (°C)	C_d (fF/ μm²)	J_A (A/cm²)	VCC (ppm/V²)	Radius (mm)	Bend cycles
HfO_2 [50]	PI	Sol-gel	250	10.35	3.6×10^{-9}@5V	--	10	10^5
HfO_2 [51]	PI	Inkjet print	220	0.953	3×10^{-7}	18-452	--	--
ZrO_2 [52]	PEN	ALD	200	10.3	5.3×10^{-6}@1V	--	10	6×10^2
ZrO_2 [53]	PEN	ALD	150	17.8	1.3×10^{-9}@1V	--	7	10^3
Nb_2O_5 [54]	PET	Sputtering	50	11	--	557	10.43	2.5×10^3
BSA [55]	PET	Spin coating	40	0.261	1.2×10^{-7}@1V	-0.008	7.5	3×10^2
Graphene oxide [56]	PET	Drop casting	RT	4	--	250	10.43	3×10^2
NCTF [57]	PI	Sol-gel	RT	2.4	10^{-9}@1V	--	20	10^4
Al-Ti-Zn-O [21]	PET/PI/ Epoxy	ALD	RT	6.2-7.4	3×10^{-8}@3V	422-456	8.2	4×10^3
Al-Ti-Sn-O Our work				4.2-5.6	$0.7\text{-}8.7\times10^{-5}$	21-76	8.2	4×10^3

Remark: BSA=Bovine Serum Albumin; NCTF=1-bromoadamantane-P123 nanoscale composite thin-film.

comprehensive performances with super bendability. Al-Ti-Zn-O displays larger C_d, much smaller J_A, and reasonable VCC, whereas Al-Ti-Sn-O shows very low VCC value, with moderate C_d and J_A. ALD-derived devices at low T/RT exhibit better or comparable electrical properties, compared to other flexible capacitors prepared at higher temperature, indicating a lower thermal budget and shorter run time. Furthermore, RTALD increases the flexible substrate selection, such as paper, silk, or other biological substrates. In addition, ALD has superb compatibility with ICs, proved by mass manufacture and brilliant performances. It should have great potential in future flexible and wearable electronics.

12.7 CONCLUSION AND OUTLOOK

In summary, a series of metal oxide-based high-K dielectric systems on rigid Si and flexible polymer substrates have been prepared by ALD and investigated thoroughly. It is found that the PEALD-derived $Zr_{0.43}Sn_{0.096}Ti_{0.48}O_2$ MIM capacitors on Si exhibit the best electrical performances of the C_d of 11.2fF/μm², the J_A of 8.6×10^{-8} A/cm², the voltage linearity of -158 ppm/V², the breakdown voltage of 6.8 V, and the temperature linearity of 95.1 ppm/°C. These data have approached the key requirements of high-density MIM capacitors proposed by ITRS, however it is still one tough challenge how to achieve the trade-off between the large C_d and low J_A, due to the intrinsic contradiction of high-K and band gap, along with the complicated effects on VCC value. The development of elaborately-modulated composite high-K materials and device architecture might be a promising route to solve this issue. In addition, the electrical properties of several high-K nanolaminated MIM capacitors prepared on polymer substrates at low temperature or room temperature by PEALD/thermal ALD, such as Al-Ti-Zn-O, have tended to the level on Si substrate. Based on the performance and the bending resistance, PET-based high-K nanolaminated MIM capacitors undoubtedly show better comprehensive performances, followed by epoxy-based MIM capacitors. There is much room for the improvement of PI-based MIM capacitors, due to the surface roughness problem of PI substrates. For flexible MIM capacitors, the processing temperature is a strict limitation. Low-temperature/RT ALD technology offers a promising avenue for producing high-performance polymer-based MIM capacitors due to its compatibility with semiconductor processing. With the fast expansion of flexible electronics and wearable devices, the demand of flexible MIM capacitors will also be growing exponentially. Extensive and intensive study on ALD novel high-K materials systems is urgently needed on flexible substrates, such as the precursor choice, surface-pre-treatment, deposition temperature, and reaction mechanism because of the ALD window (Li, 2016). Above all, our work indicates that ALD-derived metal oxide high-K nanolaminated thin films are competitive candidates for next-generation high-density MIM capacitors applied in ICs and flexible electronics.

ACKNOWLEDGMENTS

This work was funded by the Natural Science Foundation of China (52073142, 51721001) and Jiangsu Province (BK20201252). It was also supported partially by the Huawei Innovation Research Program under Grant YB2015110047. The authors thank Mr. Chang Fang and Ms. Mei Wang, Ping Han, and Jin Lei for their contributions and efforts in this work.

REFERENCES

1. Ishikawa, T., Kodama, D., Matsui, Y., Hiratani, M., Furusawa, T., and Hisamoto, D. High-capacitance $Cu/Ta_2O_5/Cu$ MIM structure for SoC applications featuring a single-mask add-on process, *IEEE IEDM Tech.* (2002) 940–942. DOI:10.1109/IEDM.2002.1175991
2. Ding, S.J., Hu, H., Lim, H.F.,Kim, S.J.,Yu, X.F., Zhu, C.X., Li, M.F., Cho, B., Chan, D.S.H., Rustagi, S.C., Yu, M.B., Chin, A., and Kwong, D.L. High-performance MIM capacitor using ALD high-$KHfO_2$-Al_2O_3 laminate dielectrics, *IEEE Electron Device Lett.* 24 (2003) 730–732.DOI: 10.1109/LED.2003.820664
3. Chen, H.Y., Li, S.S.A., and Li, M.H. A low impedance CMOS-MEMS capacitive resonator based on metal-insulator-metal (MIM) capacitor structure, *IEEE Electron Device Lett.* 42 (2003) 1045–1048.DOI: 10.1109/LED.2021.3081365
4. Tsai, C.Y., Chiang, K.C., Lin, S.H., Hsu, K.C., Chi, C.C., and Chin, A. Improved capacitance density and reliability of high-K $Ni/ZrO_2/TiN$ MIM capacitors using laser-annealing technique, *IEEE Electron Device Lett.* 31 (2010) 749–751. DOI: 10.1109/LED.2010.2049636
5. Chen, S.B., Lai, C.H., Chin, A., Hsieh, J.C., and Liu, J. High-density MIM capacitors using Al_2O_3 and $AlTiO_x$ dielectrics, *IEEE Electron Device Lett.* 23 (2002) 185–187. DOI: 10.1109/55.992833
6. Hudec, B., Husekova, K., Tarre, A., Han, J.H., Han, S., Rosova, A., Lee, W., Kasikov, A., Song, S.J., Aarik, J., Hwang, C.S., and Frohlich, K. Electrical properties of TiO_2-based MIM capacitors deposited by $TiCl_4$ and TTIP based atomic layer deposition processes, *Microelectron. Eng.* 88 (2011) 1514–1516. DOI: 10.1016/j.mee.2011.03.059
7. Lee, S.K., Kim, K.S., Kim, S.W., Lee, D., Park, J.S.J., and Kim, S. Characterizing voltage linearity and leakage current of high density $Al_2O_3/HfO_2/Al_2O_3$ MIM capacitors, *IEEE Electron Device Lett.* 32 (2010) 384–386. DOI: 10.1109/LED.2010.2099200

8. Zhu, B., Liu, W. J., Wei, L., Zhang, D.W., Jiang, A.Q., and Ding, S.J. Voltage linearity modulation and polarity dependent conduction in metal-insulator-metal capacitors with atomic-layer-deposited Al_2O_3/ZrO_2/SiO_2 nano-stacks, *J. Appl. Phys.* 118 (2015) 63. DOI: 10.1063/1.4923349
9. Zhang, L., He, W., Chan, D.S.H., and Cho, B.J. High-performance MIM capacitors using HfLaO-based dielectrics, *IEEE Electron Device Lett.* 31 (2009) 17–19. DOI: 10.1109/LED.2009.2034545
10. Sun, J.K., Cho, B.J., Li, M.F., Ding, S.J., Zhu, C., Yu, M.B., Narayanan, B., Chin, A., and Kwong, D.L. Improvement of voltage linearity in high-K MIM capacitors using HfO_2-SiO_2 stacked dielectric, *IEEE Electron Device Lett.* 25 (2004) 538–540. DOI: 10.1109/LED.2004.832785
11. Rogers, J.A. Nanomesh on-skin electronics, *Nat. Nanotechnol.* 12 (2017) 839–840. DOI: 10.1038/nnano.2017.150
12. Ramasauskas, L.,Meskys, R., and Ratautas, D. Real-time glucose monitoring system containing enzymatic sensor and enzymatic reference electrodes, *Biosens. Bioelectron.* 164 (2020) 112338. DOI: 10.1016/j.bios.2020.112338
13. Maity, K.,Garain, S., Henkel, K., Schmeißer, D., and Mandal, D. Self-powered human-health monitoring through aligned PVDF nanofibers interfaced skin-interactive piezoelectric sensor, *ACS Appl. Polym. Mater.* 2 (2020) 862–878.DOI: 10.1021/acsapm.9b00846
14. Semiconductor Industry Association (SIA). The International Technology Roadmap for Semiconductors, USA (1999), http://www.itrs.net/reports.html
15. Sherman, A. *Atomic layer deposition for nanotechnology*, Ivoryton: Ivoryton Press, (2008).
16. Pinna, N., and Knez, M. *Atomic Layer Deposition of Nanostructured Materials*,Wiley-VCH Verlag & Co. KGaA, Weinheim (2012).
17. George, S.M. Atomic Layer Deposition: An Overview, *Chem. Rev.* 110 (2010) 111–131.DOI: 10.1021/cr900056b
18. Li, A.D. *Atomic Layer Deposition Technology: Fundamental and Applications*, Beijing: China Science Press, (2016).
19. Li, A.D. Atomic Layer Deposition for Novel Nonvolatile Memory Devices, Chap. 3, pp. 85–126 in Book *Atomic Layer Deposition (ALD) Fundamentals, Characteristics and Industrial Applications*, New York: Nova Science Publishers, (2015)
20. Profijt, H.B., Potts, S.E., van de Sanden, M.C.M. and Kessels, W.M.M. Plasma-assisted atomic layer deposition: basics, opportunities, and challenges, *J. Vac. Sci. Technol. A* 29 (2011) 050801–050826. DOI: 10.1116/1.3609974
21. Lei, J., Fang, C., Liu, C., Wu, D., and Li, A.D. Flexible Al-Ti-Zn-O MIM capacitors fabricated by room temperature atomic layer deposition and their electrical performances, *J. Alloy. Compd.* 870(2021) 159391. DOI: 10.1016/j.jallcom.2021.159391
22. Smitha, P.S., Babu, V.S., and Shiny, G. Critical parameters of high performance metal-insulator-metal nanocapacitors: A review. *Mater. Res. Express.*, 2019, 6: 122003. DOI: 10.1088/2053-1591/ab5511
23. Fang, C., Wang, M., Han, P., Cao, Y.Q., Wu, D., and Li, A.D. High-performance MIM capacitors using Zr-Sn-Ti-O dielectrics derived from atomic layer deposition, *IEEE Electron Device Lett.* 40 (2019) 682–685. DOI: 10.1109/LED.2019.2907981
24. Han, P. Study on high-K composite dielectrics-based charge-trapping memory cells and high-density metal-insulaotor-metal capacitors by atomic layer deposition, Master's Thesis of Nanjing University. (2018), 57–67.
25. Wang, M. Study on Si-based/Polymer-based High-density Capacitors and HfO_2-based Ferroelectric Capacitors Prepared by Atomic Layer Deposition, Master's Thesis of Nanjing University. (2018) 45–87.
26. Wilk, G.D., Wallace, R.M., and Anthony, J.M. High-K gate dielectrics: Current status and materials properties considerations, *J. Appl. Phys.* 89 (2001) 5243–5275. DOI: 10.1063/1.1361065
27. Lee, S.K., Kim, K.S., Kim, S.W., Lee, D.J., Park, S.J., and Kim, S. Characterizing voltage linearity and leakage current of high density Al_2O_3/HfO_2/Al_2O_3 MIM capacitors, *IEEE Electron Device Lett.* 32 (2010) 384–386. DOI: 10.1109/LED.2010.2099200
28. Kim, H.,McIntrye, P., and Saraswat, K. Effects of crystallization on the electrical properties of ultrathin HfO_2 dielectrics grown by atomic layer deposition, *Appl. Phys. Lett.* 82 (2003) 106–108. DOI: 10.1063/1.1533117
29. Li, M., Qian, X., Li, A.D., Cao, Y.Q.,Zhai, H.F., and Wu, D. A comparative study of growth and properties of atomic layer deposited transparent conductive oxide of Al doped ZnOfilms from different Al precursors, *Thin Solid Films.* 646 (2018) 126–131.DOI: 10.1016/j.tsf.2017.11.039

30. Iddles, D.M., Bell, A.J., and Moulson, A.J. Relationships between dopants, microstructure and the microwave dielectric properties of ZrO_2-TiO_2-SnO_2 ceramics, *J. Mater. Sci.* 27 (1992)6303–6310. DOI: 10.1007/BF00576276
31. Dover, R.B.V.L.,Schneemeyer, F., and Fleming, R.M. Discovery of a useful thin-film dielectric using a composition-spread approach, *Nature*, 392 (1998) 162–164. DOI: 10.1038/32381
32. Wang, M. Study on Si-based/polymer-based high-density capacitors and HfO_2-based ferroelectric capacitors prepared by atomic layer deposition, Master's Thesis of Nanjing University. (2018) 45–87.
33. Bertaud, T., Bermond, C., Blonkowski, S., Vallée, C., Lacrevaz, T., Farcy, A., GrosJean, M., and Fléchet, B. Electrical characterization of advanced MIM capacitors with ZrO_2 insulator for high-density packaging and RF applications, *IEEE Trans. Compon. Packag. Manuf. Technol.* 2 (2012) 502–509. DOI: 10.1109/TCPMT.2011.2182611
34. Wu, J.R., Wu, Y.H., Lin, C.C., Ou, W.Y., Wu, M.L., and Chen, L.L. Effect of nitrogen passivation on the performance of MIM capacitors with a crystalline-TiO_2/SiO_2 stacked insulator, *IEEE Electron Device Lett.* 33 (2012) 878–880. DOI: 10.1109/LED.2012.2192253
35. Kwak, H.Y., Kwon, H.M., Jung, Y.J., Kwon, S.K., Jang, J.H., Choi, W., Ha, M.L., Lee, J., Lee, S.J., and Lee, H.D. Characterization of Al_2O_3–HfO_2–Al_2O_3 sandwiched MIM capacitor under DC and AC stresses, *Solid-State Electron.* 79 (2012) 218–222. DOI: 10.1016/j.sse.2012.09.007
36. Ho, C.S., Chang, S.J., Chen, S.C.,Liou, J.J., and Li, H. A reliable Si_3N_4/Al_2O_3-HfO_2 stack MIM capacitor for high-voltage analog applications, *IEEE Trans. Electron Devices.* 61 (2014) 2302–2308. DOI: 10.1109/TED.2014.2332046
37. Padmanabhan, R., Mohan, S., Morozumi, Y., Kaushal, S., and Bhat, N. Performance and reliability of TiO_2/ZrO_2/TiO_2 (TZT) and AlO-doped TZT MIM capacitors, *IEEE Trans. Electron Devices.* 63 (2016) 3928–3935. DOI: 10.1109/TED.2016.2601419
38. Lukosius, M., Kaynak, C.B., Wenger, C., Ruh, G., and Rushworth, S. Electrical characteristics of Ti–Ta–O based MIM capacitors, *J. Vac. Sci. Technol. B.* 29 (2011) 1–6. DOI: 10.1116/1.3534020
39. Zhang, Q.X., Zhu, B., Ding, S.J., Lu, H.L., Sun, Q.Q., Zhou, P., and Zhang, W. Full ALD Al_2O_3/ZrO_2/SiO_2/ZrO_2/Al_2O_3 stacks for high-performance MIM capacitors, *IEEE Electron Device Lett.* 35 (2014) 1121–1123. DOI: 10.1109/LED.2014.2359195
40. Hourdakis, E., and Nassiopoulou, A.G. Reaching state-of-the art requirements for MIM capacitors with a single-layer anodic Al_2O_3 dielectric and imprinted electrodes, *Appl. Phys. Lett.* 111 (2017) 033503. DOI: 10.1063/1.4993898
41. Meen, J.S., Chu, M.C., Kuo, S.W., Chang, F.C., and Ko, F.H. Improved reliability from a plasma-assisted metal-insulator-metal capacitor comprising a high-K HfO2 film on a flexible polyimide substrate, *Phys. Chem. Chem. Phys.* 12 (2010) 2582–2589. DOI: 10.1039/b917604g
42. Mackus, A.J.M., Garcia-Alonso, D., Knoops, H.C.M., Bol, A.A., and Kessels, W.M.M. Room-temperature atomic layer deposition of platinum, *Chem. Mat.* 25 (2013) 1769–1774. DOI:10.1021/cm100274n
43. Kot, M., Das, C., Wang, Z., Henkel, K.,Rouissi, Z., Wojciechowski, K., Snaith, H.J., and Schmeisser, D. Room-temperature atomic layer deposition of Al_2O_3: Impact on efficiency, stability and surface properties in perovskite solar cells, *ChemSusChem*, 9 (2016) 3401–3406. DOI:10.1002/cssc.201601186
44. Shao, Y., Wu, X., Zhang, M.N., Liu, W.J., and Ding, S.J. High-performance a-InGaZnO thin-film transistors with extremely low thermal budget by using a hydrogen-rich Al_2O_3 dielectric, *Nanoscale Res. Lett.* 14 (2019) 122. DOI: 10.1186/s11671-019-2959-1
45. Fang, C. Fabrication and electrical properties of flexible polymer-based MIM capacitors with high-K nanolaminate dielectrics by atomic layer deposition, Master's Thesis of Nanjing University.(2020),37–77.
46. Lin, R.X., Xiao, K., Qin, Z.Y., Han, Q.L., Zhang, C.F., Wei, M.Y.,Saidaminov, M.I., Gao, Y., Xu, J., Xiao, M., Li, A.D., Zhu, J., Sargent, E.H. and Tan, H.R. Monolithic all-perovskite tandem solar cells with 24.8% efficiency exploiting comproportionation to suppress Sn(ii) oxidation in precursor ink, Nature *Energy.* 4 (2019) 864–873. DOI: 10.1038/s41560-019-0466-3
47. Pilz, J., Perrotta, A.,Leising, G., and Coclite, A.M. ZnO thin films grown by plasma-enhanced atomic layer deposition: material properties within and outside the "Atomic Layer Deposition Window", *Phys. Status Solidi A-Appl. Mat.* 217 (2020) 1900256. DOI: 10.1002/pssa.201900256
48. Wang, L.G., Cao, Z.Y., Qian, X., Zhu, L., Cui, D.P., Li, A.D., and Wu, D. Atomic layer deposited oxide-based nanocomposite structures with embedded CoPtx nanocrystals for resistive random access memory applications, *ACS Appl. Mater. Interfaces.* 9 (2017) 6634–6643. DOI: 10.1021/acsami.6b16098

49. Dupin, J.C., Gonbeau, D., Vinatier, P., and Levasseur, A. Systematic XPS studies of metal oxides, hydroxides and peroxides, *Phys. Chem. Chem. Phys.* 2 (2000) 1319–1324. DOI: 10.1039/a908800h
50. Meen, J.S., Chu, M.C., Kuo, S.W., Chang, F.C., and Ko, F.H. Improved reliability from a plasma-assisted metal-insulator-metal capacitor comprising a high-K HfO_2 film on a flexible polyimide substrate, *Phys. Chem. Chem. Phys.* 12 (2010) 2582–2589. DOI: 10.1039/b917604g
51. Vescio, G., López-Vidrier, J., Leghrib, R., Cornet, A., and Cirera, A. Flexible inkjet printed high-K HfO_2-based MIM capacitors, *J. Mater. Chem.* C. 4 (2016) 1804–1812. DOI: 10.1039/C5TC03307A
52. Guo, T., Zhang, G., Su, X., Zhang, H., Wan, J., Chen, X., Wu, H., and Liu, C. Transparent and flexible capacitors with an ultrathin structure by using graphene as bottom electrodes, *Nanomaterials*, 7 (2017) 418. DOI:10.3390/nano7120418
53. Guo, T., Zhang, G.Z., Zhang, H., Su, X., Chen, X., Wan, J.X., Wu, H., and Liu, C. Highly flexible, stable and transparent capacitors with enhanced performances by composite electrodes of AZO and metallic nanomeshes, *J. Alloy. Compd.* 819 (2020) 152973. DOI:10.1016/j.jallcom.2019.152973
54. Hota, M.K., Bera, M.K., and Maiti, C.K. Flexible metal–insulator–metal capacitors on polyethylene terephthalate plastic substrates, *Semicond. Sci. Technol.* 27 (2012) 105001. DOI: 10.1088/0268-1242/27/10/105001
55. Kumar, P., Guhathakurata, S., Choudhury, A., Sharma, A., Tripathy, A.R., Kumar, S.S., Pancham, P.P., Das, P., Mahato, S.S., Mahata, S., and Mallik, S. Flexible BSA MIM capacitor with negative voltage coefficient for RF applications, *Appl. Phys. Lett.* 116 (2020) 171904. DOI: 10.1063/1.5139494
56. Bag, A., Hota, M.K., Mallik, S., and Maiti, C.K. Graphene oxide-based flexible metal–insulator–metal capacitors, *Semicond. Sci. Technol.* 28 (2013) 055002. DOI: 10.1088/0268-1242/28/5/055002
57. Tiwari, J.N., Meena, J.S., Wu, C.S., Tiwari, R.N., Chu, M.C., Chang, F.C., and Ko, F.H. Thin-film composite materials as a dielectric layer for flexible metal-insulator-metal capacitors, *ChemSusChem*, 3 (2010) 1051–1056. DOI: 10.1002/cssc.201000118

13 Polymer-based High-K Dielectrics for Energy Application

Lipsa Priyadarshini, L. Biswal, Priyambada Mallick, S. K. Satpathy, and Srikanta Moharana

13.1 INTRODUCTION

Currently, all across the globe, fossil fuels have been the major source of energy. Excessive use of these fossil fuels has resulted in environmental pollution and degradation. Also, deposits of fossil fuels are getting depleted daily. For this purpose, researchers in academics and industry are focusing on using renewable energy sources to produce green energy. They are in search of convenient and cost-effective methods to harvest energy from these renewable sources, transmit harvested energy over long distances, and store in high-performance energy storage devices [1–3]. Scientists are paying more attention to exploring green energy from renewable energy sources like wind and sun. A major chunk of researchers is focusing on the design and development of efficient power energy storage systems. From portable devices to large industrial equipment, high-performance and cost-effective electrical energy storage devices (EESD) are needed. Scientists and engineers in industries are trying to manufacture efficient EESDs with high energy density and the required power density. Depending on the mechanism involved, these EESDs are grouped into different categories, such as dielectric capacitors, electrochemical capacitors, supercapacitors, batteries, and solid oxide fuel cells [4]. These EESDs in current use have their advantages and disadvantages. Due to the fast charge-discharge capability of dielectric capacitors, they have high power density, thus, are used in many pulsed discharge and power conditioning electronic applications [5]. However, these dielectric capacitors have low energy density. On the other hand, batteries and solid oxide fuel cells have high energy density, but low power density. The electrochemical capacitors and supercapacitors have energy density and power density that fall between those of conventional capacitors and batteries [6]. Accordingly, the electrical energy storage device for specific applications is decided and designed.

Dielectric capacitors with high power density, low loss, and higher operating voltage are required in various applications, like pulsed power generation systems, high-frequency inverters, power factor correction, electric vehicles, etc [7]. In these applications, dielectric capacitors are superior to supercapacitors and electrochemical capacitors. The functional component of these dielectric capacitors is the dielectric material used in them. The important property of a good dielectric material is its ability to support an electric field with dissipation of minimum energy. Most dielectric materials used in dielectric capacitors are solids made up of ceramics, paper, solid electrolytes, or polymers. Compared to ceramic or electrolytic dielectric materials, polymer dielectrics can be easily prepared at low temperatures. These polymers can be easily processed with low cost and have important properties, like flexibility, high thermal stability, and high breakdown strength, in comparison to ceramics and other materials [8, 9]. However, the major problem with traditional polymer dielectrics is that they have low dielectric constant and low breakdown strength [9]. Many countries are trying to develop electric vehicles or hybrid vehicles with low CO_2 emissions. The dielectric capacitors used in these vehicles require a large volume and weight [5]. This need increases the cost of

DOI: 10.1201/9781003352365-13

vehicles, as well as reduces the efficiency. To reduce the volume and weight of dielectric capacitors, polymer materials with high energy density are needed. This can be achieved by developing high-K polymer dielectrics with low loss, high dielectric breakdown strength, and good thermal, chemical, and mechanical stability.

Currently, the research community is looking for functional polymer composite dielectrics (FPCD). These FPCDs, with novel properties and high dielectric constant, can be easily fabricated and have been found to be more useful in dielectric capacitors than traditional polymers. Now, FPCD-based dielectric capacitors are used in high-powered and flexible electronics, electric transport, advanced combat vehicles, construction of artificial skins and muscles, etc. [10–12]. These functional polymer composites have high-power density, better stability, high efficiency, and comparatively higher energy density. Much work is going on to synthesize new inorganic or polymer-based, advanced functional materials for their use in dielectric capacitors. To improve the required properties, researchers are working on polymer-based composite systems, with a polymer matrix containing organic, ceramic, and carbon-based fillers that may disperse randomly or in an ordered manner. Based on the nature of the filler used, the FPCDs can be categorized into two types: (i) dielectric-dielectric composites (DDC) and (ii) conductor-dielectric composites (CDC). In the first category, high dielectric constant and low loss ceramic particles are dispersed in the polymer matrix in a mono-dispersive or poly-dispersive manner, resulting in enhancement of the dielectric constant. In the second category, the polymer matrix contains electrically conducting nano particles, resulting in a remarkable increase in dielectric constant value. However, the major challenge to overcome for these FPCDs is that they have low breakdown strength. In order to acquire high breakdown strength, multi-layer structures are designed. This also enhances the dielectric constant, which is required for energy applications.

This chapter starts with an overview of fundamental principles and mechanisms involved at the microscopic level, which are responsible for deciding the dielectric constant, dielectric loss, and dielectric breakdown of polymer dielectrics. The understanding of these theories and mechanisms helps us to tailor and tune the properties required for high-performance polymer dielectric capacitors, making them suitable for specific applications. Then we have listed some polymer dielectrics that are currently in use in EESD, along with highlighting their limitations. Some recent developments while trying to overcome the existing challenges are discussed, citing some current research works. We have also discussed novel methods to fabricate high-K, low loss, and high-dielectric breakdown strength polymer (composite) dielectrics required for energy applications. Lastly, the future perspective in this field is summarized.

13.2 FUNDAMENTAL THEORIES AND MECHANISMS INVOLVED

Dielectric materials are easily polarizable materials when bringing them under the action of the external electric field. In these materials, an electric field can persist. On the other hand, insulators are the materials that prohibit current flow within it. The dielectric property is quantified by measuring its dielectric constant, whereas the insulating property is quantified by measuring its electrical resistivity. Though good dielectrics are low-conductivity materials, yet the dielectric property can be seen more or less in all materials in real use.

When a dielectric material is put under the influence of an external electric field (**E**), the net dipole moment is induced in it, due to (i) separation of charge carriers or (ii) orientation of existing dipole moments in the material in response to the external electric field [13]. This results in polarization of the dielectric material and the extent of this polarization depends on the different processes occurring microscopically in the material, the band gap, the crystallinity and defect density in the material, the strength of the external electric field applied and the physical conditions in which the dielectric material exists [14]. Below, we have discussed briefly the major mechanisms occurring in the material in response to the external field and the influence of physical conditions in which the dielectric is kept, deciding the dielectric behavior. Also, we have defined some quantities that allow the performance of dielectrics in energy applications to be quantified.

13.2.1 Microscopic Processes Responsible for Polarization

In dielectrics, polarization or the dipole moment is induced when electric field is applied. This polarization comes because of one or more of five different types of polarization processes that occur: (i) Electronic polarization, (ii) Atomic or ionic polarization, (iii) Dipolar polarization, (iv) Spontaneous polarization and (v) Interface or space charge polarization [13]. The electronic and atomic polarization results, due to the displacement of electron clouds and lattice vibration within the atoms or molecules that constitute the material, thus, is an intra-molecular phenomenon. The restoring force against the displacement resulting in deformation is relatively insensitive to temperature. On the other hand, dipolar polarization and spontaneous polarization come under the category of orientation polarization. The orientation polarization comes from the inelastic (rotational) movement of molecules or particles possessing permanent dipole moments. This orientation polarization is an inter-molecular phenomenon and is strongly dependent on the physical conditions in which the material is kept.

Based upon the polarization processes that occur in materials inducing the net dipole moment, dielectric materials are categorized as (1) nonpolar, (2) polar, and (3) dipolar dielectrics. In nonpolar dielectrics, the external electric field can result only in elastic displacement of electron clouds, hence, these materials can have electronic polarization only. Polar dielectrics are the materials in which an external electric field causes elastic displacement of the valence electrons as well as elastic displacement of the relative positions of ions. Thus, these materials can have both electronic and ionic polarization components. The materials in which all three fundamental polarizations (i.e., electronic, ionic, and orientation) are seen are called dipolar materials. Dipolar materials constitute molecules having permanent dipole moments (even in the absence of an external electric field) due to their asymmetric structure, in which the center of positive charge and negative charge do not coincide. The external electric field exerts torque on these dipole moments, resulting in ordered orientation of these permanent dipoles.

Application of an electric field results orientation of the permanent dipoles spatially parallel to the electric field, resulting in net orientation polarization and, with the removal of the electric field, this net polarization vanishes because of the thermal energy of the molecules that randomize the alignment of the dipoles. But there exists one class of dipolar dielectric material, in which permanent polarization persists even in the absence of the electric field. This is because of a special kind of polarization, the spontaneous polarization seen in these materials. The unit cells of these special classes of dielectrics are noncentrosymmetric in nature. Hence, the molecules possess permanent dipole moment. Spontaneous polarization is a short chain reaction process, in which each unit cell possesses a reversible electric dipole moment spontaneously oriented parallel to the dipole moment of neighboring cells. This gradual building up of dipole moments pointing along one crystal axis forms a domain of a certain size, giving a net large dipole moment. All the dipoles in one domain are directed in the same direction. The dipole moments of all such domains in the material align themselves in such a way as to reduce the free energy of the material to a minimum. Spontaneous polarization is seen in materials with ordered crystalline structures and, hence, occurs only in single crystals or crystallites in polycrystalline materials with a noncentrosymmetric structure. These materials possess very high dielectric constant in comparison to the normal dielectrics up to a critical temperature (called curie temperature), after which a transition from an ordered to disordered phase occurs, thereby destroying spontaneous polarization in the material.

The dielectrics having spontaneous polarization possess novel properties like piezoelectric, pyroelectric, ferroelectric, electro-optic, and electro-mechanical properties, which are absent in ordinary dielectrics. These novel properties make the dielectric an active or functional dielectric. In these materials, the application of mechanical stress generates electricity, and the electric field produces mechanical stress. This effect is called as piezoelectric effect, and the materials are known as piezoelectric materials. Pyroelectric materials are a subgroup of piezoelectric materials, in which heat can be generated by changing the state of polarization of the material and vice-versa. Further, ferroelectric materials are a subgroup of pyroelectric materials, which possess reversible spontaneous polarization over a certain temperature range. These ferroelectric materials show nonlinear electrical behavior (i. e., the net polarization induced in these materials changes with the applied electric field in a nonlinear manner). They exhibit hysteresis loops and are associated with mechanical strain. Some of these ferroelectric materials exhibit birefringence and interesting electro-optic effects.

Except for these four kinds of polarization processes seen in the dielectric materials, there is another kind of polarization process called interfacial polarization. The real dielectric materials are in amorphous or polycrystalline form. They contain a large number of various traps. Real dielectric materials are also seen to be conducting (due to vacancies or mobile ions), involving charge carriers injected from electrical contacts. These charge carriers are trapped in the bulk or at the interfaces or may be impeded from being discharged or replaced at electrical contacts. Hence, space charges are formed, resulting in distortion of field distribution affecting the dielectric constant. Hoping polarization (polarization due to the hoping of ions and vacancies, electrons and holes from one site to a neighboring site) and interfacial polarization (resulting because of the separation of mobile positively and negatively charged particles under an applied electric field), which form positive and negative space charges in the bulk of the material or at the interfaces between different materials, come under this category.

Each type of the above discussed polarization processes require different time intervals for its set up. Hence, the total polarization depends on the frequency of the applied field. The chemical structure of the material, imperfections (defects) present in it, and the physical conditions (like pressure, temperature, etc.) have a large effect on the degree of polarization of these basic components. The polarizability (α), which give an estimation of the net polarization in the material under given conditions, has components contributed from all the above types. Among all these polarizations, the electronic and ionic polarizations are the deformational polarization processes, which involve vibrating modes. Hence, these are called resonance polarization processes. The other polarization and depolarization processes, like orientation and space charge polarization, require quite a long time. Thus, they are called relaxation polarizations or processes. In contrast to electronic and ionic polarization, dipolar polarization occurs at lower frequencies and is thus important because it can greatly affect the capacitive and insulating properties of dielectric materials for low-frequency applications.

13.2.2 Evaluation of Dielectric Response

The quantification of the dielectric response of the material, and hence its performance evaluation, is carried out through some quantities like dielectric constant (ε_r), loss tangent $(Tan\delta)$, dielectric breakdown strength (E_b), electrical conductivity, etc., and their dependence on various factors like physical conditions, thickness, temperature and chemical stability, scalability, etc.

The dielectric constant (ε_r) of a material gives a measure of its induced polarization and the ability to store electrical energy. When a dielectric material is placed in a time-varying electric field, depending upon the frequency of the electric field, some polarization processes are activated and some are relaxed out. At very low frequencies, all types of polarization mechanisms are involved. Hence, we get high value of dielectric constant, though the completion of all types of polarization does not occur instantaneously. This delay in setting up of polarization in response to an applied electric field is attributed to (1) kind of viscosity or resistance of the medium to the localized and long-range motion of charge carriers and (2) the inertia of the dipoles during their orientation to the applied electric field [13]. These resistive effects are associated with the dissipation of energy. With an increase in frequency, the different polarization processes relax out one by one. In fact, the term relaxation refers to delay of system to follow the external stimulus and is always associated with dissipation of energy. As is already pointed out, with increase of frequency, different polarization processes are relaxed out one by one. The relaxation phenomenon of each polarization process is estimated by a quantity called relaxation time (τ). The relaxation time of different polarization mechanisms in dielectric material is different. For dipolar materials, the relaxation time depends on the measure of the interaction strength of the dipoles and the molecular environment. In solid materials, the relaxation may arise due to different types of charge species involved, disorderness in the material, and relaxation of permanent and induced dipole moments. Hence, the dielectric relaxation in a material is characterized by a distribution of relaxation times.

When the applied electric field is below a certain critical field, for most materials, the dielectric constant is seen to be independent of electric field strength. At or above the critical field, injections of carriers into the material through the electrodes become significant. Dielectric loss tangent (tan δ) is related with the ability of the dielectric material to support the applied electric field. It also gives a measure of the dissipation of electric field energy in the dielectrics. These dielectric constants and dielectric losses are found to be function of the (i) frequency of applied electric field, (ii) the physical conditions (like temperature, pressure, etc.) in which the dielectric is kept, (iii) the concentration of defects present in the material (iv) the disorderedness present in the material and (v) the strength of interaction between molecules. The value of dielectric constant and dielectric loss is also dependent on the possible polarization processes occurring in the material.

The electrical response of a dielectric material is found to be neither purely resistive nor purely reactive. Hence, this electrical response (at any particular frequency of applied electric field and under given physical conditions) is many times modeled by a series or parallel combinations of a resistance and a capacitance. Such an equivalent circuit analysis is done by introducing a complex dielectric constant $\left(\varepsilon^* = \varepsilon_r - j\varepsilon_i\right)$. This helps one to incorporate simultaneous formulations of both the responses. The real part $\left(\varepsilon_r\right)$ measures the strength of polarization that can is result in the material (proportional to the out-of-phase component of the current) and the imaginary part $\left(\varepsilon_i\right)$ measures loss due to conducting current (proportional to the in-phase-component of the current). Then, the loss factor $\left(\tan\delta\right)$ is expressed as $\tan\delta = \frac{\varepsilon_i}{\varepsilon_r}$. The physical meaning of the loss tangent can be ascribed to the ratio of dissipated energy to stored energy per charging cycle. The two quantities ε_r and $\tan\delta$ largely decide the usefulness of the dielectric material in device applications.

Dielectric strength $\left(E_b\right)$ of the material estimates the maximum voltage that the dielectric can withstand without breakdown. When the applied electric field exceeds a critical value, the electrical resistance is reduced drastically and the material becomes a conductor. This results in breakdown of the dielectric property of the material, and the material acts as a conductor. The maximum electric field up to which the dielectric does not breakdown is called as dielectric breakdown field (E_b). This dielectric breakdown strength depends on band gap and permittivity of the dielectric material. Many other factors, like defects, porosity, grain size, phase distribution, sample thickness, etc., largely affect the dielectric breakdown strength. The dielectric breakdown strength E_b limits the energy storage capacity of the dielectric capacitor.

Electrical conductivity $\left(\sigma\right)$ gives a measure of the electrical conduction in dielectrics due to the ordered motion of weakly bound charged particles in response to the electric field. Studies of electrical conductivity help one to determine the structure of crystal, to obtain information on the lattice energy, as well as on defects and impurities, etc.

13.2.3 Evaluation of Energy Storage Performance

There are some factors that strongly affect the parameters like dielectric constant, loss tangent, electric breakdown strength, and conductivity that evaluates the energy storage performance of dielectrics. Among them are temperature, pressure, frequency of applied electric field, chemical composition, molecular structure, crystalline/amorphous nature, defect density that are more significant.

Dielectrics are suitable materials for electrical energy storage because of their ability to polarize. If a dielectric can be polarized to a high value without breakdown, higher capacitance and charge storage can be achieved. Hence, a larger amount of energy can be stored. But along with this, the amount of energy that can be recovered per cycle is also very important.

The performance of a dielectric capacitor for energy storage applications device is evaluated by estimating its (i) stored and recoverable energy density, (ii) energy storage efficiency, (iii) charging and discharging time, (iv) dielectric breakdown field, (v) fatigue endurance and (vi) chemical, thermal, and mechanical stability.

(i) Stored and recoverable energy density:
The stored energy density is the energy stored per unit volume in a polarized dielectric. It is given by:

$$u_s = \int_0^{D_{max}} EdD \tag{13.1}$$

Where E stands for applied electric field and D stands for electric displacement. The electric displacement is related to electric field by: $D = \varepsilon_0 E + P = \varepsilon E = \varepsilon_0 \varepsilon_r E$. Here, ε is the permittivity of the dielectric material and ε_r is the dielectric constant.

It is obvious that the strength of an applied electric field has strong influence on energy density. The stored energy density increases with an increase in the electric field. However, electric field strength is decided by the working conditions of the capacitor. On the other hand, dielectric constant is the material property. Hence, large energy density can be achieved by using high dielectric constant material in the capacitor.

If the permittivity is independent of the external field, which is the case of linear dielectrics, then,

$$u_s = \varepsilon \int_0^{E_{max}} EdE = \frac{1}{2} \varepsilon E_{max}{}^2$$

In case of high permittivity dielectrics, a small electric field can result n large polarization, hence, the electric displacement can be taken approximately equal to the polarization vector (P).

Then

$$u_s = \int_0^{P_{max}} EdP$$

The stored energy is recovered during the depolarization process. The recoverable energy density during the depolarization process is given by:

$$u_r = -\int_{P_r}^{P_{max}} EdP$$

Where P_r and P_{max} are the remnant and maximum polarization.

From the above expressions, it is clear that a dielectric material with high permittivity and breakdown strength can store more energy. Also, the material should be a low conducting material; otherwise, the stored energy will be dissipated soon.

(ii) Energy storage efficiency:
The second parameter, which is very important in energy storage applications of dielectrics is: energy storage efficiency (η). During the discharging process, part of stored energy is dissipated due to various factors like hysteresis loss. So the stored energy cannot be

recovered completely for applications. The storage efficiency per cycle of a dielectric is estimated as:

$$\eta = \frac{u_r}{u_s} = \frac{u_r}{u_s + u_l}$$

Where u_l is the energy loss per volume/mass, due to hysteresis loop.

To achieve greater energy storage efficiency, the dielectric needs to have low hysteresis loss, low remnant polarization, and delayed saturation polarization. This can be achieved if the dielectric has a thin hysteresis loop, high dielectric constant, and high breakdown strength.

(iii) Charging discharging time
The charging/discharging time of dielectric capacitors ranges from microsecond to millisecond. The quantification of charging and discharging time is done using relaxation time (τ) for polarization/depolarization processes. This relaxation time depends upon permittivity, geometry of the dielectric and the load resistance to which the dielectric capacitor is connected. For pulsed power applications, the load resistance is so selected that the relaxation time is maintained as small as possible.

(iv) Dielectric breakdown field:
As is already discussed, the energy density of dielectric can be increased by applying a strong electric field. But when the applied field exceeds a critical value, dielectric material collapses, and then it is converted to a conductor. Hence, dielectric breakdown strength (E_b) also plays a significant role in selecting dielectric materials for energy storage applications.

(v) Fatigue endurance
In energy storage applications, during charging and discharging, the dielectric capacitor passes through repetitive cycling. Passing through repetitive charging-discharging cycles, crystallographic changes occur many times due to lattice strain relaxation, long range diffusion, aggregation of defects, domain wall pinning, and formation of low permittivity dead layers at the dielectric/electrode interface. The fatigue that develops degrades the energy storage performance of dielectric materials. Therefore, dielectric materials with high fatigue resistance are essential.

(vi) Chemical, mechanical and thermal stability
In electronic applications, when capacitors are used for a long time, heat is generated due to dielectric loss and leakage current. This heat is managed by dissipation. Many times, capacitors are used in a wide range of temperatures, not just high temperatures. If the dielectric parameters change significantly with temperature, then energy storage performance is largely affected. Hence, thermally stable dielectric materials need to be designed with low leakage current, and low dielectric loss. Moreover, the dielectric capacitors should bear mechanical shocks and should be nonreactive. Chemically stable dielectrics with thermally stable polarization are better candidates for energy storage applications.

13.3 DIELECTRIC RESPONSE OF POLYMER AND ITS COMPOSITES

Most polymers and their matrix composites come under the category of insulators, due to very low concentrations of free charges in them. All these polymers and their composites possess dielectric properties. In a polymer or its matrix composites, polarization comes from macromolecular chains. The arrangement of polymer molecules, segments of macromolecular chain, twisting in and around C-C bonds, rotation of side groups, etc., have their contribution to net polarization. In some polymers, all types of polarization mentioned earlier are seen. Due to their dielectric nature, polymers and their composites show dielectric relaxation phenomena. The frequency and temperature-dependent dielectric response of these materials help one to distinguish the different types of polarization mechanisms present in the material.

Some polymers have permanent dipole moment, hence, are polar/dipolar in nature. Some polymers like PVDF also possess a ferroelectric property. In these polymers with permanent dipole moment, the extent of orientation of dipoles in response to an applied electric field depends on their local environment. The alignment of dipoles connected to flexible chains is easier than the dipoles connected to stiffer chains. This orientation of dipoles in response to an applied field also depends on the structure of the polymer. In these polymers, both permanent and induced dipoles of different sizes and mass contribute. Due to all these complications, the dielectric spectra of polymers are complex in nature. The dielectric response of polymers is temperature dependent. One way to separate contributions from different regimes of polarization processes is to look into the temperature dependent dielectric loss graph at a constant frequency (as shown in Figure 13.1). But this is possible only if the relaxation times involved for different polarization mechanisms are separated far apart. Then, one obtains peaks indicating the relaxation of different polarization mechanisms, with each peak corresponding to the relaxation of one type of dipole. But if the relaxation times of different polarization mechanisms are not significantly different, the peaks corresponding to different relaxations overlap. In the temperature-dependent plot of dielectric loss at a constant frequency, different relaxation processes corresponding to different peaks are labeled by lowercase Greek letters α, β, γ, δ,…etc. where relaxation corresponds to relaxation at the highest temperature and so on. With the rise in temperature, intermolecular forces within polymer chains weaken, resulting in increased thermal agitation. The dipole moments of the dipolar component become free to orient in response to a changing electric field. On the other hand, at low temperatures, the motion of segments within the chain freezes out, reducing the dielectric constant. Also, at very high temperatures, thermal random energy disturbs the orientation, which in turn decreases the dielectric constant. On the other hand, the electronic and atomic polarizations are spontaneously maintained at all frequencies and temperatures. So the relaxation time of the dipole polarization is affected by temperature, and at higher temperatures, the dipole moment follows in phase with the oscillating electric field. Segmental and chain motions in polymeric material result in peaks in temperature-dependent plots of dielectric loss, each peak indicating relaxation of one type of dipole moment [15]. The Micro-Brownian motion of the whole chain, also indicated as glass transition, is named as α relaxation. The rotations of polar groups about the C-C bond result in a peak termed as β relaxation. Similar other relaxation peaks are observed, termed as γ relaxation and so on. The relaxation processes that occur from local

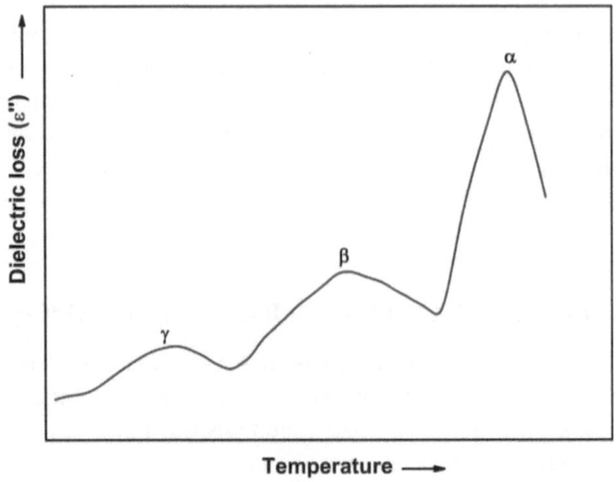

FIGURE 13.1 Variation of dielectric loss with temperature at a fixed frequency. Peak positions represent relaxation.

motions of side groups are called secondary relaxation processes. β and γ relaxation come under this category. Secondary relaxations have a lower value of relaxation time and shorter activation energy.

Composites of polymers are heterogeneous mixtures of two or more components, out of which the matrix component is the polymer. Because of the coexistence of amorphous and crystalline phases, the predominant interfacial polarization effect is seen in polymer matrix composites. In these types of materials, interfacial polarization (also called Maxwell-Wagner polarization) contributes largely to total dielectric response, due to the large interface area available between different phases. When an external electric field is applied, the unbound charges, which are formed during manufacturing of composite polymer, migrate and are trapped at the interfaces of different phases, thus producing large dipole moments. In fact, in all kinds of heterogeneous systems, such interfacial polarization contributes largely to the total dielectric response. But these unbound charges involved in this polarization process are relaxed out as the applied field frequency approaches the radio frequency regime. Consequently, the Maxwell-Wagner effect is restricted to the low frequency regime only. In IR and UV frequencies, only the electronic polarization, ionic polarization, and the dipolar polarization contributes.

13.4 POLYMERS AS DIELECTRIC MATERIALS IN ENERGY STORAGE DEVICES

Polymer materials and/or matrix composites with a high dielectric constant, low dielectric loss and leakage current, high dielectric breakdown strength, good thermal and chemical stability, and high fatigue endurance can be fabricated and used in electrical energy storage devices. Based on the polarization mechanisms involved in polymers and their composites, they can be categorized as low-K and high-K polymer dielectric materials. High-K polymer dielectric materials are prominent candidates for energy storage applications. To increase energy and power storage efficiency, researchers are focusing on the fabrication of high-K polymer dielectrics, due to their ease of processing, low cost, mechanical flexibility, and scalability.

Most Polymers are paraelectric and are capable of providing stable capacitance about temperature and frequency. They also have good mechanical properties, ease of processing, a low dissipation factor, and high dielectric breakdown strength. Also, a group of polymers with a polar backbone, which are ferroelectric, poses higher values of dielectric constant. But then, even the dielectric constant of polymers is not so significant for energy storage applications. To increase the discharge energy density of polymer dielectrics, its spatial efficiency is to be enhanced. This can be achieved by synthesizing/fabricating high-K polymer dielectrics or its matrix composites or by following any other novel method. In this direction, some researchers have tried introducing high permittivity ceramic particles like $Pb(Zr_{1-x}Ti_x)O_3$, $CaCu_3Ti_4O_{12}$, $Na_{1/3}Ca_{1/3}Bi_{1/3}Cu_3Ti_4O_{12}$, $BaTiO_3$, $(Ba_{1-}xSr_x)TiO_3$, etc. or conducting fillers like metal particles, carbon nano-tube, grapheme, etc., into the polymer matrix [11, 16–18] This no doubt enhances permittivity of the polymer composites significantly, but the dielectric breakdown strength is reduced. Another way to enhance the dielectric constant is through interface engineering of filler particles and the polymer matrix [5]. Substantial increase in energy storage capability with high breakdown strength is still not achieved through these procedures. Hence, this area is still open and has become a burning topic among researchers, as efforts are ongoing to enhance both the dielectric constant and dielectric breakdown strength. People are trying to design polymer dielectrics with novel structures to achieve all these goals. We are going to discuss certain state-of-the-art high-K dielectric polymers that are in use or may be found to be promising candidates to be used in energy storage devices by forming their composites or by modifying their structural design.

13.5 LINEAR DIELECTRIC POLYMERS AND THEIR COMPOSITES

Linear polymer dielectrics can be divided into two groups, polar or nonpolar. Dielectric polymers polyetherimide (PEI), Poly(methyl methacrylate) PMMA, polysulfone (PSF), poly thiourea (PTU), etc., belong to a polar dielectric category, whereas polymer dielectrics polypropylene (PP) and polystyrene (PS) belong to the nonpolar category. Many polymer dielectrics, for example, poly(ethylene 2,6-naphthalate) (PEN), polycarbonate (PC), poly(ethylene terephthalate) (PET), poly(phenylene sulfide) (PPS) have been developed and are in commercial applications [19–24].

In comparison to their ferroelectric counterpart, linear dielectric polymers have comparatively low dielectric constant values. But they have advantages like low loss, good thermal stability, and high efficiency. With structural designing/engineering or making their composites, their dielectric constant can be substantially enhanced without much affecting their dielectric loss [25]. As is already discussed, the dielectric constant of linear dielectric polymers can be significantly enhanced using inorganic, conducting, or nanofillers. Inorganic fillers such as ceramic particle(s) of high dielectric constant can be mono or poly-dispersed into the polymer matrix, resulting in a multiphase composite material [26]. Hence, high dielectric constant and good energy storage characteristics can be achieved. But such efforts are associated with reduction in dielectric breakdown strength and enhance the dielectric loss factor [27]. It is also observed that the inorganic ceramic particles and the polymer matrix have poor compatibility. To increase the dielectric constant, the concentration of inorganic fillers is to be increased, which in turn reduces the flexibility of the composite formed, giving a poor performance [28]. This is why it is more appropriate to use organic fillers to increase dielectric constant and energy storage performance.

Polyimide (PI) is such a dielectric polymer. It has good thermal stability and also has high glass transition temperature (T_g). It has high-K (K = 7.8) and also has high energy density (15 J/cm^3) [29]. Over the years many other linear dielectrics with good thermal stability and high T_g have been synthesized. Zhang et al. [30] could successfully synthesize a dielectric polymer based on sulfonylated poly (2,6-dimethyl-1,4-phenylene oxide) (SO_2-PPO) by post-polymer fictionalization and showed that it has a high dielectric constant (6-8, depending on temperature), high T_g (T_g = 220°C), high discharged density (22 J/cm^3), low dissipation factor ($\tan\delta \sim 0.003$) and the discharge efficiency at 800 MV/m is 92%. Li et al. [31] presented a novel design of organic polymers to achieve high energy density and high efficiency. They introduced flexible segments into the rigid polar aromatic polymers, thus achieved enhancing the flexibility of dipoles. They designed three new polymers using three base polymers, one from the polythiourea family named PDTC-HK511, the second from the polyurea family named TDI-EDR148, and third from the polyimide family named BTDA-HK25, through addition reactions. They found multiple sub-glass transition peaks associated with low activation energy, indicating enhanced free movement of dipoles, thus increasing the dielectric constant by 46 % compared to base polymers. They could successfully achieve high energy densities above 11 J/cm^3 with above 90% efficiency. Zhang et al. [32] reported organo-soluble polymers with a high dielectric constant (~6), low dielectric loss, and good heat resistance in a temperature window of 50-200°C. They introduced solphonyl side groups (4.5 D) and achieved in getting sulfonylated PIM (SO_2-PIM), in which solphonyl dipole could freely rotate in response to applied electric field, thus enhancing dielectric constant. They could achieve optimal balance between dielectric constant and low loss and found that the discharged energy density reached to 17 J/cm^3. The discharge efficiency was 94% to 88%, depending on temperature. Wu et al. [33] fabricated a series of polyimide (PI) films through the introduction of self-synthesized aniline trimmer (ACAT) with a conjugated structure to enhance electronic polarization. The conjugated ACAT in the main chains of PI increased the electron polarization effect and also the carrier mobility of polyimide chains. This increased the dielectric constant (4.4–7.4) with a low dissipation factor (0.002–0.013). They observed stable dielectric behavior up to a temperature of 200 °C along with excellent mechanical properties (70–105 MPa). They could achieve a maximal energy density of value 3.6 J cm^{-3} under the field of 426 kV/mm. Yang Feng, Shengtao Li, and co-workers [34] reported a new amphiphilic block copolymer with clear phase separation. This polymer dielectric showed high energy density of 29.8 J cm^{-3}, which they attributed to the coiled block molecule in the interfacial region, which could substantially increase the breakdown strength from 652 MV/m to 1166 MV/. They observed that in this way, the dielectric constant is maintained at a high value (5) with a low dielectric loss.(< 0.01). Polyetherimide (PEI) is a modified form polymer of PI characterized by an imide ring on the main chain, and ether bonds are added to increase the thermo-plasticity [35]. It has low tan δ value and also has good thermal stability. But it has a low discharge energy density (~ 0.5 J/cm^3), though it has a good discharge efficiency (90%). Zhang et al., for mixed PEI and poly(ether methyl ether urea) (PEMEU), in various mass ratios increased the dielectric constant and reduced dielectric loss [36]. Similarly, researchers have tried blending and filling of nonpolar polymers. Polypropylene (PP) is a nonpolar polymer and is made up of hydrocarbons. Though it has high breakdown strength (700 MV/m), its dielectric constant is less (2.25 at 1 kHz). The energy storage efficiency of PP based polymers can be significantly enhanced by incorporating polar groups into the PP molecular chain. Similarly conducting fillers or nano fillers can be incorporated into PP molecular chain forming composites. Through this process, dielectric constant and energy storage efficiency can be significantly enhanced [37].

Layered structures of composite polymer dielectrics or surface-engineered polymer dielectrics are also prepared to enhance the dielectric constant and energy density. Li et al. [38] fabricated layered nanocomposite filers using polystyrene as the polymer matrix. They filled one layer of polystyrene with a wide band gap of Al_2O_3 and the next layer by nano-sized TiO_2. This composite layered structure showed enhanced energy density (4.43 J/cm^3 at 550 MV/m). Y. Li and coworkers [39] enhanced the dielectric constant and dielectric breakdown strength of Poly(methylmethacrylate) (PMMA) through surface modifications, achieving a high energy density of 17.6 J/cm^3 and discharge efficiency of greater that 87% (measured at 800 MV/m). Normally, the increase in dielectric constant of composite polymers is attributed to interfacial polarization.

Even if linear dielectric polymers and their composites have a moderate dielectric constant and energy storage efficiency, nevertheless, because of their advantages like high dielectric strength, low loss, long service life, low cost, ease in processing, scalability, many of them are used commercially in modern energy and power systems.

13.6 CHALLENGES AND FUTURE PERSPECTIVE

High-K-polymer dielectrics with high energy density have been a new interest for present research and study, owing to their multifunctionality and flexibility. In order to cater to the need for energy storage materials, tremendous achievements have been made in this field. To develop next-generation energy storage devices and achieve their best performance, the top attributes of materials that must be delivered are low energy losses, high permittivity, high breakdown strength, long lifetime, and high-temperature rating. Because of the contradictory changes between dielectric loss and permittivity, the aim should be to achieve moderate energy density for polymer-based dielectrics.

For achieving the required properties and performance of the polymer dielectrics and polymer-nanocomposites the following issues and challenges must be addressed with greater attention.

- Current nanoparticles and polymer interface models are of a hypothetical nature. In recent developments, observation of the polymer-nanoparticles interface and measurement of the local dielectric parameters through electrostatic force microscope is quite impressive. The electrode and dielectric interface and band gap structure should also be analyzed properly in order to minimize the charge injection and subsequently reduce the conduction loss. Thus, combined the study of theoretical and experimental results should be pursued to focus on issues and design the required devices.
- Most dielectrics are now reported for a small electrode surface area in the range of mm^2. Thus, the scope in the future should cater to the use of polymer dielectrics in a larger surface area to ensure reliability and reproductivity.
- Most of the polymers become soft at elevated temperatures. To make the polymer dielectric feasible for electronic devices, the working temperature range of the materials should be extensively studied. To overcome this challenge, either new polymers and functional fillers

can be synthesized with high thermal stability or the nature of interface between the polymer matrix and fillers can be improved.

- The charge-discharge efficiency under operational conditions of the dielectrics is related to the dielectric loss. Thus, the future investigations should be concentrated on decreasing the dielectric loss.
- Operation of dielectric polymer under the action of applied electric field results in dielectric loss, and hence, the temperature of the material increases. Therefore, enhanced thermal conductivity of polymer based dielectric materials is extremely necessary for achieving multiple applications in pulsed power generation, field-activated actuators to energy storage and mechanical energy harvesting, etc.

REFERENCES

1. Qiu, J.; Gu, Q.; Sha, Y.; Huang, Y.; Zhang, M.; Luo, Z., "Preparation and application of dielectric polymers with high permittivity and low energy loss: A mini review", *J. Appl. Polym. Sci.*, 2022, vol. 139 (24), 1–18, https://doi.org/10.1002/app.52367
2. Kausar, A., "Overview on conducting polymer in energy storage and energy conversion system", *J. Macromol. Sci. A*, 2017, 54:9, 640–653, https://doi.org/10.1080/10601325.2017.1317210
3. Stanley Whittingham, M., "History, evolution, and future status of energy storage", *Proc. IEEE*, 2012, *100*,1518–1534, https://doi.org/10.1109/JPROC.2012.2190170
4. Sherrill, S.A.; Banerjee, P.; Rubloff, G.W.; Lee, S.B., "High to ultra-high power electrical energy storage" *Phys. Chem. Chem. Phys.*, 2011, *13* (46), 20714–20723, https://doi.org/10.1039/c1cp22659b
5. Chen, Q.; Shen, Y.; Zhang, S.; Zhang, Q.M., "Polymer-based dielectrics with high energy storage density", Annu. Rev. Mater. Res. 2015 45:1, 433–458, https://doi.org/10.1146/annurev-matsci-070214-021017
6. Veerapandiyan, V.; Benes, F.; Gindel, T.; Deluca, M., "Strategies to improve the energy storage properties of perovskite lead-free relaxor ferroelectrics: a review", *Materials* 2020, *13*, 5742. https://doi.org/10.3390/ma13245742
7. Palneedi, H.; Peddigari, M.; Hwang, G.-T.; Jeong, D.-Y.; Ryu, J., "High-performance dielectric ceramic films for energy storage capacitors: progress and outlook", *Adv. Funct. Mater.*,2018, 1803665, https://doi.org/10.1002/adfm.201803665
8. Zhang, X.; Jiang, J.; Shen, Z.; Dan, Z.; Li, M.; Lin, Y.; Nan, C.-W.; Chen, L.; Shen, Y., "Polymer nanocomposites with ultrahigh energy density and high discharge efficiency by modulating their Nanostructures in Three Dimensions", *Adv. Mater.* 2018, 1707269, https://doi.org/10.1002/adma.201707269
9. Chu, B.J.; Zhou, X.; Ren, K.L.; Neese, B.; Lin, M.R.; Wang, Q., et al., "A dielectric polymer with high electric energy density and fast discharge speed", Science 2006, 313, 334 336. https://doi.org/10.1126/science.1127798
10. Anderson, I.A.; Gisby, T.A.; McKay, T.G.; O'Brien, B.M.; Calius, E.P., "Multi-functional dielectric elastomer artificial muscles for soft and smart machines", J. Appl. Phys. 112, 041101 (2012); https://doi.org/10.1063/1.4740023
11. Dang, Z.-M.; Yuan, J.-K.; Zha, J.-W.; Zhou, T.; Li, S.-T.; Hu, G.-H., "Fundamentals, processes and applications of high-permittivity polymer–matrix composites", *Prog. Mater. Sci.*, 57(4), 2012, 660–723, https://doi.org/10.1016/j.pmatsci.2011.08.001
12. García Núñez, C.; Manjakkal, L.; Dahiya, R. "Energy autonomous electronic skin". *npj Flex Electron* **3**, 1 (2019). https://doi.org/10.1038/s41528-018-0045-x
13. Chi Kao, K., "Dielectric phenomena in solids", (Elsevier Academic Press, Amsterdam, New York, USA, 2004).
14. Wu, X.; Chen, X.; Zhang, Q.M.; Tan, D.Q., "Advanced dielectric polymers for energy storage", *Energy Storage Mater.*, Volume 44, 2022, Pages 29–47, https://doi.org/10.1016/j.ensm.2021.10.010
15. Young, R.J., *Introduction to Polymers*, (1989) Chapman & Hall Ltd.
16. Yao, Z.H.; Song, Z.; Hao, H.; Yu, Z.Y.; Cao, M.H.; Zhang, S.J., et al., Homogeneous/inhomogeneous-structured dielectrics and their energy storage performances, *Adv. Mater.* 29 (2017) 1601727.

17. Kum-onsa, P.; Thongbai, P.; Maensiri, S. et al. Greatly enhanced dielectric permittivity in poly(vinylidene fluoride)-based polymeric composites induced by $Na_{1/3}Ca_{1/3}Bi_{1/3}Cu_3Ti_4O_{12}$ nanoparticles. *J Mater Sci: Mater Electron* **27**, 9650–9655 (2016). https://doi.org/10.1007/s10854-016-5023-2
18. Dang, Z.M.; Zheng, M.S.; Zha, J.W., 1D/2D carbon nanomaterial-polymer dielectric composites with high permittivity for power energy storage applications, *Small*, 2016, 12(13), 1688–1701.
19. Guan, F.-X.; Wang, J.; Pan, J.-L.; Wang, Q.; Zhu, L. Time and poling history dependent energy storage and discharge behaviors in poly(vinylidene flfluoride-co-hexaflfluoropropylene) random copolymers. *Chin. J. Polym. Sci.* 2010, *29*, 65–80. https://doi.org/10.1007/s10118-010-1020-8
20. Tan, D.; Zhang, L.; Chen, Q.; Irwin, P. High-Temperature Capacitor Polymer Films. *J. Electron. Mater.* 2014, *43*, 4569–4575. https://doi.org/10.1007/s11664-014-3440-7
21. Mannodi-Kanakkithodi, A.; Treich, G.M.; Huan, T.D.; Ma, R.; Tefferi, M.; Cao, Y.; Sotzing, G.A.; Ramprasad, R. Rational co-design of polymer dielectrics for energy storage. *Adv. Mater.* 2016, *28*, 6277–6291. https://doi.org/10.1002/adma.201600377
22. Ho, J.S.; Greenbaum, S.G. Polymer capacitor dielectrics for high temperature applications. *ACS Appl. Mater. Interfaces* 2018, *10*, 29189–29218, https://doi.org/10.1021/acsami.8b07705
23. Singh, M.; Apata, I.E.; Samant, S.; Wu, W.; Tawade, B.V.; Pradhan, N.; Raghavan, D.; Karim, A. Nanoscale strategies to enhance the energy storage capacity of polymeric dielectric capacitors: review of recent advances. *Polym. Rev.* 2021, *23* 1–44. https://doi.org/10.1080/15583724.2021.1917609
24. Luo, J.; Mao, J.; Sun, W.; Wang, S.; Zhang, L.; Tian, L.; Chen, Y.; Cheng, Y. Research progress of all organic polymer dielectrics for energy storage from the classifification of organic structures. *Macromol. Chem. Phys.* 2021: *222*, 1–15. https://doi.org/10.1002/macp.202100049
25. Han, Z.; Wang, Q.. Recent progress on dielectric polymers and composites for capacitive energy storage. *iEnergy*, *1*(1): 50–71, March 2022, https://doi.org/10.23919/IEN.2022.0008
26. Li, D.; Zeng, X.; Li, Z.; Shen, Z.-Y.; Hao, H.; Luo, W.; Wang, X.; Song, F.; Wang, Z.; Li, Y. Progress and perspectives in dielectric energy storage ceramics. *J. Adv. Ceram.* 2021, 10, 675–703. https://doi.org/10.1007/s40145-021-0500-3
27. Wei, X.; Yan, H.; Wang, T.; Hu, Q.; Viola, G.; Grasso, S.; Jiang, Q.; Jin, L.; Xu, Z.; Reece, M.J. Reverse boundary layer capacitor model in glass/ceramic composites for energy storage applications. *J. Appl. Phys.* 2013, 113, 1–6. https://doi.org/10.1063/1.4775493
28. Huang, X.; Sun, B.; Zhu, Y.; Li, S.; Jiang, P. High-K polymer nanocomposites with 1D fifiller for dielectric and energy storage applications. *Prog. Mater. Sci.* 2019, 100, 187–225. https://doi.org/10.1016/j.pmatsci.2018.10.003
29. Ma, R., Baldwin, A. F., Wang, C. C., Offenbach, I., Cakmak, M., Ramprasad, R.Sotzing, G. A., Rationally designed polyimides for high-energy density capacitor applications. ACS Appl. Mater. Interfaces, 2014, *6*: 10445–10451. https://doi.org/10.1021/am502002v
30. Zhang, Z. B., Wang, D. H., Litt, M. H., Tan, L. S., Zhu, L. High-temperature and high-energy-density dipolar glass polymers based on sulfonylated poly(2,6-dimethyl-1,4-phenylene oxide). *Angew. Chem., Int. Ed.*, 2018, *57*: 1528–1531. https://doi.org/10.1002/anie.201710474
31. Li, Z. Z., Treich, G. M., Tefferi, M., Wu, C., Nasreen, S., Scheirey, S. K., Ramprasad, R., Sotzing, G., Cao, Y. High energy density and high efficiency all-organic polymers with enhanced dipolar polarization. *J. Mater. Chem. A.*, 2019, *7*: 15026–15030. https://doi.org/10.1039/C9TA03601F
32. Zhang, Z. B., Zheng, J. F., Premasiri, K., Kwok, M. H., Li, Q., Li, R. P., Zhang, S. B., Litt, M. H., Gao, X. P. A., Zhu, L. (2020). "High- K polymers of intrinsic microporosity: A new class of high temperature and low loss dielectrics for printed electronics". *Mater. Horiz.*, *7*: 592–597, https://doi.org/10.1039/C9MH01261C
33. Wu, Z. Q., Guo, Q., Liu, Y., Zhou, H. H., Zheng, H., Lei, X. F., Gong, L., Chen, Y. H., Liu, Z. G., Zhang, Q. Y. (2021). Excellent polyimide dielectrics containing conjugated ACAT for high-temperature polymer film capacitor. *Macromol. Mater. Eng.*, *306*: 2100456. https://doi.org/10.1002/mame.202100456
34. Feng, Y., Jiang, L. H., Yang, A. Q., Liu, X., Yang, L. Q., Lu, G. H., Li, S. T. (2022). Interfacial effect on dielectric properties of self-as sembled polythiourea-based copolymers for ultrahigh energy storage. *Macromol. Rapid Commun.*, *43*: 2100700. https://doi.org/10.1002/marc.202270013
35. Liu, J.; Su, J.; Li, R.; Zhao, L.; Lu, Y.; Liu, X. Characterization of cavities in polyetherimide and the inflfluence on insulation performance of ultra-high voltage pulse. *Mater. Lett.* 2019, *240*, 153–156. https://doi.org/10.1016/j.matlet.2019.01.003
36. Zhang, Q.; Chen, X.; Zhang, T.; Zhang, Q.M. Giant permittivity materials with low dielectric loss over a broad temperature range enabled by weakening intermolecular hydrogen bonds. *Nano Energy* 2019, *64*: 1–7. https://doi.org/10.1016/j.nanoen.2019.103916

37. Yuan, X.; Matsuyama, Y.; Chung, T.M. Synthesis of functionalized isotactic polypropylene dielectrics for electric energy storage applications. *Macromolecules* 2010, 43: 4011–4015. https://doi.org/10.1021/ma100209d
38. Li, H., Yao, B., Zhou, Y., Xu, W. H., Ren, L. L., Ai, D., Wang, Q. (2020). Bilayer-structured polymer nanocomposites exhibiting high breakdown strength and energy density via interfacial barrier design. *ACS Appl. Energy Mater.*, *3*: 8055–8063. https://doi.org/10.1021/acsaem.0c01508
39. Li, Y. P., Yin, J. H., Feng, Y., Li, J. L., Zhao, H., Zhu, C. C., Yue, D., Liu, Y. P., Su, B., Liu, X. X. (2022). Metal-organic framework/Polyimide composite with enhanced breakdown strength for flexible capacitor. *Chem. Eng. J.*, *429*: 132228, https://doi.org/10.1016/j.cej.2021.132228

14 Core-Shell Structured High-K Dielectrics

Yogita Batra and Puspender Singh Beniwal

14.1 INTRODUCTION

In today's electronic and electrical industry, polymer and high-K ceramic nanoparticle-based nanocomposites are gaining popularity for dielectric and energy storage applications. The high dielectric constant (high-K) filler brings along electric field distortion. An increase in the dielectric constant typically comes at the expense of a significant reduction in the breakdown strength of nanocomposites. High-K polymer nanocomposites with high breakdown strength have been created using hybrid nanoparticles. Traditional ferroelectric ceramics have a high dielectric constant. However, their use is restricted by their poor processing characteristics and low breakdown strength. The method for achieving the desired dielectric properties in the next generation of energy storage devices is to incorporate high-K ceramics into a polymer matrix [1]. Core-shell nanoparticles are highly functional materials, with tunable properties. They can be tailored for different applications. Generally, they are distinguished by the presence of a core (inner material) and a shell (outer material). Concentric multilayered nanoparticles, also known as 'core-shell' particles, have been developed to improve the properties of semiconductor materials. Core and shell materials are highly influenced by end-use and usage. Changing the constituents or the core-shell ratio changes properties such as reactivity and thermal stability, resulting in unique properties. The core particle coating serves a variety of functions, including improved functionality, surface modification, reduced use of valuable materials, and increased stability [2]. In addition to having a high dielectric constant, a high-K material should also have a low dielectric loss, high breakdown strength, and good processability. Surface electric stress can be significantly reduced by high-K materials and kept below the air's breakdown strength, potentially eliminating surface discharge [3]. The material's magnetic and electric energies are coupled to create a high-K core-shell particle that can be used in energy storage devices. Small devices benefit from nanomaterials with ferromagnetic and ferroelectric properties. Many techniques are based on the interplay of electrical and magnetic properties. Magnetoelectric (ME) coupling is a well-known phenomenon in core-shell nanoparticles. Core-shell structures with spherical interfaces connecting ferroelectric and antiferromagnetic materials on an atomic scale would be excellent candidates for future applications. ME coupling is caused by strain transfer at boundaries. The high surface area-to-volume ratio of nanomaterials results in improved ME binding. And research supports stronger coupling of spherical, rectangular, and cylindrical core-shell nanostructures [4]. Many researchers have used surface modifications such as surface coating to form the heterogeneous core-shell structure to improve the dielectric properties of functional materials. For example, $Ba_{0.75}Sr_{0.25}TiO_3$ (BST) powders coated with $Mg_{0.9}Zn_{0.1}O$ to form a "core-shell structure" could prevent BST particle growth during sintering and significantly improve dielectric tuning performance. Complex composites coated with metal oxides as a shell structure on the surface of the core can significantly improve the dielectric temperature performance by inhibiting the formation of the second phase [5]. When we add other compounds to the core material, it is difficult to achieve a uniform distribution of additives using the solid-state reaction method. On the other hand, the chemical coating can form a homogeneous shell after selecting the proper additives that react with the core material. Coated particles are perfectly spherical and quite small. Chemical coating ensures the formation of the desired core-shell structure by forming a uniform layer of doped additives over the

DOI: 10.1201/9781003352365-14

core material. Coated core-shell materials have good dielectric properties, such as high permittivity at room temperature and low dielectric loss. The dielectric loss decreases significantly with temperature and also increases as the element content used as an additive for shell formation increases [6]. The chemical precipitation method can be used to create a core-shell structure in ceramics by coating the core material with suitable elements. The ions of the element used for coating replace the atoms in the core material, forming a solid solution to a certain mole percentage limit. As the additive content increases, it begins to form a shell over the formed particle of solid solution, and a significant decrease in grain size is observed. We obtain a core-shell structure in ceramics by coating above the limit of solid solution, which significantly improves the dielectric properties of these core materials. We face the problem of temperature stability near Curie temperature when using dielectric ceramics independently. To overcome such challenges, dopants (additives) are coated over the material, forming a core-shell structure after reactions and atom substitution. Various preparation methods have been shown to have a significant impact on the substitution behavior of additive ions in the core material. When compared to the traditional solid-state method, the chemical precipitation method produces a uniform çoating layer and an inhomogeneous "core-shell" structure. The dielectric polarization and relaxation mechanisms influence the permittivity of nanocomposite materials. Permittivity is seen to be monotonically decreasing with increasing frequency in typical behavior [7].

Materials having high dielectric constant (high-K) and low dielectric losses are sought after for use in wireless communication. High-K materials offer a chance to create antennas that are smaller, lighter, and more effective in terms of signal quality. Energy storage and microwave communication are both simple applications for these materials. The composition of the core material, how the core and shell materials interface, and how evenly they are distributed all play a significant role in determining how well the core-shell material performs. This is referred to as the material's third phase [8, 9]. Ferroelectric random access memories and electronic devices can both benefit from the use of composite nanofibers with excellent dielectric and ferroelectric properties, in addition to good mechanical properties [10]. Table 14.1 shows the dielectric properties of various core-shell structured nanocomposites.

14.2 TYPES OF CORE-SHELL STRUCTURES

14.2.1 Core-Shell Nanostructures Based on Core Shapes

Although core-shell particles are frequently spherical, numerous other core-shell nanostructures can take the form of nanotubes, nanowires, nanorods, nanorings, or nanostars depending on the core shape. The cores of core-shell particles can be dense, porous, or eccentric, and hollow nanostructures can be created when the core material is completely removed from the structure, through calcination or solvent dissolution. There are various kinds of shells, including discontinuous and particulate shells, continuous and porous shells, and continuous and dense shells. More advanced nanostructures, are as the following: Core nanoparticles with a bilayer, composite shell, or multiple layers [33, 34]. With the development of new synthesis methods, in addition to the symmetrical (spherical) shape, nanoparticles can now be created in a variety of other shapes, such as cube, prism, hexagon, octahedron, disc, wire, rod, and tube. [33].

14.2.1.1 Nanorod

At low energies, the nanorods reduce the horizontal localized surface plasmon resonance (LSPR) perpendicular to the rod length and the longitudinal LSPR, due to the geometry generated perpendicular to the rod length. Silver (Ag) has better plasmonic properties than gold (Au) due to its lower plasmon resonance damping. Adding an Ag layer to Au nanorods enhances their plasmonic properties and expands their surface applications. As a result, coating the Au core with a thin palladium layer can improve the catalytic properties while preserving most of the plasmonic properties. The composition and morphology of the surface and subsurface layers greatly affect the optical and catalytic properties

TABLE 14.1
Dielectric Properties of Various Core-shell Structured Nanocomposites

Material	Dielectric Constant (k)	Dielectric Loss	Reference
Ag@TiO_2 (25wt. %)	~53	0.53	[11]
Ag@TiO_2 (20vol. %)	379	0.15	[11]
Ag@C/PVDF	295	0.08	[11]
Ag@polydopamine/PVDF	53		[12]
Polyetherimide/ $BaTiO_3$@Al_2O_3	3.7	0.002	[13]
Polyetherimide/ $BaTiO_3$@TiO_2 (150°C)	3.6	0.005	[13]
Polyetherimide /$BaTiO_3$@AlN	4.2	0.0025	[13]
$BaTiO_3$-polydopamine-Ag/ Epoxy	8.4	0.024	[9]
$BaTiO_3$/Epoxy (20% vol)	6.67		[9]
$BaTiO_3$/Epoxy (40% vol)	15	0.06	[9]
FR-406	3.92	0.017	[9]
RT/Duriod 5870	2.33 ± 0.02	0.001	[9]
PDMS-$NdTiO_3$ (20 vol%)	~6.5	0.015	[9]
Epoxy-$BaTiO_3$ (20 vol%, 300 nm)	7.03	0.045	[9]
Silicone-$SrTiO_3$ (31.2 vol%)	8.4	0.017	[9]
Epoxy-$BaTiO_3$ (40 vol%, 100 nm)	~10.8	0.04	[9]
Zn-ZnO/PVDF	60	0.07	[14]
$BaTiO_3$/epoxy	-	0.035	[15]
PMN–PT/P(VDF–TrFE)	-	0.1	[16]
Bimodal $BaTiO_3$/epoxy	-	0.03	[17]
$BaTiO_3$/P(VDF–HFP)	-	<0.07	[18]
CCTO/P(VDF–TrFE)	-	0.26	[19]
CCTO/PI	-	0.2	[20]
Ag flake/epoxy	1000	0.02	[8]
Ag@C/epoxy	>300	<0.05	[21]
Al/epoxy	109	0.02	[22]
Ni–$BaTiO_3$/PVDF	300	0.5	[23]
CB/epoxy	13,000	3.5	[24]
Al–Ag/epoxy	160	0.045	[25]
Ag/epoxy	300	0.05	[26]
Ag-CB/epoxy	2260	0.45	[27]
CNT	152	0.06	[28]
DVB-GMA@MWNTs/epoxy	152	0.035	[28]
$BaTiO_3$/PVDF (30 vol%)		0.090	[29]
PVDF@$BaTiO_3$/PVDF		0.086	[29]
PS@$BaTiO_3$/PVDF		0.887	[29]
PEN/$BaTiO_3$@PANI	14.0	0.025	[30]
Nano-Si-Epoxy	257.0	0.11	[31]
$BaTiO_3$/HCu-Phthalocyanine	17	0.054	[32]

of bimetallic nanorods. The metal reduction potential is critical for successful core-shell growth, as electrochemical displacement can occur when the core material is oxidized to give a metallic structure during the shell metal reduction process. The extrinsically formed smooth metallic shell is a general consequence of metal bloom with metals with similar crystal structures and low lattice constants and surface energies, as evidenced by the hatched growth of Ag on Au. Thermal annealing is used to fine-tune the morphology and crystallographic orientation of the metallic shell. The mesoporous silica coating served as a dimensionally stable template that kept the nanorods spherical at high temperatures and prevented aggregation. The mesoporous silica shell surrounding the bimetallic nanorods prevented strong plasmonic coupling between the nanorods and allowed access to the nanorod surface, thus allowing us to compare different core-shell bimetallic nanorods. The main factors affecting the morphology and crystal structure of the resulting core-shell nanoparticles are the mechanism of metallic shell growth [35]. Such data are important for extending synthetic methods and predicting the metal overgrowth behavior of various metal combinations.

14.2.1.2 Nanowhiskers

Diffusion is used to create crystallographically perfect single crystals. The basis of this method (MBE) is the carbon-mediated synthesis of core-shell-metal nanowhiskers using molecular beam epitaxy. The outer Au layer of the Ag-Au core-shell nanowhisker epitaxially covers the single-crystalline Ag whiskers on the substrate surface, while the spaces between the whiskers are nano-crystalline. This allows the original 'core' nanostructure to 'sculpt' the shape, size, and morphology of the internal voids. Nanowhiskers gouge at a constant velocity, proportional to the radius of the sphere. Ag atoms diffuse along the inner surface of the growing Au nanotube from the remaining (filled) portion of the Ag core to the root of the nanowhisker. The kinetic bottleneck of the process is the migration of Ag atoms from the bottom inner surface of the nanowhisker to the nearest grain boundary (GB) of the Au film. Ag atoms easily diffuse out of the nanowhiskers when they reach the grain boundaries of the Au film. Since Ag is difficult to dissolve in carbon, the Au-carbon interface can be thought of as a "reflector" that diffuses Ag atoms. The solution resistance effect reduces the mobility of Au-GBs that have been enriched with Ag. As a result, as one moves away from the root of a nanowhisker, the average grain size of the Au film increases. Near-GB is a dynamic drilling model and the dynamic bottleneck is assumed to be the diffusion of Ag atoms along the Au-carbon interface from the inner nanowhisker core. The obtained drilling rate equation can be interpreted as a competition between the decrease in energy driving the drilling process and the increase in energy due to the creation of new surfaces [35].

14.2.1.3 Nanoparticles

Core-shell nanomaterials are a type of nanoparticle (NP) that consists of a core and one or more shells. The nucleus's properties must be stabilized and protected from outside influences by the shell, which also serves this purpose. The characteristics and capabilities of NP can be changed using this structure. A variety of new opportunities for novel materials and multifunctionality are made possible by the combination of two or more materials in a single NP. For the synthesis of core-shell material's NP, numerous methods are being developed, including sol-gel processes, co-precipitation, layer-by-layer (LBL) methods, metal reduction, etc.[36]. Core-shell nanostructures are advantageous materials, due to their high thermal and chemical stability, low toxicity, high solubility, and enhanced permeability to specific target cells. Several physicochemical and eco-friendly techniques have been developed for the synthesis of nanostructures, in contrast to conventional approaches that require the use of plants or their extracts. Core-shell nanoparticles are a hybrid class of nanoparticles composed of two, three, or more different types of individual nanocomponents [37].

14.2.1.4 Nanotube

On the other hand, polymer-based thermally conductive composites have received a lot of attention in the past ten years in a variety of fields, particularly in electronic devices for heat dissipation.

Electronic devices require quick heat dissipation due to their smaller size, higher power consumption, and numerous functions. Electronic devices' dependability, performance, and lifetime can all be impacted by even the slightest difference in operating temperature. Due to their unusual combination of qualities—excellent mechanical strength and extremely high electrical and thermal conductivity—carbon nanotubes (CNTs) have received a lot of attention [38]. Because of their mechanical stability and high conductivity, carbon nanotubes (CNTs) are promising modification materials for microbial fuel cell (MFCs) improvement.[39]

14.3 BASED ON THE NATURE OF CORE AND SHELL

14.3.1 Organic-Inorganic

Organic-inorganic (O-I) nanocomposites synthesized under sustainable conditions offer the optimal balance between durability, mechanical properties, and other characteristics such as density, permeability, color, and hydrophobicity. These composites are designed using both organic and inorganic substances, and can be used to create polypeptides. The potential functionality of hybrid O–I interfaces arises from the catalytic activation of organic materials confined in the interstices or porosity of inorganic catalysts, such as clays and zeolites. The development of numerous mixed O-I nanocomposites is made possible by work on glass and ceramics, using molecules and polymers [40].

14.3.2 Organic-Organic

In this group, the particles' shells and core are both made of polymers or other organic materials. These types of particles are also known as "smart particles". Catalysis, drug delivery, bio-sensing, chemical separations, and biomaterials are just a few of the industries that use them. A polymer coating on another polymer can change the material's physical properties, such as toughness and glass transition temperature. The glass transition temperature is an important property of polymers because, below this temperature, the polymer is in the "glassy state". When the temperature of the polymer exceeds the glass transition temperature, the mechanical properties of the material change from brittle to elastic. Inorganic-inorganic core-shell particles are also widely used in controlled drug delivery system applications, due to their excellent biodegradability and drug encapsulation properties [41, 42].

14.3.3 Inorganic-Organic

Metals, metal compounds, metal oxides, silica, polymers, or other dense organic materials can be used as the core and shell of inorganic-organic core-shell nanoparticles. The organic coating on the inorganic material has numerous benefits. The metal core's increased oxidation stability prevents the metal core's surface atoms from oxidizing to metal oxide, in a typical environment. Catalysis, additives, pigments, paints, cosmetics, and inks are just a few of the applications of inorganic materials with polymer coatings. Magnetic nanoparticles with arbitrary polymer coatings are used for magnetic recording, magnetic sealing, electromagnetic shielding, and MRI, especially in biological fields such as drug targeting, and magnetic cell separation. Wet chemical processes are often used to make magnetic core particles because of their advantages of good crystallinity and magnetization. The stability of magnetic nanoparticles under highly applied magnetic fields is of great importance for applications in biology and other fields. Therefore, magnetic nanoparticles are coated with various organic materials to reduce aggregation and improve the stability of particle suspensions [43, 44].

14.3.4 Inorganic-Inorganic

The most significant category of core-shell nanoparticles is comprised of inorganic-inorganic particles. These kinds of particles are frequently employed in a variety of applications, including quantum dots, biological labeling, optoelectronics, optical bio-imaging, information storage, and semiconductor efficiency improvement. The shells and cores are both made of silica, other inorganic substances, metal, metal oxide, or metal. A core particle's silica coating has several benefits. The silica coating is the most fundamental advantage over other inorganic metals or metal oxides. It lowers bulk conductivity and improves the core particles' suspension stability. Additionally, because silica is the most chemically inert substance on the market, the core surface can be clogged without affecting the redox reactions taking place there. Since silicon dioxide is optically transparent, it can be used to modify the position and intensity of surface plasmon absorption bands. This allows spectroscopic investigation of chemical reactions on the core surface. Numerous researchers have looked into the noble metals coated in silica, such as gold (Au) and silver (Ag), as potential metal cores [45, 46].

14.3.5 High-K Ceramic Composite

Ceramics, especially ferroelectric ceramics, have a high dielectric constant but are brittle and have a low dielectric strength, while polymers have a high dielectric strength but a very low dielectric constant. As a result, considerable effort has been expended in developing ceramic-polymer composites formed by suspending zero-dimensional ceramic powders in a three-dimensional continuous polymer matrix. Recently developed composite materials combine the advantages of ceramics and polymers, creating a new type of material that is flexible, easy to process and has a relatively high dielectric constant and high dielectric strength [19, 47]. Polymer-ceramic composites, combining substrate fabrication capabilities with thermal stability, can be developed based on new concepts for large-scale integrated (LSI) chip assembly. Polymer-ceramic composites combine the low processing temperature of polymers and the high dielectric constant of ceramic and can therefore easily be used for capacitors embedded in organic substrates [48]. SiCN ceramic made from polymers, a non-oxide material, has been demonstrated in numerous investigations to be a structural ceramic beneficial in high-temperature applications. The rational design of SiCN ceramic microspheres (Figure 14.1) is currently a very competitive research area because of the large range of related applications, such as the light-filler in high-temperature thermal insulation coating and the infrared opacifier in inorganic aerogels. The most promising candidates for use in high-temperature thermal insulation are core-shell structured composites, with SiCN as the shell or core composition. The fact that core-shell ceramic microspheres are made using a straightforward emulsion approach and polymer-derived ceramics (PDCs) technology makes them highly valuable for a variety of applications [49, 50].

14.3.6 Epoxy-based Hybrid Materials

Epoxy resins, also known as poly-epoxides, are epoxide-containing reactive prepolymers and polymers. Epoxide functional groups are also referred to as epoxy. GBN/epoxy composites, with boron

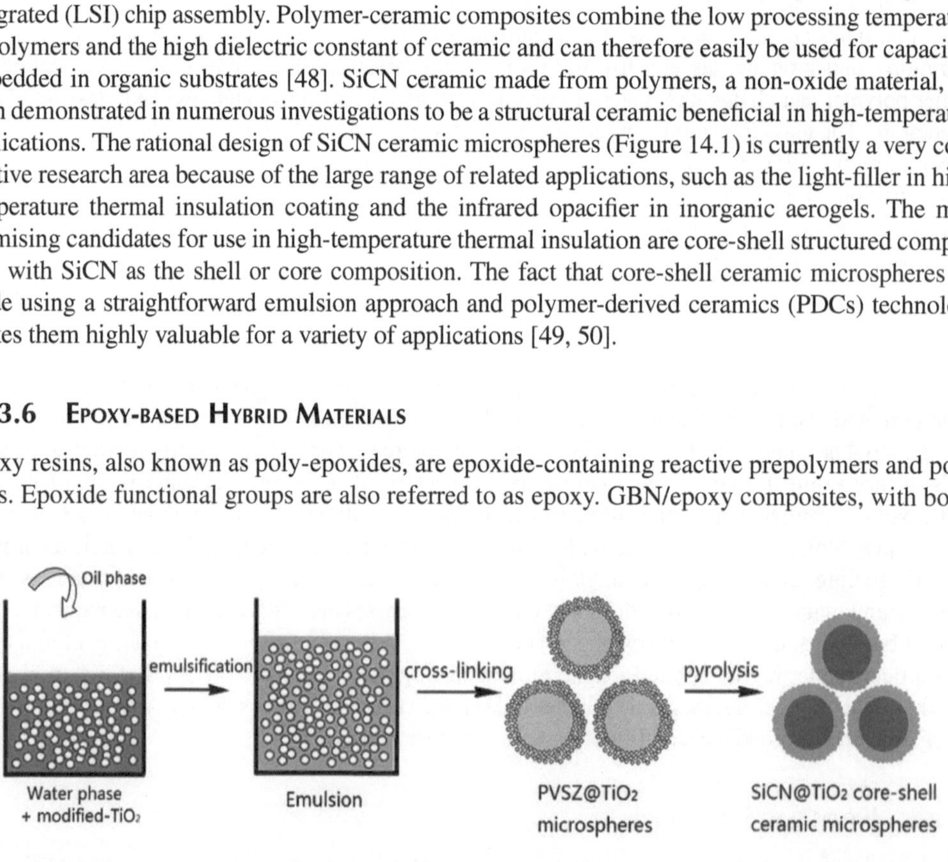

FIGURE 14.1 Schematic of the fabrication process for SiCN@TiO_2 core–shell ceramic microspheres. [49]. (Copyright 2016. Reproduced with permission from Elsevier.)

nitride nanoparticle concentrations as low as 5 wt%, exhibit significantly improved thermal conductivity and dielectric properties. Due to graphene's high electrical conductivity, composites suffer from poor electrical insulation and significant dielectric loss. Composites made of GBN and epoxies are created by mixing and curing. Electron transport between adjacent graphene is hindered by hybrid nanoparticles with thermal and insulating properties, which also create phonon transmission channels. As a result, the thermal conductivity of the composites increases significantly [51]. The stable dielectric constants at 3-18 GHz at 40 vol% for the epoxy-based composites with $BaTiO_3$ nanoparticles of 100 nm and 200 nm are 10.8 and 15, respectively. High filler loading, on the other hand, exacerbates processing challenges and results in significant dielectric loss. Even with a relatively small percentage of ceramic fillers, it remains challenging to obtain the desired dielectric properties. The dielectric properties of the $BaTiO_3$/epoxy composite may be improved as a result of improved dispersion and increased interfacial adhesion with the matrix [9].

14.4 SYNTHESIS

14.4.1 Surface Oxidation

This method allows for the production of core-shell particles with good shell adhesion and controllable composition and morphology by utilizing fundamental knowledge of gas dynamics and kinetics. Under specific circumstances, the exothermic surface reaction of silicon nanoparticles with oxygen results in the formation of a thin SiO_2 shell, as depicted in Figure 14.2. The chosen process parameters, such as silicon particle size, oxygen partial fraction, temperature, pressure, and reaction time, determine the thickness. It is simple to create core-shell nanostructures by oxidizing the surface of core particles [52]. Silica (SiO_2) is a frequently used material for the formation of core-shell particles because of the weak Van der Waal attractive forces between the particles and the medium. Particles need to be evenly dispersed in the medium for a variety of reasons, and this can be accomplished by coating them in silica to create an encapsulating shell [34].

14.4.2 Flame Synthesis

Core-shell nanostructures can also be created in low-pressure flame reactors. The choice of precursors is crucial for the successful fabrication of core-shell particles, and the reaction rate coefficient of the shell precursor must be greater than that of the core precursor. On one end of a low-pressure reaction chamber, there is an axially mobile, cooled, sintered bronze plate burner. The precursors or mixture of argon (Ar) is added to the burner and burns up in flames. The obtained core-shell particles are then separated from the gas stream using a variety of techniques, including molecular beam probing, filtration, and thermophoresis. At room temperature, magnetic particle cores show super-paramagnetic behavior. This is explained by the separation of the magnetic cores from one another and their exchange-decoupling using the environment, which causes the particle cores to behave as distinct magnetic entities. However, compared to the bulk magneto-crystalline anisotropy constant, the effective uniaxial anisotropy constant calculated from the super-paramagnetic blocking temperature is larger. This difference can be attributed to additional contributions from the core-matrix interface. Additionally, dipolar interactions between the particles cannot be ruled out [53].

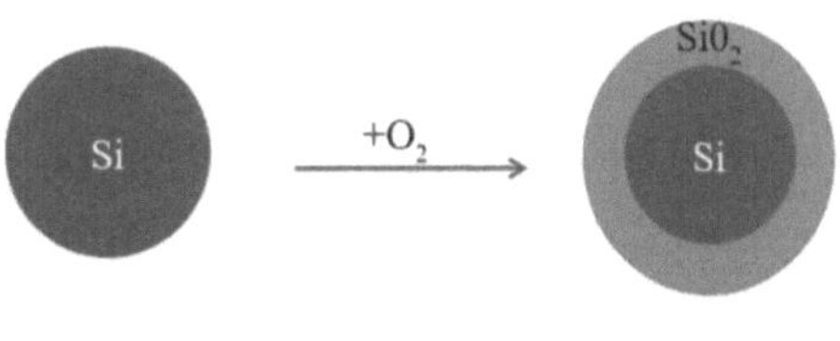

FIGURE 14.2 Schematic for illustration of surface oxidation of silicon nanoparticle [34, 52].

14.4.3 Spray Pyrolysis

A spray pyrolysis procedure can also be used to create core-shell nanostructures. Aerosols are a fluid mixture of extremely small precursor droplets and carrier gas. The spray pyrolysis method relies on the formation of particles through reaction or precipitation, solvent evaporation from droplets, and coating of the particles. This procedure is typically carried out in a reactor made up of two reaction zones or two subsequent reactors. When core and shell particles are created by ultrasonic spray pyrolysis, the properties of the core and shell can be changed by modifying the process parameters. This makes it possible to customize core-shell particles for particular applications. The stream of shell material that was solidified was combined with the aerosol produced at an ultrasonic frequency. The desired core-shell particles were created by passing this mixture through the pyrolysis zone [54, 55].

14.4.4 Chemical Deposition on Core Particles

This method of core-shell particle synthesis in the liquid phase is also known as the deposition-precipitation method. Amorphous or crystalline layers can be directly formed on salt core particles using this method, which is desirable. Heterogeneous nucleation of the solid shell phase occurs at the surface of the core particle. The processing conditions determine whether the precursors react only in the surrounding liquid phase by homogeneous nucleation to form particles or form particles directly on the surface of the core particles by chemical deposition. Chemical vapor deposition produces spherical core-shell particles with very narrow size distribution and varying shell thickness. Through the hydrolysis and condensation of a precursor shell material, a shell is created around a core particle [56]. Core nanoparticles with narrow size distribution and various compositions can be created using a variety of methods, such as precipitation from solutions, microemulsions, aerosol-vapor methods, spray pyrolysis, laser pyrolysis, hydrothermal methods, etc. The main issue with the synthesis of core-shell particles is that ferrite nanoparticles have a strong tendency to agglomerate because of their high surface energies and large surface-to-volume ratio. This problem can be resolved by stabilizing and surface-modifying the newly created nanoparticles, typically using citric acid. In the entire pH range, citric acid generates a negative particle charge and shifts potential toward more negative values. The surface of the ferrite nanoparticles, combined with -OH or -C-OH groups, is used to chemically deposit a layer of silica in the following step [57].

14.5 ASSEMBLY OF PREFORMED NANOPARTICLES BY ELECTROSTATIC INTERACTIONS

By assembling the previously formed nanoparticles on the core particle's oppositely charged surface, core-shell structures can be created. The assembly process is typically carried out by homogeneously nucleating nanoparticles from the liquid phase in which the core particles are homogeneously dispersed. The pH must be set at a level where negatively charged core particles will attract positively charged shell nanoparticles for the assembly process to work. Only if nanoparticle agglomeration is prevented during the deposition process can uniform core-shell structure be obtained. By producing periodic local chemical composition fluctuations, it is possible to significantly alter the properties of functional materials. Ferroelectric ceramic oxides, like polycrystalline $BaTiO_3$, which are widely used as MLCC dielectrics in the electronic industry, serve as a good illustration [52, 58].

14.6 GRAFTING-FROM AND GRAFTING-TO METHOD

(a) **Grafting-From Method**: The secret to success with this strategy is to add enough initiation sites to the nanoparticles' surfaces. During radical atom transfer polymerization and reversible addition-fragmentation chain transfer polymerization, the shell layer that is deposited on the nanoparticle surfaces prevents nanoparticle aggregation. By applying a shell layer to the nanoparticle surfaces, this method stops the aggregation of nanoparticles. Additionally, core-shell nanoparticles can be directly used to create nanocomposites using the shell layer as a matrix, producing high-quality, and highly filled nanocomposites. The concentration of functionalized monomer and initiator nanoparticles can be altered by adjusting the feed ratio. Prepared nanocomposites displayed significant improvement because they had a higher breakdown strength, high dielectric constant, and capacity to store energy [3].

(b) **Grafting-To Method**: As shown in Figure 14.3., the reaction between the polymer end-groups and the functional groups on the nanoparticle surface grafts the primed polymer chains onto the surface of the nanoparticles to form a nanocomposite. Composition of polymer chains to achieve the desired performance of finished nanocomposites [3].

14.7 FACTORS AFFECTING DIELECTRIC PROPERTIES

14.7.1 Particle Dispersion

Due to their large specific surface areas, nanoparticles exhibit significantly higher surface tension and frictional forces than microparticles. The nanoparticles become more cohesive as a result, which makes dispersing them more challenging. A sufficiently large particle can disrupt and locally amplify the electric field, due to the filler's and the matrix material's different conductivities and permittivities [59]. To perform specific modifications on a ZnSe-$CoSe_2$ core-shell particle, external elements, such as Carbon (C), Nitrogen (N), etc., are added in a specific stoichiometry. The core-shell structure's N-doped carbon framework's uniform distribution of ZnSe and $CoSe_2$ nanoparticles is supported by the uniform distribution of C, N, and Se elements in the composite. Modified ZnSe and $CoSe_2$ nanoparticles on the nanocomposite framework, in particular, can preserve the composite's distinctive structure and increase the material's conductivity, which improves the electrochemical performance [60].

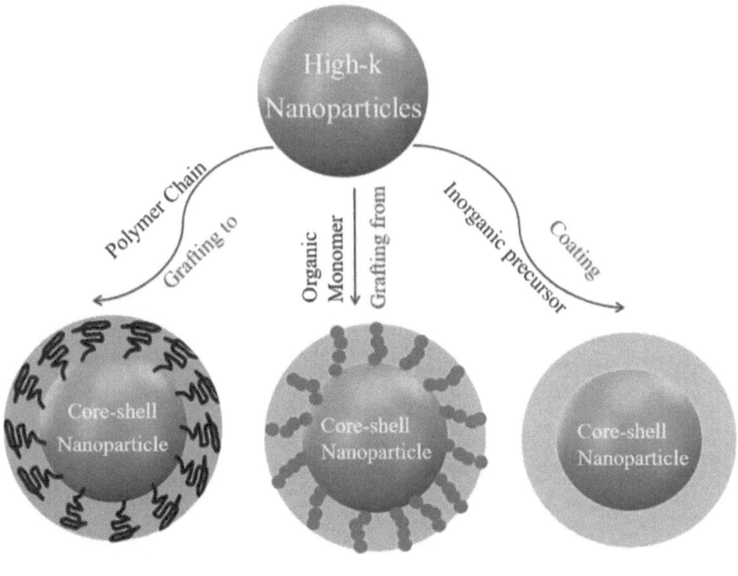

FIGURE 14.3 Schematic of the design and construction of core-shell nanoparticles for high-K polymer nanocomposites [3].

14.7.2 Interfacial Interactions

Unique physicochemical characteristics that cannot be attained by a component acting alone are exhibited by polymer nanocomposites. The polymer matrix has a significant impact on processability, transverse tensile and compressive properties, shear properties, heat resistance, and resistance to environmental media. The continuous phase acts as a binding agent for the nanofillers. The uniform dispersion of the nanofillers and the strong interfacial interactions between the matrix and the nanofillers are primarily responsible for the noticeably improved mechanical properties. It is thought that the best way to raise the interfacial shear stress and enhance stress transfer is to create covalent bonds between the fillers and the matrix. Interfacial interactions between nanoparticles and the polymer matrix improve stability by reducing mechanical stress during redox reactions [61].

14.7.3 Interphase Formation

The study aims to investigate how the particle interphase affects the dielectric properties of nanocomposite materials by employing various combinations of core-shell nanoparticles. There are two distinct zones where the particles and host interact: the particle interphase and the polymer interphase. In the polymer interphase, when moving from the particle interface to the layer of the polymer interface, it is claimed that the properties change instantly, rather than gradually, and that this interaction exists in the region where the two bulk phases interact, as shown in Figure 14.4. The polymer interphase is the name given to this area of interaction. Nearer to the nanoparticle is where the positively or negatively charged particles have located, causing diffusion [7]. Nanoparticles have a lot of defects on their surfaces because of their large surface area. These flaws might be the result of surface irregularities, such as foreign atoms, or surface geometry. These flaws cause the creation of new localized energy levels that are located in the space between the valence and conduction bands. Because mobile charges are present in the conduction or valence band, either electrons or holes, they prefer to move through the particle interphase. These could transition from one particle

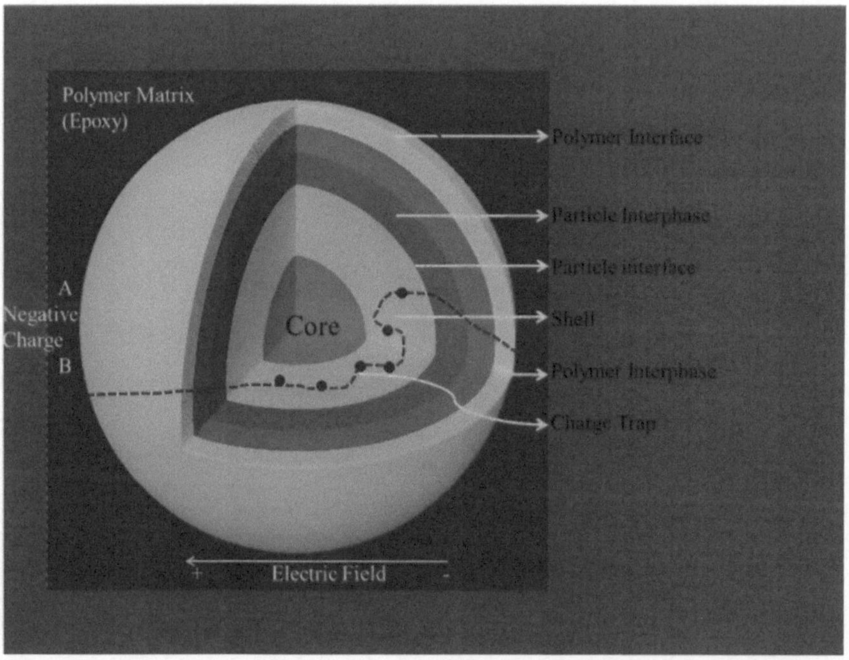

FIGURE 14.4 Possible ways of particle interface interaction with the charge movement. (a) Dielectric – energy barrier constraining charge movement. (b) Conductive – facilitating charge movement [7].

interphase to another, making it easier for charges to move around in nanocomposites and weakening their electrical insulating capabilities. The particle interphase, and subsequently the entire particle, will be perceived by the charge carriers as a restriction on their motion, increasing conduction paths. This demonstrates a potential mechanism that enhances the dielectric properties of nanocomposites [59].

14.7.4 Morphological Changes

Due to their inherent fragility, processing challenges, and weak dielectric breakdown strength, high-performance applications hardly ever use inorganic materials with high K. Ceramic and polymeric materials have lower dielectric constants and thermal conductivity issues, but they also have excellent mechanical flexibility, high dielectric strength, light weight, and ease of fabrication. When the composition and morphology of the composites are properly designed and controlled, composites are an efficient way to combine the advantageous properties of polymers and ceramics and solve these problems [33].

14.7.5 Shell thickness

The investigation of poly-vinylidene fluoride, or PVDF, filled with silicon dioxide-coated β-silicon carbide whiskers (β-SiC_w@ SiO_2) as core-shell nanofillers, demonstrates that the core-shell architecture and the shell thickness affects the thermal and the electrical conductivity of composites. SiO_2 layers can suppress the dielectric constant (K) of polymer composites, as well as significantly reduce their dielectric loss. An increase in the shell thickness further reduces the dielectric loss without changing K. Composites show a simultaneous increase in dielectric constant and thermal conductivity with controlled SiO_2 shell thickness. Outside of the pure β-SiC_w whisker, a thin, uniformly thick amorphous SiO_2 layer forms, and as the SiO_2 content rises, so does the layer's thickness. As the SiO_2 content rises, the effect of SiO_2 shell thickness on K is less pronounced than it was when the shell was first forming [33]. By creating an insulating SiO_2 shell, it is possible to significantly reduce both K and loss. However, increasing shell thickness only has a small impact on K. The electrons need more time to jump to another nanofiller, which explains why the conductivity and dielectric loss continuously decrease, and gradually, the transition frequency moves toward the low-frequency end as the shell thickness rises. We take it a step further and demonstrate that short-range charge oscillation is not significantly impacted by shell thickness in the suppression of long-range electron transport. As a result, while maintaining a similar filler polarization level, modified fillers can significantly reduce the dielectric loss caused by conductivity DC as the shell thickness increases. This finding suggests that high core-shell filler and thick insulating shell polymer composites can benefit from high K value and low dielectric loss [51].

14.7.6 Polarization

Infiltrative composites which have high dielectric constant is a result of both fast intraparticle and slow interparticle polarization. These two polarizations are possible due to the formation of an insulating shell, which is typically coupled in conventional infiltrative composites, to be controlled separately. By encouraging polarization within particles and suppressing polarization between particles, core-shell nanocomposites can achieve a high dielectric constant (K) and low dielectric loss at the same time. The insulating shell not only reduces dielectric loss but also serves other purposes. While intraparticle polarization refers to electron shift within a single filler, interparticle polarization describes electron migration across a collection of sparsely spaced fillers. The two polarizations cannot be controlled separately because they are inseparable from the typical infiltrating polymer composites. To simultaneously achieve a high dielectric constant and a low dielectric loss, the core-shell nanocomposites must suppress interparticle polarization and promote intraparticle polarization [62]. The slow polarization takes longer because the electrons must travel through the polymer and filler shell in between the nanoparticles. The dielectric properties of nanocomposites can also be characterized by the three variables of intraparticle polarization, interparticle polarization, and dielectric mass of polyvinylidene fluoride (PVDF), which has high thermal stability and excellent chemical resistance. The fact that PVDF and particle-to-particle polarization occur in the same frequency range raises the possibility that chain motion plays a role in particle-to-particle charge transport. In the same frequency range, the dielectric loss rises as the dielectric constant does. Due to their high relaxation strength from polarization between particles, polymer nanocomposites should not have their dielectric constant increased, as this causes significant energy loss and a rise in temperature in dielectric devices. Instead, it is preferable to maximize the contribution of intraparticle polarization to simultaneously achieve a high dielectric constant and a low loss. The insulating shell allows for separation and independent control of the internal and intergranular polarization instead of reducing dielectric loss. Nanocomposites with core-shell structures can achieve high dielectric constant while maintaining low dielectric loss, thanks to optimized shell thickness and charge concentration [33, 63].

14.7.7 Charge Transfer Mechanism

The insulating shell's blocking effect on charge transfer between the conductive fillers benefits the core-shell filler directly, allowing for a significant decrease in conductivity and related dielectric loss. A charge diffusion model accounts for the contribution of electron transport and the corresponding charge shift along the filler network. The enhanced charge separation is accompanied by the charge migration over long distances, which also explains the high K value and loss. Long-range charge migration can be effectively prevented by enclosing the conductive fillers with an insulating shell, and as shell thickness is increased, the suppression effect gradually becomes stronger, along with a steady decline in the dielectric loss. For fillers with a core-shell structure, this short-range charge transfer mainly affects the polarization of the fillers, since most of the charge is confined to the boundary between the conductive core and the insulating shell. The level of charge shift is not significantly altered by this dynamic, although some charge can still be released through the shell and the polymer gap because it moves much more slowly than the transient charge separation within the conducting whisker. With decreasing frequency, the electrons can move to nearby nanofillers. Then the motion of the charge can be considered random motion due to the random orientation of the charge [64, 65].

14.8 APPLICATIONS

14.8.1 Applications in Energy Storage and Conversion

By properly adjusting shells and cores of various kinds of materials with core-shell structures, It is possible to achieve advantageous properties essential for energy storage and conversion. The current energy crisis may have a workable solution in the form of core-shell materials. Due to their potential use as gate dielectrics, capacitors with large charge storage capacities, and electronically active materials, high-K materials are growing in popularity. Multi-component nanomaterials can be combined to enhance their properties, and variations in their structure, and composition can expand their potential applications. Nanomaterials with core-shell structures are the most promising candidates for these applications. A specific surface structure (based on porosity, surface area, etc.) is provided by the shell supported by the core, resulting in excellent energy storage and conversion performance [66].

14.8.2 Applications in the Biomedical Field

Core-shell nanoparticles have a wide range of fascinating biomedical applications. Core-shell nanoparticles are primarily used in the biomedical sector for applications such as tissue engineering, bioimaging, cell labeling, biosensors, and controlled drug delivery [67].

14.8.2.1 Targeted Drug Delivery

Recent developments in nanotechnology have made it possible to deliver medications in a controlled manner to a specific organ or area of the body. Overall effectiveness can be raised by using the right nanoparticles, which have a higher selectivity toward the target tissue. To enable the drug's selective binding to a particular cell surface, the drug is first contained in mesoporous nanoparticles with an appropriate surface coating. When the nanocarrier reaches the target cell, it dissolves or activates chemically attached supramolecular gates to release its cargo. Heat or light from the outside can trigger this release, but taking advantage of the target's local chemical environment is preferable [67]. Active targeting is a more focused method of drug delivery that enables selective receptor adsorption by target cells and involves the binding of specialized ligands to the drug-enveloping nanoparticles. With passive targeting, the drugs' appropriate surface modification allows them to enter the body, causing them to accumulate only with the targeted cells as a result of physicochemical or pharmacological factors [68].

14.8.2.2 Bioimaging

Optical and magnetic resonance imaging methods are most suitable as they exploit the unique luminescent and magnetic properties of nanoparticles. Dye-doped and quantum dots (QDs), which could be used for optical imaging, have the drawbacks of being toxic, having a propensity for photo-oxidation, and being poorly soluble in water. However, these restrictions can be reduced by covering them with appropriate materials. Then, those middle-shell nanoparticles can be applied to optical imaging now [69]. The surface plasmonic properties of Ag metals can be exploited for tumor cell imaging and safe local delivery at tumor sites by coating them with smart gel polymers. While Ag is used for cellular imaging, pH-controlled drug delivery is performed in copolymer hydrogels, as shown in Figure 14.5. When debris is placed near the tumor cells and exposed to light, the fluorescent fabric emits light that allows for the detection of the tumor cells. Here, the best shell coating is combined with iron oxide, Fe, Ni, Co, or high-quality paramagnetic iron oxide (SPION) center particles magnetic nanoparticles, to increase contrast for applications [70, 71].

14.8.2.3 Sensors

Sensors as a device are used to measure physical characteristics and convert them into signals that can be analyzed by a person or an instrument, whether analog or digital. As sensors in products, nanoparticles are used to detect DNA, damaged cells, RNA, cholesterol, glucose, and other substances. A sensor could be made out of a magnetic fabric covered in fluorescence. Here, fluorescent dyes track the particle's position and a magnetic shell uses magnetic excitation to transfer heat to the region [72]. Magnetic-based nanocomposites are used for bio-analytical sensors after coating with any other material, such as metal, fluorescent, polymer, or silica [73].

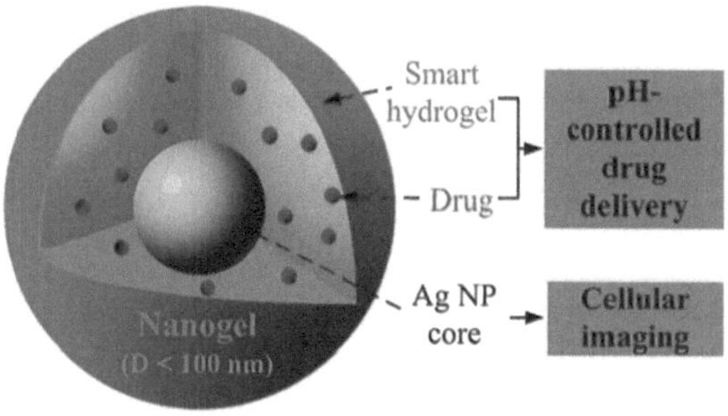

FIGURE 14.5 Schematic illustration of drug-loaded core/shell hybrid nanogels with the smart gel shell immobilized on the Ag NP core for integration of tumor cell imaging and pH-triggered drug delivery [71]. (Copyright 2010. Reproduced with permission from ACS Publications.)

14.9 APPLICATIONS IN THE MICROELECTRONIC FIELD

When magnetic scrap is coated with a useful material, such as a precious metal, a semiconductor, or a suitable oxide, the physical properties of the combined particles improve compared to pure core scrap. Embedded passive technology has yet to be commercialized for digital packages due to fabric and system issues. To enable the commercialization of embedded passive technology, materials that meet fabrication requirements, as well as electrical and mechanical performances, must be developed. Dielectrics with K values greater than 1000 and low dielectric loss pose material challenges. Radiofrequency applications require a very low dissipation factor to avoid signal losses. Much higher values can be tolerated in decoupling applications [8]. Nanopolymer-ceramic composites can introduce high dielectric constant with a low-temperature thick film fabrication technique because it combines the advantages of the high dielectric constant of a ceramic part and the main processability body of the polymer element. Polymer-ceramic composites have been extensively studied due to their unique advantages, and they have been selected as prime candidates for the development of capacitors embedded in natural substrates [74]. Even though ceramic-polymer composites have demonstrated some commercial success, their dielectric permittivity ($K<100$) is insufficient for many applications in next-generation electronics. To address this issue, disruptive innovation in polymer excessive-K composites has been investigated. One promising technique is high-K polymer-conductive filler composites based primarily on percolation mechanisms. According to the scaling principle, high dielectric permittivity can be obtained when the real filler loading is very close to the percolation threshold. Because the filler loading in percolative composites is lower than in ceramic-polymer composites, improved material processibility and adhesion strength can be anticipated [75], [48]. However, the electrical and dielectric properties of a percolative composite are extremely sensitive to the composition, due to its percolation nature. A minor composition trade will result in a massive exchange of fabric houses, posing a significant challenge to producing materials with reproducible properties [76].

14.10 APPLICATIONS IN THE ELECTRICAL ENGINEERING FIELD

High-K polymer composites, one of the new purposeful materials, are used in various ways in electrical engineering. For example, excessively elastic, high dielectric constant rubber matrix composites are becoming increasingly popular, especially in electrical engineering programs and cable accessories, due to their ease of processing, low temperatures, and versatility. With fillers such as metals, carbon black, carbon fibers, and carbon nanotubes, the permittivity of the composites increases sharply near the percolation threshold. In this case, the dielectric loss of the composites typically increases dramatically, while the resistivity decreases sharply. As a result, if bendy rubber-matrix composites are to be used as electric cable add-ons, the aforementioned issues must be addressed [77, 78]. Mobile electronics, stationary power systems, hybrid electric vehicles, and pulse power applications all depend heavily on electrical energy storage. The dielectric nanocomposites showed minimal temperature dependence and weak frequency dependence as prepared. As a result, the resulting PVDF/Ag@C nanocomposites have a wide range of potential applications in the electrical and electronic sectors, including printed circuit boards. Additionally, this research provides a new method for creating flexible nanocomposites with a low percolation threshold and superior dielectric qualities [65].

14.11 APPLICATION IN MULTILAYER CERAMIC CAPACITORS (MLCCs)

A material that exists in several crystal structures, the modification of material properties, is quite simple. Because of their high dielectric constant, low leakage current, small dielectric loss, and large dielectric breakdown, polyatomic ceramic materials are widely used as multilayer ceramic capacitors (MLCCs). MLCCs are in high demand in electrical industries, due to surface mounting technology and the need for portable devices. The development of MLCCs was accomplished by increasing the number of active dielectric layers to achieve higher capacitance, lower cost, and smaller size. To be used as MLCCs, the microstructure of grains must be properly designed so that the obtained dielectric layers are quite thin. MLCCs are widely used in electronics that require a dielectric ceramic with exceptional dielectric properties that is also cost-effective and environment friendly. A uniform coating layer of dopant would significantly slow down material's transport while also deactivating the sintering kinetics. It produces an element gradient distribution to form a core-shell structure, which significantly improves the dielectric properties and reliability of ceramics used as dielectric in MLCCs co-doped with rare earth elements [51].

14.12 CONCLUSIONS

Compared with the material being used as the core, the coating of the shell by certain chemical or physical methods shows dramatically enhanced dielectric properties. The composites' dielectric properties exhibit excellent thermal stability and high energy density. We want to highlight how the field of core-shell nanocomposites research has recently expanded into a new dimension. This field will ultimately lead to breakthrough discoveries regarding the synthesis, property modification, and applications of this new class of nanomaterials. Future advances in core-cell nanoparticles will lead to better-applied parts, with several different properties and improved performance. Core-shell nanoparticles are known for their high stability and tendency to protect the core material from the environment, enhanced chemical and physical characteristics, enhanced dielectric properties, ease of bio-functionalization, and so on. Recent exciting discoveries demonstrate that this opens up enormous opportunities for biomedical applications, including targeted drug delivery, bio-imaging, controlled drug delivery and release, diagnostics, biosensors, and many others. This paper confirms that various types of core-shell nanoparticles, due to their numerous uses in biomedicine and electronics, inorganic/inorganic nanoparticles coated with other inorganic materials, are currently the subject of extensive research. Magnetic or semiconductor cores are frequently coated with a polymer or another inert inorganic material to increase biocompatibility for biological applications, such as bioimaging or controlled drug delivery. Core materials can be selectively coated with shell materials that are comparably suitable, to increase the effectiveness of a core-shell particle. Due to modifications in some of their physical and chemical characteristics, some organic/organic and organic/inorganic core/shell nanoparticles are noteworthy and have a wider range of industrial applications. The most popular inorganic shell materials are metals (Au, Ag) and oxides (SiO_2, TiO_2) because they increase the core particle's biocompatibility, mechanical strength, oxidation resistance, thermal stability, and colloidal stability. In the past decade, interest has shifted from the synthesis and characterization of various core-shell nanoparticles to the use of these particles in industries like biomedicine, catalysis, and electronics, as well as the synthesis of more complex particles. In addition to spherical particles, core-shell nanoparticles with a variety of shapes have been the focus of research in recent years. They are typically made by controlling the growth of a shell over a core material or by using a core particle with a similar shape, as a template. Unlike the symmetric growth of the shell material on the core surface of spherical core-shell nanoparticles, the shell coatings for other core-shell nanoparticle forms are not uniform. These have novel properties that might be useful for particular applications, in addition to being of academic interest. To help with the formation of different core-shell nanoparticles, some new mechanisms, like ion exchange or charge transfer mechanisms, have been developed recently. From an application standpoint, it is clear that there is a diverse set of applications that can be broadly classified as biomedical applications, electronics, biomedical sensors, energy storage, electrical engineering, and so on.

Advanced models for dielectric composites need to be refined or developed. In particular, when used in harsh conditions, it is crucial to develop high-K materials with low frequency or temperature dependence of dielectric permittivity. Future research will also reveal new synthesis-facilitating mechanisms in addition to the development of new core-shell materials. The significance of process scalability from the laboratory to the commercial scale must be understood. Lastly, with these advancements in experimental work, more theoretical studies are anticipated to improve understanding and ease experimental endeavors.

14.13 FUTURE OUTLOOK

Advanced models for dielectric composites need to be refined or developed. It is critical to develop high-K materials with low frequency or temperature dependence of dielectric permittivity, especially when used in extreme conditions. In addition to the creation of new core-shell materials, new mechanisms to facilitate synthesis will also be revealed in the future. It is important to recognize the importance of process scalability from the laboratory to the commercial scale. And finally, with such improvements in experimental work, more theoretical studies are anticipated to improve understanding and ease experimental endeavors.

REFERENCES

[1] K. Yang, X. Huang, J. He and P. Jiang, Strawberry-like Core-Shell Ag@Polydopamine@BaTiO $_3$ Hybrid Nanoparticles for High- k Polymer Nanocomposites with High Energy Density and Low Dielectric Loss, *Adv. Mater. Interfaces* 2 (2015), pp. 1500361.

[2] M. Kurian and S. Thankachan, Structural diversity and applications of spinel ferrite core - Shell nanostructures- A review, *Open Ceramics* 8 (2021), pp. 100179.

[3] X. Huang and P. Jiang, Core-Shell Structured High- k Polymer Nanocomposites for Energy Storage and Dielectric Applications, *Adv. Mater.* 27 (2015), pp. 546–554.

[4] M. Reaz, A. Haque and K. Ghosh, Synthesis, Characterization, and Optimization of Magnetoelectric BaTiO3–Iron Oxide Core–Shell Nanoparticles, *Nanomaterials* 10 (2020), pp. 563.

[5] H. Wang, M. Cao, T. Mao, J. Fu, W. Pan, H. Hao et al., Fabrication of BaTiO3@FeO core-shell nanoceramics for dielectric capacitor applications, *Scripta Materialia* 196 (2021), pp. 113753.

[6] Y. Wang, B. Cui, Y. Liu, X. Zhao, Z. Hu, Q. Yan et al., Fabrication of submicron La 2 O 3 -coated BaTiO 3 particles and fine-grained ceramics with temperature-stable dielectric properties, *Scripta Materialia* 90–91 (2014), pp. 49–52.

[7] S. Chaudhary, T. Andritsch and A.S. Vaughan, Effect of core-shell particles on the dielectric properties of epoxy nanocomposites, in 2019 *IEEE Conference on Electrical Insulation and Dielectric Phenomena (CEIDP)*, 2019, pp. 729–732.

[8] Z.-M. Dang, J.-K. Yuan, J.-W. Zha, T. Zhou, S.-T. Li and G.-H. Hu, Fundamentals, processes and applications of high-permittivity polymer–matrix composites, *Progress in Materials Science* 57 (2012), pp. 660–723.

[9] M. Zhang, P. Xu, H. Peng and F. Qin, A rational design of core-shell-satellite structured BaTiO3 fillers for epoxy-based composites with enhanced microwave dielectric constant and low loss, *Composites Part B: Engineering* 215 (2021), pp. 108764.

[10] A. Mayeen, M.S. Kala, S. Sunija, D. Rouxel, R.N. Bhowmik, S. Thomas et al., Flexible dopamine-functionalized BaTiO3/BaTiZrO3/BaZrO3-PVDF ferroelectric nanofibers for electrical energy storage, *Journal of Alloys and Compounds* 837 (2020), pp. 155492.

[11] X. Xiao, H. Yang, N. Xu, L. Hu and Q. Zhang, High performance of P(VDF-HFP)/Ag@TiO $_2$ hybrid films with enhanced dielectric permittivity and low dielectric loss, *RSC Advance* 5 (2015), pp. 79342–79347.

[12] L. Zhang, S. Yuan, S. Chen, D. Wang, B.-Z. Han and Z.-M. Dang, Preparation and dielectric properties of core–shell structured Ag@polydopamine/poly(vinylidene fluoride) composites, *Composites Science and Technology* 110 (2015), pp. 126–131.

[13] X. Wu, G. Song, W. Zhang, H. Feng, Y. Liu, E. Huang et al., Atomic layer deposition fabricated core–shell nanostructures for enhanced polyetherimide composite dielectrics, *J. Mater. Chem. A* 10 (2022), pp. 13097–13105.

[14] D. He, Y. Wang, X. Chen and Y. Deng, Core–shell structured BaTiO3@Al2O3 nanoparticles in polymer composites for dielectric loss suppression and breakdown strength enhancement, *Composites Part A: Applied Science and Manufacturing* 93 (2017), pp. 137–143.

[15] H.-W. Choi, Y.-W. Heo, J.-H. Lee, J.-J. Kim, H.-Y. Lee, E.-T. Park et al., Effects of Ni particle size on dielectric properties of PMMA-Ni-BaTiO3 composites, *Integrated Ferroelectrics* 87 (2007), pp. 85–93.
[16] Y. Bai, Z.-Y. Cheng, V. Bharti, H.S. Xu and Q.M. Zhang, High-dielectric-constant ceramic-powder polymer composites, *Appl. Phys. Lett.* 76 (2000), pp. 3804–3806.
[17] V. Bobnar, B. Vodopivec, A. Levstik, M. Kosec, B. Hilczer and Q.M. Zhang, Dielectric Properties of Relaxor-like Vinylidene Fluoride–Trifluoroethylene-Based Electroactive Polymers, *Macromolecules* 36 (2003), pp. 4436–4442.
[18] Jiongxin Lu, Kyoung-Sik Moon and C.P. Wong, Development of Novel Silver Nanoparticles/Polymer Composites as High K Polymer Matrix by In-situ Photochemical Method, in *56th Electronic Components and Technology Conference 2006*, 2006, pp. 1841–1846.
[19] M. Arbatti, X. Shan and Z.-Y. Cheng, Ceramic–Polymer Composites with High Dielectric Constant, *Adv. Mater.* 19 (2007), pp. 1369–1372.
[20] Z.-M. Dang, T. Zhou, S.-H. Yao, J.-K. Yuan, J.-W. Zha, H.-T. Song et al., Advanced Calcium Copper Titanate/Polyimide Functional Hybrid Films with High Dielectric Permittivity, *Adv. Mater.* 21 (2009), pp. 2077–2082.
[21] Y. Shen, Y.H. Lin and C.-W. Nan, Interfacial Effect on Dielectric Properties of Polymer Nanocomposites Filled with Core/Shell-Structured Particles, *Adv. Funct. Mater.* 17 (2007), pp. 2405–2410.
[22] M.T. Sebastian and H. Jantunen, Polymer-Ceramic Composites of 0-3 Connectivity for Circuits in Electronics: A Review, *International Journal of Applied Ceramic Technology* (2010).
[23] Z.-M. Dang, Y. Shen and C.-W. Nan, Dielectric behavior of three-phase percolative Ni–BaTiO3/polyvinylidene fluoride composites, *Appl. Phys. Lett.* 81 (2002), pp. 4814–4816.
[24] S. Ramesh, B.A. Shutzberg, C. Huang, Jie Gao and E.P. Giannelis, Dielectric nanocomposites for integral thin film capacitors: materials design, fabrication and integration issues, *IEEE Trans. Adv. Packag.* 26 (2003), pp. 17–24.
[25] D.-D. Feng, K. Yu and H. Wang, Dielectric Properties of CCTO/Poly(vinylidene fluoride) *Composites for Energy Storage Applications, J Adv Phys* 4 (2015), pp. 384–387.
[26] Y. Shen, Y. Lin, M. Li and C.-W. Nan, High Dielectric Performance of Polymer Composite Films Induced by a Percolating Interparticle Barrier Layer, *Adv. Mater.* 19 (2007), pp. 1418–1422.
[27] J. Lu, K.-S. Moon, J. Xu and C.P. Wong, Synthesis and dielectric properties of novel high-K polymer composites containing in-situ formed silver nanoparticles for embedded capacitor applications, *J. Mater. Chem.* 16 (2006), pp. 1543.
[28] H. Tang, G.-X. Chen and Q. Li, Epoxy-based high-K composites with low dielectric loss caused by reactive core-shell-structured carbon nanotube hybrids, *Materials Letters* 184 (2016), pp. 143–147.
[29] J. Ma, U. Azhar, C. Zong, Y. Zhang, A. Xu, C. Zhai et al., Core-shell structured PVDF@BT nanoparticles for dielectric materials: A novel composite to prove the dependence of dielectric properties on ferroelectric shell, *Materials & Design* 164 (2019), pp. 107556.
[30] Y. You, Y. Wang, L. Tu, L. Tong, R. Wei and X. Liu, Interface Modulation of Core-Shell Structured BaTiO3@polyaniline for Novel Dielectric Materials from Its Nanocomposite with Polyarylene Ether Nitrile, *Polymers* 10 (2018), pp. 1378.
[31] Z. Wang, Y. Cheng, M. Yang, J. Huang, D. Cao, S. Chen et al., Dielectric properties and thermal conductivity of epoxy composites using core/shell structured Si/SiO2/Polydopamine, *Composites Part B: Engineering* 140 (2018), pp. 83–90.
[32] W. Xu, G. Yang, L. Jin, J. Liu, Y. Zhang, Z. Zhang et al., High- k Polymer Nanocomposites Filled with Hyperbranched Phthalocyanine-Coated $BaTiO_3$ for High-Temperature and Elevated Field Applications, *ACS Appl. Mater. Interfaces* 10 (2018), pp. 11233–11241.
[33] D. Cao, W. Zhou, M. Yuan, B. Li, T. Li, J. Li et al., Polymer composites filled with core–shell structured nanofillers: effects of shell thickness on dielectric and thermal properties of composites, *J Mater Sci: Mater Electron* 33 (2022), pp. 5174–5189.
[34] S. Kalele, S.W. Gosavi, J. Urban and S.K. Kulkarni, Nanoshell particles: synthesis, properties and applications, *Current Science* 91 (2006), pp. 16.
[35] S. Baylan, G. Richter, M. Beregovsky, D. Amram and E. Rabkin, The kinetics of hollowing of Ag–Au core–shell nanowhiskers controlled by short-circuit diffusion, *Acta Materialia* 82 (2015), pp. 145–154.
[36] S. Dembski, C. Schneider, B. Christ and M. Retter, Core-shell nanoparticles and their use for in vitro and in vivo diagnostics, in *Core-Shell Nanostructures for Drug Delivery and Theranostics*, Elsevier, 2018, pp. 119–141.

[37] M. Khatami, H. Alijani, M. Nejad and R. Varma, Core@shell Nanoparticles: Greener Synthesis Using Natural Plant Products, *Applied Sciences* 8 (2018), pp. 411.

[38] M.C. Vu, Y.H. Bae, M.J. Yu, Md.A. Islam and S.-R. Kim, Core-shell structured carbon nanotube-poly(methylmethacrylate) beads as thermo-conductive filler in epoxy composites, *Composites Part A: Applied Science and Manufacturing* 109 (2018), pp. 55–62.

[39] Y.-C. Liu, Y.-H. Hung, S.-F. Liu, C.-H. Guo, T.-Y. Liu, C.-L. Sun et al., Core–shell structured multiwall carbon nanotube–graphene oxide nanoribbon and its N-doped variant as anodes for high-power microbial fuel cells, Sustainable *Energy Fuels* 4 (2020), pp. 5339–5351.

[40] C. Sanchez, P. Belleville, M. Popall and L. Nicole, Applications of advanced hybrid organic–inorganic nanomaterials: from laboratory to market, *Chem. Soc. Rev.* 40 (2011), pp. 696.

[41] J. Jang, J. Ha and S. Kim, Fabrication of polyaniline nanoparticles using microemulsion polymerization, *Macromol. Res.* 15 (2007), pp. 154–159.

[42] M. Ballauff and Y. Lu, "Smart" nanoparticles: Preparation, characterization and applications, *Polymer* 48 (2007), pp. 1815–1823.

[43] M. Salavati-Niasari, F. Davar and M. Farhadi, Synthesis and characterization of spinel-type CuAl2O4 nanocrystalline by modified sol–gel method, *J. Sol-Gel Sci. Technol.* 51 (2009), pp. 48–52.

[44] Th. Batzilla and A. Tulke, Preparation of encapsulated aluminum pigments by emulsion polymerization and their characterization, *J. Coatings Tech.* 70 (1998), pp. 77–83.

[45] Y. Yu and P. Mulvaney, Synthesis and characterization of Au/TiO2 core-shell structure nanoparticles, *Korean J. Chem. Eng.* 20 (2003), pp. 1176–1182.

[46] J. Lee, Y. Lee, J.K. Youn, H.B. Na, T. Yu, H. Kim et al., Simple Synthesis of Functionalized Superparamagnetic Magnetite/Silica Core/Shell Nanoparticles and their Application as Magnetically Separable High-Performance Biocatalysts, *Small* 4 (2008), pp. 143–152.

[47] Y. Yang, H. Sun, D. Yin, Z. Lu, J. Wei, R. Xiong et al., High performance of polyimide/$CaCu_3Ti_4O_{12}$ @Ag hybrid films with enhanced dielectric permittivity and low dielectric loss, *J. Mater. Chem. A* 3 (2015), pp. 4916–4921.

[48] M. Kakimoto, A. Takahashi, T. Tsurumi, J. Hao, L. Li, R. Kikuchi et al., Polymer-ceramic nanocomposites based on new concepts for embedded capacitor, *Materials Science and Engineering: B* 132 (2006), pp. 74–78.

[49] H. Liu, H. Zhang, J. Li, W. Kang and P. Chu, A novel SiCN@TiO 2 core–shell ceramic microspheres derived from a polymeric precursor, *Ceramics International* 42 (2016), pp. 7135–7140.

[50] J. Feng, Y. Yan, D. Chen, W. Ni, J. Yang, S. Ma et al., Study of thermal stability of fumed silica based thermal insulating composites at high temperatures, *Composites Part B: Engineering* 42 (2011), pp. 1821–1825.

[51] J. Ren, Q. Li, L. Yan, L. Jia, X. Huang, L. Zhao et al., Enhanced thermal conductivity of epoxy composites by introducing graphene@boron nitride nanosheets hybrid nanoparticles, *Materials & Design* 191 (2020), pp. 108663.

[52] V. Srdic, B. Mojic, M. Nikolic and S. Ognjanovic, Recent progress on synthesis of ceramics core/shell nanostructures, *PAC* 7 (2013), pp. 45–62.

[53] C. Janzen, J. Knipping, B. Rellinghaus and P. Roth, Formation of Silica-Embedded Iron-Oxide Nanoparticles in Low-Pressure Flames, *Journal of Nanoparticle Research* 5 (2003), pp. 589–596.

[54] S. Stopic, B. Friedrich, M. Schroeder and T.E. Weirich, Synthesis of TiO2 core/RuO2 shell particles using multistep ultrasonic spray pyrolysis, *Materials Research Bulletin* 48 (2013), pp. 3633–3635.

[55] Y.S. Chung, S.B. Park and D.-W. Kang, Magnetically separable titania-coated nickel ferrite photocatalyst, *Materials Chemistry and Physics* 86 (2004), pp. 375–381.

[56] A.-H. Lu, E.L. Salabas and F. Schüth, Magnetic Nanoparticles: Synthesis, Protection, Functionalization, and Application, *Angew. Chem. Int. Ed.* 46 (2007), pp. 1222–1244.

[57] B. Mojić, K.P. Giannakopoulos, Ž. Cvejić and V.V. Srdić, Silica coated ferrite nanoparticles: Influence of citrate functionalization procedure on final particle morphology, *Ceramics International* 38 (2012), pp. 6635–6641.

[58] M.T. Buscaglia, M. Viviani, Z. Zhao, V. Buscaglia and P. Nanni, Synthesis of $BaTiO_3$ Core–Shell Particles and Fabrication of Dielectric Ceramics with Local Graded Structure, *Chem. Mater.* 18 (2006), pp. 4002–4010.

[59] F.N. Alhabill, R. Ayoob, T. Andritsch and A.S. Vaughan, Introducing particle interphase model for describing the electrical behaviour of nanodielectrics, *Materials & Design* 158 (2018), pp. 62–73.

[60] C. Pan, K. Kou, Y. Zhang, Z. Li, T. Ji and G. Wu, Investigation of the dielectric and thermal conductive properties of core–shell structured HGM@hBN/PTFE composites, *Materials Science and Engineering: B* 238–239 (2018), pp. 61–70.
[61] C. Yang, H. Wei, L. Guan, J. Guo, Y. Wang, X. Yan et al., Polymer nanocomposites for energy storage, energy saving, and anticorrosion, *J. Mater. Chem. A* 3 (2015), pp. 14929–14941.
[62] W. Zhou, T. Li, M. Yuan, B. Li, S. Zhong, Z. Li et al., Decoupling of inter-particle polarization and intra-particle polarization in core-shell structured nanocomposites towards improved dielectric performance, *Energy Storage Materials* 42 (2021), pp. 1–11.
[63] P. Wang, B. Sun, Y. Liang, H. Han, X. Fan, W. Wang et al., Stretchable and super-robust graphene superhydrophobic composite for electromechanical sensor application, *J. Mater. Chem. A* 6 (2018), pp. 10404–10410.
[64] B. Li, C.I. Camilli, P.I. Xidas, K.S. Triantafyllidis and E. Manias, Structured Polyethylene Nanocomposites: Effects of Crystal Orientation and Nanofiller Alignment on High Field Dielectric Properties, *MRS Advances* 2 (2017), pp. 363–368.
[65] Z. Chen, H. Li, G. Xie and K. Yang, Core–shell structured Ag@C nanocables for flexible ferroelectric polymer nanodielectric materials with low percolation threshold and excellent dielectric properties, *RSC Advances* 8 (2018), pp. 1–9.
[66] H. Feng, L. Tang, G. Zeng, Y. Zhou, Y. Deng, X. Ren et al., Core-shell nanomaterials: Applications in energy storage and conversion, *Advances in Colloid and Interface Science* 267 (2019), pp. 26–46.
[67] R. Ghosh Chaudhuri and S. Paria, Core/Shell Nanoparticles: Classes, Properties, Synthesis Mechanisms, Characterization, and Applications, *Chem. Rev.* 112 (2012), pp. 2373–2433.
[68] V.P. Torchilin, Passive and Active Drug Targeting: Drug Delivery to Tumors as an Example, in *Drug Delivery*, M. Schäfer-Korting, ed., Springer Berlin Heidelberg, Berlin, Heidelberg, 2010, pp. 3–53.
[69] I.D. Grosu, M.A. Toms and S.A. Toms, Nanoimaging and neurological surgery, *WIREs Nanomed Nanobiotechnol* 2 (2010), pp. 601–617.
[70] N. Sounderya and Y. Zhang, Use of Core/Shell Structured Nanoparticles for Biomedical Applications, *BIOMENG* 1 (2008), pp. 34–42.
[71] W. Wu, T. Zhou, A. Berliner, P. Banerjee and S. Zhou, Smart Core–Shell Hybrid Nanogels with Ag Nanoparticle Core for Cancer Cell Imaging and Gel Shell for pH-Regulated Drug Delivery, *Chem. Mater.* 22 (2010), pp. 1966–1976.
[72] W. Schärtl, Current directions in core–shell nanoparticle design, *Nanoscale* 2 (2010), pp. 829.
[73] L. Stanciu, Y.-H. Won, M. Ganesana and S. Andreescu, Magnetic Particle-Based Hybrid Platforms for Bioanalytical Sensors, *Sensors* 9 (2009), pp. 2976–2999.
[74] Y. Rao and C.P. Wong, Material characterization of a high-dielectric-constant polymer-ceramic composite for embedded capacitor for RF applications, *J. Appl. Polym. Sci.* 92 (2004), pp. 2228–2231.
[75] T. Tanaka, G.C. Montanari and R. Mulhaupt, Polymer nanocomposites as dielectrics and electrical insulation-perspectives for processing technologies, material characterization and future applications, *IEEE Trans. Dielect. Electr. Insul.* 11 (2004), pp. 763–784.
[76] Z.-M. Dang, B. Peng, D. Xie, S.-H. Yao, M.-J. Jiang and J. Bai, High dielectric permittivity silver/polyimide composite films with excellent thermal stability, *Appl. Phys. Lett.* 92 (2008), pp. 112910.
[77] Z.-M. Dang, B. Xia, S.-H. Yao, M.-J. Jiang, H.-T. Song, L.-Q. Zhang et al., High-dielectric-permittivity high-elasticity three-component nanocomposites with low percolation threshold and low dielectric loss, *Appl. Phys. Lett.* 94 (2009), pp. 042902.
[78] G. Gallone, F. Carpi, D. De Rossi, G. Levita and A. Marchetti, Dielectric constant enhancement in a silicone elastomer filled with lead magnesium niobate–lead titanate, *Materials Science and Engineering: C* 27 (2007), pp. 110–116.

15 Hybrid High-K Dielectrics for Energy Storage Applications

Nibedita Nath, Subhendu Chakroborty, and Laxmi Mallik

15.1 INTRODUCTION

With the expanded functionality of current power systems and power electronic devices [1, 2], high power and high energy storage capacitors are a key research and development goal. Electrical and electronic engineering pulsed power devices employ dielectric capacitors, due to their quick charging and discharging rates, low density, and light weight. Polymer-based capacitors outperform inorganic ceramic capacitors in breakdown strength, dielectric loss, production, and cost [3]. When compared to electrochemical capacitors and batteries, dielectric capacitors offer superior operating voltage and power density, longevity, and cycle stability. [4–8]. Dielectric capacitors are useful for distributed power systems and renewable energy storage because of all of these advantages. Dielectric capacitors' characteristics are virtually entirely determined by their dominant component. High-K dielectrics have gained interest in addition to dielectric capacitors as helpful materials for artificial muscles [9, 10], cable insulation [14, 15], electrocaloric cooling [11–13], and other applications. A dielectric capacitor consists of a layer of dielectric material sandwiched between two electrically conducting layers, or electrodes (Figure 15.1a). Electric polarization takes place within the dielectric under an applied electric field (E app), and charges build up on the electrodes. This is termed the charging process (Figure 15.1b).

Polymer-based capacitors have several advantages, such as easy processing and manufacture, low dielectric loss, high breakdown strength, and low cost. Microwave communications, underground gas and oil exploration, power transmission, microelectronic devices, wind power generation, and hybrid electric vehicles are all interesting areas for their application [16–21]. The development of dielectrics with a large capacity for storing energy has emerged as a major research focus. Biaxially orientated polypropylene (BOPP), currently the most widely used dielectric material, offers development hurdles due to its poor energy storage density (1–2 J/cm^3). Additionally, temperature changes have a negative impact on the efficiency and density of its energy storage [22–25]. This chapter summarizes the various polymer-based high-K dielectrics for energy application.

15.2 ELECTRICAL PROPERTIES OF POLYMER

Due to their low dielectric constant, pure polymers are inappropriate for applications requiring high energy densities [26]. High dielectric constant electro-ceramic nanoparticles are utilized as filler

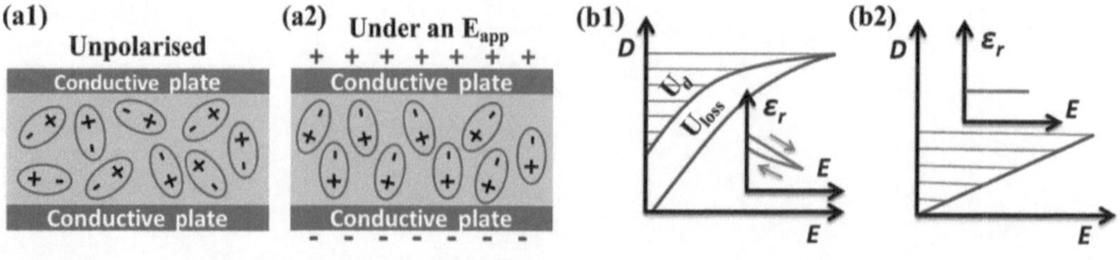

FIGURE 15.1 Diagrams of a dielectric capacitor (a) before and (b) after charging are shown. (Reprinted from ref [1] with permission).

DOI: 10.1201/9781003352365-15

materials to get over low dielectric constant restrictions. The creation of charge carriers often occurs by thermal excitation because semiconductors have a tiny band gap, but dielectric materials have a very large band gap, which causes charge carriers to be introduced from electrical contacts or external sources. In dielectrics, switching from one band to another needs considerable energy [27]. The dielectric phenomena, that interact with applied external fields like an electric or magnetic field, is caused by electrons within the valence shell of an atom. Despite the introduction of electrically conductive nanofillers, high-K ceramics are still used as filler materials in the creation of nanocomposites. Although this device architecture has demonstrated greater dielectric response, there are several drawbacks to take into account, including poor breakdown strength and high dissipation factor [28]. Ceramic materials are regarded as effective filler materials for applications in energy storage because of their high dielectric constant.

Only a perfect dielectric substance can displace charges, causing polarization. In a standard capacitor setup, where the voltage is carried by the current by a phase angle of $\pi/2$, an alternating E is applied to an ideal dielectric, which absorbs no power and has no dielectric loss ($\tan \delta = 0$). (i.e., 90°). Dielectric losses are linked to electrical polarization, ion vibration, migration, and deformation. Dielectric loss in ceramics is generally caused by the movement of ions [29]. We find that tan δ is significantly affected by both frequency and temperature. Compared to dielectric ceramics, polymer dielectrics are preferable for transistor applications because to their lower K and tan values. Their low leakage current and strong breakdown strength make them excellent for nanoscale layers that enable smaller electrical devices. Table 15.1 lists polymer breakdown strengths. Ionic polarization caused by asymmetric migration of the core metal cation in the crystal lattice gives ceramics high dielectric constants. Polymers have ionic, electronic, and orientation polarization, which dictates their dielectric constants [30, 31]. Dielectrics often have at least one type of polarization mechanism present, depending on the material's makeup and the field to which it is subjected.

The dielectric properties of polymer nanocomposites are affected by a wide range of factors, such as filler material type, concentration, size, and form, and the polymer matrix itself. Silicon carbide (SiC), molybdenum sulphide (MoS_2), zinc oxide (ZnO), and titanium dioxide (TiO_2), are some of the semiconductive fillers that have been studied for high energy density uses. There is a lot of hope for these fillers as replacements for high voltage polymer dielectric uses [26]. Polymer nanocomposites have had their dielectric and energy-storage properties studied and enhanced by including dielectric ceramics with perovskite structures as filler materials.Table 15.2 lists several polymers and ceramic fillers, along with their corresponding dielectric constants.

15.3 VARIOUS ELECTRICAL PARAMETERS OF POLYMER

The dielectric constant measures how much electrical energy a dielectric can store in an electric field and utilize to polarize it. High-K polymer composites are made using two methods. The percolation

TABLE 15.1
Commonly Used Polymers' Dielectric Strengths [22]

Polymer used	Dielectric strength (V/μm)
Polyvinylidene fluoride	10.2
Epoxy Resin	25–45
Polystyrene	200
Polyethylene	200
Polycarbonate	252
Polyimide	280
Polyester	300

TABLE 15.2
Commonly Used Polymer and Filler for Energy Storage Application

Polymer	Dielectric Constant	Filler	Dielectric Constant
Polystyrene	2.6	ZrO_2	25
Polynapthalene	2.2	TiO_2	80
Silsesquioxane	2–8–3.0	$SrTiO_3$	2000
Poly(arylene ether)	2.9	$BaTiO_3$	1700
Fluorinated polyimide	2.6–2.8	PLZT	2590
Nonfluorinated aromatic polyimide	3.2–3.6	$CaCu_3Ti_4O_{12}$	~60000
Poly(ether ketone ketone)	~3.5	$La_{1.8}Sr_{0.2}NiO_4$	~100000

hypothesis [28] mixes conductive filler with an insulating polymer matrix to create composites with a power law dielectric constant (kc) [29].

$$kc \propto (pc - p) - s, pc > p \tag{15.1}$$

Critical exponent s and percolation threshold pc. Equation (15.1) indicates that the dielectric constant of a percolative composite can grow dramatically as the volume percent of the conductive filler approaches the percolation threshold. Even with a filler percentage below the percolation threshold, this composite's breakdown strength is affected by the matrix and filler's severe electrical mismatch [30]. Duxbury and Beale devised the breakdown strength model for dielectrics with conductive particles [31], where L is the dielectric diameter.

$$Eb \approx \frac{(pc - p)v}{\ln(L)} \tag{15.2}$$

Equation (15.2) indicates that conductive particle-filled dielectrics have negligible breakdown strength when p is near to pc. Interfacial polarization and high leakage currents also provide conductive filler composites, and exceptional frequency-dependent dielectric properties. Conductive filler composites have substantial dielectric loss [32–38]. Based on the shape, the dielectric constant, and orientation of the filler particles, volume fraction ratio between the matrix and filler, and the filler/matrix interaction, one or more of the equations can estimate the dielectric constant of composites [39–42]. Numerous investigations show that high-K composite materials prefer higher dielectric constant particles.

15.3.1 Dielectric Loss and Electrical Conductivity

Dielectric loss is dielectric power loss. Relaxation and electrical conduction produce dielectric loss in a single dielectric. The dielectric constant/conductivity mismatch causes interfacial polarization, which increases dielectric loss [43, 44]. Avoid dielectric loss in energy storage and dielectric applications. The following elements affect low-loss polymer composite preparation. Integrating low-loss ceramic particles with a low-loss polymer matrix, increasing interfacial compatibility since an unsuitable interface will create microvoids where charge carriers have increased mobility when an electric field is applied, and minimizing the introduction of contaminants, especially ions that can pass through the dielectric. Polymer composite dielectric loss depends on filler placement and alignment. Figure 15.2 shows Randall's study on high-K polymer composite loss. Composite dielectric loss rises with the electric field. The 1–3 parallel (z-aligned) composites had the largest dielectric

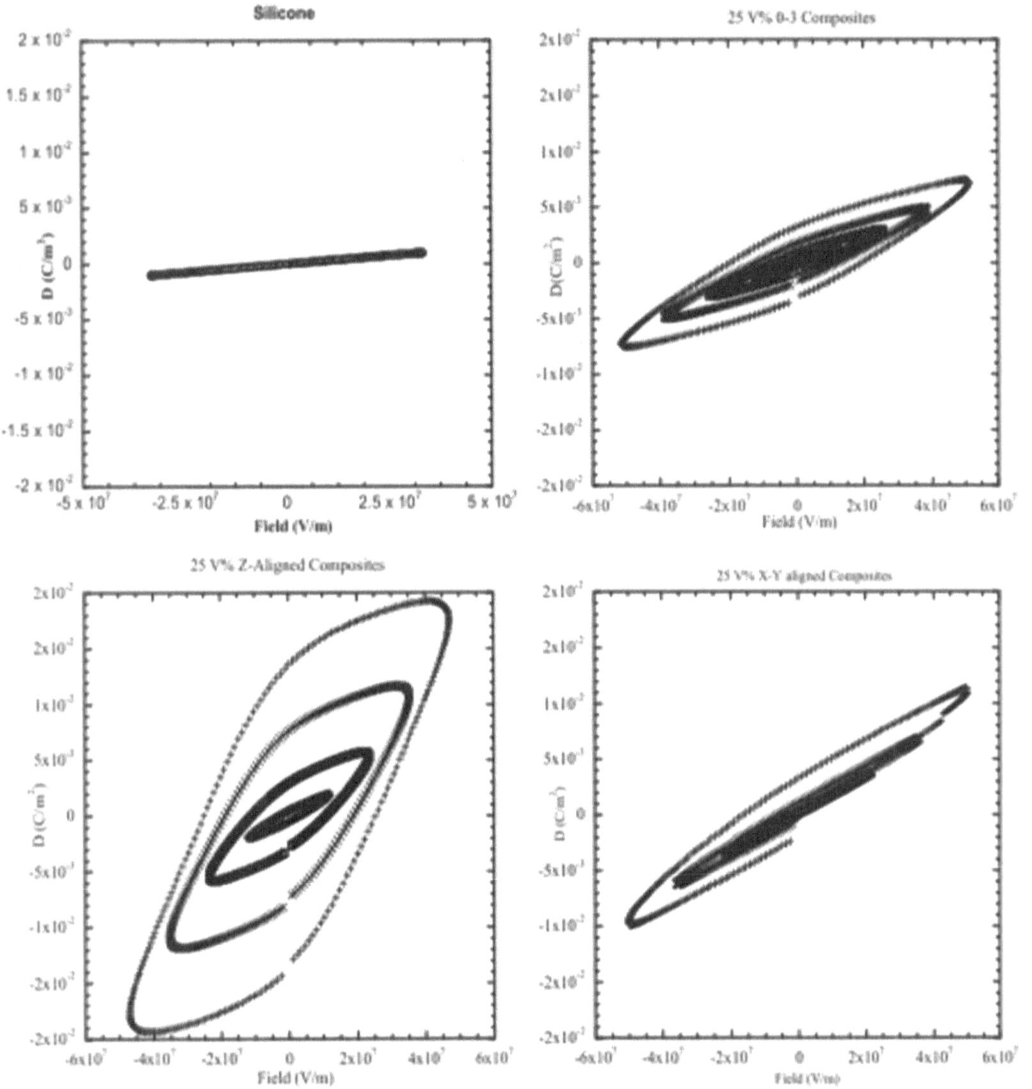

FIGURE 15.2 Shows the D-E loops of polydimethylsiloxane (PDMS) and its composites with 25 vol% $BaTiO_3$ NPs as a function of the electric field for pure PDMS, 0–3 composites, 1–3 (z-aligned), and 1–3 (x-y-aligned) composites. (Reprinted from ref [46] with permission).

loss due to their upright bipolar loops, whereas the 1–3 perpendicular (x-y aligned) composites had the lowest. Local electric field amplification and high conduction loss increase dielectric losses in Z-aligned composites [45]. Higher particle chain spacing in X-y-aligned composites could reduce dielectric losses [46].

15.3.2 Dielectric Nonlinearity

Dielectric substances store electrical energy (U) based on the applied electric field and dielectric displacement (D) [47].

$$U = \int_{0}^{E_{BD}} E.dD = \varepsilon_0 \int_{0}^{E_{BD}} E.k(E)dE \tag{15.3}$$

Where $\varepsilon_0 = 8.85 \times 10^{-12}$ F m-1 is the vacuum dielectric constant and k(E) depends on the electric field.

Dielectric breakdown field (Ebd) constrained fields reduce energy loss in linear dielectrics. Dielectric nonlinearity increases energy loss (electric field-dependent electric displacement area). Due to nonlinearity, dielectric material energy storage density is best calculated by integrating the charge curve over the polarization axis [48]. Ferroelectric-filled composites display dielectric nonlinearity under high electric fields [49–52]. Unlike ferroelectric ceramics, dielectric polymer composites with high-K filler increase in dielectric constant with the electric field, decreasing partial discharge in power electronic modules [50] and electric stress in high voltage bushings [53] and cable terminations[52, 54]. Robertson et al. reported dielectric nonlinearity in acrylic resin/$BaTiO_3$ composites [111, 116]. Due to their bigger percentage dielectric constant variation with the electric field, composites containing little $BaTiO_3$ exhibit higher nonlinearity. Larger $BaTiO_3$ particles create composites with larger dielectric constants [51]. Recent study recommends controlling power equipment's electric field distribution by simultaneously employing the non-linear electrical conductivity and dielectric constant [55–57].

15.3.3 Breakdown Strength

Voltage frequency and control, sample thickness, ambient medium, humidity electrode geometry and contact, temperature, and all impact dielectric breakdown strength [58, 59–61]. If these factors are controlled, thermal breakdown, intrinsic electric breakdown, and electromechanical breakdown cause a short-term solid dielectric breakdown [62]. A low-temperature electron avalanche causes intrinsic electric breakdown. Electrons might come via electrode injection emission through impurities or collision ionization. Thermal breakdown happens when joule heating and dielectric loss produce far more heat than conduction and convection loss. The current and thermal continuity equation describes heat generation-dissipation instability.

$$C_v \frac{dT}{dt} - div(\lambda.gradT) = \sigma E^2 \tag{15.4}$$

Where Cv and T are specific heat per unit volume and temperature. Equation (15.4) says breakdown strength diminishes exponentially with temperature. Electromechanical breakdown is caused by electrostatic compressive forces that attract negative and positive charges separated by the dielectric deforming elastic organic materials or breaking inorganic materials. Electromechanical failure can be assessed.

$$\frac{\varepsilon_0 k V^2}{2d^2} = Yln\left(\frac{d_0}{d_r}\right) \tag{15.5}$$

$$E_b = 0.6\left(\frac{y}{\varepsilon_0\, k}\right)^{1/2} \tag{15.6}$$

Where d_o and d are the specimen's original and lowered thicknesses and Y is Young's modulus. The breakdown strength of a heterogeneous material, such as a polymer composite, depends on several variables, including filler size, electrical contrast, microstructure (e.g., filler distribution and dispersion), and shape (e.g., flakes, particles, fibers), mechanical property (e.g., Young's modulus), filler/matrix interface, thermal conductivity, and defects (e.g., pores, voids,). The Joule heating energy density exponentially increases with temperature, accelerating breakdown strength. Heating breaks down polymer nanocomposites; however, reducing electrical conductivity is more effective than increasing thermal conductivity [63]. Several dielectric fillers can boost polymer and polymer composite breakdown strength by blocking high-energy charge carriers, dispersing them, or increasing mechanical characteristics [64, 65, 66]. 2D boron nitride nanosheets (BNNSs) can increase the breakdown strength of polymers like polyethylene (PE), poly(phenylene sulphide), and polypropylene, even at high temperatures (vinylidene fluoride-chlorotrifluoroethylene) (PVDFCTFE).

15.4 POLYMER-BASED HIGH-K DIELECTRICS

Dielectric polymers are monomer-based, high-molecular-mass organic molecules. The D-E loop and ferroelectric analyzer can distinguish antiferroelectric, relaxor ferroelectric, ferroelectric, and linear dielectric materials by dielectric properties and polymer polarity at high electric fields. Table 15.3 summarizes polymer dielectric properties [67]. Three structures filling, mixing, and multilayer describe their benefits and potential. Electric fields do not change linear dielectric constants. Polar and nonpolar linear polymers differ. Polystyrene (PS) and polypropylene (PP) are nonpolar, while poly(methyl methacrylate) and polyetherimide (PEI) are polar (PS). PC, poly(phenylene sulphide), poly(ethylene 2,6-naphthalate), and PET are capacitor dielectrics [68–73]. Most nonlinear polymers are ferroelectric. Ferroelectric polymers have high dielectric constants but limited breakdown energy storage and strength, making them unsuitable capacitor dielectrics [74]. Ferroelectric polymers' vast domains and highly linked dipoles shift dipole orientation, causing polarization hysteresis. Discharge energy density, dielectric loss, and charge-discharge efficiency decrease. Due to their simplicity, versatility, and low cost, high-K dielectrics have promising energy storage and dielectric applications. Polymer-based High-K dielectrics for inverters, humidity, energy, organic field-effect transistors (OFETs), electro-optics, and temperature sensors are being studied. High-K materials have a K value greater than 3.9 (SiO_2). Dielectric polymers form the matrix, while inorganic/organic fillers enhance polymer dielectric nanocomposites. Polymers have high breakdown strengths and HED, whereas fillers, especially dielectric ceramics, have high dielectric constants (K). Polymers and fillers affect dielectric characteristics [75].

The researchers first focused on improving breakdown strength and dielectric constant (ε_r) of the dielectric capacitor (Eb). Polymers with high breakdown strength and strong polariZation are difficult to produce. Thus, the key challenges in designing new high-performance polymer dielectrics are the close relationship between breakdown strength and dielectric constant from a scientific perspective and service life, from an engineering one. Many capacitor energy storage research studies have found that filled composite dielectrics, stacked composite dielectrics, and new polymers boost energy storage density [76–83].

Filled composite dielectrics are made by adding metal oxides, ceramics, and conductive elements to polymers. During filling, the composites' breakdown strength decreases, interface polarization improves, and the dielectric constant increases [84–87]. Filling determines interface area, requiring smaller fillers. Nanofillers outweigh micro fillers. It's hard to get enough inorganic filler for composites' dielectric constant. Inorganic metal oxides reduce polymer flexibility and processing [88–91]. Sandwich polymers make sandwich dielectrics. Functions between layers superimpose strong breakdown strength and high dielectric constant layers. The delamination action sets each layer's conductivity and dielectric constant. Despite the fact that the layered interface could be employed to reduce carrier migration and enhance insulation performance, polarization performance is not significantly enhanced [92–97].

15.5 POLYMER-BASED HIGH-K DIELECTRICS FOR ENERGY STORAGE APPLICATION

Some polymer materials with outstanding dielectric properties, like extremely high permittivity, strong breakdown field, and low dielectric loss at the operating voltages and frequencies, would be

TABLE 15.3
Various Polymer and Their Dielectric Properties

Polymer used	Dielectric Constant at 25°C (1kHz)	Loss tangent at 25°C (1kHz)	Glass transition temperature (°C)	Dielectric breakdown strength (MV/m)	Energy density (J/cm³)	Reference
Polycarbonate	~ 2.25	~0.0001	~150–160	528@150 °C	1.29@400MV/m 150_C and 10 Hz	[98–100]
Poly(arylene ether urea)	~ 4.7	~0.0015	>250	400@150°C	2.9 @400MV/m 150_C and 10 Hz	[101]
Poly(ether ether ketone)	~3.2	~0.0015	~150	290@150°C	0.5 @250MV/m 150_C and 10 Hz	[100, 102–103]
Poly(phthalazinone ether ketone)	~3.5	~0.0063	~250	470@25°C	2.1 @450MV/m 190_C and 10 Hz	[104]
Poly(ether ketone ketone)	~3.6	~0.003	155	468@70°C	2.37 @450MV/m 140_C and 10 Hz	[105]
poly(vinylidene fluoride)	10	0.04	-40	150–500 @25°C	14@500 MV/m 25_C and 10 Hz	[106]
Poly(vinylidene fluoride-hexafluoropropylene)	15	0.05	-90	~750@25°C	25 @700 MV/m 25_C and 10 Hz	[107]
Poly(methyl methacrylate)	~3	0.0418	107	240@25°C	1.5 @330 MV/m 25_C and 10 Hz	[108–109]

helpful as electrical and electronic technologies advance [110, 111–118]. Despite having a high breakdown strength, polymer materials often have low dielectric constants [119]. The polymeric chain architecture has been changed, polymers have been blended with other polymers, and conducting nanoparticles and ceramics have been dispersed in polymers to increase the intrinsic dielectric constants of polymeric material. Because of the structural alteration, various studies have discovered that polymers can significantly enhance their dielectric and energy storage performances [120]. In this section, we'd like to highlight a few successes that epitomize the outcomes of recent decades.

Fluoro-polymer functionalized RGO and polydopamine coated RGO (PDA-RGO) were prepared easily and ecologically (PF-PDA-RGO). The nanocomposites' low percolation threshold (fc = 1.06 vol percent) and good flexibility were due to the PF-PDA-RGO nanosheets' superior dispersion in a ferroelectric polymer matrix (P(VDF-HFP)) and strong interfacial adhesion. Unlike the RGO, P(VDF-HFP)/PF-PDA-RGO nanocomposites, with a low dielectric loss and high dielectric constant, performed best. P(VDF-HFP) nanocomposite with 1.0 vol percent PF-PDA-107.9 RGO's dielectric constant and 0.070 dielectric loss at 1000 Hz make it promising for dielectric applications [121]. Hanani et al. created high-K nanocomposites utilizing polylactic acid (PLA), a biodegradable polymer matrix, and various inorganic nanofillers with different morphologies and dielectric properties. D-E loops measured the energy storage of PLA-based nanocomposites with spherical, rod, and wire particles. Rod-like filler nanocomposites showed the greatest Wrec and Wtot, 1.80 J cm^{-3} and 3.63, respectively. Thus, PLA-based nanocomposites with high dielectric permittivity and high aspect ratio nanofillers store energy well [122]. In the chosen solvent, block copolymers containing terpyridyl-metal ion pendants self-assembled into nanostructures after ring-opening metathesis polymerization. Metal ions were added to polymer side chains to improve dielectric properties. These polymers have a higher dielectric constant in the broadband range of 100 Hz to 1 MHz, due to metal ion-induced electrical and interfacial polarization, and the self-assembled nanostructure reduces dielectric loss. Electronic and interfacial polarization from metal ions and nanostructure

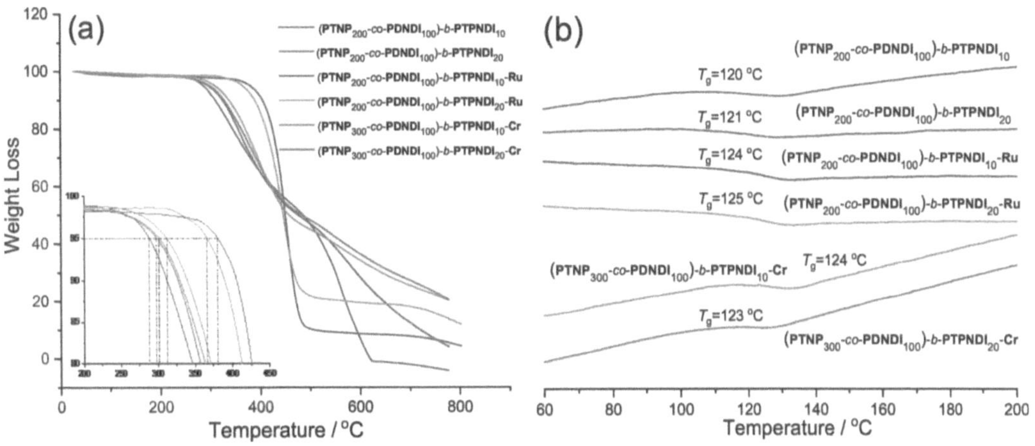

FIGURE 15.3 Dielectric constant (a) and dielectric loss (b). (Reprinted from ref [123] with permission.)

boosted energy storage density to 5.2 J cm^{-3} at 325 mV cm^{-1} and charge-discharge efficiency to 83.3%. Ru(II)-containing block copolymers have a high dielectric constant of 32.5. Figure 15.3a and b show the polymers' dielectric constant and dielectric loss [123].

Zhang et al. described an all-organic sandwich-structured composite with polyetherimide layers and a PMMA/PVDF mix layer. As temperature rises, PMMA increases the intermediate layer's permittivity faster than PVDF. XRD patterns for PMMA/PVDF sandwich-structured composites, single-layer composites, and pure PEI give 18.4 and 20.0 diffraction peaks on PVDF's (020) and (110) crystal planes. PVDF-phases' crystalline structure causes the (110) and (200) planes, which create the peak at 20.4 [124, 125]. The sandwich-structured dielectric composites and single-layer PMMA/PVDF composites have comparable XRD patterns as PMMA concentration increases. As PMMA percentage increases, PVDF crystallinity decreases in blended composites. Energy storage requires permittivity, which gauges polarization capacity. See Figure 15.4 for frequency-dependent permittivity (ε') and dielectric loss tests of sandwich-structure composites and blended PMMA/PVDF composites. In Figure 15.4(a), blended PMMA/PVDF composites decrease as PMMA content increases. Two factors impact the polymer's polarization in an electric field. The first is local dipole distribution and direction, while the second is polymer chain mobility and freedom. PMMA reduces the intermediate layer's crystallinity, accelerating permittivity increase with temperature. Electric field distribution must be sensitively self-adaptive to temperature change to prevent composite failure at high temperatures. In P-30 percent -P at 100 °C, Eb is 486.05 MV/m and polarization is 4.48 C/cm^2. U_d is 8.65 J/cm^3, higher than at normal temperature. This finding is crucial for high-performance polymer dielectric energy storage research [126].

Pristine PVDF is an electroactive flexible polymer having dielectric, ferroelectric, and piezoelectric properties. Mixing filler with pure PVDF can increase its device application functionality. Hafnia-Zirconia (HZ) ceramic filler ($0.19HfO_2$–$0.81ZrO_2$) affects the dielectric, micro structural, electrical characteristics, and structural of pure PVDF matrix. Flexible polymer-ceramic composites are made via environmentally friendly and cost-effective solution casting. Room-temperature XRD spectra demonstrate flexible composite phase formation. Filler increases crystallinity. 5% ceramic filler offers the greatest -phase crystallinity. As predicted, these two ceramics-polymer composites with 5 and 10 wt percent HZ ceramics filler had an enhanced dielectric constant of 53 and 20 at 100 Hz, due to polar-phase enrichment. Percolation modelling showed how filler content affects dielectric properties [127].

1-Ethyl-3-methylimidazolium methyl sulphate [EMIM][$MeSO_4$], an imidazolium-based ionic liquid (IL), affects the structural, dielectric, ion transport, and thermal characteristics of Chitosan (CS), poly(vinyl pyrrolidone (PVP), sodium iodide (NaI), and polyethylene glycol (PEG),

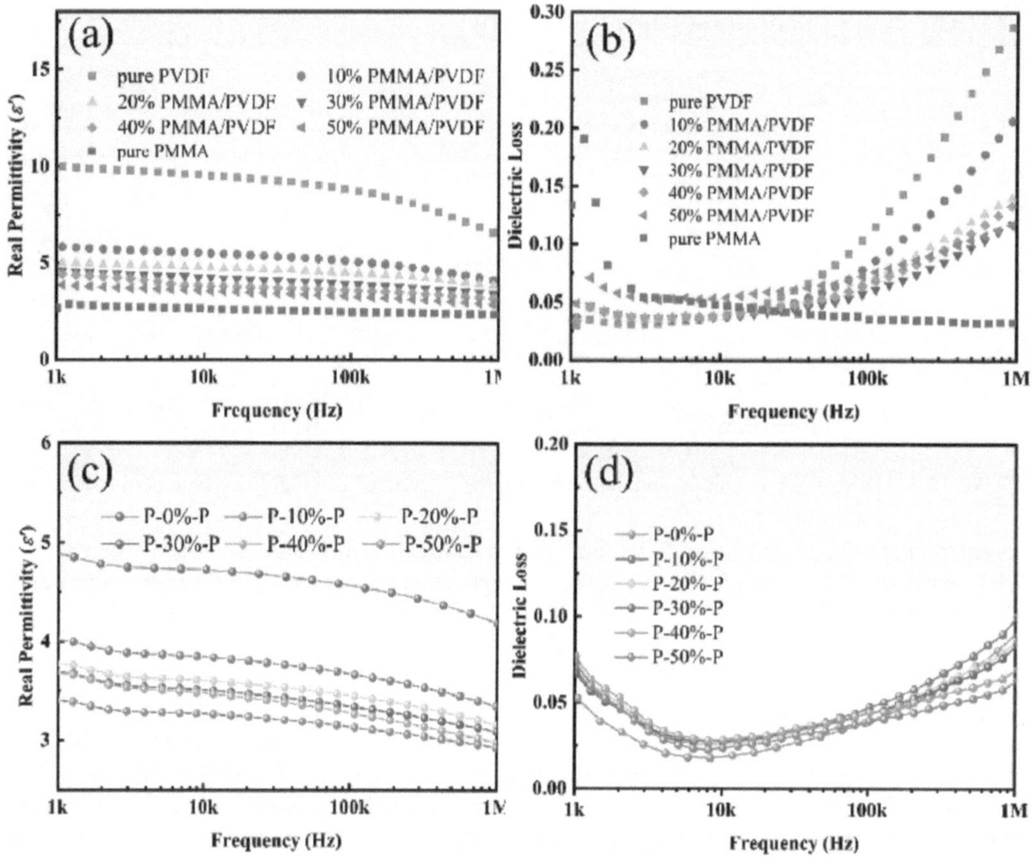

FIGURE 15.4 The ε' and dielectric loss for sandwich-structured dielectric composites (c) and (d) and PMMA/PVDF composites (a) and (b), respectively. (Reprinted from ref [126] with permission.)

bio-polymers. Solution casting created BPBE flicks. IL concentration promotes BPBE film amorphousness, according to XRD studies. TGA data shows IL affects BPBE film thermal stability. IL concentration lowers surface roughness according to optical microscopy. SEM analysis supported this result. All samples show Arrhenius-type conductivity fluctuation with temperature. Ea1 and Ea2 of the 50IL sample are 0.248 and 0.012 eV, respectively. Dielectric permittivity and electric modulus showed charge storage and conductivity relaxation. The ITN says ions drive the present BPBE system's conduction. LSV and CV tests show 50 IL's ESW varies from -3.13 V to +2.65 V [128].

Solution casting was used to make polymer nanocomposite (PNCs) films reinforced with solid-state $SrTiO_3$ nanoparticles (NPs). Figure 15.5 shows SEM images of pure PVA/CMC mix and varied $SrTiO_3$ NP loadings. Figure 15.5a displays the blend picture's uniformly rough, fissure-free surface. This snapshot supports Morsi et al. [129] and Saadiah [130]. In Figure 15.5b–d, $SrTiO_3$ NP filling changed the PVA/CMC blend's surface morphology. PNC with equally distributed $SrTiO_3$ NPs (1.0 wt%) improves surface roughness Figure 15.5b. Figure 15.5c indicates some aggregation at 3.0 wt% filling. The nanocomposite (6 wt%) sample in Figure. 15.5d shows that the $SrTiO_3$ NPs have a concentrated structure and are agglomerated, spherical, and randomly dispersed, with varying sizes over the surface of the PVA/CMC mix.

The white dots indicate the clusters of NPs due to the high filling level. Following the filling procedure, the pure (PVA/CMC) must undergo a structural change, as well as a change in its dielectric characteristics. This is demonstrated by the variations in the filled films' dielectric loss ($\varepsilon' = \sigma'\varepsilon_0\omega$) and dielectric constant (ε' = cd εoa) curves (Figure. 15.6a, 6b). Figure 15.6 shows how the

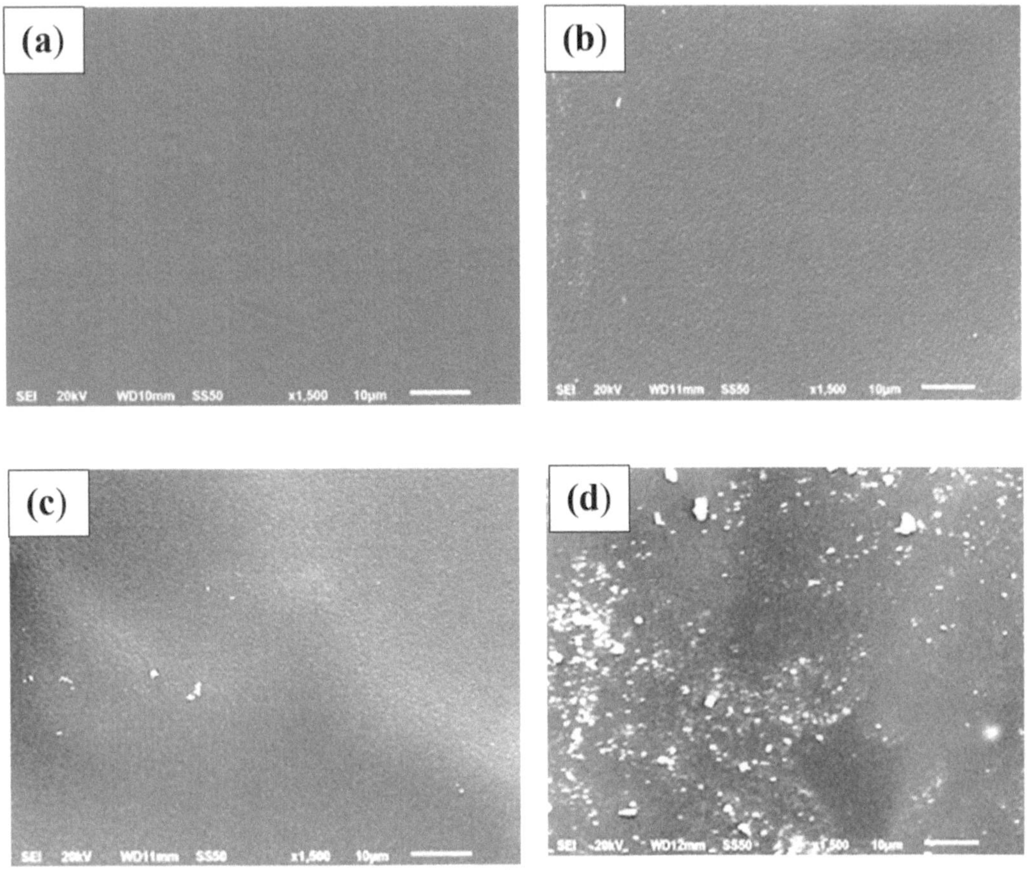

FIGURE 15.5 Shows SEM image of (a) pristine PVA/CMC blend, (b) 1.0% $SrTiO_3$, (c) 3.0% SrTiO3, and 6.0% $SrTiO_3$. (Reprinted from ref [131] with permission.)

PVA/polar CMC's nature, together with $SrTiO_3$ NPs' efficiency, increases its capacity to store energy. The increase in resulting dipoles aligned with the field brought on by the electrostatic interaction between the polar groups of the host matrix and the $SrTiO_3$ NPs can be used to explain this [131].

Interfacial characteristics may alter dielectric polymer nanocomposites' electrical energy storage. This section examines the interfacial region's polarity function. $BaTiO_3$ (BT) nanoparticles are coated with three polar polymers: polyglycidyl methacrylate, polymethyl methacrylate (PMMA), and polymethylsulfonyl ethyl methacrylate (PMSEMA). Shell polymer polarity affects dielectric polymer nanocomposite energy storage and dielectric. PMSEMA@BT, the most polar shell polymer, reduces nanocomposites' breakdown strength and dielectric loss, limiting energy storage. PMMA@BT, the least polar shell polymer, may shatter nanocomposites more violently. PMMA@BT stores the most electricity. D-E loops investigate nanocomposites' energy storage and electric polarization. Nanoparticle loading enhances nanocomposite electrical displacement because BT has a much higher dielectric constant than P(VDF-HFP). The nanocomposite of BT and a low-polarity shell (PMMA) exhibits lower dielectric loss, higher breakdown strength, efficiency, and discharged energy density than the others. These findings show that coupling contact between the shell and polymer matrix improves dielectric nanocomposite nanocomposites' energy storage and may influence future design and fabrication [132].

Cheng and his coworker found high-temperature energy storage in a polymer dielectric sandwiched between layers of medium-electrical-conductivity, medium-dielectric-constant, and medium-bandgap nanoscale deposition. We demonstrate that the dielectric constant should be

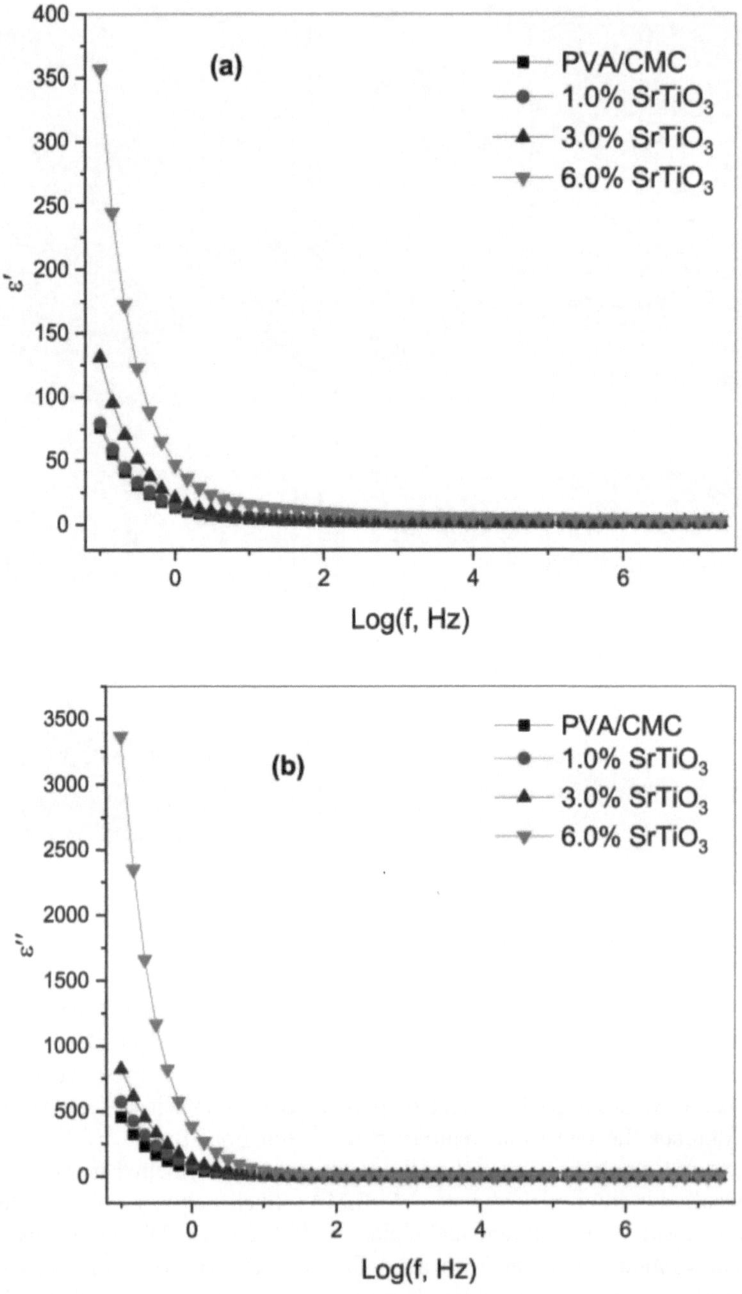

FIGURE 15.6 Frequency variation of (a) ε' and (b) ε'' for PVA/CMC- $SrTiO^3$ PNC films at 300 K. (Reprinted from ref [131] with permission.)

addressed while selecting deposition materials, since it can boost performance. Energy densities above 90% at 150 °C and 200 °C are 2.8 J cm^3 and 3.5 J cm^3, respectively, which are 133%and 67% greater than the best values of polymer dielectrics covered with low-dielectric-constant insulating materials [133].

Adding nanosized -Al_2O_3 with a cubic defect spinel structure and α-Al_2O_3 with a wider bandgap to a heat-resistant dielectric polymer creates two nanocomposites. The TSDC study discovered that nanoscale -Al_2O_3 with a unique cubic defect spinel structure may generate deeper traps in the polyetherimide (PEI) matrix, lowering leakage current and enhancing efficiency, energy storage density, and breakdown strength. The -Al_2O_3/PEI-0.3% nanocomposite can attain 450 MV/m with over 90% efficiency and 550 MV/m with 4.2 J/m^3. Nanomaterials with defective spinel topologies may boost polymer dielectric energy storage in hazardous environments [134].

Polyimide (PI) composite films with surface-inserted strontium oxide (SrO) nanoparticles were made in-situ. SrO has not been shown to protect polymers' electrical properties. Schottky, ISPD, PEA, and finite element simulations were used to study the breakdown and conduction of the PI film covered with SrO layer at high fields. The SrO nanolayer increases the energy level of localized states of PI film and decreases surface molecular conformations. The SrO layer boosts the breakdown strength, dielectric stability, and energy storage of PI films from 30 to 250°C [135].

15.6 CONCLUSION AND OUTLOOK

Given the drive toward downsizing and high performance in electronic systems, high-energy-storage dielectric materials are essential. Dielectric polymers are recommended for smart electrical devices and pulsed power due to their energy storage and flexible sensing potential. Polymer coupling with ceramics, carbon compounds, or conductive metals improves energy storage efficiency and mechanical qualities. However, energy loss control is limited in most dielectric polymer composites. Researchers are interested because multilayer thin film technology in multipolarization systems reduces polymer material dielectric loss.

High-energy-density polymer dielectrics for applications and energy storage have improved, but many obstacles remain. First, nanofillers help nanocomposites retain energy, but not enough to utilize electrically. 0D nanoparticles with high packing volume fraction increase er but decrease breakdown strength and capacitor energy density when employed as nanofillers. High-aspect ratio 1D nanofibers have the maximum polarizability and breakdown strength, but they also lose energy and have worse charge-discharge efficiency. 2D nanosheets outperform 0D and 1D nanofillers in breakdown strength and dielectric loss. Another issue with 2D fillers is their high conductivity, poor polarization, or excessive thickness. As a result, more optimization of nanocomposites' composition is currently required to achieve outstanding energy storage capability. The developments are also hampered by the nanocomposite films preparation techniques. Therefore; there is still room for improvement in the preparation techniques for applications. Overall, tight multidisciplinary collaboration between chemists, materials scientists, mechanical engineers, and other specialists is necessary to further the development of polymer dielectrics.

REFERENCES

1. Hu, Jin, Shufen Zhang, and Bingtao Tang. "2D filler-reinforced polymer nanocomposite dielectrics for high-K dielectric and energy storage applications." *Energy Storage Materials* 34 (2021): 260–281.
2. Li, Qi, Guangzu Zhang, Feihua Liu, Kuo Han, Matthew R. Gadinski, Chuanxi Xiong, and Qing Wang. "Solution-processed ferroelectric terpolymer nanocomposites with high breakdown strength and energy density utilizing boron nitride nanosheets." *Energy & Environmental Science* 8, no. 3 (2015): 922–931.
3. Thakur, Vijay Kumar, and Raju Kumar Gupta. "Recent progress on ferroelectric polymer-based nanocomposites for high energy density capacitors: synthesis, dielectric properties, and future aspects." *Chemical Reviews* 116, no. 7 (2016): 4260–4317.
4. Hao, Xihong. "A review on the dielectric materials for high energy-storage application." *Journal of Advanced Dielectrics* 3, no. 01 (2013): 1330001.
5. Zhang, Haibo, Mohsin Ali Marwat, Bing Xie, Malik Ashtar, Kai Liu, Yiwei Zhu, Ling Zhang, Pengyuan Fan, Chanatip Samart, and Zuo-guang Ye. "Polymer matrix nanocomposites with 1D ceramic nanofillers for energy storage capacitor applications." *ACS Applied Materials & Interfaces* 12, no. 1 (2019): 1–37.
6. Yang, Letao, Xi Kong, Fei Li, Hua Hao, Zhenxiang Cheng, Hanxing Liu, Jing-Feng Li, and Shujun Zhang. "Perovskite lead-free dielectrics for energy storage applications." *Progress in Materials Science* 102 (2019): 72–108.

7. Sun, Zixiong, Zhuo Wang, Ye Tian, Ge Wang, Wen Wang, Mengdie Yang, Xiaoying Wang, Fanghui Zhang, and Yongping Pu. "Progress, outlook, and challenges in lead-free energy-storage ferroelectrics." *Advanced Electronic Materials* 6, no. 1 (2020): 1900698.
8. Palneedi, Haribabu, Mahesh Peddigari, Geon-Tae Hwang, Dae-Yong Jeong, and Jungho Ryu. "High-performance dielectric ceramic films for energy storage capacitors: progress and outlook." *Advanced Functional Materials* 28, no. 42 (2018): 1803665.
9. Qiu, Yu, Elric Zhang, Roshan Plamthottam, and Qibing Pei. "Dielectric elastomer artificial muscle: materials innovations and device explorations." *Accounts of Chemical Research* 52, no. 2 (2019): 316–325.
10. Brochu, Paul, and Qibing Pei. "Dielectric elastomers for actuators and artificial muscles." In *Electroactivity in polymeric materials*, pp. 1–56. Springer, Boston, MA, 2012.
11. Huang, Houbing, Guangzu Zhang, Xingqiao Ma, Deshan Liang, Jianjun Wang, Yang Liu, Qing Wang, and Long-Qing Chen. "Size effects of electrocaloric cooling in ferroelectric nanowires." *Journal of the American Ceramic Society* 101, no. 4 (2018): 1566–1575.
12. Qian, Jianfeng, Penghao Hu, Chen Liu, Jianyong Jiang, Zhenkang Dan, Jing Ma, Yuanhua Lin, Ce-Wen Nan, and Yang Shen. "High electrocaloric cooling power of relaxor ferroelectric BaZrxTi1–xO3 ceramics within broad temperature range." *Science Bulletin* 63, no. 6 (2018): 356–361.
13. Weyland, Florian, Thorsten Eisele, Sebastian Steiner, Till Froemling, George A. Rossetti Jr, Juergen Roedel, and Nikola Novak. "Long term stability of electrocaloric response in barium zirconate titanate." *Journal of the European Ceramic Society* 38, no. 2 (2018): 551–556.
14. Huang, Xingyi, Bin Sun, Yingke Zhu, Shengtao Li, and Pingkai Jiang. "High-K polymer nanocomposites with 1D filler for dielectric and energy storage applications." *Progress in Materials Science* 100 (2019): 187–225.
15. Huang, Xingyi, and Pingkai Jiang. "Core–shell structured high-K polymer nanocomposites for energy storage and dielectric applications." *Advanced Materials* 27, no. 3 (2015): 546–554.
16. Thakur, Vijay Kumar, and Raju Kumar Gupta. "Recent progress on ferroelectric polymer-based nanocomposites for high energy density capacitors: synthesis, dielectric properties, and future aspects." *Chemical Reviews* 116, no. 7 (2016): 4260–4317.
17. Yao, Zhonghua, Zhe Song, Hua Hao, Zhiyong Yu, Minghe Cao, Shujun Zhang, Michael T. Lanagan, and Hanxing Liu. "Homogeneous/inhomogeneous-structured dielectrics and their energy-storage performances." *Advanced Materials* 29, no. 20 (2017): 1601727.
18. Li, Qi, Fang-Zhou Yao, Yang Liu, Guangzu Zhang, Hong Wang, and Qing Wang. "High-temperature dielectric materials for electrical energy storage." *Annual Review of Materials Research* 48 (2018): 219–243.
19. Liu, Yang, Haibibu Aziguli, Bing Zhang, Wenhan Xu, Wenchang Lu, J. Bernholc, and Qing Wang. "Ferroelectric polymers exhibiting behaviour reminiscent of a morphotropic phase boundary." *Nature* 562, no. 7725 (2018): 96–100.
20. Johnson, R. Wayne, John L. Evans, Peter Jacobsen, James R. Thompson, and Mark Christopher. "The changing automotive environment: high-temperature electronics." *IEEE transactions on electronics packaging manufacturing* 27, no. 3 (2004): 164–176.
21. Li, Qi, Kuo Han, Matthew Robert Gadinski, Guangzu Zhang, and Qing Wang. "High energy and power density capacitors from solution-processed ternary ferroelectric polymer nanocomposites." *Advanced Materials* 26, no. 36 (2014): 6244–6249.
22. Chen, Qin, Yang Shen, Shihai Zhang, and Q. M. Zhang. "Polymer-based dielectrics with high energy storage density." *Annual Review of Materials Research* 45, no. 1 (2015): 433–458.
23. Huan, Tran Doan, Steve Boggs, Gilbert Teyssedre, Christian Laurent, Miko Cakmak, Sanat Kumar, and Rampi Ramprasad. "Advanced polymeric dielectrics for high energy density applications." *Progress in Materials Science* 83 (2016): 236–269.
24. Kim, Se Hyun, Kihyon Hong, Wei Xie, Keun Hyung Lee, Sipei Zhang, Timothy P. Lodge, and C. Daniel Frisbie. "Electrolyte-gated transistors for organic and printed electronics." *Advanced Materials* 25, no. 13 (2013): 1822–1846.
25. Zhu, Lei. "Exploring strategies for high dielectric constant and low loss polymer dielectrics." *The Journal of Physical Chemistry Letters* 5, no. 21 (2014): 3677–3687.
26. Li, Qi, Kuo Han, Matthew Robert Gadinski, Guangzu Zhang, and Qing Wang. "High energy and power density capacitors from solution-processed ternary ferroelectric polymer nanocomposites." *Advanced Materials* 26, no. 36 (2014): 6244–6249.

27. Gerasimov, G. N., and L. I. Trakhtenberg. "Vapor deposited composite films consisting of dielectric matrix with metal/semiconductor nanoparticles." *Thin Films and Nanostructures* 34 (2007): 523–579.
28. Yang, Ke, Xingyi Huang, Liyuan Xie, Chao Wu, Pingkai Jiang, and Toshikatsu Tanaka. "Core–shell structured polystyrene/BaTiO3 hybrid nanodielectrics prepared by in situ RAFT polymerization: a route to high dielectric constant and low loss materials with weak frequency dependence." *Macromolecular Rapid Communications* 33, no. 22 (2012): 1921–1926.
29. Kong, L. B., Li, S. Zhang, T. S., Zhai, J. W., Boey, F. C., and Ma, J. "Electrically tunable dielectric materials and strategies to improve their performances." *Progress in Materials Science* 55, no. 8 (2010): 840–893.
30. Riggs, Brian C., Shiva Adireddy, Carolyn H. Rehm, Venkata S. Puli, Ravinder Elupula, and Douglas B. Chrisey. "Polymer nanocomposites for energy storage applications." *Materials Today: Proceedings* 2, no. 6 (2015): 3853–3863.
31. Naeem, Abdul, Asad Mahmood, Yaseen Iqbal, Asad Ullah, Tahira Mahmood, and Muhammad Humayun. "Dielectric and impedance spectroscopic studies on (Ba0. 5Sr0. 5) Mnx (Ti0. 95Fe0. 05) 1– xO3 ceramics synthesized by using sol–gel method." *Journal of Alloys and Compounds* 645 (2015): 290–296.
32. Barber, Peter, Shiva Balasubramanian, Yogesh Anguchamy, Shushan Gong, Arief Wibowo, Hongsheng Gao, Harry J. Ploehn, and Hans-Conrad Zur Loye. "Polymer composite and nanocomposite dielectric materials for pulse power energy storage." *Materials* 2, no. 4 (2009): 1697–1733.
33. Jia, Qingchao, Xingyi Huang, Guanyao Wang, Jinchao Diao, and Pingkai Jiang. "MoS2 nanosheet superstructures based polymer composites for high-dielectric and electrical energy storage applications." *The Journal of Physical Chemistry C* 120, no. 19 (2016): 10206–10214.
34. Nan, Ce-Wen. "Physics of inhomogeneous inorganic materials." *Progress in Materials Science* 37, no. 1 (1993): 1–116.
35. Nan, Ce-Wen, Yang Shen, and Jing Ma. "Physical properties of composites near percolation." *Annual Review of Materials Research* 40, no. 1 (2010): 131–151.
36. Coppard, R. W., J. Bowman, L. A. Dissado, S. M. Rowland, and R. T. Rakowski. "The effect of aluminium inclusions on the dielectric breakdown of polyethylene." *Journal of Physics D: Applied Physics* 23, no. 12 (1990): 1554.
37. Beale, Paul D., and Philip M. Duxbury. "Theory of dielectric breakdown in metal-loaded dielectrics." *Physical Review B* 37, no. 6 (1988): 2785.
38. He, Linxiang, and Sie Chin Tjong. "High dielectric permittivity and low loss tangent of polystyrene incorporated with hydrophobic core–shell copper nanowires." *RSC Advances* 5, no. 48 (2015): 38452–38459.
39. Brosseau, Christian. "Generalized effective medium theory and dielectric relaxation in particle-filled polymeric resins." *Journal of Applied Physics* 91, no. 5 (2002): 3197–3204.
40. Calame, J. P. "Finite difference simulations of permittivity and electric field statistics in ceramic-polymer composites for capacitor applications." *Journal of Applied Physics* 99, no. 8 (2006): 084101.
41. Sihvola, Ari H., and Jin Au Kong. "Effective permittivity of dielectric mixtures." *IEEE Transactions on Geoscience and Remote Sensing* 26, no. 4 (1988): 420–429.
42. Zhang, L. I. N., and Z-Y. Cheng. "Development of polymer-based 0–3 composites with high dielectric constant." *Journal of Advanced Dielectrics* 1, no. 04 (2011): 389–406.
43. Zhu, Lei. "Exploring strategies for high dielectric constant and low loss polymer dielectrics." *The journal of physical chemistry letters* 5, no. 21 (2014): 3677–3687.
44. Fattoum, A., F. Gmati, N. Bohli, M. Arous, and A. Belhadj Mohamed. "Effects of the matrix molecular weight on conductivity and dielectric relaxation in plasticized polyaniline/polymethylmethacrylathe blends." *Journal of Physics D: Applied Physics* 41, no. 9 (2008): 095407.
45. Liu, Suting, Ming Tian, Liqun Zhang, Yonglai Lu, Tung W. Chan, and Nanying Ning. "Tailoring dielectric properties of polymer composites by controlling alignment of carbon nanotubes." *Journal of Materials Science* 51, no. 5 (2016): 2616–2626.
46. Tomer, V., and C. A. Randall. "High field dielectric properties of anisotropic polymer-ceramic composites." *Journal of Applied Physics* 104, no. 7 (2008): 074106.
47. Chen, Qin, Yang Shen, Shihai Zhang, and Q. M. Zhang. "Polymer-based dielectrics with high energy storage density." *Annual Review of Materials Research* 45, no. 1 (2015): 433–458.
48. Huang, X., Sun, B., Zhu, Y., Li, S., Jiang, P., High-K Polymer Nanocomposites with 1DFiller for Dielectric and Energy Storage Applications, *Progress in Materials Science* (2018), 187–225.

49. Varlow, B. R., and Ke Li. "Non-linear AC properties of filled resins." *IEE Proceedings-Science, Measurement and Technology* 150, no. 2 (2003): 75–82.
50. Wang, Ningyan, Ian Cotton, Jeff Robertson, Sarah Follmann, Kim Evans, and David Newcombe. "Partial discharge control in a power electronic module using high permittivity non-linear dielectrics." *IEEE Transactions on Dielectrics and Electrical Insulation* 17, no. 4 (2010): 1319–1326.
51. Robertson, J., and B. R. Varlow. "Non-linear ferroelectric composite dielectric materials." *IEEE Transactions on Dielectrics and Electrical Insulation* 12, no. 4 (2005): 779–790.
52. Strumpler, Ralf, Jakob Rhyner, Felix Greuter, and Petra Kluge-Weiss. "Nonlinear dielectric composites." *Smart Materials and Structures* 4, no. 3 (1995): 215.
53. Paredes-Olguin, M., Carlos Gomez-Yañez, Fermin Pascual Espino-Cortes, and E. Ramirez. "Electric stress grading on bushings of combined instrument transformers using high permittivity polymeric composites." *IEEE Transactions on Dielectrics and Electrical Insulation* 20, no. 6 (2013): 2335–2342.
54. Ye, Hanyu, Tobias Fechner, Xianzhang Lei, Yi Luo, Mingyu Zhou, Zhengyi Han, Haitian Wang, Qikai Zhuang, Ruoyu Xu, and Duo Li. "Review on HVDC cable terminations." *High Voltage* 3, no. 2 (2018): 79–89.
55. Robertson, J., and D. A. Hall. "Nonlinear dielectric properties of particulate barium titanate–polymer composites." *Journal of Physics D: Applied Physics* 41, no. 11 (2008): 115407.
56. Yang, Xiao, Xiaolei Zhao, Qi Li, Jun Hu, and Jinliang He. "Nonlinear effective permittivity of field grading composite dielectrics." *Journal of Physics D: Applied Physics* 51, no. 7 (2018): 075304.
57. He, Jinliang, Jingcheng Xie, and Jun Hu. "Progress of nonlinear polymer composites for improving nonuniform electrical fields." *High Voltage Engineering* 40, no. 3 (2014): 637–647.
58. Coppard, R. W., J. Bowman, L. A. Dissado, S. M. Rowland, and R. T. Rakowski. "The effect of aluminium inclusions on the dielectric breakdown of polyethylene." *Journal of Physics D: Applied Physics* 23, no. 12 (1990): 1554.
59. Huang, Xingyi, Liyuan Xie, Pingkai Jiang, Genlin Wang, and Yi Yin. "Morphology studies and ac electrical property of low density polyethylene/octavinyl polyhedral oligomeric silsesquioxane composite dielectrics." *European Polymer Journal* 45, no. 8 (2009): 2172–2183.
60. Ieda, M., M. Nagao, and M. Hikita. "High-field conduction and breakdown in insulating polymers. Present situation and future prospects." *IEEE Transactions on Dielectrics and Electrical Insulation* 1, no. 5 (1994): 934–945.
61. Li, Shengtao, Guilai Yin, George Chen, Jianying Li, Suna Bai, Lisheng Zhong, Yunxa Zhang, and Qingquan Lei. "Short-term breakdown and long-term failure in nanodielectrics: A review." *IEEE Transactions on Dielectrics and Electrical Insulation* 17, no. 5 (2010): 1523–1535.
62. Laghari, Javaid R., and W. James Sarjeant. "Energy-storage pulsed-power capacitor technology." *IEEE Transactions on Power Electronics* 7, no. 1 (1992): 251–257.
63. Shen, Zhong-Hui, Jian-Jun Wang, Jian-Yong Jiang, Yuan-Hua Lin, Ce-Wen Nan, Long-Qing Chen, and Yang Shen. "Phase-field model of electrothermal breakdown in flexible high-temperature nanocomposites under extreme conditions." *Advanced Energy Materials* 8, no. 20 (2018): 1800509.
64. Li, Qi, Lei Chen, Matthew R. Gadinski, Shihai Zhang, Guangzu Zhang, Haoyu U. Li, Elissei Iagodkine et al. "Flexible high-temperature dielectric materials from polymer nanocomposites." *Nature* 523, no. 7562 (2015): 576–579.
65. Li, Qi, Kuo Han, Matthew Robert Gadinski, Guangzu Zhang, and Qing Wang. "High energy and power density capacitors from solution-processed ternary ferroelectric polymer nanocomposites." *Advanced Materials* 26, no. 36 (2014): 6244–6249.
66. Li, Qi, Guangzu Zhang, Feihua Liu, Kuo Han, Matthew R. Gadinski, Chuanxi Xiong, and Qing Wang. "Solution-processed ferroelectric terpolymer nanocomposites with high breakdown strength and energy density utilizing boron nitride nanosheets." *Energy & Environmental Science* 8, no. 3 (2015): 922–931.
67. Cheng, Rui, Yifei Wang, Rujia Men, Zhipeng Lei, Jiancheng Song, Yuanyuan Li, and Meiqing Guo. "High-energy-density polymer dielectrics via compositional and structural tailoring for electrical energy storage." *Iscience* (2022): 104837.
68. Guan, Fang-xiao, Jing Wang, Ji-lin Pan, Qing Wang, and Lei Zhu. "Time and poling history dependent energy storage and discharge behaviors in poly (vinylidene fluoride-co-hexafluoropropylene) random copolymers." *Chinese Journal of Polymer Science* 29, no. 1 (2011): 65–80.
69. Luo, Jiaming, Jiale Mao, Wenjie Sun, Shuang Wang, Lei Zhang, Tian Liliang, Chen Yu, and Yonghong Cheng. "Research progress of all organic polymer dielectrics for energy storage from the classification of organic structures." *Macromolecular Chemistry and Physics* 222, no. 11 (2021): 2100049.

70. Singh, Maninderjeet, Ikeoluwa E. Apata, Saumil Samant, Wenjie Wu, Bhausaheb V. Tawade, Nihar Pradhan, Dharmaraj Raghavan, and Alamgir Karim. "Nanoscale strategies to enhance the energy storage capacity of polymeric dielectric capacitors: Review of recent advances." *Polymer Reviews* 62, no. 2 (2022): 211–260.
71. Mannodi-Kanakkithodi, Arun, Gregory M. Treich, Tran Doan Huan, Rui Ma, Mattewos Tefferi, Yang Cao, Gregory A. Sotzing, and Rampi Ramprasad. "Rational co-design of polymer dielectrics for energy storage." *Advanced Materials* 28, no. 30 (2016): 6277–6291.
72. Ho, Janet S., and Steven G. Greenbaum. "Polymer capacitor dielectrics for high temperature applications." *ACS Applied Materials & Interfaces* 10, no. 35 (2018): 29189–29218.
73. Tan, Daniel, Lili Zhang, Qin Chen, and Patricia Irwin. "High-temperature capacitor polymer films." *Journal of Electronic Materials* 43, no. 12 (2014): 4569–4575.
74. Li, Qi, and Qing Wang. "Ferroelectric polymers and their energy-related applications." *Macromolecular Chemistry and Physics* 217, no. 11 (2016): 1228–1244.
75. Mahmood, Asad, Abdul Naeem, and Tahira Mahmood. "High-K polymer nanocomposites for energy storage applications." *Polymer Dielectrics: Properties and Applications of* (2017): 23.
76. Li, He, Ding Ai, Lulu Ren, Bin Yao, Zhubing Han, Zhonghui Shen, Jianjun Wang, Long-Qing Chen, and Qing Wang. "Scalable polymer nanocomposites with record high-temperature capacitive performance enabled by rationally designed nanostructured inorganic fillers." *Advanced Materials* 31, no. 23 (2019): 1900875.
77. Jiang, Jianyong, Zhonghui Shen, Xingke Cai, Jianfeng Qian, Zhenkang Dan, Yuanhua Lin, Bilu Liu, Ce-Wen Nan, Longqing Chen, and Yang Shen. "Polymer nanocomposites with interpenetrating gradient structure exhibiting ultrahigh discharge efficiency and energy density." *Advanced Energy Materials* 9, no. 15 (2019): 1803411.
78. Li, Yushu, Yao Zhou, Yujie Zhu, Sang Cheng, Chao Yuan, Jun Hu, Jinliang He, and Qi Li. "Polymer nanocomposites with high energy density and improved charge–discharge efficiency utilizing hierarchically-structured nanofillers." *Journal of Materials Chemistry A* 8, no. 14 (2020): 6576–6585.
79. Luo, Suibin, Yanbin Shen, Shuhui Yu, Yanjun Wan, Wei-Hsin Liao, Rong Sun, and Ching-Ping Wong. "Construction of a 3D-BaTiO 3 network leading to significantly enhanced dielectric permittivity and energy storage density of polymer composites." *Energy & Environmental Science* 10, no. 1 (2017): 137–144.
80. Luo, Suibin, Junyi Yu, Shuhui Yu, Rong Sun, Liqiang Cao, Wei Hsin Liao, and Ching Ping Wong. "Significantly enhanced electrostatic energy storage performance of flexible polymer composites by introducing highly insulating ferroelectric microhybrids as fillers." *Advanced Energy Materials* 9, no. 5 (2019): 1803204.
81. Wang, Huan, Haian Xie, Shan Wang, Zhaodongfang Gao, Chuanbin Li, Guo-Hua Hu, and Chuanxi Xiong. "Enhanced dielectric property and energy storage density of PVDF-HFP based dielectric composites by incorporation of silver nanoparticles-decorated exfoliated montmorillonite nanoplatelets." *Composites Part A: Applied Science and Manufacturing* 108 (2018): 62–68.
82. Zhou, Tao, Jun-Wei Zha, Rui-Yao Cui, Ben-Hui Fan, Jin-Kai Yuan, and Zhi-Min Dang. "Improving dielectric properties of BaTiO3/ferroelectric polymer composites by employing surface hydroxylated BaTiO3 nanoparticles." *ACS Applied Materials & Interfaces* 3, no. 7 (2011): 2184–2188.
83. Li, He, Feihua Liu, Baoyan Fan, Ding Ai, Zongren Peng, and Qing Wang. "Nanostructured ferroelectric-polymer composites for capacitive energy storage." *Small Methods* 2, no. 6 (2018): 1700399.
84. Nan, Ce-Wen, Yang Shen, and Jing Ma. "Physical properties of composites near percolation." *Annual Review of Materials Research* 40, no. 1 (2010): 131–151.
85. Chu, Baojin, Xin Zhou, Kailiang Ren, Bret Neese, Minren Lin, Qing Wang, F. Bauer, and Q. Zhang. "A dielectric polymer with high electric energy density and fast discharge speed." *Science* 313, no. 5785 (2006): 334–336.
86. Zha, Jun-Wei, Ming-Sheng Zheng, Ben-Hui Fan, and Zhi-Min Dang. "Polymer-based dielectrics with high permittivity for electric energy storage: A review." *Nano Energy* 89 (2021): 106438.
87. Shipman, Joshua, Brian Riggs, Scott Grayson, and Douglas Chrisey. "First principles modeling of nanoparticle–polymer surface functionalizations for improved capacitive energy storage." *Journal of Materials Science* 55, no. 33 (2020): 15813–15825.
88. Thakur, Yash, Tian Zhang, C. Iacob, Tiannan Yang, J. Bernholc, L. Q. Chen, J. Runt, and Q. M. Zhang. "Enhancement of the dielectric response in polymer nanocomposites with low dielectric constant fillers." *Nanoscale* 9, no. 31 (2017): 10992–10997.

89. Fenouillot, Françoise, Philippe Cassagnau, and J-C. Majesté. "Uneven distribution of nanoparticles in immiscible fluids: Morphology development in polymer blends." *Polymer* 50, no. 6 (2009): 1333–1350.
90. Ren, Lulu, Lijun Yang, Siyu Zhang, He Li, Yao Zhou, Ding Ai, Zongliang Xie et al. "Largely enhanced dielectric properties of polymer composites with HfO_2 nanoparticles for high-temperature film capacitors." *Composites Science and Technology* 201 (2021): 108528.
91. Luo, Hang, Xuefan Zhou, Christopher Ellingford, Yan Zhang, Sheng Chen, Kechao Zhou, Dou Zhang, Chris R. Bowen, and Chaoying Wan. "Interface design for high energy density polymer nanocomposites." *Chemical Society Reviews* 48, no. 16 (2019): 4424–4465.
92. Feng, Mengjia, Yu Feng, Tiandong Zhang, Jinglei Li, Qingguo Chen, Qingguo Chi, and Qingquan Lei. "Recent Advances in Multilayer-Structure Dielectrics for Energy Storage Application." *Advanced Science* 8, no. 23 (2021): 2102221.
93. Wang, Yifei, Yi Li, Linxi Wang, Qibin Yuan, Jie Chen, Yujuan Niu, Xinwei Xu, Qing Wang, and Hong Wang. "Gradient-layered polymer nanocomposites with significantly improved insulation performance for dielectric energy storage." *Energy Storage Materials* 24 (2020): 626–634.
94. Jiang, Jianyong, Zhonghui Shen, Jianfeng Qian, Zhenkang Dan, Mengfan Guo, Yue He, Yuanhua Lin, Ce-Wen Nan, Longqing Chen, and Yang Shen. "Synergy of micro-/mesoscopic interfaces in multilayered polymer nanocomposites induces ultrahigh energy density for capacitive energy storage." *Nano Energy* 62 (2019): 220–229.
95. Jiang, Yanda, Xin Zhang, Zhonghui Shen, Xinhui Li, Jingjing Yan, Bao-Wen Li, and Ce-Wen Nan. "Ultrahigh breakdown strength and improved energy density of polymer nanocomposites with gradient distribution of ceramic nanoparticles." *Advanced Functional Materials* 30, no. 4 (2020): 1906112.
96. Wang, Yifei, Jin Cui, Qibin Yuan, Yujuan Niu, Yuanyuan Bai, and Hong Wang. "Significantly enhanced breakdown strength and energy density in sandwich-structured barium titanate/poly (vinylidene fluoride) nanocomposites." *Advanced Materials* 27, no. 42 (2015): 6658–6663.
97. Wang, Yao, Yafang Hou, and Yuan Deng. "Effects of interfaces between adjacent layers on breakdown strength and energy density in sandwich-structured polymer composites." *Composites Science and Technology* 145 (2017): 71–77.
98. Zhu, Yingke, Hao Yao, Pingkai Jiang, Jiandong Wu, Xi Zhu, and Xingyi Huang. "Two-dimensional high-K nanosheets for dielectric polymer nanocomposites with ultrahigh discharged energy density." *The Journal of Physical Chemistry C* 122, no. 32 (2018): 18282–18293.
99. Rabuffi, Maurizio, and Guido Picci. "Status quo and future prospects for metallized polypropylene energy storage capacitors." *IEEE Transactions on Plasma Science* 30, no. 5 (2002): 1939–1942.
100. Li, He, Yao Zhou, Yang Liu, Li Li, Yi Liu, and Qing Wang. "Dielectric polymers for high-temperature capacitive energy storage." *Chemical Society Reviews* 50, no. 11 (2021): 6369–6400.
101. Zhang, Tian, Xin Chen, Yash Thakur, Biao Lu, Qiyan Zhang, J. Runt, and Q. M. Zhang "A highly scalable dielectric metamaterial with superior capacitor performance over a broad temperature." *Science Advances* 6, no. 4 (2020): eaax6622.
102. Ho, Janet, and Richard Jow. *Characterization of high temperature polymer thin films for power conditioning capacitors*. Army Research Lab Adelphi Md Sensors And Electron Devices Directorate, 2009.
103. Li, Qi, Guangzu Zhang, Feihua Liu, Kuo Han, Matthew R. Gadinski, Chuanxi Xiong, and Qing Wang. "Solution-processed ferroelectric terpolymer nanocomposites with high breakdown strength and energy density utilizing boron nitride nanosheets." *Energy & Environmental Science* 8, no. 3 (2015): 922–931.
104. Pan, Jilin, Kun Li, Sunanta Chuayprakong, Tim Hsu, and Qing Wang. "High-temperature poly (phthalazinone ether ketone) thin films for dielectric energy storage." *ACS Applied Materials & Interfaces* 2, no. 5 (2010): 1286–1289.
105. Pan, Jilin, Kun Li, Junjun Li, Tim Hsu, and Qing Wang. "Dielectric characteristics of poly (ether ketone ketone) for high temperature capacitive energy storage." *Applied Physics Letters* 95, no. 2 (2009): 022902.
106. Li, Wenjing, Qingjie Meng, Yuansuo Zheng, Zhicheng Zhang, Weimin Xia, and Zhuo Xu. "Electric energy storage properties of poly (vinylidene fluoride)." *Applied Physics Letters* 96, no. 19 (2010): 192905.
107. Zhou, Xin, Xuanhe Zhao, Zhigang Suo, Chen Zou, James Runt, Sheng Liu, Shihai Zhang, and Q. M. Zhang. "Electrical breakdown and ultrahigh electrical energy density in poly (vinylidene fluoride–hexafluoropropylene) copolymer." *Applied Physics Letters* 94, no. 16 (2009): 162901.
108. Gross, Silvia, Daniele Camozzo, Vito Di Noto, Lidia Armelao, and Eugenio Tondello. "PMMA: A key macromolecular component for dielectric low-κ hybrid inorganic–organic polymer films." *European Polymer Journal* 43, no. 3 (2007): 673–696.

109. Xie, Liyuan, Xingyi Huang, Yanhui Huang, Ke Yang, and Pingkai Jiang. "Core@ double-shell structured BaTiO3–polymer nanocomposites with high dielectric constant and low dielectric loss for energy storage application." *The Journal of Physical Chemistry C* 117, no. 44 (2013): 22525–22537.
110. Qenawy, Shaymaa A., Loai S. Nasrat, Hanafy M. Ismail, and Jeannette N. Asaad. "Evaluation of dielectric strength of SiR/TiO_2 composites using feed-forward neural network." *IET Nanodielectrics* 3, no. 3 (2020): 74–80.
111. Thostenson, Erik T., and T-W. Chou. "Carbon nanotube networks: sensing of distributed strain and damage for life prediction and self healing." *Advanced Materials* 18, no. 21 (2006): 2837–2841.
112. Homes, C. C., T. Vogt, S. M. Shapiro, S. Wakimoto, and A. P. Ramirez. "Optical response of high-dielectric-constant perovskite-related oxide." *science* 293, no. 5530 (2001): 673–676.
113. Carlson, C. M., T. V. Rivkin, P. A. Parilla, J. D. Perkins, D. S. Ginley, A. B. Kozyrev, V. N. Oshadchy, and A. S. Pavlov. "Large dielectric constant (ε/ε 0> 6000) Ba 0.4 Sr 0.6 TiO_3 thin films for high-performance microwave phase shifters." *Applied Physics Letters* 76, no. 14 (2000): 1920–1922.
114. Scott, J. F. "High-dielectric constant thin films for dynamic random access memories (DRAM)." *Annual Review of Materials Science* 28, no. 1 (1998): 79–100.
115. Kotecki, David E. "A review of high dielectric materials for DRAM capacitors." *Integrated Ferroelectrics* 16, no. 1–4 (1997): 1–19.
116. Rao, Yang, S. Ogitani, Paul Kohl, and C. P. Wong. "Novel polymer–ceramic nanocomposite based on high dielectric constant epoxy formula for embedded capacitor application." *Journal of Applied Polymer Science* 83, no. 5 (2002): 1084–1090.
117. Robertson, John. "High dielectric constant oxides." *The European Physical Journal-Applied Physics* 28, no. 3 (2004): 265–291.
118. Pelrine, Ron, Roy Kornbluh, and Guggi Kofod. "High strain actuator materials based on dielectric elastomers." *Advanced Materials* 12, no. 16 (2000): 1223–1225.
119. Jiang, S. L., Y. Yu, and Y. K. Zeng. "Novel Ag–BaTiO3/PVDF three-component nanocomposites with high energy density and the influence of nano-Ag on the dielectric properties." *Current Applied Physics* 9, no. 5 (2009): 956–959.
120. Liu, Yongbin, Jinghui Gao, Yan Wang, Jun Zhou, Liang Cao, Zhixin He, Yang Zhang, Chao Tang, and Lisheng Zhong. "Enhanced temperature stability of high energy density ferroelectric polymer blends: the spatial confinement effect." *Macromolecular Rapid Communications* 40, no. 21 (2019): 1900406.
121. Yang, Ke, Xingyi Huang, Lijun Fang, Jinliang He, and Pingkai Jiang. "Fluoro-polymer functionalized graphene for flexible ferroelectric polymer-based high-K nanocomposites with suppressed dielectric loss and low percolation threshold." *Nanoscale* 6, no. 24 (2014): 14740–14753.
122. Hanani, Zouhair, Daoud Mezzane, M'barek Amjoud, Mohammed Lahcini, Matjaž Spreitzer, Damjan Vengust, Arash Jamali, Mimoun El Marssi, Zdravko Kutnjak, and Mohamed Gouné. "The paradigm of the filler's dielectric permittivity and aspect ratio in high-K polymer nanocomposites for energy storage applications." *Journal of Materials Chemistry C* 10, no. 30 (2022): 10823–10831.
123. Ma, Fengwang, Cuihong Ma, Yu Zhang, Ruyi Sun, Xiaojuan Liao, Wei Song, and Meiran Xie. "Dielectric and energy storage performance of polymers bearing terpyridyl group and metal ion complex by ring-opening metathesis polymerization." *Reactive and Functional Polymers* 165 (2021): 104944.
124. Chi, Qingguo, Yinhua Zhou, Chao Yin, Yue Zhang, Changhai Zhang, Tiandong Zhang, Yu Feng, Yongquan Zhang, and Qingguo Chen. "A blended binary composite of poly (vinylidene fluoride) and poly (methyl methacrylate) exhibiting excellent energy storage performances." *Journal of Materials Chemistry C* 7, no. 45 (2019): 14148–14158.
125. Meng, Qingjie, Wenjing Li, Yuansuo Zheng, and Zhicheng Zhang. "Effect of poly (methyl methacrylate) addition on the dielectric and energy storage properties of poly (vinylidene fluoride)." *Journal of Applied Polymer Science* 116, no. 5 (2010): 2674–2684.
126. Zhang, Tianran, Qinzhao Sun, Fang Kang, Zepeng Wang, Rong Xue, Jiping Wang, and Lixue Zhang. "Sandwich-structured polymer dielectric composite films for improving breakdown strength and energy density at high temperature." *Composites Science and Technology* 227 (2022): 109596.
127. Sahoo, Sushrisangita, Yadav, A., K. P. Andryushin, and R. N. P. Choudhary, "Influence of (0.19 HfO2–0.81 ZrO2) ceramics filler content on structural and dielectric properties of PVDF polymer." *Synthetic Metals* 287 (2022): 117097.
128. Rawat, Pooja, and Achchhe Lal Saroj. "Effect of ionic liquid on plasticized CS-PVP-NaI based bio-polymer blend electrolytes: Structural, thermal, dielectric and ion transport properties study." *Materials Science and Engineering: B* 288 (2023): 116215.

129. Morsi, M. A., A. H. Oraby, A. G. Elshahawy, and R. M. Abd El-Hady. "Preparation, structural analysis, morphological investigation and electrical properties of gold nanoparticles filled polyvinyl alcohol/carboxymethyl cellulose blend." *Journal of Materials Research and Technology* 8, no. 6 (2019): 5996–6010.
130. Saadiah, M. A., D. Zhang, Y. Nagao, S. K. Muzakir, and A. S. Samsudin. "Reducing crystallinity on thin film based CMC/PVA hybrid polymer for application as a host in polymer electrolytes." *Journal of Non-Crystalline Solids* 511 (2019): 201–211.
131. Al-Muntaser, A. A., Rami Adel Pashameah, Sharma, K., Alzahrani, E., Hameed, S. T., and Morsi, M. A. "Boosting of structural, optical, and dielectric properties of PVA/CMC polymer blend using SrTiO3 perovskite nanoparticles for advanced optoelectronic applications." *Optical Materials* 132 (2022): 112799.
132. Li, Hongfei, Liwei Wang, Yingke Zhu, Pingkai Jiang, and Xingyi Huang. "Tailoring the polarity of polymer shell on $BaTiO_3$ nanoparticle surface for improved energy storage performance of dielectric polymer nanocomposites." *Chinese Chemical Letters* 32, no. 7 (2021): 2229–2232.
133. Cheng, Sang, Yao Zhou, Yushu Li, Chao Yuan, Mingcong Yang, Jing Fu, Jun Hu, Jinliang He, and Qi Li. "Polymer dielectrics sandwiched by medium-dielectric-constant nanoscale deposition layers for high-temperature capacitive energy storage." *Energy Storage Materials* 42 (2021): 445–453.
134. Fu, Jing, Mingcong Yang, Rui Wang, Sang Cheng, Xiaoyan Huang, Shaojie Wang, Junluo Li, Manxi Li, Jinliang He, and Qi Li. "Improvement of high-temperature energy storage performance in polymer dielectrics by nanofillers with defect spinel structure." *Materials Today Energy* 29 (2022): 101101.
135. Dong, Jiufeng, Renchao Hu, Yujuan Niu, Li Li, Shuai Li, Liang Sun, Yuqi Liu et al. "Scalable in-situ surface-coated polymer dielectrics with significantly enhanced high-temperature breakdown strength." *Materials Today Energy* 30 (2022): 101158.

16 Multipurpose Polymer Dielectric Materials and Composites for Advanced Energy Applications

D. Vignesh and Ela Rout

16.1 INTRODUCTION

Energy is the primitive entity that integrates multifarious processes in the universe. Urbanization and population surge have drastically imposed alarming concerns over energy storage and conversion devices. Economical and eco-friendly technological advancements are stimulated with extended applications from mechanical to electronic devices [1, 2]. A gradual shift from conventional to renewable energy sources, although complementing source replenishment, establishes discrete technical and economical constraints. The synergy of vital challenges has impelled researchers and practitioners to probe into energy storage and conversion devices for smart energy applications. Standalone energy storage devices with pure and derived dielectric materials have been attempted over the years, yet significant disparity persists between energy and power density, which hinders the overall operational efficiency. Polymer dielectric materials (PDMs) are thus a novel class of materials for electrical insulation and energy storage applications. The prominent characteristics of high dielectric strength, low dielectric loss, and low series resistance, alongside high voltage endurance, personify their monopoly amongst other classes of materials in recent years [3]. Unlike ceramics, the flexibility of PDMs results in mechanical deformation when subjected to the external electric field. High-temperature operation alongside electrification of transport presents structural inflexions which limit lucrative implications of PDMs for advanced energy applications and thus provoke one of the vital ethos of research progression [4]. Biaxially oriented polypropylene (BOPP) is a low-temperature (< 105 °C) operative polymer under an applied electric field, regulating the compositional benefits of capacitive energy storage under harsh environments [3]. Despite a high dielectric breakdown (> 6000 MV m^{-1}) and latent low dielectric loss, poor energy density is profoundly obstructed by the low implicit dielectric strength ($\varepsilon < 2.2$) [5]. A steep decline in the capacitive performance of PDMs is further associated with the exponential surge in the leakage current activated by thermal and electric fields. The cooperative influence of high temperature and an electric field increases the conductivity, yet simultaneously enforces Joule's heating effects. The lower thermal conductivity of PDMs obstructs prominent heat dissipation constituting thermal instability and resulting in thermal breakdown [6, 7]. Thus, high-temperature energy harvesting with suppressed conduction using PDMs projects as one of the primary materials–oriented challenges. Additional challenges with high ε PDMs include threshold voltage (V_{th}) instability during functioning. Theoretically, a threshold voltage (V_{th}) shift observed from the hysteresis is attributed to the charge trapping effect in the existing trapping centers within the host PDMs [8]. Figure 16.1 shown below demonstrates the wide applications of PDMs in the energy sector.

Diverse polymers comprise C—H covalent bonds and grant a feeble increase in polarization of the host material, which reflects over equivalent increase in the dielectric strength. Composite polymers with multiphase polymer matrix incorporated by fillers on the other hand escalate the

DOI: 10.1201/9781003352365-16

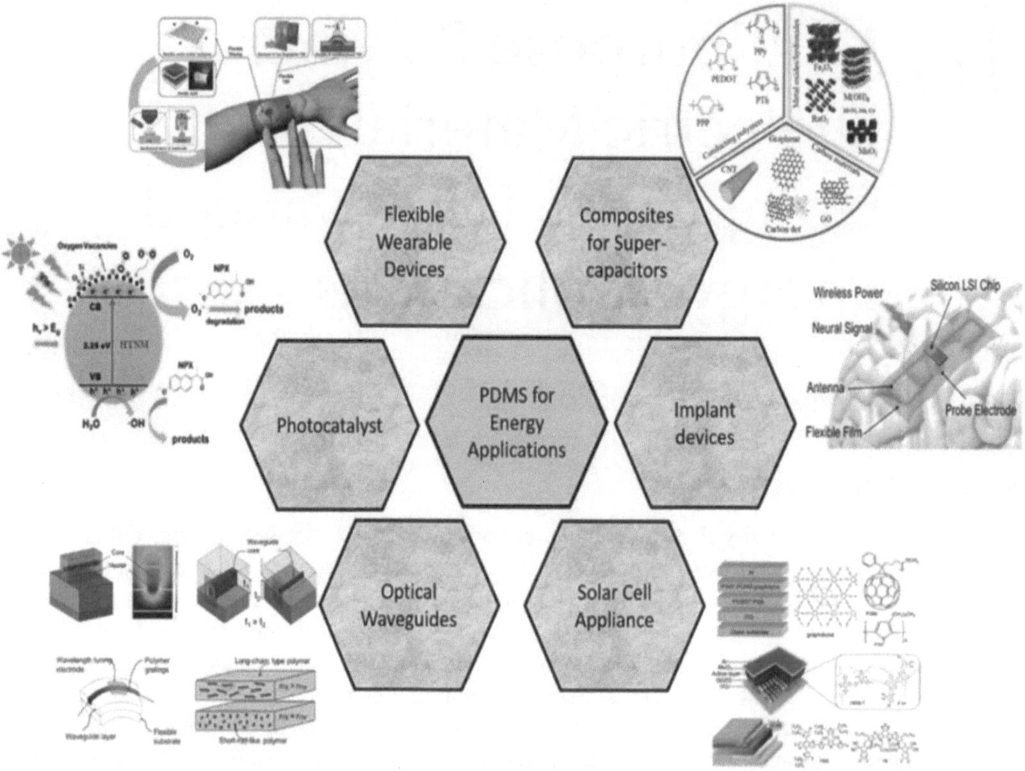

FIGURE 16.1 Schematic representation of diverse energy applications of PDMs.

interfacial and dipolar polarization [9–12]. Practical synthesis of high-end hybrid PDMs establishes immense material modeling hindrance. Fractionally larger inorganic non-metals impart mechanical instability, while conductive fillers (metals/carbon) amplify the leakage current, due to the percolation phenomenon [13]. The disclosed concern establishes an overview of improvement in dielectric properties at the expense of mechanical durability. Novel material improvisations involve suitable polar segments, amelioration of crystalline and amorphous regions and the co-extrusion of processed thin films by suitable polymer pairs respectively [13]. A simple blend of multiphase polymers by co-extrusion processing resolves the incompatibility of multiple polymer architectures and simultaneously accomplishes high dielectric strength. The co-extruded multiplayer polymers and composites embed alternative layers of polymers with high dielectric strength [14]. Dielectric strength is one of the elusive characteristics which governs long-term voltage endurance alongside short-term dielectric breakdown respectively. Such implicit compositional behavior vanquishes the material's architecture and hinders the volumetric power density, dielectric thickness, and compositional tuning [15, 16]. Recent material engineering has traced the coupling of high polar molecular groups with linear polymers to revamp the energy density. High volumetric density, with the dipolar coupling of polar molecular groups, enables the host polymer matrix to demonstrate high dielectric strength with the low dielectric loss [17–19]. Meanwhile, polymer composites have gradually driven the intellect of researchers. The deposition of conducting particles on the HPD surface has illustrated exemplary results over the dielectric properties of the parent material. Such conducting particles (Ni, Ag) potentially promise as an alternative filler to the polymer composites preserving mechanical durability. Hybrid polymer-conducting particles ($Ag–Na_{0.5}Bi_{0.5}Cu_3Ti_4O_{12}$) from the outcomes of Kum-onsa et al. [20] claimed the dominant contribution of conducting particles over enhanced dielectric properties and the resulting fine microstructure of the principal composition.

Although the hybrid conducting oxides and high permittivity oxides are predominantly opted fillers in the polymeric composite matrix, the oxidation of conducting particles alters the dielectric properties of the ternary composition and the consequent capacitance of the capacitor. Limiting the oxidation of metallic particles in the PDMs further presents additional constraints over material optimization. A recent study by Silakaew et al. [21] reports Ruthenium oxide (RuO_2) as an equivalent filler to polymer dielectric composites. The surface deposition of RuO_2 over $BaTiO_3$ incorporated into polyvinylidene fluoride (PVDF) as a three-phase HPD composite showcased an improved dielectric constant, with a 50% volumetric infusion of RuO_2-$BaTiO_3$ within the host composition. The disclosed overview suggests the collapse of physical, mechanical and chemical integrity, due to PDMs thermodynamical variability. The evolution of thermally generated voids in the composite polymer matrix ameliorates the material's dielectric strength. The chapter is intended to emphasize material limitations associated with dielectric polymers and assimilate to highlight the plausible efforts and outcomes to enhance the dielectric properties of the host composition with progressive research ideas.

16.2 FOUNDATION OF DIELECTRICS

Dielectrics are a utilitarian class of materials with the ability to polarize under the influence of an external electric field. The dielectric materials (DMs) are classified distinctly depending on the nature of relaxation mechanisms and hysteresis behavior. Numerous dielectric properties, including dielectric breakdown, dispersion, dielectric loss, ferroelectric phase transition, and many more, can be annotated exclusively without the atomic and molecular properties. DMs encompass atoms and molecules with closed or nearly closed electronic shells, which are markedly distinguished from metals with free electrons in the conduction band. The non-centrosymmetric displacement of tightly bound electrons to the atomic cores frames the backbone of the linear response theory of dielectrics, as shown in Eq. (16.1).

$$p = \alpha \varepsilon_0 E \tag{16.1}$$

Where,

p = induced electric dipole moment
α = polarizability of the atom
ε_0 = permittivity in free space
E = externally applied electric field

Principally, DMs are heavily relied on and investigated for energy applications, due to high dielectric breakdown and low remanent electrical displacement (D_τ) The amplitude of energy loss in DMs intensifies with equivalent nonlinearity in the dielectric behavior [22]. However, feeble residual polarization with discrete phenomena exists amongst nonlinear dielectrics, even after a prominent coupling between dipoles during successive charging and discharging cycles. Recognized energy storage devices comprise of capacitors, batteries, combustion engines, and more. The relevant figure of merit of energy storage capacity is weighed using the device's energy density (J). The energy storage principle in a parallel plate capacitor initiates charge accumulation across the electrodes under an applied electric field. The charging process cultivates an internal electric field and nullifies the effect of an external electric field, with a consequent increase in charge accumulation across the electrodes [23]. Such a depolarizing effect is a geometry-dependent property. The effect is prominent in diverse geometric DMs relative to ellipsoidal DMs. The depolarizing factor along three orthogonal directions (N_x, N_y, N_z) for ellipsoidal symmetries satisfies the relation shown in Eq. (16.2):

$$N_x + N_y + N_z = 1 \tag{16.2}$$

The real-world elementary ellipsoidal shapes with prolate and oblate configurations include needle and disc respectively. The nonlinear distribution of depolarizing factors among different semi-axes consequently reflects the geometry's dielectric strength. While one of the depolarizing components is low for prolate ellipsoids, oblate configurations on the other hand display two low depolarizing components. The symmetric distribution of a depolarizing factor (1/3) among all the components in spherical symmetry establishes high depolarization, and a consequent high internal electric field [24].

The charge displacement by the external electric field and electrostatic energy is stored in the dielectric layer and is evaluated using Eq. (16.3):

$$W = \int_0^{Q_{max}} V dq \tag{16.3}$$

where,

W = Work done to store energy
V = applied potential
Q_{max} = Maximum charge accumulated on the electrodes during the charging process
dq = incremental charge

In principle, all dielectric materials establish a threshold toward the exposed electric field, which imposes irreversible modifications in the medium, escorted by the onset of an acute and disrupted charge diffusion. A steep gradient in dielectric loss and the resulting material's insulation properties is termed as dielectric breakdown. The frequency-dependent dielectric performance of the host material is accessed using
Eqs. (16.4) and (16.5):

$$\varepsilon(\omega) = \varepsilon'(\omega) + j\varepsilon''(\omega) \tag{16.4}$$

$$tan\delta = \frac{\varepsilon''(\omega)}{\varepsilon'(\omega)} \tag{16.5}$$

where,

$\varepsilon(\omega)$ = total dielectric constant
$\varepsilon'(\omega)$ = real part of dielectric constant
$\varepsilon''(\omega)$ = imaginary part of dielectric constant
$tan\delta$ = loss tangent factor

Time-dependent dielectric variability, also termed as aging, in the absence of stress and isothermal conditions, accounts for one of the significant obstructions for practical appliances [25]. The rate of aging varies between different classes of materials, yet multiple ceramics undergo vital electromechanical perturbations and modified dielectric behavior. Diverse permutations of DMs have been probed to pacify diverse disclosed effects. Material remedy in terms of optimized PDMs has emerged as a novel gateway for the research community in recent years. Dipoles in dielectric macromolecules can exist in different plausible orientations (side groups, along the backbone, and end groups). The macromolecular displacement within the free volume of the dielectric polymer matrix constitutes the implicit mechanical flexibility. The controlled response of macromolecular displacement to the external field presents a scope of engineering the mechanical durability of PDMs. In accordance with the studies of Picci et al. [26] and Chen et al. [27], most PDMs display considerable resistance to dielectric breakdown accounting to hover between ($10^6 - 10^8$ V m^{-1}) respectively. However, structural differences are reflected in the associated low-dielectric-constant dielectric characteristics. Material modeling via foreign substituents impairs the long-range ferroelectric ordering of the electrical domains and constitutes broad polar nano regions, which expands and flattens the dielectric isotherm. While in the case of PDMs, those elements with a plausible tendency to form a double bond with a sp^2 hybridized carbon, preserve a rotational degree of freedom at high temperatures and demonstrate temperature-dependent polarization [28]. The compiled limitation of thermos-mechanical instability, aging, performance inflexions and sustainability imposes a rationale to introspect appropriate PDMs and HPDMs respectively.

16.3 A REVISIT TO DIELECTRIC BREAKDOWN

Dielectric breakdown (E_b) corresponds to a transition of DMs from insulating characteristics to a more conductivity counterpart. The characteristic variability emerges upon exceeding the critical electric field limit of the DMs [29]. The magnitude of dielectric breakdown strength determines the electrostatic storage density in DMs [30]. The breakdown mechanisms rely solely upon the compositional complexity. Among distinct breakdowns, free volume breakdown, electromechanical breakdown, and thermal breakdown are profoundly favorable at high temperatures [31]. The analogous perspective of dielectric breakdown is recognized as the aftermath of microstructural degradation, due to defect cluster formation under the influence of an external electric field [29]. Trivially, the theory of dielectric breakdown is categorized into three groups viz, thermal, avalanche, and intrinsic breakdown respectively. The dominant contribution amongst which emerges from the avalanche effect induced by the non-uniformity of the field strength produced by the charge collision ionization [32, 33]. In the low-temperature limit, the dominant avalanche effect demonstrates negative thickness alongside positive temperature dependence of the applied electric field [34]. The proportionality of defect probability over the electrode area of a dielectric film prioritizes the impact of the electrode area over the magnitude of the breakdown field [30]. The mechanism of breakdown is further classified into three groups, namely the wear-out phase, the onset of breakdown, and the breakdown propagation. The dimension of DMs has a consequent effect on the breakdown mechanism. The effects become more prominent at the nano-scale. While an abrupt gradient in the conduction mechanism is perceived with DMs having thickness > 5 nm, a progressive loss of their insulating characteristics becomes dominant with thickness < 5 nm respectively. Such a progressive dielectric breakdown is associated with the prolonged surge in the current across the dielectric unit up to a threshold for operational conditions [29]. The barrier effect is an established concept applied for breakdown voltage engineering [35–37]. The presence of a barrier (material with different permittivity and conductivity) in a polymer matrix induces a space charge layer at the interface of the barrier when subjected to an electric field. The synergy causes the electric field redistribution across the barrier-matrix interface [38]. Figure 16.2 shown below demonstrates different breakdown mechanisms as a function of time and temperature: Dielectric breakdown in PDMs is associated with multiple parameters, including temperature, residual stress, defects and impurities, exposure time to an applied voltage, and the nature of charge transfer. The intrinsic breakdown property within a polymer matrix relies on the structure, electronic states, and bandgap energy at the atomic scale under standard atmospheric conditions. Multiple kinds of literature illustrate a proportional impact of dielectric breakdown over the order of crystallinity at lower temperature regimes [39].

The segmental motion of polymeric chains freezes and results in reduced dielectric strength of the overall PDM. Prominent chain and segmental motions occur in PDMs that are classified as follows:

1. α relaxations: They are formally labeled as glass transitions, due to micro-Brownian displacement of the polymeric chains. [40]
2. β relaxations: appears due to rotation of polar groups about the C—C bond alongside a conformational flip of the cyclic unit
3. γ relaxations: eventuates due to constrained segmental chain displacement and redeemed phenyl ring. Such relaxations predominantly occur in aromatic PDMs.

In the perspective of low temperatures, γ relaxations illustrate weighted probability due to facile perturbations of smaller phenyl rings and C—H units. Successive order is followed by β and α relaxations in relevance to the scale of segmental motion. Figure 16.3 shown below demonstrates the different relaxation behavior of PDMs as a function of temperature [41]. Equivalent impacts over crystallinity and polymeric structures are also detected at high-temperature regimes. Thermal agitations at high temperatures escalate due to a steep increase in intermolecular forces between polymeric chains. This provides flexibility among polar groups to re-orient along different crystallographic axes emulating the track of varying electric fields [41]. Control over the breakdown parameters and a subsequent improvisation becomes a tedious process, with standalone polymers and hybrid polymers due to complex molecular chains, interfaces, and surfaces respectively. The prevalence of the charge-trapping effect across the interface and interstitial sites within a polymer matrix curtails the dielectric strength of the polymeric composition [42]. One of the vital features of dielectric breakdown in PDMs is the evolution of hollow conducting channels within the polymer matrix due to partial discharges. Such hollow conducting channels develop from the top electrode to the bottom and simulate an inverted tree-like structure and hence are termed as electric trees [43–47]. The growth of an electrical tree amplifies under divergent electric stress and originates within the polymeric matrix embedding voids, defects, and impurities in a confined space. There exist two distinct approaches to understand the dielectric breakdown mechanism. In the first case, the dc voltage ramps are applied to a dielectric polymer matrix until the breakdown. The resulting breakdown voltages are thereafter analyzed using Weibull statistics to deduce the nominal breakdown field [48]. The second case deals with the determination of a specific breakdown field defined as the voltage gradient, corresponding to the mean value of time delay between two successive breakdowns subjected to electrical stress tending to infinity [49].

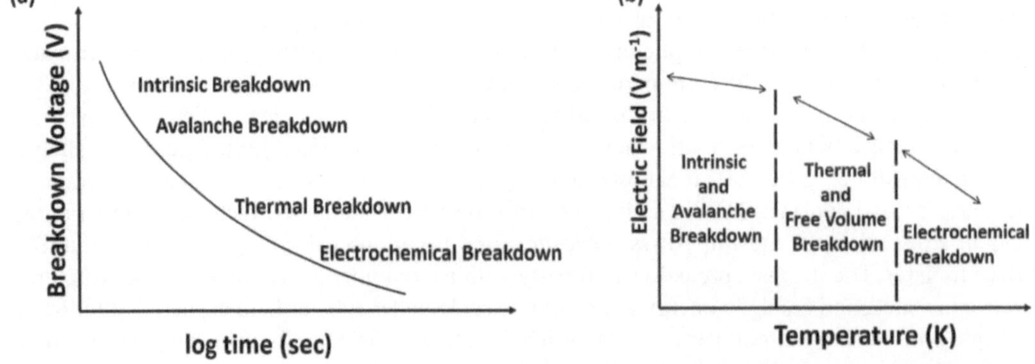

FIGURE 16.2 Distinct dielectric breakdown mechanism as a function of (a) breakdown voltage and time, (b) electric field and temperature.

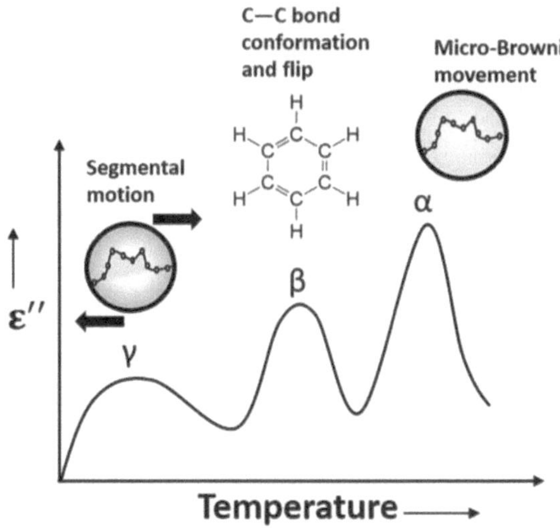

FIGURE 16.3 Temperature-dependent relaxations of hybrid high-K polymer dielectrics.

It is now familiar that physical voids, free volume in the molecular chain and structural disorders are all polymeric defects. Hence, reduction of micro and macrostructural defects via modification of a polymer molecular chain is an essential requisite to minimize the structural defects and enhance the breakdown strength [50, 51]. Diverse polymeric compositions include block co-polymer for suitable interface architecture, layered polymers, aromatic polymers, aliphatic polymers, nanofillers for voltage endurance, and many more [17, 52, 53]. The implication of nanoparticles in the high ε polymer matrix establishes a variable permittivity between the nanofiller and the polymer matrix and resolves the nanoparticle dispersibility under an intense electric field. Intermediate temperature-centric applications using PDMs, conventionally illustrate free volume breakdown where the high accelerated electrons, due to breakdown voltage, breach the polymeric chains and initiate the breakdown process [31]. Crosslinked polymeric chains are thus adopted in a large scale to counteract the free volume breakdown process and enhance the breakdown voltage at high temperatures respectively. With numerous interdependent factors governing discrete breakdown mechanisms, it becomes an elemental entity to investigate, discern, and distinguish different breakdown mechanisms for optimized PDMs.

16.4 HIGH-K PDMs FOR ADVANCED ENERGY APPLICATIONS

Energy storage, conversion and utilization are of vital importance for eco-friendly energy applications. Modern-day electronics ranging from micro-grid to automobiles demonstrate large-scale reliability over dielectric materials as a backbone for power systems. Traditional dielectric ceramics and inorganic materials demand tedious material processing, which over the years has shifted to polymer dielectrics. The flexible processing techniques, low compositional reactivity to chemically active atmospheres and economical facets have enriched the appliance of PDMs as smart materials for advanced energy applications [2]. An ideal high-K PDM ($k > 3.9$) is recognized with the ability to resist the free flow of charges and permit the displacement of charges, with a resultant polarization. Polymer dielectrics flourish high or low dielectric constants, which are proportionally associated with the permittivity (polarization response to the external electric field). In the early 90s, aromatic [41] and aliphatic fluoropolymers [54] were assertively used as dielectric materials for energy storage applications.

Although PDMs typically feature low dielectric strength, they are still commonly used in electronic devices due to their inability to provide high electrical insulation [55]. As a consequence, distinct attempts are executed to enhance the dielectric properties of PDMs via polymeric chain architecture. Highly polar aromatic rings and chalcogens are employed as a means to escalate the dielectric characteristics. The dielectric characteristics of host polymeric compositions are extensively investigated through the use of high-dielectric-constant polymers, conducting filers, and other material engineering techniques. Polymers can be distinguished as polar and nonpolar, susceptible to the electronic dipole arrangements and the resulting polymer geometry. Typical polar polymers include poly(methyl methacrylate) PMMA [56], polyvinyl chloride (PVC) [57], and many others, while the nonpolar counterpart includes polytetrafluoroethylene (PTFE) [58]and other fluoropolymers that have unique dielectric properties in terms of the relaxation mechanism. Figure 16.4 shown below demonstrates a polar and nonpolar molecule alongside different polarization mechanisms.

The response to alternating electric fields is different for different classes of PDMs. The independence of dielectric relaxations to the applied alternating field by nonpolar molecules arises due to instantaneous electron polarizations. The extended polymeric configuration of PTFE with high dipole moment -CF_2 units attached to the alternate carbon backbone nullifies the net dipole moment vector sum and restrains the $k < 3$. While PVC on the other hand exhibits enforced dielectric permittivity, due to the parallel orientation of dipoles with non-zero net dipole moment [41]. The dielectric constant illustrates a spike prior to a symmetrical drop about a frequency range. Such inflexions depict the optimal polarization in phase with oscillating frequency. Table 16.1 shown below delineates the dielectric strength of different PDMs.

From Table 16.1, low dielectric constants of polar and nonpolar PDMs can be enhanced via different techniques. A blend of different PDMs to form a hybrid high-K polymer matrix besides the use of Electroceramics and conductive nanofiller is one of the modern strategies utilized for energy storage applications. While inorganic polymers illustrate a high dielectric constant over organic

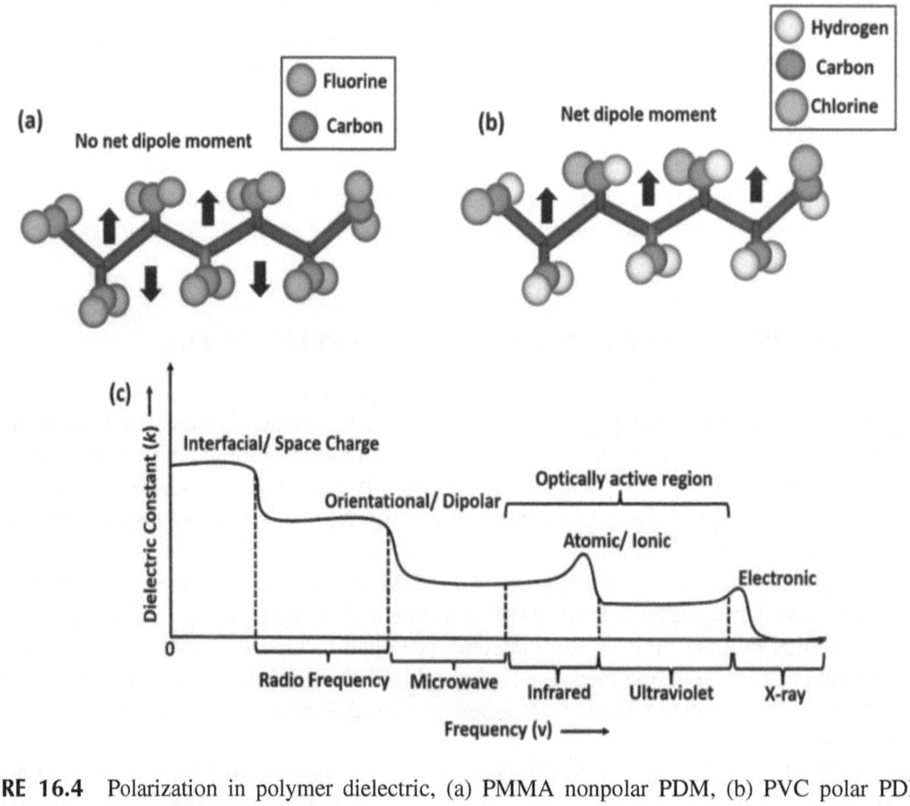

FIGURE 16.4 Polarization in polymer dielectric, (a) PMMA nonpolar PDM, (b) PVC polar PDM, (c) polarization response at different frequency.

TABLE 16.1
Dielectric Properties of Polar and Nonpolar PDMs

Polymer Dielectric Materials (PDMs)	Dielectric Constant (ε)	Ref
	Polar dielectric polymers	
Polyimide (PI)	4.2	[59]
Biaxially oriented polypropylene (BOPP)	2.2	[60]
Polystyrene	2.6	[61]
	Nonpolar dielectric polymers	
Polypropylene (PP)	2.3–2.8	[62]
Polyquinoline (PQ)	2.8	[63]
Poly (methyl methacrylate) PMMA	3.12	[64]

polymers, yet tedious material synthesis forbids the compatible integration of organic-inorganic hybrid polymer matrix [65]. In compliance with the hybrid polymer dielectrics, the study conducted by Toomeyet al. [66] illustrates the conglomeration of the synthesized material responsible for the disordered behavior of the polymer matrix. Polymer precursor incompatibility and gel formation over time are two momentous factors that constitute the compromised dielectric properties of the synthesized hybrid compositions.

High dielectric constant, low dielectric loss, and electrical conductivity are requisite criteria to accomplish potent energy storage materials with high energy density. Prominent loss occurs at the interface of the polymer matrix demonstrated by the mid-point of infinite and static dielectric constant in the Nyquist plot respectively. The non-Debye relaxations in large hybrid PDMs display high viscosity, with an entanglement between polymeric chains. The synergy of effects stands contrary to polymeric systems with different relaxation mechanisms, where the mobility-dependent relaxation time fluctuates under local environmental entropy [41]. The multi-component hybrid PDMs follows interfacial charge reorganization via Maxwell-Wagner-Sillars (MWS) interfacial polarization (IP). The charges accumulate across the surface of different dielectric interfaces with distinct retention time. Such heterogeneous polymeric configurations are an effective approach for high-K hybrid polymers. However, inflexions in the inhomogeneous charge distribution demonstrate dielectric failure and demand optimization of performance parameters [67].

16.5 CURRENT SCENARIO IN POLYMER DIELECTRICS

Dielectrics are capaciously employed for diverse applications. The insulation property of dielectrics is exploited to isolate the electrical conducting surfaces and charged elements. Technological advancements have driven the immense attention of researchers and practitioners to upgrade the dielectric strength via suitable material modeling and synthesis routes. A promising impedance over the potential application of PDMs for commercialization lies in the low energy density relative to energy conversion electrochemical devices. Figure 16.5 shown below demonstrates a relative comparison of energy and power density disparity among different energy storage and conversion devices: In the context of low energy density, depressed dielectric strength with capacitors underlines the structural and processability issues as the prime concerns [60, 68]. However, the ability to enhance the dielectric strength via hybrid PDMs is accompanied by the expense of reduced dielectric breakdown strength [69]. Substantial studies have illustrated the inability to enhance the dielectric permittivity with the use of a single polymeric substance. In a relative aspect, the dielectric constant of organic PDMs is significantly lower than the inorganic counterpart, due to the variability in the nature of covalent molecular bonds [70]. Structural and microstructural diversity produced by differing molecular chains hinders the dielectric properties of the PDMs. Dielectric anisotropy has thus been introduced to examine the dielectric strength of PDMs as a function of variable morphology and spatially aligned nanofillers [5]. Computational techniques and simulations have been utilized in recent years for the rational design and architecture of repeated macromolecular chains, with high polarizability on the basis of electronic characteristics [15, 71]. Studies conducted by Haqueet al. [72] claim an effective role of defect cluster morphology over the dielectric performance of the PDMs. Internal, interstitial, and corona partial discharge are conventionally encountered defects within standalone and hybrid polymeric structures that perturb the polymeric insulations for high voltage endurance. A discharge at the void surface deposits conductive layers due to chemical decomposition reactions around the void boundary. The charge carriers thus get trapped at the polymeric surface, due to the modified boundary, and revamp the polymer dielectric properties.

Copious material engineering is executed to enhance the dielectric characteristics of the host PDM. Copolymerization of two or more polymers is one of the frequently adopted strategies to tailor the dielectric properties of unique hybrid PDMs and PDM composites. The synergy of covalent bonding of polymeric structures attributes to the evolution of hybrid polymeric chains. The covalently bonded structures during the cross-linking process induce chain rigidity, which is proportional to the cross-link density and alters the relaxation time [41]. An increase in cross-link density congruently escalates the covalent bond rigidity and actuates a longer time for bond relaxations at the equilibrium configuration. In reference to the prior discussion, the former parameter foists considerable effects over different relaxations (α, β, γ) in hybrid and polymer composites. Discrete relaxations thus vary in terms of kinetics due to wavering large-scale polymeric chain segmental motions. The variability in relaxation time is utilized as one of the parameters to monitor the chemical kinetics across the polymer interface in hybrid and composite PDMs. Copolymerization at the nano-scale to form hybrid polymer dielectric nanocomposites has been administered in recent years, which has incited a rapid pace due to the potential advancement in dielectric properties. Studies by Liet al. [73], Siva et al. [74] and many other established works of literature illustrate a valuable

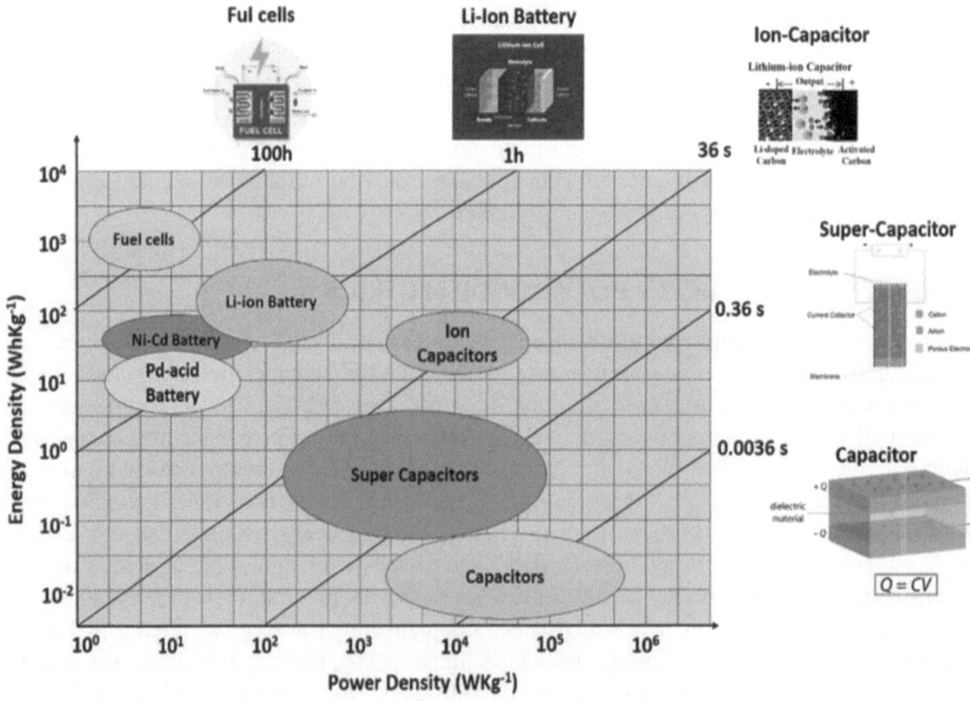

FIGURE 16.5 Energy and power density of energy conversion and storage devices with operational durability for electrochemistry.

dielectric response (3–3.4 times) of hybrid and PDM nanocomposites at higher frequency signals (~ 1MHz) over neat polymers [75]. A plausible account for the observed attribute is associated with the nature of fillers employed at the polymer interface. Statistically, the dielectric constant of the hybrid PDMs escalates proportionally with an equivalent rise in the filler concentrations. Further enrichment of dielectric characteristics is featured by sufficient compatibility of organic and inorganic fillers achieved using the ultrasonication technique [76]. There are numerous permutations executed towards such strategy, Afzal et al. [77]. and Kuo et al. [78] asserted the embodiment of titania (TiO_2), barium titanate ($BaTiO_3$), zirconia (ZrO_2) and alumina (Al_2O_3) as susceptible fillers across the polymer interface. However, a major concern with hybrid polymers, polymer composites, and nanocomposites is phase heterogeneity, which constitutes cracks and voids across the polymer interface and boundary. Such an effect is termed the Maxwell-Wagner effect and comprises the dielectric constant of the polymeric composition. Manifold models such as Maxwell-Wagner, Bruggeman Model and Logarithmic Mixing Law were proposed to examine the dielectric properties of the composite structures [79]. Such models provide flexibility to optimize the composition ratio, filler volume, filler sizes, and more. However, the proposed model illustrates a poor fit for highly complex polymeric systems. The perceived characteristic is accounted for by the inconsistent interface interactions within a polymer matrix that incites disparity in material engineering. A more serious concern with hybrid and composite PDMs is the misinterpretation of dipolar polarization at low-frequency sweep.

16.6 COMMERCIALLY AVAILABLE HIGH-K HYBRID PDMs

Polymers are conventionally organic compounds with high molecular weight due to repetitive structural units, synthesized via natural or synthetic processing. Non-conducting materials with the ability to store charges are termed dielectrics. Thus, a dielectric is a physical model that illustrates the response of a material to an applied electric field due to microscopic dipole reorientation [80]. High chemical compatibility at high temperatures with potential dielectric properties accords a larger fraction of organic polymers in high-K hybrid and composite PDMs [41]. The early exercise of dielectric polymers for capacitive applications was initiated with aliphatic and aromatic fluoropolymers, with chalcogens attached in branched polymeric structures. The high electronegative chalcogens were used as a means to enhance the dielectric properties of the host polymeric compositions [72]. Dielectric polymeric films can be fairly synthesized via spin coating and solution casting, with immersion in an organic substrate and glow discharge techniques respectively. Numerous fabrication procedures grant low-temperature processing flexibility [72]. Both high-K and low-K polymer dielectrics find their vitality in modern-day electronics and advanced energy storage applications. While low-K PDMs are predominantly employed as insulators, fast signal propagation and crosstalk in high-density power systems and are also termed as passivation materials, high-K PDMs, on the other hand, are used for capacitive energy storage, cooling agent, gate dielectrics, memory cell dielectrics and more [81]. However, charge-trapping effects in high-K dielectrics obstruct their efficient functionality and commercialization. High dielectric constant, thermal stability, large electronic bandgap, and band offsets are some of the requisites for the lucrative appliance of high-K hybrid and composite PDMs. Table 16.2 shown below indicates different commercially available polymer dielectrics with their corresponding dielectric constant.

Besides adequate insulation, nonpolar PDMs can be contrived polar by incorporating impurities in small concentrations. This enhances the material's ability to store considerable charges under small, applied signals. An early instance of such features was observed with polyvinylidene fluoride incorporated with chlorotrifluoroethylene impurities [87]. The presence of fluorinated substituents as impurities, fillers, and nano-additives are progressively attached to diverse aliphatic and aromatic PDMs to amplify the dielectric response and characteristics to external electric fields for energy storage applications. Diverse works in recent years have been executed with perovskite-based ceramic nanoparticles as conducting nanofillers and inorganic fillers, in addition to modified core-shell structures and interface in the hybrid polymer matrix to escalate the resulting dielectric

TABLE 16.2
Different Commercially Available Polymer Dielectric Material

Polymer Dielectric Material	Dielectric Constant (*k*)	Ref
Polyvinyl chloride (PVC)	3.4	[82]
Biaxially oriented polypropylene (BOPP)	2.2	[83]
Polystyrene (PS)	2.6	[61]
Silsesquioxane	2.8 – 3.0	[84]
Polynaphthalene	2.2	
Poly(tetrafluoroethylene) (PTFE)	2.1	[85]
Polyimide (PI)	3.2	[86]
Polyvinylidene Fluoride (PVDF)	12	
Polypropylene (PP)	2.2	
Polycarbonate (PC)	2.8	

characteristics [88–90]. Chemical and thermoelectric trade-off and perturbations demand optimization of PDM constituents for high operational efficiency.

16.7 ALIPHATIC STANDALONE AND HYBRID HIGH-K FLUOROPOLYMERS

Fluoropolymers are a class of dielectric polymers from the family of fluorinated compounds. Aliphatic fluoropolymers are pragmatic macromolecules with attractive dielectric characteristics. They are highly desired due to chemical inertness and mechanical and thermal stability alongside promising dielectric properties [91]. The global market of aliphatic fluoropolymers grew steeply over the years. with a significant share of polytetrafluoroethylene (PTFE) or Teflon due to their considerable hydrophobic characteristics [92]. The advent of Teflon is used widely in real-life components. Adequate mechanical toughness for Teflon equipment arises from the elementary hydrocarbon structure. Among the C—C and C—F structural backbone, C—F contributes immensely due to rigid covalent ionic bonds, and the high-grade fluorination imparts appreciable hydrophobic effects [93]. Novelty in PTFE films, with a thickness (< 5 μm), have been fabricated via the heat treatment process of water emulsion PTFE nanoparticles coated copper foils [94]. Subsequent thermal annealing allows PTFE films to soften, expand, and sequentially level to yield a compact film with fiber-like networks. The dense interconnected fibrous microstructure permits heat treatment up to 380°C, with a plausible breakdown field of 410 MV m^{-1}examined at room temperature, which decrements with a consequent rise in temperature from 395 MV m^{-1}at 150°C to 350MV m^{-1} at 200°C, respectively. Figure 16.6 illustrates the epoxy resin-coated polymer interconnects.

Although temperature resistance from 300–350°C proves beneficial, a major limitation evolves in terms of low dielectric constant and material processing, due to insoluble crystalline helical structures. The electrical breakdown of PTFE films relies on surface flatness. The effect increases during ultrathin film processing via the coating process and high-temperature calcination due to interconnected fibrous structures [95]. Material improvisations with Teflon and other aliphatic fluoropolymer dielectric derivatives have been carried over the years in either chemical or physical ways, with the chemical means allowing the addition of more hydrogen contents, while the physical means grant the incorporation of diverse heteroatoms respectively. The substitution of one fluorine atom with a heteroatom eased the processability, with a more distorted helix at the expense of a reduced dielectric constant. Luoet al. [95] adopted the use of an epoxy resin coating to smooth the interconnected fibrous units with a PTFE film. The optimized implication (0.5 wt%) of epoxy resin decreases the local electric field perturbations at the electrodes and simultaneously increases the surface flatness.

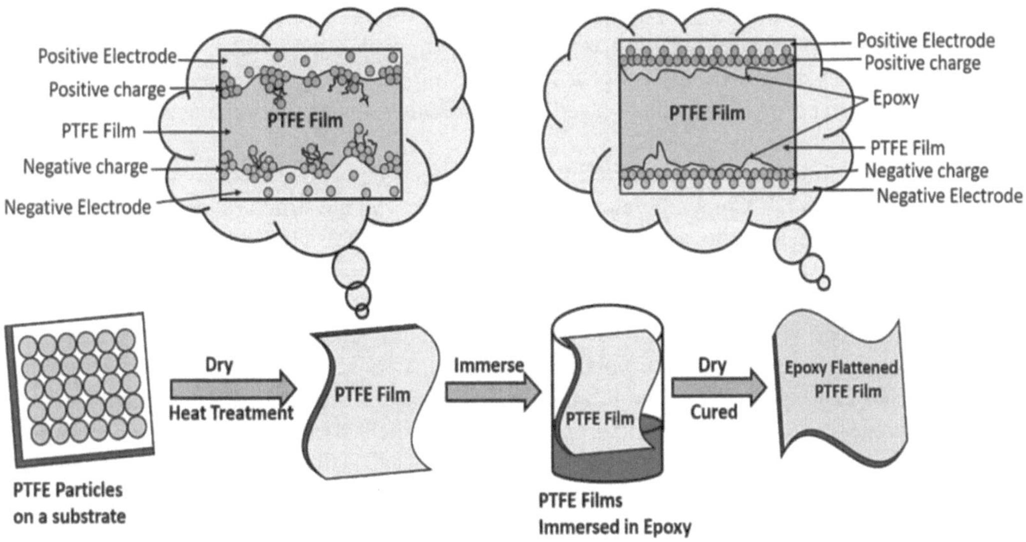

FIGURE 16.6 Sequential process of epoxy resin coating over polymer interconnects with pre and post-epoxy-coated PTFE films (inset).

Given energy storage applications and the consequent requisite dielectric constant, aliphatic fluoropolymers were redesigned with different plausible functional groups and heteroatoms via copolymerization of commercially accessible dielectric fluoropolymers that gained priority in recent years [96–98]. The inductive effect of high electronegative chalcogens like fluorine and chlorine was exploited in most polymeric units. The studies conducted by Kraytsberg et al. [99], Ghosh et al. [100], and Banerjee et al. [101] perceived the enhanced electron density among the neighboring functional group with the presence of fluorine atoms at the polymeric backbone or on a spacer linked to the backbone, which could enhance the dielectric properties of the host polymer. Additionally, the fluoropolymers were intentionally incorporated with Phosphorus atoms to escalate the charge dynamical features to extend the application to energy conversion devices (fuel cells) for charge diffusion. Wehbi et al. [91] synthesized and examined the conductivity of polymer electrolyte membranes processed via phosphonic acid fluoropolymers. The resulting membrane displayed improved proton conductivity over the former existing membrane at 100% relative humidity. The observed attribute is associated with the presence of difluoromethylene adjacent to a phosphonic acid functional group. The study of such model compounds escalated with upgraded charge diffusion kinetics through the bulk electrolyte. Compared to amorphous and nonpolar PDMs, ferroelectric PDMs are characterized by switchable and implicit permanent dipoles with an overall high dielectric constant [102]. Notable ferroelectric PDMs include PVDF [103], hexafluoropropylene [104], and their associated copolymer derivates permit tunable dielectric characteristics. The semi-crystalline nature of PVDF with five distinct crystal structures depends on the chain conformers (trans (T) or gauche (G)) with molecular packing in the crystalline phase [105]. The fundamental chemical structures of PVDF are represented via $-\left(CF_2 - CH_2\right)_n -$ units. The polar β-phase with a molecular chain, in all-trans configuration while direct crystallization from solid solution, generates a non-ferroelectric α phase with TGTG' conformation and low dielectric properties. Thus, electrical poling and mechanical stretching are some of the diversely opted strategies for α to β phase transformations. High dielectric constant and ferroelectric hysteresis of β-phase PVDF compromises the energy density of polymeric composition. The sequential investigation of the energy density of different phases of PVDF by Xiaet al. [106] claimed the highest of 14 Jcm^{-3} for the γ-phase over the others, with an electric field threshold for phase transitions. The three trans coupling to a gauche (T_3GT_3G') configuration directs to small polarizability and constrained remnant displacement, favors highly discharged energy efficiency [107]. The stability criteria for α-phase are constrained to 300 MV m^{-1} field limit, and an α to γ phase transition appears. This contravenes the fundamental conception of large ferroelectric hysteresis by high β-phase PVDF compositions. Such anomalous behavior is attributed to the erratic reversible small-sized polar structures under the influence of an external electric field.

It is visceral to eliminate the ferroelectric hysteresis effects and simultaneously enhance the discharge efficiency of PVDF. Conglomerate strategies with incorporate structural defects and impurities have been increased to produce high-K relaxor ferroelectric PDMs [108, 109]. Even so, potential concerns with relaxor ferroelectric PDMs emerge in terms of saturated polarization at low-intensity electric fields. The goal is to maximize the energy density and confer requisite properties for considerable energy storage applications [110]. The outcome of the study showed moderate dielectric constant with an enhanced saturation polarization limit that is beneficial over their counterpart. High-K aliphatic PDMs like PVDF and P(VDF-TrFE) display considerable dielectric constant (k ~ 10) at room temperature, while dielectric properties, such as dielectric and energy loss, display profound oscillations and instability. The study conducted by Hoet al. [16] and Wei et al. [34] expounded the reason for formerly disclosed characteristics with associated parameters. Most aliphatic PDMs with low glass transition temperatures (T_g < 120 °C) and strong chain segmental motion manifest large energy dissipation.

Material improvisations in view of high-K hybrid and composite PDMs inhibit high crystallinity due to polycrystalline molecular chain displacement and enable the resultant dielectric compositions to withstand high thermal-dependent dielectric response. Electrostatic energy storage applications of aliphatic hybrid polymer nanocomposites are widely implemented in modern-day supercapacitors and hybrid ionic capacitors respectively under diverse configurations. The study over TiO_2 dielectric aliphatic polymer nanocomposites conducted by Tomara et al. [111] reported a superior dielectric permittivity over their standalone polymer counterpart, yet the constrained insulation effect at high temperatures due to active charge mobility weakens the dielectric performance of polar aliphatic nanocomposites with strong polarization effects. Interfacial chemistry, modified core-shell structures, and many other structural, chemical, and thermodynamical parameters need to be configured and optimized appropriately, with extended insight into aromatic fluoropolymers and their hybrid derivatives for further advancements. Meanwhile, it is significant to introduce uniformity and pacify the heterogeneity of a dielectric constant at the interface of high-K hybrid PDMs for adequate energy storage density and operational efficiency. Lower fractions of appropriate epoxy resin with nanostructured polymer additives, such as conductive fillers in complex aromatic branched polymeric units, prompt the current and progressive material advancements with potential scope for lucrative applications.

16.8 AROMATIC AND COMPLEX HIGH-K HYBRID PDMs

Polymer dielectric materials with adequate dielectric response and high-temperature resistance are desirable for next-generation energy storage applications [16, 60, 112]. PDMs for high-temperature applications must acquire high T_g and high melting points to ensure considerable operational stability. Although high dielectric constants with standalone and hybrid aliphatic fluoropolymers are attractive, crystalline and amorphous dipole mobility stimulates the ionic and electronic mobility, which constitutes a steep dielectric loss [113]. Progressive research with PDMs shifted toward aromatic compounds due to alluring dielectric characteristics. High-K aromatic PDMs demonstrate high temperature and voltage durability. Such a class of compounds finds their application as interlayer thin film dielectrics in microelectronics, due to adequate tensile strength and easy microelectronic fabrications [114]. The backbone of aromatic PDMs is linked via phenyl rings with C—H, C—C, and C——C bonds in the backbone structures that embrace the polymer membrane with substantial mechanical, chemical, and thermal strengths [115, 116]. However, a hydrophobic group forces the absorption of excessive water molecules. The resultant ionic and hydrogen bonds of void-occupied water molecules in the polymer matrix constitute swelling, mechanical deterioration, and oxidation instability, alongside reduced dielectric permittivity of the reactant precursors [117, 118]. The effects and characteristics of aromatic PDMs are essential to adopt for electrochemical energy conversion devices and energy storage applications. The feasibility of an enhanced intrinsic dielectric constant with aromatic PDMs is plausible with a consequent rise in the total polarizability and rigid aromatic chain stacking. Aromatic polymers are thus prioritized for a high π-π stacking force, which constitutes large repeatable units. Material modeling via highly polar and electronegative branches using aliphatic and aromatic chains are of ultimate significance [119]. However, certain dielectric parameters remain ambiguous when applied to polymeric systems, such as the poor correlation of the dielectric constant with density and dipole moment, and demand further clarity.

Polyimides (PI) are a class of polymer fabricated by employing two monomers viz; dianhydride and diamine. The synthesis involves a two-step approach, with a nucleophilic attack, followed by a cyclodehydration rection of reactant precursors and thermal imidization of soluble polyamic acid (PAA) intermediates with H_2O evolution to form insoluble PI [120]. Monomers with discrete characteristics determine the overall property of the polymeric composition. Large functional groups enclosed within principal aromatic rings produce a large increase in adhesivity, stability, and dielectric response [72]. The aromatic PIs showcase a linear, symmetric dimide structure with an effective implicit net dipole moment for repeated units. Moisture absorbance tendency by PIs enhances the PI layers and increases the dielectric constant. The fluorinated chains in the aromatic PI backbone in the polymer matrix hinder the moisture absorption property, decrease the dielectric constant, and increase the responsivity of the host polymer toward chemical attacks. The undesirable features of PI are eliminated via material modifications and synthesis of novel aromatic hybrid high-K PDMs and their nanocomposites. The study conducted by Zhanget al. [121] over the dielectric response of three and four-layered PI films illustrates the superior insulation property of three-layered films. The dielectric response of four-layered PI films remained constant while a gradient in three-layered PI films was obtained between a frequency sweep of 50 -10^6 Hz respectively. The opposing response is devoted to the reduced porosity in three-layered films. Inflexions in the dielectric loss of three-layered films decreased and increased with a frequency progression. While the prominent contribution from relaxation polarization loss occurs at lower frequencies, conductivity loss dominates at high-frequency regimes, thus summing up the total dielectric loss within a polymer dielectric matrix. Although the dielectric loss in four layered PI films demonstrated frequency independence, the weak relaxation peaks corresponding to (~ 10^2 – 10^3 Hz) are associated with broad interface polarization relaxation time.

Besides PI, polyurea and polythiourea are well known for their fascinating dielectric, ferroelectric, and piezoelectric properties. Sufficient dielectric constant (~ 4.2) at 690 MV m^{-1} breakdown field and low leakage current in aromatic PDM films via vapor deposition polymerization techniques [122, 123]. The perceived high energy density (9 Jcm^{-3}) is attributed to the structural frameworks of the polymer matrix. The direct unification of highly polar urea groups with hydrophobic aromatic rings facilitates the delocalization of electrons, improving the dielectric response and regulating the dipole-dipole coupling to cultivate low dielectric loss. Recent computations from the study of Plania et al. [124, 125] claimed the metal ions incorporated polymeric structures via covalent bonding present a potential scope to inflate the dielectric properties of host PDMs. The resulting infusion allows fine dispersion of metals through the polymer backbone and eliminates undesirable segregation effects. Additional benefits of metal-heteroatom bonds in the polymer matrix reflect over increased dipolar and atomic polarization, due to metal-heteroatom inductive effects over the carbon-heteroatom counterpart.

A significant priority is rendered to hybrid materials with a symbiotic aggregate of organic and inorganic constituents at the molecular scale. State-of-the-art energy storage of electrochemical devices, including supercapacitors, employ electrochemical double layer capacitance and pseudocapacitive proposition for promising grid-scale applications [126]. The optimized synthesis of high-K

PDMs thus becomes an essential criterion, with a cautious scaling of foreign substituents. Eco-friendly hybrid materials for energy storage applications are of high interest. The synthesis routes of hybrid materials are broadly classified into ex-situ and in-situ techniques, with adequate reliability provided to in-situ routes over the latter counterpart, due to the interplay of nanomaterials over the polymer matrix and their consequent effect on the composition's performance. The ex-situ synthesis involves the incorporation of a blend of one or more substituents from the molten or solvent state. In such a procedure, the end material exists in a final shape.

Nonetheless, the material architecture with diverse substituents is exemplified for future research and development. Appropriate concerns over the concentration of blending components are to be provided to preserve compositional homogeneity despite the explicit incorporation of nanocomposites. In concurrence with former artifacts, Puliet al. [127]synthesized barium zirconium titanate-barium calcium titanate (BZT-BCT) and PMMA nanocomposites. The resulting composite, due to the incorporation of BZT-BCT within the PMMA backbone, revealed enhanced dielectric properties, due to increased compatibility upon grafting. On the other hand, an inorganic filler-modified hybrid polymer matrix is contemplated to serve as a potential dielectric material for advanced energy storage technology. Huang et al. [128] addressed the fabrication of TiO_2nanowires-modified barium titanate nanoparticles via a solvothermal process. The outcome of examined $BaTiO_3$@TiO_2 poly (vinylidene fluoride-co-hexafluoro propylene) (PVDF-HFP) composites demonstrated considerable dielectric strength alongside interfacial polarization, due to robust interactions between $BaTiO_3$@TiO_2 particles with the host polymer matrix. The featured highlights are associated with the inflated interfacial areas between the fillers and polymer matrix. Figure 16.7 shown below demonstrates different strategies used to synthesize hybrid polymer dielectric materials: Maji et al. [129] fabricated PMMA-$NiFe_2O_4$ nanocomposite via an in-situ polymerization technique. The amalgamation of $NiFe_2O_4$ within the host polymer improved the

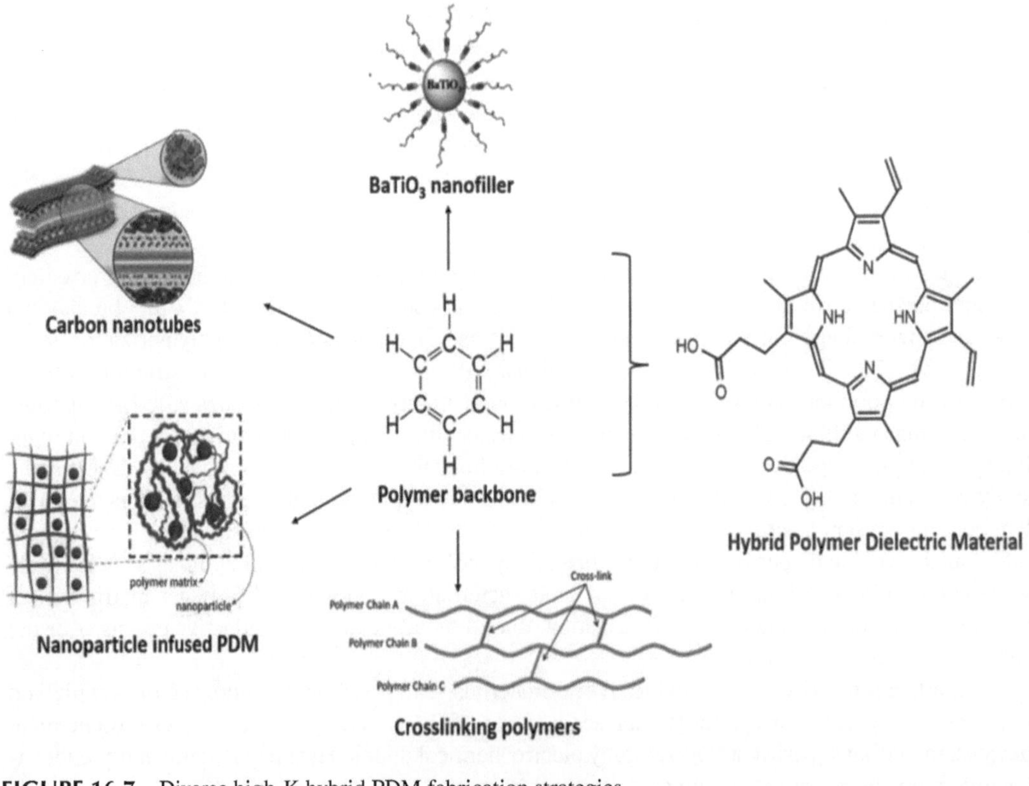

FIGURE 16.7 Diverse high-K hybrid PDM fabrication strategies.

thermal stability facilitating the extended scope for high catalytic activity. Meanwhile, the Fe_3O_4@ TiO_2 epoxy nanocomposites prepared by Nardi et al. [130] demonstrated a gradient in permittivity with relaxed electric stress across the electrode-insulator interface. Considerable dielectric constant by aliphatic polymer dielectric illustrates an inverse interplay over the isothermal response and structural stability. The crystalline and amorphous dipole mobility stimulates the ionic and electronic mobility, which constitutes a steep dielectric loss. High π-π stacking force constitutes large repeatable units in aromatic polymeric structures. Infusion of polar branches and electronegative substituents alongside fillers and nanoparticles in the host polymeric backbone are different strategies employed to enhance the dielectric responsivity under diverse thermodynamic constraints. The idea of polymerization of two or more highly polar polymeric rings with suitable thermal and chemical stability in a resultant high-K hybrid polymeric structure widens the scope of dielectric research for energy storage. Even so, a reasonable choice of highly polar functional groups and polymeric units with appropriate macromolecular structures in terms of main and side chain units is essential to escalate the dielectric feedback of the overall compositions.

16.9 ROLE AND PROPERTIES OF HIGH-K AND HYBRID PDMs

A hybrid polymeric material comprises two or more foreign substituents integrated within a single polymer matrix. The principal motive of hybrid polymeric structures arises to exploit the synergic properties of the participating standalone polymeric units. The formation of a hybrid polymer involves multiple intermittent interfacial interactions. Synthesis techniques strongly rely on the components of the hybrid polymers, and the end product may illustrate a diverged trait relative to the employed precursors. However, the derived hybrid PDMs rely on the relative abundances and prominence of individual constituents. The blend of heteroatoms with discrete base components to synthesize a thermal and chemically stable high dielectric polymer has attracted immense scientific attention. The amplitude of interfacial interactions determines the reinforcing effects of the constituents on the properties of the hybrid polymer matrix. This imposes requisite criteria to establish a considerable interfacial area between the fillers and polymer matrix [131]. One of the critical factors involves reducing the electrical breakdown (E_b) of the dielectric composite. The approach to accomplish the former motive is categorized into three groups, viz, the addition of nanosized fillers, core-shell composites, and multilayered structures respectively [132–137]. The E_b of hybrid PDMs relies upon the dispersion of fillers and dispersants (free small molecules) in the polymer matrix. Improved E_b and mechanical properties are attributed to low filler concentration (< 10 vol%) due to electron scattering effects and cultivate material processing issues under large-scale productions [138–141]. Recent works in hybrid PDM revealed a prominent effect of T_g over the polymeric chain dynamics. The inverse interplay between T_g and the dielectric constant equivalently decreases the dielectric breakdown. Low T_g polymers are thus most favorable with extended dielectric properties as hybrid polymer constituents within the matrix. The flexible chains twine around the nanoparticle and model a strong interacting interface (soft interface). Opposingly, the hard interfaces (non-flexible chains with high T_g) diminish the breakdown performance.

In the early stages, ferroelectric ceramic fillers, such as $BaTiO_3$(BT), Pb[(Zr, Ti)O_3] (PZT) and many others, gained prominent study due to high dielectric constants and polarizability with non-centrosymmetric structures. Liet al. [142] first examined the BT particles within a PVDF composite polymer matrix, while Muralidhar et al. [143] reported the variations between real and imaginary dielectric constants in the former BT-PVDF composite PDM against a large range of frequency sweep and modified volume fraction of filler contents. The outcome of the study provided an increase in permittivity of the composite to 133 at 10Hz and 30 °C, corresponding to a proportional increase in filler content of 70 vol% in the polymer matrix. The large surface area of BT nanoparticles was specifically incorporated within the host polymer matrix to enhance the interfacial interactions. The computational study was conducted by Wang and Tan [144] on the local electric field distribution over the core-shell-filled composite polymer matrix. The distinction between the dielectric constants of participating constituents in the matrix, the vertical component of electric field (E_y) is steeply enhanced along the applied field. The following characteristic claims local electric distortions near the polarized particle poles. While a feeble electric field enhancement is obtained in the core-shell particles, the depolarization effect nullifies the former effect due to accumulated charges in the filler/composite matrix interface. Conducting fillers with metal particles demonstrate divergent electrical mismatch with different host polymers and compromises the dielectric constant of the resulting hybrid PDMs. However, nanosized conductive fillers begin to epitomize nano effects, such as quantum confinement and coulomb blockade [145]. The ultra-small filler nano-particles are utilized effectively to minimize the dielectric breakdown relative to ceramic nanofillers [146, 147]. A substitute to conventional nanofillers was attempted by Hao and her co-workers [148] with the use of ultrafine nanocrystals. The dimensions of the nanocrystals were regulated under 10nm and obtained fruitful results. The nanocrystals effectively elevated the dielectric strength and electrical displacement of the PDM composite, due to comparatively large interfaces. The successful observations illustrate the significance of filler dimensionality. The smaller dimension of fillers constitutes an effective thin film dielectric, which enhances the capacitance per unit volume in any conventional dielectric capacitor.

Numerous organic-inorganic hybrid polymer dielectrics have been fabricated via distinguished techniques, with pre-synthesized fillers opted frequently in the polymer matrix. Besides different structural customization, the final hybrid PDMs display conglomeration, which implants inconsistent dielectric response. The polymer aging and a plausible gel formation over time additionally endow the compromise [66]. The emerging issue of dielectric loss concerns the commercialization of dielectric nanocomposites for energy storage applications. An artificially synthesized dielectric layer with fused conductive nanoparticles pacifies the loss factor at the expense of reduced dielectric constant and high leakage current, due to conductive networks or tunneling effects between conductive nanoparticles respectively. A feasible unfold of the former exposition is associated with the homogeneous distribution of fillers in the host hybrid polymer matrix to avoid inter-filler contact [149]. An additional "all polymer" approach using polar polymer blends is considered a favorable method to enhance the dielectric constant at a reduced dielectric loss. A systematic advancement to the former approach is the "percolative dielectric composite," where the conducting organic domains are dispersed within the host hybrid polymer matrix, which results in space segregation across the interface and dielectric domains. High dielectric and polarization response in such cases is attributed to the "super dipole" constructive pattern formed by the conductive domains. The dielectric performance elevates progressively with a consequent increase in the volume fraction of the conductive domains until a percolation threshold. However, the exercise of a hybrid organic-inorganic domain results in high leakage current and poor dielectric breakdown.

A multistructured and multilayer polymeric structure is a novel strategy employed to enhance the dielectric properties of any hybrid PDM. This method is an attempt to use the properties of different polymer constituents for a thick hybrid polymer dielectric layer or film. However, processing discrepancies emerge due to the oversized dimensionality of overall hybrid polymeric compositions [150, 151]. Arbatti et al. [152] probed the influence of compositional architecture over the resulting dielectric properties of calcium copper titanate (CCT)copolymerized vinylidene fluoro-trifluoroethylene (P(VDF-TrFe)) multilayered composite synthesized via layer hot press method. The observations suggested the tendency to enhance the dielectric constant of the resultant hybrid PDM via incorporating additional layers followed by high-temperature annealing. The vitality of the appropriate synthesis route was demonstrated with considerable dielectric constant, showcased by sol-gel synthesized CCT-PVDF composite over the conventional solid-state reaction technique. Such disparity among the properties of the same compositions with different synthesis procedures reflects in the particle size distribution. The smaller and fine particle size achieved via sol-gel over the latter synthesis route befits and alters the dielectric behavior and response of the host polymer composites. Regulated spatial distribution of nanosized fillers is thus arguably one of the strategies to optimize the microscopic response. Fillers with discrete length scales are synthesized via anisotropic dielectrophoretic assembly in association with flow/stress-gradient induced alignment routes. The filler alignment within the host matrix can considerably alter the dielectric, conducting behavior with a proportional effect over recoverable energy density.

The compact outline suggests the importance of advancements in standalone polymer dielectrics to more complex hybrid structures via the implication of fillers, modified core-shell structures and multilayered and multistructural approaches. Although development in dielectric constant within the broad spectrum of dielectric properties accompanies at the expense of an equivalent decrease in other associated parameters (leakage current, T_g). Collective recovery of dielectric response and performance stands as the prime concerns among different research groups. In compilation with prior studies, homogeneously distributed conductive nanosized fillers over the organic-inorganic counterpart are considered more favorable under optimized volume fractions to avoid interfacial aggregations. The poor compatibility between organic and inorganic nanosized fillers obstructs the growth of high-K hybrid polymer dielectrics. Besides, the high leakage current due to spatial perturbations and inflexions in filler content within the polymer matrix raises the poor dielectric response. Significant variability in permittivity among different polymeric constituents distorts the electric field inside the hybrid polymer matrix and imparts adequate dielectric breakdown. Dimensionality constraints of nanosized fillers are to be taken care of to avoid quantum effects (coulomb blockade and quantum confinement). The hybrid polymeric matrix must henceforth combine the constituents with a controlled difference between the dielectric properties to avoid interfacial breakdowns.

16.10 ENTERPRISE TO ENHANCE DIELECTRIC CHARACTERISTICS

The current understanding and progress of high-K polymer and polymer composites for energy storage applications lies in the consequent advancements in capacitors, supercapacitors, batteries, and many other compact technological devices. In accordance with the prior investigations, standalone and hybrid polymers showcase hindrances to requisite dielectric characteristics. Any enterprise to enhance an individual dielectric property is accompanied by an equivalent decline in the other, due to significant interdependence. The intent of polymer nanodielectrics emerged with a broad motive to infiltrate the bandwagon of nanotechnology, a multi-component dielectric with hybrid polymeric structures. The dimensional priority in terms of polymer nanodielectrics and nanocomposites is associated with the differences in material engineering techniques and demands [153]. The conventional hybrid polymeric structures and composites incorporate high micro-sized filler concentrations $(\geq 30\%)$ which scale down to lower volume fractions under nano-dimensions. Nanometric manipulations of the host polymer backbones introduce significant variations in the dielectric properties and response. Additionally, the stark distinction with polymer nanocomposites lies in interfacial chemistry. The architecture of a polymer dielectric nanocomposite involves the application of an extended range of filler materials classified under zero (nanoparticles), one (nanofibers) and two dimensions (nanosheets) respectively. Such fillers are activated to modify the interfacial chemistry (bond strength and dielectric characteristics) within the host hybrid polymer matrix. Figure 16.8 shown below illustrates the interfacial gap between different-sized fillers within a polymer matrix.

From Figure 16.8, the available interfacial region differs between different-sized fillers within the host polymer matrix. Preserving the same volume fraction of fillers, it is evident from the visual perspective that more interfacial region is available with nano-sized fillers than with their micro-sized counterpart, due to an increase in inter-particle distance. The infusion of nanoparticles alters the epoxy of the in-contact polymer matrix, and the extended surface area architects the nature of the host polymer composite. The increment in the interfacial area around the surface of the nano-fillers implants a gradient in the dielectric characteristics, due to morphological variations and charge chemistry. The vitality of interfacial chemistry is devoted to the additional phase (three phases) contribution by the nanoparticles, distinguishable from the pillar polymer material [73]. In general, melt stretching [154], hot pressing [155], and spin coating [156] are numerous nanocomposite synthetic strategies. Surface functionalization is the prior approach before the progression of any composite synthesis to establish

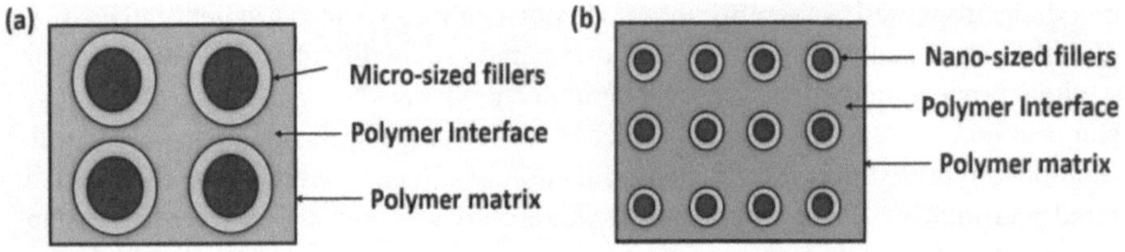

FIGURE 16.8 Interfacial chemistry of fillers in a polymer matrix, (a) micro-sized fillers, (b) nano-sized fillers.

better compatibility between the filler and the host polymer matrix. Epoxy resin is a prevalent potting material and a technique applied for suitable electrical insulation. Pure epoxy showcases feeble thermal conductivity (~ 0.2 $Wm^{-1}K^{-1}$) and conservative dielectric properties. However, nanofiller-incorporated epoxy resin uncovers the possibility of improving the former discrepancies. Alumina (Al_2O_3) is a conventional epoxy resin used to encapsulate the electrification apparatuses [157].

Zhou et al. [158] addressed a comparative analysis of the dielectric properties of PVDF polymer dielectric nanocomposite with standalone $BaTiO_3$ (BT) nanoparticles. The outcome illustrated permittivity surge at low temperatures and positive frequency response with nanocomposites synthesized via hydroxylated BT nanoparticles (h-BT). The observed features are attributed to the strong interactions between the h-BT nanoparticles and the host polymer matrix, and constitute high energy density. Additionally, Ehrhardt et al. [159] employed n-octyl phosphonic acid (OPA) and pentafluorobenzyl phosphonic acid (PFBPA) for surface modification of BT nanoparticles. The subsequent nanocomposites were fabricated via spin coating of dispersed BT nanoparticles in Poly(vinylidene fluoride-co-hexafluoropropylene) poly(VDF-co-HFP) copolymer solution. The fallout illustrates a superficial increment in the dielectric permittivity of phosphonic acid with a fluorinated organic host over the n-octyl phosphonic acid incorporated polymer matrix. The analysis presents fair evidence of high chemical compatibility offered by pentafluorobenzyl phosphonic acid with the host poly(VDF-co-HFP) respectively. However typical limitations include homogeneous nanoparticle dispersion and interface engineering to achieve requisite properties. Nanoparticle surface modifications executed via diverse techniques refrain from showing full potential functioning of high-K hybrid PDMs, due to the insignificant role of modifiers for property appraisal [160–162].

In compliance with former kinds of literature, it is now familiar that the most reliable method to enhance the dielectric constant of the host polymer includes fillers with high volume fractions (> 20 vol%). The richness of high-K is accompanied at the expense of reduced dielectric breakdown. The nanofiller contacts become inevitable, causing a thin layer of polymer/organic molecules to sufficiently allow the charge carrier migration amid the conductive fillers. The prominent interfacial barrier effect at high filler concentrations and aggregation enforces a conductive pathway upon established contacts and constitutes degraded dielectric breakdown [163]. The invoked dilemma can be resolved with matrix-free core-shell structured polymer composites. The framework comprises a direct grafting of the polymer matrix with the nanoparticles and the resultant composite system obtained via coupled grafting of nanoparticles at elevated temperatures [164]. This forces a thick and uniformly spaced grafter polymer layer between the nanofillers. Nanocomposite synthesis via core-shell strategies is categorized into four groups viz:

i. Direct account of core-shell nanoparticles fabricated via "grafting from" [165]
ii. Direct account of core-shell nanoparticles fabricated via "grafting to" [166, 167]
iii. Employed organic-inorganic nanofillers [141, 168]
iv. Other nanofillers [145, 169, 170]

A double core-shell approach presents an improvised strategy to augment the dielectric strength of hybrid PDMs. The technique involves a simple blend of a filler incorporated polymer with a unique polymer shell into a nano-composite forming a double-core structure. The end hybrid PDM is obtained after executing one of the former disclosed composite synthetic strategies. However, traditional nanocomposites with embedded zero-dimensional nanoparticles suffer a dominant reduction in dielectric breakdown, due to local electric field concentration upon a mismatch of dielectric permittivity of filler and the host polymer matrix. Efforts to subsidize the local electric field effects involve surface functionalization, core-shell, and multiplayer polymer architecture. Multilayer PDMs comprise a central layer with promising dielectric breakdown strength and an outer layer with high permittivity. Such interleave composites pacify the compromise between permittivity, dielectric breakdown, and dielectric loss. Multilayer and multistructural PDMs, in recent years, have monopolized polymer-based energy materials for advanced energy storage applications, with a considerable dielectric response. High-quality coextruded defect-free multilayer hybrid PDMs with alternative high energy density polymeric structures have exemplified giant dielectric breakdown strength over nominal dielectric films. Rationally designed multilayer PDMs are much more successful in integrating the suppressed electrode/interface effects on charge injection and transport to elevate the energy density [171].

The aggregate of the section presents distinct feasible scope to enhance the dielectric characteristics of hybrid PDMs. The choice of nanosized fillers over their micro-sized counterpart establishes the impact of the interfacial chemistry between different filler constituents within the host polymer matrix. The venture to elevate the material's dielectric response via high filler volume fractions decreases the breakdown strength, due to filler aggregation across the interface. Different kinds of literature have presented a surface modification of nanofillers for homogeneous dispersion within the polymer matrix and a resultant uniform PDM composite film. Distinct synthetic procedures, including hot pressing, grafting, and many others, were reported with standalone benefits. On the other hand, core-shell strategies eliminate the interfacial nanoparticle segregations and discard the electrical percolation transition in PDM nanocomposites incorporated with conducting nanofillers. The synergy of double-core shell structures improves the overall dielectric characteristics of the host high-K hybrid PDMs. Optimum filler concentrations in conjunction with multi-layered structures, is essential to regulate the electric field distribution and establish adequate electrical insulation for enhanced energy storage applications. Tuning of stacked layers proposes a positive impact over the dielectric interface and modified dielectric performance. The consequence of multi-layered hybrid PDMs allows the polarized ions in the dielectric interface to develop an active depolarizing electric field to pacify the charge carrier displacement across the interface and refrain from the propagation of electrical trees. The coalition renders escalated dielectric breakdown and energy density.

16.11 NANOFILLERS AND NANOPARTICLES IN HIGH-K HYBRID PDMs

The universal figure of merit for the energy storage property of any dielectric material is associated with the extent of dielectric breakdown and loss. Pure and derived ceramic dielectrics, although they manifest copious amounts of dielectric constant, still lags within formerly established parameters. Dielectric polymers with an economical synthetic approach exhibit chemical and mechanical stability with a sufficient dielectric breakdown at the expense of a low dielectric constant [51, 172]. Material engineering of polymeric units followed by co-polymerizations with effective ceramic, conductive, organic, inorganic, and hybrid nanofillers have raised the standards of energy storage devices in recent years. The birth of the nanocomposite approach affords the unified advantages of nanofillers and the host polymer matrix. The synergetic effect assists the cooperative enhancement of the dielectric constant and breakdown alongside energy storage density with controlled nanofiller features and interfacial interactions. Substantial research with variable chemical compositions and dimensions (0D (nanoparticles), 1D (nanowires), and 2D (nanosheets)) of nanofiller blended ferroelectric polymeric composites have been attempted over the last two decades to upgrade the dielectric performance [22, 173–175]. The energy density of a hybrid PDM nanocomposite is evaluated as the volumetric summation of individual constituents, besides the role of interfacial regions [176]. $BaTiO_3$ is one of the well-known and widely accessible dielectric nanofillers incorporated in multiple polymer matrices [176–181]. Untreated nanoparticles are prone to interfacial aggregation and constitute poor dielectric films with reduced breakdown strengths. Surface modifications using organic ligands and polymers are frequently employed to mitigate dielectric disparity across the polymer-nanofiller interface. Meanwhile, the large filler loading ratios adversely introduce mechanical instability with poor tensile strength and processibility. The demand for controlled nanofiller content and homogeneous dispersion is rendered a giant priority to subsidize the trade-off between different dielectric parameters. Recent progress has shown the possibility of hierarchical nanofiller structures with complex material design, layered nanocomposites, and multi-component nanocomposites for enhanced energy storage properties.

Luo et al. [180] addressed a high energy density of 8.13 Jcm^{-3} by hydantoin epoxy resin-coated BT-P(VDF-HFP) nanocomposite. The surface functionalization certified a homogeneous nanofiller dispersion even at elevated loading ratios. The resultant 20 vol% BT nanofillers within the polymer matrix, although they demonstrated desirable outcomes, yet the goal of high-K was accomplished at the compromise of dielectric breakdown strength (< 100 MV m^{-1}) and loss. The addressed issue is made known based on molecular interactions via strong hydrogen and chemical bonds between polymer and nanoparticle constituents [177]. The nanofiller-polymer matrix interface is a critical parameter to monitor the energy storage performance of the resulting hybrid PDMs. The distribution of nanofillers within fluorinated PDMs results are problematic due to lower surface energy. The agglomeration of nanoparticles constitutes low-field-driven electronic conduction, with abominable porosity and dielectric failure [177, 182]. The infusion of high dielectric nanofillers within the host hybrid polymer matrix escalates the breakdown strength gradually to a threshold after which steeply depreciates due to the emergence of conductive paths, space charge formation, and particle-particle interactions under an applied external electric field [183–186]. Figure 16.9 illustrates the temperature dependent space charge formation under applied voltage and surface modification of nanofillers.

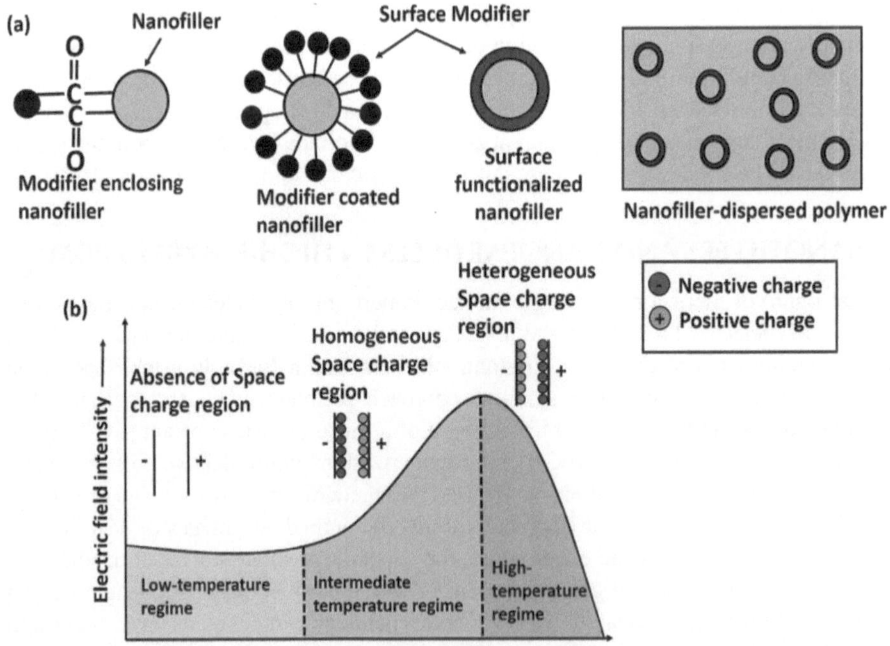

FIGURE 16.9 Surface functionalization and space charge agglomeration across the filler-matrix interface, (a) surface functionalization process, (b) space charge agglomeration at different temperature regimes.

Studies of Ding et al. [187] and Luo et al. [140] illustrate the vitality of polymer dielectric nanocomposites over the conductive paths and resulting interfacial chemistry. The improved dielectric breakdown and energy density of surface modified Zinc sulphide-PVDF (ZnS-PVDF) nanocomposite over the standalone PVDF counterpart attributes to the delayed construction of conductive paths by the charged interfaces and a regulated applied electric field, due to an implicit micro electric field. The surface engineering of ZnS allowed the formation of a non-conductive junction that substantially increased the breakdown strength and the consequent energy density. The weakening of the local electric field and the relaxed space charge accumulation was hypothesized due to the capture and confinement of charge carriers within the hybrid polymer composites through a potential barrier modeled at the heterojunction of ZnO-ZnS nanofillers [3].

Besides the nanofiller concentrations within the polymer matrix, the geometry of nanofillers additionally imposes constraints over dielectric optimizations. Numerous studies suggest a proportional impact over percolation threshold and nanofiller dimensions, apart from sharing an indirect correlation between the flexible concentrations within any polymer nanocomposites. The outcomes also highlight the relative ease to achieve a percolation threshold for larger dimensional nanofillers over their remaining counterparts under congruent volume fractions [188, 189]. With a motive to augment the dielectric constant of hybrid polymers, researchers explored different dimensional nanofillers and their corresponding core-shell structures within the host polymer matrix. Higher dimensional nanofillers (1D and 2D) pacify the concern of thermal instability, interfacial particle aggregations and breakdown inflexions [190, 191]. Luoet al. [140] incorporated 3 wt% Boron nitride (BN)-coated $BaTiO_3$ fillers into the PVDF matrix. The dielectric strength of the polymer composite equipped with BT micro-hybrid fillers enhanced the overall dielectric strength by > 160% respectively. The internal space charge density between the filler-polymer interface is significantly feeble within the same composite architecture when subjected to an electric field. The compiled outcome advocates the vitality of thermally coated BN-based 2D fillers (nanosheets) for efficient high-temperature dielectric response [3]. In agreement with former discussions, Paquin et al. [192] dispensed BN nanosheets within a polymer matrix to construct a homogenous insulation network. The robust insulation network eliminates electrochemical failure and establishes a barrier against space charge conduction and leakage current. The outcome of the investigated composite illustrated a high breakdown strength (610 MV m^{-1}) for a 12 wt% filler content. The works of Bao et al. [193] showcased a collective appraisal of dielectric breakdown and constant, with the use of negatively charged $Ca_2Nb_3O_{10}$ nanosheets alternating with a positively charged polyethyleneimine (PEI) into a PVDF matrix. The perceived feature is attributed to the delayed electronic conduction, due to the blocking effect of the reverse local electric field induced by the $Ca_2Nb_3O_{10}$ nanosheets. The synergy of events escalates the overall energy storage efficiency and increased dielectric breakdown (792 MV m^{-1}) respectively [193]. The prominence of experimental investigations was equally supported by computational techniques, simulations, and models. Linear dielectric behavior of different ceramic and polymer materials was carefully examined by the Maxwell–Garnett model, although the precision and accuracy derailed with the rising complexity among different organic, inorganic, and hybrid high-K polymer dielectrics. Satisfactory predictions for polymer nanocomposites with low loading fractions of fillers constrained their extended applications. The degree of inaccuracy varied due to overestimated inter-filler distance and filler's resistivity amongst different natures of fillers [194]. However, state-of-the-art computations, such as machine learning methods on trained experimental datasets, have been administered as a powerful tool to examine diverse structural, electrochemical, mechanical, thermal, and dielectric parameters. The multilayer perceptron models within a con-structure neural network have significantly accelerated technological advancements via a statistical approach.

The following section highlights the role of nanofillers and nanoparticles to enhance the dielectric properties of a resultant polymer matrix corresponding to the standalone polymers. The outcome from diverse pieces of literature suggests the significance of surface functionalization to eliminate the dielectric disparity between different polymeric constituents across the interface. Different surface modification techniques viz, covalent and non-covalent methods have been widely implied to improve the interfacial interactions via surfactants and strong chemical bonding. Recent progress with high-K hybrid PDMs involves the exercise of hierarchical nanofiller structures with complex material design, layered nanocomposites and multi-component nanocomposites for enhanced energy storage properties. Regulated nanofiller concentrations via an established control over the percolation threshold were achieved with different nanofiller dimensionality. Higher-dimensional nanofillers (1D and 2D) regulate the electronic conduction via blocked conductive channels, due to the local reverse electric field. The following feature uplifts the dielectric breakdown strength with a simultaneous surge in the permittivity. The novel interconnected fibrous networks within a polymer nanocomposite aid to control the nanofiller concentrations and percolation threshold and establish strong insulation properties.

16.12 INTERFACIAL CHEMISTRY OF HIGH-K HYBRID PDMs

A polymer interface is a sub-micron to a nanometer system with an array of polymeric units on a 2D o 3D structural framework. Symmetric interfaces occur with a contact of two or more identical polymeric structures, while asymmetric interfaces bind dissimilar polymeric systems with randomly oriented nanofillers. The polymer interface establishes a prominent role in the dielectric characteristics of hybrid PDMs [195]. Interfacial polarization (also termed as the Maxwell-Wagner-Sillars polarization) ensues due to the interfacial charge accumulation within a homogeneous dielectric polymer matrix [196, 197]. Numerous factors (impurities and incomplete polymeric contacts) govern interfacial chemistry within the host polymer matrix. The effect of interfacial polarization is realized at lower frequency regions. The utilization of interfacial polarization is possible with the account of high-K nanofillers and multi-layered dielectric polymer films [34, 60]. The interfacial regions differ between nanofillers and multi-layered films, wherein the interfacial region for multi-layered structures occurs at the contact regions of different layers, due to variable dielectric and

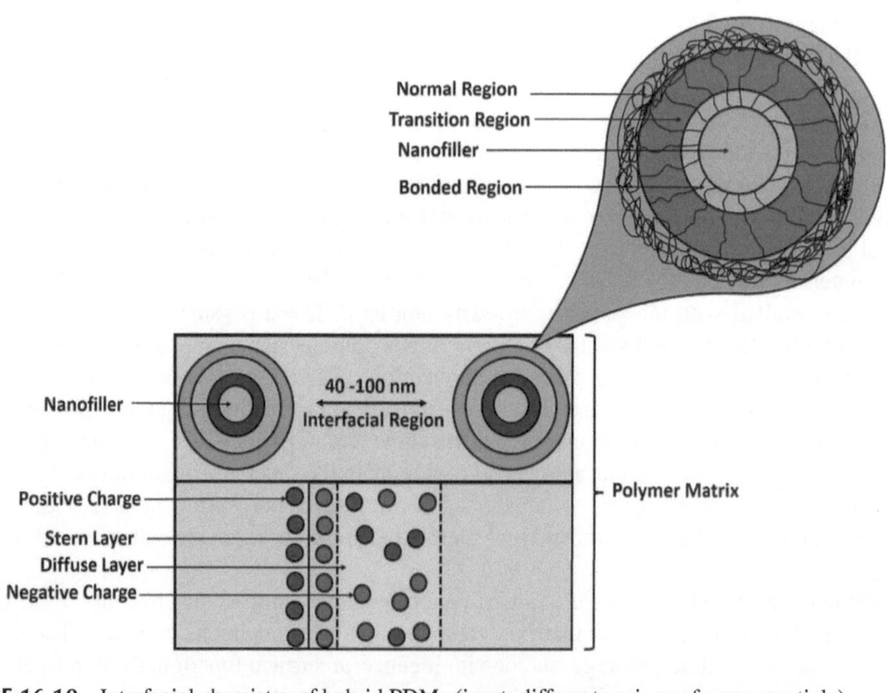

FIGURE 16.10 Interfacial chemistry of hybrid PDMs (inset: different regions of a nanoparticle).

conductivity properties of individual constituents. The interfacial region within a polymer matrix is classified into three zones (from interior to exterior) as shown in Figure 16.10.

From Figure 16.10, the primary region is referred to as the bonded region. In this region, a fraction of counter ions is absorbed by the coulombic interaction around the nanoparticles and additional forces, due to dissimilar Fermi energy between organic and inorganic polymeric constituents within a polymer matrix. The disparity among electronic energy levels between diverse polymeric constituents within the host matrix introduces steep traps with high density, due to a strong potential gradient in the bonded region. The intensity of potential and the depth of the trapping centers decrement with a shift from the interior to an exterior region and constitute shallow traps. Space charge accumulation therefore increases rapidly due to the applied electric field and promotes dielectric failure across the bonded region. The molecular chains in the translational region of the polymer matrix are rigidly associated with the bonded region and the nanofiller surface. The molecular chains adopt a symmetric arrangement such that each molecular chain endures an average force under applied electrochemical stress. The translational region is directed by a strong cohesive energy density (CED) of the host matrix with molecular chain flexibility due to feeble CED fractional values and demands higher energy for molecular bond dissociations. The resulting variations in the molecular chain displacement and segmental motion, besides perturbed crystallinity, alter the polymer nanocomposite characteristics. The transitional region preserves molecular chain periodicity and retains a higher thickness corresponding to the bonded region. Even so, perturbations in the crystallinity are proportionally associated with the surface functionalization of the nanofillers within the hybrid polymer matrix. The exterior normal region is congruent to the matrix, with different enfolded polymer chains around the randomly distributed nanofillers [198]. A diffuse electrical double-layer model proposed by Lewis et al. [195]was used to elucidate the interfacial chemistry within a PDM nanocomposite. The outcomes suggest a suppressed charge accumulation across the filler-polymer interface. The consequent attribute arises due to the charged nanoparticle surface as a result of a gradient in Fermi level and chemical potential of the nanofiller and the host polymer matrix. The positively charged nanofillers constitute a positive potential on one of the layers (Gouy–Chapman layer) due to static charged impurities, trapped ions, mobile electrons, and holes within the nanoparticles, respectively. Also, the negatively charged immobile layer (Stern layer) closer to the nanoparticle surface is formed, due to the presence of absorbed ions, small molecules, and solvated ions respectively. The interfacial charges within a polymer matrix generate a Gouy–Chapman and Stern layer at the nanoparticle interface and govern the dielectric constant, besides the interfacial polarization [199]. Additionally, the Gouy–Chapman diffused layer around the Stren layer potentially operates as an interaction zone composed of oppositely charged ions. This zone becomes responsible for fluctuations in the electric and dielectric properties of the resultant polymer matrix.

The dielectric properties of any hybrid PDM are raised with the incorporation of high-volume fractions of high-K dielectric nanofillers, even though, from prior discussions, high nanofiller loading ratios result in agglomeration across the polymer interface and reduce the dielectric constant due to subsequent pore formation. The homogeneous dispersion of nanofillers at lower concentrations demands optimized interfacial chemistry between the nanoparticles and the host polymer matrix [67]. Customarily, interfacial polarization is regarded as the elemental mechanism in polymer nanocomposites. The energy density of devices escalates with suitable modifications of interfacial chemistry between the nanofillers and surface functionalization of filler-polymer matrix [183]. The addressed issue is improvised via the augmentation of surface modifiers to the nanofillers. Conventionally, surface modifiers subsist to fundamental components viz, anionic functional groups (OH-, SO_3^-) that adhere to the filler's surface, while the secondary component forms the macromolecular chain. As a result, the molecular structure of surface modifiers becomes critical to eliminate the emergence of voids, cracks, and unabsorbed residues in the final systems [180, 200]. Infusion of high concentrations of conducting and semi-conducting fillers enhances the thermal conductivity via conducting networks at the expense of dielectric properties. Thus, thermal interfacial resistance between fillers and the polymer matrix obstructs the homogeneous filler dispersion and subsequent enhancement in thermal conductivity at low filler volume fractions, respectively. The perceived feature is associated with the prominent scattering of phonons at the polymer matrix interface and the filler-polymer incompatibility due to restrained energy transfer [201]. The ambiguity amongst lower fractional nanofillers is experimentally visualized with the aid of dielectric spectroscopy. The powerful spectroscopic technique is utilized to examine the dynamics of polar segments at the nano-filler vicinity [202].

Besides the extensive emphasis devoted to homogeneous nanofillers' dispersion, researchers diverted their focus more toward the performance, ingenious design architecture of polymer interface, and a third phase existence in the high-K polymer nanocomposites. The feasibility of the interfacial phase to accommodate large filler volume fractions (~ 50 vol%) impacts the overall operational efficiency of the energy storage device. According to the works of Shenet al. [184], the interfacial region is estimated to be as large as 250 nm and relies on the degree of association between the polymer and nanofillers. Numerous models were proposed to examine the complex behavior (interfacial exchange coupling effects, induced local CH_2-CF_2 dipole orientation) of high-K hybrid polymer dielectrics. The Kelvin probe force microscope was utilized by Chen et al. [203] to explicitly unravel the local polarity at the polymer interface in ultra-thin dielectric films. The ambiguous interfacial chemistry and different dielectric parameters corresponding to lower filler volume fractions alongside the disparity observed between experimental and simulated results can be explained by considering the interface electric polarization. Sharma and Gupta [204, 205], in recent years, examined the dielectric constant of the epoxy or $BaTiO_3$ nanocomposite interface with the aid of electric force microscopy (EFM) and a numerical model. The EFM phase shift images reveal a higher dielectric constant of the negatively charged interfacial region around the nanofillers than the bulk epoxy. The experimental outcomes of Khare et al.[206] complement the former computational results to indicate the interfacial dielectric constant within a PDM nanocomposite besides the influence of magnitude and charge polarity upon the EFM phase shift signals respectively. Polymer interfaces on the other hand operate as charge transport barriers or charge traps. Interfaces with inorganic shells or surface coatings enhance the barrier potential and restrain the charge displacement and injection. However, interfaces composed of conductive fillers and surrounding shell materials develop into localized field sites to attract charges. The electric field-guided charges may inevitably tunnel or hop around to conduct among adjacent sites. Thin dielectric interfaces constitute the Fowler-Nordheim tunneling mechanism and render high dielectric breakdown strength with minimal dielectric loss. On the contrary, the exposure of dielectric interfaces to high-intensity electric fields releases the trapped charges to support considerable electrical conduction and dielectric failure [3].

The dielectric nanofillers are a means to enhance the dielectric constant of hybrid polymer materials. Also, high nanofiller loading ratios afford agglomeration across the polymer interface and pacifies the dielectric constant due to successive pore formation. The incorporated nanoparticles significantly alter the material's morphology. The strong hierarchy in space-time establishes a dominant challenge for hybrid high-K polymer dielectrics. Polymer interfaces in hybrid materials play a vital role where the mesoscopic dimensions (~ 10 nm). The interfacial region is classified into three different regions from interior to exterior, namely bonded region, translation region, and normal region, respectively. While a bonded region absorbs the counter ions due to coulombic interaction, the transitional region retains the molecular chain periodicity and higher thickness. Finally, the normal region resembles the host polymer matrix, with different enfolded polymer chains. The energy density of devices upgrades with the appropriate architecture of interfacial chemistry between the nanofillers and surface functionalization of the filler-polymer matrix. The molecular structure of surface modifiers thus becomes vital to prevent voids, cracks, and unabsorbed residues. Besides hard/soft interface dimensional-based classifiers, mechanical durability in terms of elastic modulus acts as a means to address the rising dielectric failures of polymer nanodielectrics. Despite the prominent role of a polymer interface upon the overall dielectric constant in several hybrid high-K PDMs, the charge trapping effect under a high-intensity external electric field at elevated temperatures imposes serious concerns. Perturbed crystallinity, due to molecular chain displacement and segmental motion, changes the overall material's characteristics. The synergy of outcomes from prior kinds of literatures suggests a need for further introspections and a clear understanding of interfacial activity within distinct hybrid polymer matrices with high-K nanofillers.

16.13 CORE-SHELL HIGH-K POLYMERIC STRUCTURES

Core-shell structures are particles with an inner core that canvases a shell with distinguished materials. The core-shell structures are essential to enhance the physical and chemical properties of the host PDMs over their single counterpart. The genesis of core-shell structures commits to alleviate the localized electric field stress across the filler-polymer interface. Specifically, the core-shell framework with two different materials (A and B) augments the stability and dispersity of the colloidal core. The shell layer enclosed upon the nanofiller surface additionally varies the surface reactions to reframe nanofiller agglomeration. However, the synthesis of polymer-based core-shell structure demands distinct underlying principles, such as the immiscibility and small interfacial energy (lattice compatibility) of two independent constituents. The overall polymer nanocomposites can be synthesized via two distinct techniques, viz., a direct and an indirect method. The direct heterogeneous nucleation via a shell-layered polymer matrix aids defect-free high-K hybrid PDMs. The robust polymer-filler bonding constitutes a rigid matrix interface. However, the optimization of the filler volume fraction occurs via tuning the monomer feed ratio and the initiator functionalized nanoparticles [172]. Different material constituents demand different material synthesis processes (e.g., a three-step process for organic polymers). Early introspections and computer simulations illustrate a feasible technique to mitigate the local electric field stress gradient and consequently enhance the dielectric breakdown. Single core and multi-shell structures with high-K nanofillers showcase a gradient decline in the dielectric constant from the shell to the polymer matrix. The shell operates as a buffer layer to pacify the localized electric field effects. Congruent studies also suggest distinct core-shell design strategies for hybrid high-K PDMs with low-shell dielectrics and a thick shell coating. The diameter of the inner core is less than the outer shell thickness due to the removal of certain nanoparticles from the core via sacrificial conversion. Multiple experimental surveys present the effectiveness of core-shell structures upon the dielectric characteristics of hybrid high-K PDMs. Traditional examples revolve around a $BaTiO_3$-based dielectric core with a low dielectric constant shell (SiO_2 and Al_2O_3) [3]. Well-dispersed nano-fillers within an epoxy composite result in a high dielectric constant. Figure 16.11 illustrates the different core-shell structures within a hybrid PDM.

A recent study by Zhou et al. [207] synthesized core-shell-based polymer nanocomposite via the calcination process. The resultant high dielectric constant of the polymer nanocomposite emerges due to substantial intra-particle and suppressed inter-particle polarization respectively. Core-shell PDMs with different dielectric materials illustrate diverse applications beyond energy storage. Organic/inorganic conductive fillers are infused into the polymer matrix up to a percolation threshold to enhance the dielectric permittivity of the overall hybrid PDMs [208]. However, conductive fillers are catastrophic for energy storage applications with high dielectric loss. Thus, the core-shell strategy proposes an effective mechanism for enhanced dielectric characteristics. Such a strategy relies upon nanocomposite formation via in-situ polymerization and surface functionalization. The vital aspect of such a strategy implies the introduction of adequate initiating sites on the nanofiller surface. Diverse recent research works associate epoxy composites with unique core-shell particles [209–213]. The outcomes claim an increase in the energy storage modulus and a subsequent thermal conductivity increase. Multilayer core-shell polymers, on the other hand, demonstrate a hike in the dielectric constant besides a rise in the former parameters [214]. However, homogeneous filler dispersion and tailoring interfacial properties remain a technical limitation to rectify [215–217]. In support of surface functionalization of nanofillers, organic modifiers are easily absorbed onto the filler surface via electrostatic interactions/hydrogen bonding, respectively. These are utilized only

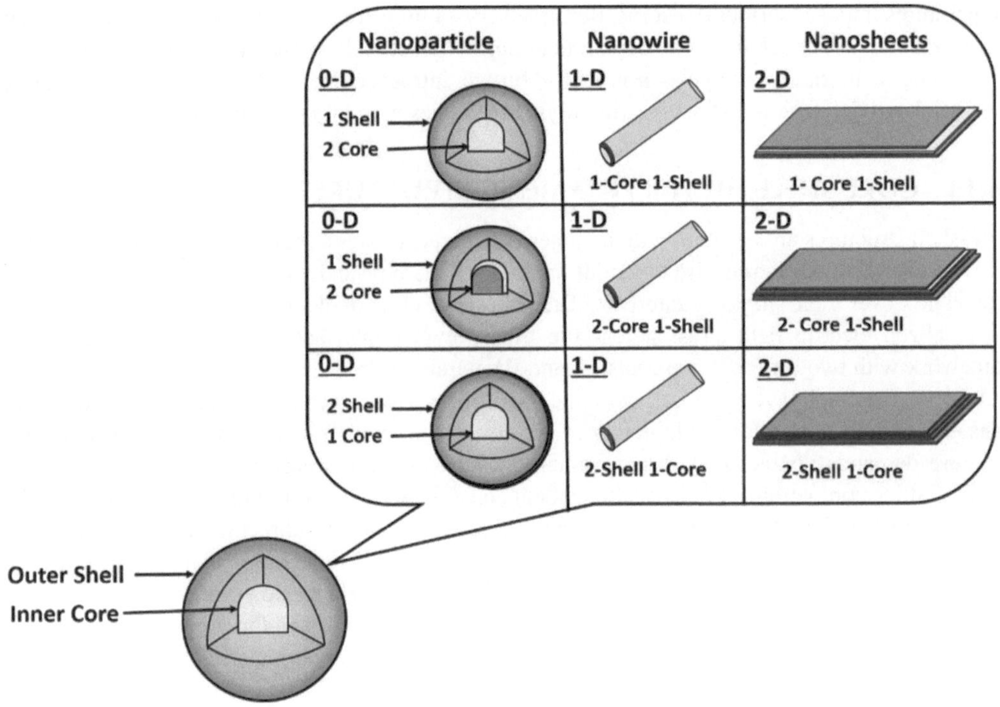

FIGURE 16.11 Schematic representation of core-shell structures and different dimensional core-shell structures (inset).

with the host hybrid polymer matrix to encounter the treatment process and solution mixing of inorganic filler constituents [218]. The constructed organic shells not only enhance the compatibility of hybrid constituents, but escalate filler dispersion and forbid implicit interfacial filler agglomeration within a hybrid PDM. A novel surface functionalization in terms of Reversible Addition–Fragmentation Chain Transfer Polymerization allows a series of core-shell structured hybrid nanoparticles with variable thickness and dimensionality. The $BaTiO_3$@ hybrid fluoro-polymers foists uniform filler distribution and progressive electrical properties of the host nanocomposites [219]. The high electronegative fluorine atoms in the modifiers and matrix curtail the disproportionate interfacial characteristics, due to strong hydrogen bonding. Fluoro-polymers thus presents a promising route for surface engineering of ceramic fillers. However, long-chain organic modifiers, on the other hand, compromise in terms of dielectric permittivity with respect to ceramic fillers. Such a steep gradient in the dielectric mismatch between organic and inorganic constituents inspires inhomogeneous electric field distribution [220, 221]. The incorporated high-K inorganic filler acts as one of the principal components of localized electric field and composite breakdown strength in the interfacial region. The tendency to subsidize the dielectric imbalance between filler and matrix accompanies the use of a low-k ceramic shell layer upon the high-K nanofiller surface. Additionally, multilayer core-shell structures with intermediate dielectric nanofillers such as TiO_2, Al_2O_3, and SiO_2, amongst other high and low-K nanofillers, are traditional preferences to enhance the energy storage density of the resultant hybrid PDMs. Consequently, former surveys establish a priority of intermediate dielectric fillers amongst numerous availabilities. The works of Chen et al. [222] claim homogeneous shell structures of Al_2O_3 upon the 2D Bi_2Te_3@Al_2O_3 nanoplates. The consequence of such uniformity is preserved to equivalent microstructural invariance. The microstructure of nanofillers critically influences the dielectric characteristics of hybrid PDMs and PDM nanocomposites. The depletion in energy density and efficiency ensues due to escalated dc conductivity and undesirable dielectric loss [223]. Efforts to uplift the energy density and storage capacity of high-K hybrid

PDMs succeed with the use of low-loss dielectric materials. However, recent reports claim a nonlinear stored energy density under an externally applied electric field. High-K PDMs are insufficient to enhance the former parameter. From the perspective of energy storage and energy density, high-K PDMs with large breakdown strength are conventionally chosen. Amongst diverse polymers, ferroelectric PDMs, such as PVDF, becomes a promising candidate, with active polarization and dipole-dipole interactions [153, 224–227]. The phase transformation (α to β phase) in ferroelectric-based PVDF polymer dielectrics leads to large ferroelectric loss and low discharge efficiency under an intense electric field. The works of Xie et al. [228] claim that a premature saturation of electric field displacement and remnant displacement strongly contributes to high energy loss in most ferroelectric polymer dielectrics, such as PVDF. To encounter the nonlinearity, high-K PDMs with non-ferroelectric and antiferroelectric traits must be adopted. Non-ferroelectric PDMs, such as PMMA, with considerable dielectric permittivity and breakdown alongside high glass transition temperature are beneficial for energy storage film. Figure 16.12 below demonstrates a comparative energy storage density between ferroelectric and non-ferroelectric PDMs. The high-K nanoparticles in core-shell structures with non-ferroelectric polymer matrices can effectively replace traditional fillers due to adequate electrical insulation and thermal conductivity with high dielectric breakdown. Double core-shell structures $BaTiO_3$@SiO_2@polydopamine PDM via SiO_2 covalent bonding and polydopamine shell onto the $BaTiO_3$ nanofiller surface. A congruent study by Tanaka et al. [229] on the extended multi-core shell structures of dielectric nanocomposites was presented. The dielectric properties of hybrid PDMs were defined in terms of interaction zones. The interfacial chemistry across different interaction zones becomes the fundamental integrity of classification. However, two major limitations of core-shell nanofillers occur in terms of low-K shells and distinct calcination performance. The divergence in lattice constants of core and shell constituents induces mechanical imperfections and instability. The synergy of variable dielectric, chemical and mechanical behavior of core-shell hybrid polymers hinders their commercial utilizations, due to significantly low discharge energy density. The scope of compatible energy storage dielectric PDMs was extended using multilayer and multistructural strategies. The former ideas are an extension of double and multi-core shell structures. However, multilayer and multistructural PDMs are devoid of the limitations of core-shell nanofiller and within hybrid PDMs.

The synopsis of the following section presents the significance of core-shell structures over other existing mechanisms to enhance the dielectric of PDMs. Diverse literature suggests high-K hybrid PDMs using conductive nanofillers. Large dielectric constants are conventionally accompanied by equivalent dielectric loss. This hinders the commercialization of dielectric composites. The interfacial effects on the dielectric and electrical properties of the hybrid polymers are cautiously examined. Core-shell strategies are a versatile technique for engineering high-K hybrid PDMs for advanced energy storage applications. This establishes a medium to control and optimize the electrical properties and nanocomposite microstructure via customized filler-matrix interfacial chemistry.

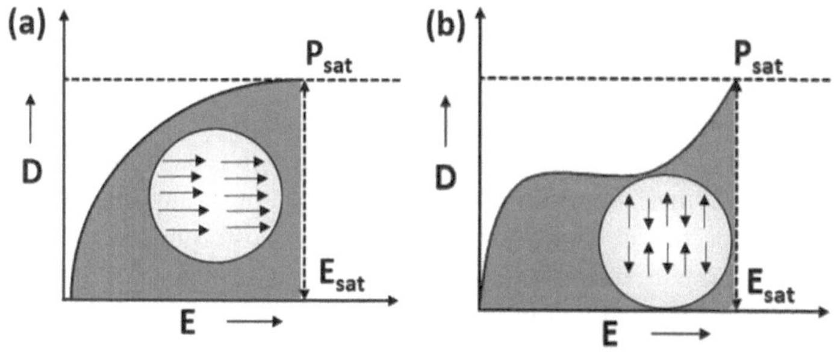

FIGURE 16.12 Energy loss of hybrid polymers with distinct nanofillers, (a) ferroelectric nanofillers (inset: arrangement of ferroelectric domains), (b) antiferroelectric nanofillers (inset: arrangement of anti-ferroelectric domains).

Core-shell methods relative to melt and solution mixing provide diverse advantages in terms of direct fabrication of high-K nanocomposites and homogeneous nanofiller dispersion with symmetric shell-matrix chemical structure. The shell layer encased on the nanofiller surface alters the surface reactions to forbid nanofiller agglomeration, which additionally acts as a buffer layer to eliminate the localized electric field gradient. Also, significant progress in dielectrics and the energy storage tendency of hybrid PDMs has been executed using single and multi-core-shell structures, competent challenges in terms of high dielectric breakdown strength and low dielectric loss. Besides, precise control upon dispersion of core-shell nanoparticles and filler-matrix interface demands advanced design strategies that need to be accomplished for compatible smart and advanced materials. On the contrary, nanoparticles with favorable nano-effects in the core-shell structures present unique freedom for advanced high-K nanocomposites.

16.14 STATE-OF-THE-ART MULTILAYER PDM COMPOSITES

Dielectric polymer blends and nanocomposites rely upon the synergetic contributions of individual material constituents. High-K polymer dielectric films are advantageous over ceramic and electrolytic films for energy storage applications. As an illustration, metalized polymer capacitors exhibit widespread applications in electronics, power systems and communications, due to high insulation, noninductive self-healing, low density, and mechanical stability respectively [3, 14, 230]. However, low dielectric constants of standalone and hybrid polymer matrix over the ceramic counterpart forbid their outright applications for advanced energy storage density and operational efficiency [231]. The primary motive involves an improvement in the dielectric permittivity of the hybrid polymeric systems. In compliance with prior discussions, several works of literature and ideas have been proposed and executed to escalate the dielectric constant of the hybrid polymer matrix. Ceramic, conductive and dielectric nanofillers, single and multi-core-shell structures, and controlled filler-polymer interfacial chemistry are diverse efforts executed to accomplish requisite dielectric properties. However, the inverse interplay between distinct dielectric characteristics reframes collaborative advancement. The former section (Sec. 4.3) featured different dimensional core-shell configurations as a motivation to enhance the dielectric properties of the host hybrid polymer matrix, due to improved filler dispersion and compatibility. However, the conventional interplay between dielectric constant and breakdown persists, despite numerous core-shell-based structural modifications. Multilayer dielectric polymeric structures, on the contrary, have proved to be a feasible and powerful technique to optimize energy storage activity. The performance of multilayer polymers can be independently designed for explicit layers. As a consequence, multilayer structures are considered to be competent to accomplish cooperative enhancement in breakdown and polarization. Multilayer polymer dielectric films with congruent filler conditions are considered as a single layer synthesized via a filling method. A multilayer dielectric structure can be equivalently thought of as multiple series capacitors with relatively small capacitance. In principle, the resultant series capacitance always turns out to be positive. However, if one or several capacitances becomes negative while the others still remain positive, an enhanced equivalence is obtained. A congruent idea was utilized with multilayer hybrid PDMs. PDM composites composed of graphite nanoparticles are dispersed within a polymer matrix to yield a negative dielectric layer with weak negative dielectric permittivity, equivalent to the dielectric constant of the positive layer. In opposition to typical multilayer hybrid polymer dielectrics, three-layer polymer composites counterbalance the competing parameters to attain remarkable dielectric properties devoid of any compromise. Different pieces of literature empirically support the former strategy to assist high energy storage density with respect to conventional multilayer polymer composites [232]. Figure 16.13 below shows a sandwiched multilayer PDM structure.

In a three-layer polymeric system, the inorganic fillers permit precise control over the dielectric characteristics of individual layers. In traditional sandwiched multilayer PDMs, the top and bottom layers typically compose high-K nanofillers to constitute an overall high dielectric constant at the

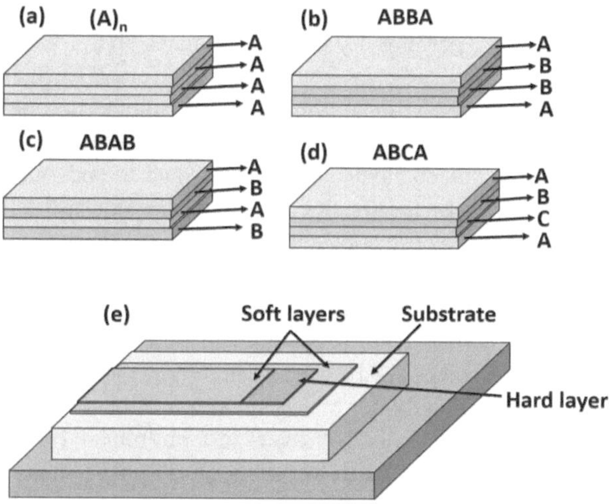

FIGURE 16.13 Multilayer high-K hybrid polymer dielectric structures, (a) A-type layers, (b) ABBA type layered structures, (c) ABAB-type layered structure, (d) ABCA-type layered structure, (e) Compiled multi-layered structure.

expense of reduced dielectric breakdown strength, while the middle layer preserves a high dielectric breakdown strength with low to intermediate dielectric permittivity. The critical factor associated with material architecture is the appropriate insulation of the middle layer to block the breakdown path of the resultant multilayer PDM composite. Nanofillers, such as $BaTiO_3$, TiO_2 and other high-K dielectric oxides, are significantly preferred for soft layers (top and bottom layers) [233–235]. The center layer, on the other hand, is doped with different dimensional filler materials, such as nanoparticles (0D), nanowires (1D), and nanosheets (2D), respectively, to accomplish high-end requisite insulation [236, 237]. The synergetic effect is visualized in terms of high dielectric breakdown, due to energy loss by the scattered free electrons under an intense applied electric field inside a Debye length within a nanofiller-composed multilayer hybrid PDM [238]. Additional artifacts for high dielectric breakdown associate with a non-uniform potential difference between different layers in a multilayer PDM. While hard layers bear considerable potential difference, soft layers on the other hand deliver a small potential difference. An increase in dielectric constant proportionally escalates with the soft layer filler concentrations, while the breakdown strength preserves nonlinear behavior. The growth in dielectric breakdown follows a subsequent decline, with congruent characteristics for energy storage density. The hard layer (middle layer) has a critical impact on the tuneable dielectric properties of high-K hybrid PDMs. In contrast to the soft-hard-soft configuration, the hard-soft-hard multi-layered sequence eases the choice of outer low-K dielectrics to enhance the resultant dielectric breakdown [71, 232, 239, 240]. It is familiar because the dielectric loss tangent under an external electric field arises from the tunneling current, which escalates exponentially with the consequent rise in electric field intensity. This declines the discharge energy density and charging/discharging efficiency. Thus, the appliance of charge blocking materials opens a gateway to minimize the leakage current and other conduction losses. One of the befitting materials with former implicit characteristics is boron nitride (BN). High mechanical strength, thermal conductivity with a wide forbidden band and low-density characteristics, BN nanosheets present an optimum filler for the outer layer, to eliminate leakage current. Also, high-K 1D oxides in the form of $SrTiO_3$ nanowires are adopted as suitable filler for intermediate soft layers within a resultant high-K hybrid PDM.

A multilayer structure with an excess of three layers in a sandwich configuration is a vital factor to explore to elucidate the impact of a number of layers on the dielectric properties of the host polymer matrix. Preserving the critical direction of interest, Wanget al. [241] fabricated and investigated the dielectric behavior of multilayer high-K nanotubes/epoxy composites. The outcomes claim a nonlinear behavior of the dielectric constant with the increase in the number of layers in a multilayer hybrid high-K PDM. The dielectric permittivity initially escalates up to four layers, and then subsequently decreases with a consequent surge in the number of layers. The four-layer high-K hybrid PDMs illustrate a four-times hike in the dielectric properties (dielectric constant and breakdown strength) with respect to their standalone elementary hybrid high-K PDMs. The consequent attribute emerges due to the absence of localized electric field effects and the nanofiller aggregation across different multilayer interfaces, with a subsequent control over the microstructural properties of the host polymer matrix. Besides the dielectric constant, multilayer PDMs showcase low dielectric dissipation relative to their single-layer counterpart, with considerable dielectric loss in the introspected frequency range. The conductive losses at the lower frequency regimes become prominent and significantly contribute to the resultant dielectric loss within a single-layer PDM, which is counterbalanced by multilayer structures. The conclusive results highlight the essentiality of structural and microstructural properties of the host polymer matrix to optimize the dielectric characteristics over a wide range. Besides examining the number of layers in a multilayer PDM, the thickness of the individual layer additionally governs the dielectric traits of the host matrix. In compliance, Feng et al. [242] synthesized sandwich-structured polymer composites using central poly(vinylidene fluoride co-hexafluoro propylene) (P(VDF-HFP)) and $BaTiO_3$-(P(VDF-HFP)) based upper and lower layers, respectively. The thickness of the middle layer was modified to introspect the influence of spin-coated three-layer sandwiched structure upon the host matrix properties. Low dielectric constant of the central layer obstructs their implication for energy storage properties. The incorporation of distinct dimensional filler constituents in terms of nanoparticles, nanowires, and nanosheets provides an additional insulation layer within a multilayer system and enhances the resultant dielectric permittivity of the overall polymer composites. A congruent study by Feng et al. [243] and Zhang et al. [244] examined the topological structures of sandwich and multilayer polymer dielectrics in terms of 2-2 composite. The outcome of such multilayer topology revealed a novel and intriguing strategy, with dispersed 2D nanofillers incorporated in an interconnected polymer matrix. Such structural advancements also establish an improvement in dielectric properties, due to the presence of an additional insulating layer. Meanwhile, the computational study by Ohshima et al. [245] illustrates the impact of low and high-intensity electric fields on the multilayer polymer matrix. The increase in the number of layers with a consequent decrease in the monolayer dielectric thickness reduces the dielectric loss due to an increase in the D-E narrow loop. This favors a resultant increase in energy storage density. Further survey establishes a reduced localized electric field gradient upon the absence of dipole flipping within the PVDF layer. Dielectric dissipation otherwise, the dielectric loss in multilayer systems principally occurs due to the dominant contribution from the ion migration over dipole switching. The genesis of the ions is because of the suspension polymerization and surfactant residues, respectively. The experimental results of the study demonstrate a decline in the low and high-field loss, with a subsequent reduction in the thickness of the PVDF layer.

The positive impact of an increase in the dielectric permittivity of overall hybrid PDMs due to the reduced thickness of multilayer structures is associated with a consequent decrease in the dielectric breakdown. The characteristic attribute arises due to the interfacial polarization and successive electronic conduction in the multilayer system. Such a phenomenon is explained using the Maxwell-Wagner Sillars model. A thick conductive layer among other layers in a multilayer structure offers large space charges for interfacial polarization. A thick insulating layer, on the other hand, reframes the internal electronic conduction. As a consequence, higher Maxwell-Wagner Sillar interfacial polarization, due to charge agglomeration, serves as an effective trap to avoid electronic conduction. A thin central layer, on the contrary, results in diminished dielectric breakdown and lifetime. The synopsis of the following section illustrates the critical role of multilayer structures on the dielectric characteristics of hybrid polymers. The inverse interplay between the dielectric constant and breakdown strength, even under single and multi-core-shell structures, is counterbalanced effectively by multi-layered sandwich polymeric structures. The regulated inorganic fillers grant the architecture of distinct dielectric layers to appropriately utilize the overall performance characteristics. In compliance with substantial analysis, contradiction persists between the demand for thin layers to subsidize the ionic conduction loss and a thick layer on the other hand to accomplish higher breakdown strength. Besides, a liner increase in the polymer content opposingly reduces the dielectric polarization and diminishes the comparative energy storage density. The principal concern of high energy storage density associates with the breakdown and polarization of different multilayer interfaces. The control and optimization of interfacial and breakdown parameters are proportionally associated with structural and microstructural optimization. Thoughtful and precise measure becomes mandatory for collective improvement in dielectric permittivity and breakdown properties.

16.15 FUTURE RESEARCH PERSPECTIVE

- Hybrid dielectric polymers with inorganic, synthetic, and bio-degradable constituents are considered green materials, which provide eco-friendly green polymer matrix with requisite dielectric behavior.
- Diverse synthesis methods with nano-foams in the polymeric structures through voids of air constitute a principal role in reduced dielectric constant.
- Organocatalytic synthesis route can be employed to achieve desirable dielectric polymer films with robust mechanical structures.
- While dominant fluorinated polymer and polymer composites constitute insulation loss via charge agglomeration and electronic conduction, semifluorinated PDMs find potential applications in state-of-the-art energy storage technologies.
- The blend of semifluorinated and dielectric polymeric constituents can be utilized to optimize the filler-polymer interface.

16.16 CONCLUSION

Polymer dielectric materials (PDMs) demonstrate extended applications in diverse domains besides energy storage technologies. Polymer dielectrics provide diverse configurations with potential applications and synthesis routes. The implicit dielectric constant of the host material is responsible for the variable energy storage density among distinct technological devices. However, the low dielectric constant and the trade-off between the former parameter and dielectric breakdown with high dielectric loss obstructs the current development of dielectric polymers. Hybrid PDMs composed of aliphatic and aromatic polymer blends illustrate the considerable improvement in dielectric properties relative to their standalone counterpart. However practical synthesis of high-end hybrid and polymer composites establishes compatibility concerns. Dielectric differences between distinct hybrid polymeric constituents disrupt the compositional homogeneity. Diverse phase relaxations (α, β and γ) and heterogeneous crystallinity due to segmental motion, physical voids and free volume in the molecular chain induce a charge-trapping effect across the interface and interstitial sites within high-K hybrid polymer composites. Charge agglomeration and electronic conduction are additional critical challenges that hinder and activate the disparity among different dielectric properties. While high electronegative atomic substituents assist dielectric customizations, strong polarizing effects and dipole mobility constitute an adequate dielectric loss. This forewarns the need to improve the dielectric permittivity and counterbalance structural, chemical, mechanical, and microstructural characteristics. Besides hybrid PDMs, novel material engineering strategies in terms of single and multi-core-shell structures, modified interfacial chemistry, organic/ceramic/inorganic nanofillers, and multilayer structures are implemented to attain desirable properties. The implication of former strategies, however, revamps individual dielectric properties, with a consistent trade-off between other dielectric parameters. High-volume fractions ($\sim > 30$ %) of conductive nanofillers form conductive networks and reduce the breakdown strength, while surface-functionalized nanofillers within multi-core-shell structures promote interstitial sites on the nanoparticle surface to enhance the energy storage modulus. Non-uniform nanofiller dispersion and ambiguous interfacial chemistry within core-shell structures hinder their effectiveness. The interfacial charge distribution and space charge movement operate as scattering points to constitute prolonged pathways for improved polarization and dielectric breakdown. As a mark of progressive advancement, multi-layered polymeric structures, on the other hand, aid simultaneous improvement in dielectric constant and breakdown strength. Material constraints in terms of multilayer dimensionality and a number of layers enforce nonlinear dielectric behavior, which needs to be encountered with suitable design engineering with pre-requisite understandings.

REFERENCES

[1] T.M. Gür, *Review of electrical energy storage technologies, materials and systems: Challenges and prospects for large-scale grid storage*, Energy Environ. Sci. 11 (2018), pp. 2696–2767.

[2] N. Meng, X. Ren, G. Santagiuliana, L. Ventura, H. Zhang, J. Wu et al., *Ultrahigh β-phase content poly(vinylidene fluoride) with relaxor-like ferroelectricity for high energy density capacitors*, Nat. Commun. 10 (2019), pp. 1–9.

[3] D.Q. Tan, *The search for enhanced dielectric strength of polymer-based dielectrics: A focused review on polymer nanocomposites*, J. Appl. Polym. Sci. 137 (2020), pp. 1–32.

[4] Y. Abdullahi Hassan and H. Hu, *Current status of polymer nanocomposite dielectrics for high-temperature applications*, Compos. Part A Appl. Sci. Manuf. 138 (2020), pp. 106064.

[5] B. Fan, M. Zhou, C. Zhang, D. He and J. Bai, *Polymer-based materials for achieving high energy density film capacitors*, Prog. Polym. Sci. 97 (2019).

[6] M. Ieda, *Electrical conduction and carrier traps in polymeric materials*, IEEE Trans. Electr. Insul. EI-19 (1984), pp. 162–178.

[7] M. Huang, Y. Zhou, Z. Zhou and B. Qi, *A combined electro-thermal breakdown model for oil-impregnated paper*, Energies 10 (2017), pp. 1–12.

[8] G. Ribes, J. Mitard, M. Denais, S. Bruyere, F. Monsieur, C. Parthasarathy et al., *Review on high-K dielectrics reliability issues*, IEEE Trans. Device Mater. Reliab. 5 (2005), pp. 5–19.

[9] Z.B. Pan, L.M. Yao, J.W. Zhai, S.H. Liu, K. Yang, H.T. Wang et al., *Fast discharge and high energy density of nanocomposite capacitors using $Ba_{0.6}Sr_{0.4}TiO_3$ nanofibers*, Ceram. Int. 42 (2016), pp. 14667–14674.

[10] M.S. Zheng, J.W. Zha, Y. Yang, P. Han, C.H. Hu and Z.M. Dang, *Enhanced breakdown strength of poly(vinylidene fluoride) utilizing rubber nanoparticles for energy storage application*, Appl. Phys. Lett. 109 (2016).

[11] S.A. Qenawy, L.S. Nasrat, H.M. Ismail and J.N. Asaad, *Evaluation of dielectric strength of SiR/TiO_2 composites using feed-forward neural network*, IET Nanodielectrics 3 (2020), pp. 74–80.

[12] X. Dong, *Low- k cross-linked polyimide for microelectronic packaging application*, (2021), pp. 21–24.

[13] J.W. Zha, M.S. Zheng, B.H. Fan and Z.M. Dang, *Polymer-based dielectrics with high permittivity for electric energy storage: A review*, Nano Energy 89 (2021), pp. 106438.

[14] Q. Chen, Y. Shen, S. Zhang and Q.M. Zhang, *Polymer-Based Dielectrics with High Energy Storage Density*, Annu. Rev. Mater. Res. 45 (2015), pp. 433–458.

[15] Z. Yao, Z. Song, H. Hao, Z. Yu, M. Cao, S. Zhang et al., *Homogeneous/Inhomogeneous-Structured Dielectrics and their Energy-Storage Performances*, Adv. Mater. 29 (2017).

[16] J.S. Ho and S.G. Greenbaum, *Polymer Capacitor Dielectrics for High Temperature Applications*, ACS Appl. Mater. Interfaces 10 (2018), pp. 29189–29218.

[17] T.D. Huan, S. Boggs, G. Teyssedre, C. Laurent, M. Cakmak, S. Kumar et al., *Advanced polymeric dielectrics for high energy density applications*, Prog. Mater. Sci. 83 (2016), pp. 236–269.

[18] Z. Wang, J. Liu, H. Gong, X. Zhang, J. Lu and Z. Zhang, *Synthesis of poly(methyl methacrylate-methallyl alcohol) via controllable partial hydrogenation of poly(methyl methacrylate) towards high pulse energy storage capacitor application*, RSC Adv. 6 (2016), pp. 34855–34865.

[19] Y. Wang, X. Huang, T. Li, Z. Wang, L. Li, X. Guo et al., *Novel crosslinkable high-K copolymer dielectrics for high-energy-density capacitors and organic field-effect transistor applications*, J. Mater. Chem. A 5 (2017), pp. 20737–20746.

[20] P. Kum-onsa, N. Chanlek, B. Putasaeng and P. Thongbai, *Improvement in dielectric properties of poly(vinylidene fluoride) by incorporation of Au–$BiFeO_3$ hybrid nanoparticles*, Ceram. Int. 46 (2020), pp. 17272–17279.

[21] K. Silakaew, E. Swatsitang and P. Thongbai, *Novel polymer composites of RuO_2@n$BaTiO_3$/PVDF with a high dielectric constant*, Ceram. Int. 48 (2022), pp. 18925–18932.

[22] H. Hu, F. Zhang, S. Luo, W. Chang, J. Yue and C.-H. Wang, *Recent advances in rational design of polymer nanocomposite dielectrics for energy storage*, Nano Energy 74 (2020), pp. 104844.

[23] L. Yang, X. Kong, F. Li, H. Hao, Z. Cheng, H. Liu et al., *Perovskite lead-free dielectrics for energy storage applications*, Prog. Mater. Sci. 102 (2019), pp. 72–108.

[24] S. Nakayama, H. Nomura, L.M. Smith and J.F. Clark, *Simultaneous estimation of intracellular free Mg^{2+} and pH by use of a new pH-dependent dissociation constant of MgATP*, Jpn. J. Physiol. 52 (2002), pp. 323–326.

[25] K. Ogino, *Review of literature on aging of dielectrics*, Ferroelectrics 87 (1988), pp. 361–377.

[26] G. Picci and M. Rabuffi, *Status quo and future prospects for metallized polypropylene energy storage capacitors*, PPPS 2001 - Pulsed Power Plasma Sci. 2001 1 (2015), pp. 417–420.

[27] T. Chen, Y. Zhao, L. Pan and M. Lin, *Insight into effect of hydrothermal preparation process of nanofillers on dielectric, creep and electromechanical performance of polyurethane dielectric elastomer/reduced graphene oxide composites*, J. Mater. Sci. Mater. Electron. 26 (2015), pp. 10164–10171.

[28] Z. Li, G.M. Treich, M. Tefferi, C. Wu, S. Nasreen, S.K. Scheirey et al., *High energy density and high efficiency all-organic polymers with enhanced dipolar polarization*, J. Mater. Chem. A 7 (2019), pp. 15026–15030.

[29] F. Palumbo, C. Wen, S. Lombardo, S. Pazos, F. Aguirre, M. Eizenberg et al., *A Review on Dielectric Breakdown in Thin Dielectrics: Silicon Dioxide, High-K, and Layered Dielectrics*, Adv. Funct. Mater. 30 (2020), pp. 1–26.

[30] H. Kumazoe, S. Fukushima, S. Tiwari, C. Kim, T.D. Huan, R.K. Kalia et al., *Hot-Carrier Dynamics and Chemistry in Dielectric Polymers*, J. Phys. Chem. Lett. 10 (2019), pp. 3937–3943.

[31] M. Ieda, *Dielectric breakdown process of polymers*, IEEE Trans. Electr. Insul. (1980), pp. 206–224.

[32] P.D. Beale and P.M. Duxbury, *Theory of dielectric breakdown in metal-loaded dielectrics*, Phys. Rev. B 37 (1988), pp. 2785–2791.

[33] O.S. Heavens, *Optical Properties of Thin Solid Films*, Courier Corporation, 1991.

[34] J. Wei and L. Zhu, *Intrinsic polymer dielectrics for high energy density and low loss electric energy storage*, Prog. Polym. Sci. 106 (2020), pp. 101254.

[35] O.S. Gefle, S.M. Lebedev and V.Y. Uschakov, *The mechanism of the barrier effect in solid dielectrics*, J. Phys. D. Appl. Phys. 30 (1997), pp. 3267–3273.

[36] S.M. Lebedev, O.S. Gefle, Y.P. Polholkov and V.I. Chichikin, Two-layer dielectrics behavior in the electric field, in *1999 Annual Report Conference on Electrical Insulation and Dielectric Phenomena (Cat. No. 99CH36319)*, 1999, pp. 265–268.

[37] D.P. Agoris, I. Vitellas, O.S. Gefle, S.M. Lebedev and Y.P. Pokholkov, *The barrier effect in three-layer solid dielectrics in quasi-uniform electric field*, J. Phys. D. Appl. Phys. 34 (2001), pp. 3485–3491.

[38] M. Singh, I.E. Apata, S. Samant, W. Wu, B. V. Tawade, N. Pradhan et al., *Nanoscale Strategies to Enhance the Energy Storage Capacity of Polymeric Dielectric Capacitors: Review of Recent Advances*, Polym. Rev. 62 (2022), pp. 211–260.

[39] H. Miyauchi and K. Yahagi, *Electronic breakdown in polyethylene film in room temperature range*, Trans. Inst. Electr. Eng. Japan. A 92 (1972), pp. 36–45.

[40] R.J. Young and P.A. Lovell, *Introduction to Polymers*, CRC Press, 2011.

[41] Z. Ahmad, *Polymeric Dielectric Materials*, (2012), pp. 3–26.

[42] T. Fukasawa, T. Sato, J. Watanabe, Y. Hama, W. Kunz and R. Buchner, *Relation between dielectric and low-frequency Raman spectra of hydrogen-bond liquids*, Phys. Rev. Lett. 95 (2005), pp. 1–4.

[43] F. Noto, N. Yoshimura and T. Ohta, *Tree initiation in polyethylene by application of DC and impulse voltage*, IEEE Trans. Electr. Insul. (1977), pp. 26–30.

[44] Y. Saito, M. Fukuzawa and H. Nakamura, *On the mechanism of tree initiation*, IEEE Trans. Electr. Insul. (1977), pp. 31–34.

[45] N. Shimizu and K. Horii, *The effect of absorbed oxygen on electrical treeing in polymers*, IEEE Trans. Electr. Insul. (1985), pp. 561–566.

[46] J. Andrianjohaninarivo, M.R. Wertheimer and A. Yelon, *Nucleation Of Electrical Trees In Polyethylene*, IEEE Trans. Electr. Insul. EI-22 (1987), pp. 709–714.

[47] T. Tanaka, Tree initiation mechanisms, in *(1991) Proceedings of the 3rd International Conference on Properties and Applications of Dielectric Materials*, 1991, pp. 18–24.

[48] W. Weibull, A.S. Distribution and W. Applicability, *Waloddi Weibull To cite this version: HAL Id: hal-03112318 A Statistical Distribution Function of Wide Applicability*, (2021), pp. 0–5.

[49] C. Laurent, E. Kay and N. Souag, *Dielectric breakdown of polymer films containing metal clusters*, J. Appl. Phys. 64 (1988), pp. 336–343.

[50] W.F. Yong and T.S. Chung, *Mechanically Strong and Flexible Hydrolyzed Polymers of Intrinsic Microporosity (PIM-1) Membranes*, J. Polym. Sci. Part B Polym. Phys. 55 (2017), pp. 344–354.

[51] L. Dou, Y.H. Lin and C.W. Nan, *An overview of linear dielectric polymers and their nanocomposites for energy storage*, Molecules 26 (2021).

[52] L. Zhu, *Exploring strategies for high dielectric constant and low loss polymer dielectrics*, J. Phys. Chem. Lett. 5 (2014), pp. 3677–3687.

[53] D.Q. Tan, *Review of Polymer-Based Nanodielectric Exploration and Film Scale-Up for Advanced Capacitors*, Adv. Funct. Mater. 30 (2020), 1808567.

[54] H. Li, Y. Zhou, Y. Liu, L. Li, Y. Liu and Q. Wang, *Dielectric polymers for high-temperature capacitive energy storage*, Chem. Soc. Rev. 50 (2021), pp. 6369–6400.

[55] S.L. Jiang, Y. Yu and Y.K. Zeng, *Novel Ag-$BaTiO_3$/PVDF three-component nanocomposites with high energy density and the influence of nano-Ag on the dielectric properties*, Curr. Appl. Phys. 9 (2009), pp. 956–959.

[56] M.S. Zafar, *Prosthodontic applications of polymethyl methacrylate (PMMA): An update*, Polymers (Basel). 12 (2020), pp. 1–35.

[57] K.F. Makris, J. Langeveld and F.H.L.R. Clemens, *A review on the durability of PVC sewer pipes: research vs. practice*, Struct. Infrastruct. Eng. 16 (2020), pp. 880–897.

[58] E. Dhanumalayan and G.M. Joshi, *Performance properties and applications of polytetrafluoroethylene (PTFE)—a review*, Adv. Compos. Hybrid Mater. 1 (2018), pp. 247–268.

[59] M. Ghosh, *Polyimides: Fundamentals and Applications*, CRC Press, 2018.

[60] Q. Li, L. Chen, M.R. Gadinski, S. Zhang, G. Zhang, H. Li et al., *Flexible higherature dielectric materials from polymer nanocomposites*, Nature 523 (2015), pp. 576–579.

[61] Berger, K. C., Brandrup, G., Brandrup, J., & Immergut, E. H. (1989). *Polymer handbook.* Wiley & Sons, New York.

[62] A.C.M. Kuo, *Poly (dimethylsiloxane)*, Polym. data Handb. (1999), pp. 411–435.

[63] K. Kamijima, M. Kaji, Z. Shinp, Y. Honda and V.J. Lee, *Thermally Stable Polyquinoline*, J. Photopolym. Sci. Technol. 10 (1997), pp. 79–82.

[64] W. Wunderlich, *Physical Constants of Poly (methyl methacrylate)*, Polym. Handb. 78 (1989).

[65] C. Zhang, H. Wang, Z. Shi, Z. Cui and D. Yan, *UV-directly patternable organic--inorganic hybrid composite dielectrics for organic thin-film transistors*, Org. Electron. 13 (2012), pp. 3302–3309.

[66] M.D. Toomey, K. Gao, G.P. Mendis, E.B. Slamovich and J.A. Howarter, *Hydrothermal synthesis and processing of barium titanate nanoparticles embedded in polymer films*, ACS Appl. Mater. Interfaces 7 (2015), pp. 28640–28646.

[67] A.N. Abougreen, A.E. Shalan, E.S.A. Serea and M.K.A. Mohammed, *Polymer Nanocomposites for Energy Storage Applications*, Eng. Mater. (2022), pp. 697–724.

[68] Z.M. Dang, J.K. Yuan, S.H. Yao and R.J. Liao, *Flexible nanodielectric materials with high permittivity for power energy storage*, Adv. Mater. 25 (2013), pp. 6334–6365.

[69] Z.H. Dai, T. Li, Y. Gao, J. Xu, J. He, Y. Weng et al., *Achieving high dielectric permittivity, high breakdown strength and high efficiency by cross-linking of poly(vinylidene fluoride)/BaTiO3 nanocomposites*, Compos. Sci. Technol. 169 (2019), pp. 142–150.

[70] Y. Shao, M.F. El-Kady, J. Sun, Y. Li, Q. Zhang, M. Zhu et al., *Design and mechanisms of asymmetric supercapacitors*, Chem. Rev. 118 (2018), pp. 9233–9280.

[71] H. Palneedi, M. Peddigari, G.-T. Hwang, D.-Y. Jeong and J. Ryu, *High-performance dielectric ceramic films for energy storage capacitors: progress and outlook*, Adv. Funct. Mater. 28 (2018), pp. 1803665.

[72] S.M. Haque, J.A.A. Rey, A.A. Mas'ud, Y. Umar and R. Albarracin, *Electrical properties of different polymeric materials and their applications: The influence of electric field*, Prop. Appl. Polym. Dielectr. InTech (2017), pp. 41–63.

[73] L. Li, Y. Zhou, Y. Liu, X. Chen, Z. Han and Q. Wang, *Perspective on scalable high-energy-density polymer dielectrics with ultralow loadings of inorganic nanofillers*, Appl. Phys. Lett. 120 (2022).

[74] V. Siva, D. Vanitha, A. Murugan, A. Shameem and S.A. Bahadur, *Studies on structural and dielectric behavior of PVA/PVP/SnO nanocomposites*, Compos. Commun. 23 (2021), pp. 100597.

[75] W. Tuichai, A. Karaphun and C. Ruttanapun, *Improved dielectric properties of PVDF polymer composites filled with Ag nanomaterial deposited reduced graphene oxide (rGO) hybrid particles*, Mater. Res. Bull. 145 (2022), pp. 111552.
[76] R.J. Sengwa and P. Dhatarwal, *PVA/MMT and (PVA/PVP)/MMT hybrid nanocomposites for broad-range radio frequency tunable nanodielectric applications*, Mater. Lett. 299 (2021), pp. 130081.
[77] A. Afzal and Y. Nawab, *Polymer composites*, Compos. Solut. Ballist. 3804 (2021), pp. 139–152.
[78] D.H. Kuo, C.C. Chang, T.Y. Su, W.K. Wang and B.Y. Lin, *Dielectric behaviors of multi-doped $BaTiO_3$/ epoxy composites*, J. Eur. Ceram. Soc. 21 (2001), pp. 1171–1177.
[79] B.K.P. Scaife, *Principles of dielectrics*, (1989).
[80] S.M.R. Billah, *Dielectric polymers*, in *Functional Polymers*, Springer, 2019, pp. 241–288.
[81] N.G. Koopman, T.C. Reiley, P.A. Totta, R.R. Tummala and E.J. Rymaszewski, *Microelectronics Packaging Handbook*, Van Reinhold 361 (1989).
[82] R. Rujiwarodom, N. Sunganun, W. Busayaporn and A. Panyanuch, *Improvement of the Electrical Insulating Property of Polypropylene by Electron Beam Irradiation*, Agric. Nat. Resour. 48 (2014), pp. 111–119.
[83] C. Han, X. Zhang, D. Chen, Y. Ma, C. Zhao and W. Yang, *Enhanced dielectric properties of sandwich-structured biaxially oriented polypropylene by grafting hyper-branched aromatic polyamide as surface layers*, J. Appl. Polym. Sci. 137 (2020), pp. 48990.
[84] H. Treichel, G. Ruhl, P. Ansmann, R. Würl, C. Müller and M. Dietlmeier, *Low dielectric constant materials for interlayer dielectric (Invited paper)*, Microelectron. Eng. 40 (1998), pp. 1–19.
[85] A.J. Osborn, K.R. Foster and M.S. Wolfe, *Dielectric properties of poly(tetrafluoroethylene) "Whiskers,"* J. Phys. Chem. 95 (1991), pp. 5915–5918.
[86] *Methods and systems of dielectric film materials for use in capacitors*. Google Patents, 2009.
[87] V. Ranjan, L. Yu, M.B. Nardelli and J. Bernholc, *Phase equilibria in high energy density PVDF-based polymers*, Phys. Rev. Lett. 99 (2007), pp. 1–4.
[88] S.H. Xie, B.K. Zhu, J.B. Li, X.Z. Wei and Z.K. Xu, *Preparation and properties of polyimide/aluminum nitride composites*, Polym. Test. 23 (2004), pp. 797–801.
[89] L. Liu, B. Liang, W. Wang and Q. Lei, *Preparation of polyimide/inorganic nanoparticle hybrid films by sol-gel method*, J. Compos. Mater. 40 (2006), pp. 2175–2183.
[90] H. Li, G. Liu, B. Liu, W. Chen and S. Chen, *Dielectric properties of polyimide/Al_2O_3 hybrids synthesized by in-situ polymerization*, Mater. Lett. 61 (2007), pp. 1507–1511.
[91] M. Wehbi, A. Mehdi, C. Negrell, G. David, A. Alaaeddine and B. Améduri, *Phosphorus-Containing Fluoropolymers: State of the Art and Applications*, ACS Appl. Mater. Interfaces 12 (2020), pp. 38–59.
[92] J.E. Marshall, A. Zhenova, S. Roberts, T. Petchey, P. Zhu, C.E.J. Dancer et al., *On the solubility and stability of polyvinylidene fluoride*, Polymers (Basel). 13 (2021), pp. 1–31.
[93] Y. Yang, Y. Hong and X. Wang, *Utilizing the diffusion of fluorinated polymers to modify the semiconductor/dielectric interface in solution-processed conjugated polymer field-effect transistors*, ACS Appl. Mater. Interfaces 13 (2021), pp. 8682–8691.
[94] S. Luo, J. Yu, T.Q. Ansari, S. Yu, P. Xu, L. Cao et al., *Elaborately fabricated polytetrafluoroethylene film exhibiting superior high-temperature energy storage performance*, Appl. Mater. Today 21 (2020), 100882.
[95] S. Luo, T.Q. Ansari, J. Yu, S. Yu, P. Xu, L. Cao et al., *Enhancement of dielectric breakdown strength and energy storage of all-polymer films by surface flattening*, Chem. Eng. J. 412 (2021), pp. 128476.
[96] B. Ameduri, *From vinylidene fluoride (VDF) to the applications of VDF-Containing polymers and copolymers: Recent developments and future trends*, Chem. Rev. 109 (2009), pp. 6632–6686.
[97] T. Cai, K.-G. Neoh and E.-T. Kang, *Functionalized and functionalizable fluoropolymer membranes*, Handb. Fluoropolymer Sci. Technol. (2014), pp. 149–181.
[98] J.T. Goldbach, R. Amin-Sanayei, W. He, J. Henry, W. Kosar, A. Lefebvre et al., *Commercial synthesis and applications of poly (vinylidene fluoride)*, Fluorinated Polym. 2 (2016), pp. 127–157.
[99] A. Kraytsberg and Y. Ein-Eli, *Review of advanced materials for proton exchange membrane fuel cells*, Energy and Fuels 28 (2014), pp. 7303–7330.
[100] A. Ghosh and S. Banerjee, *Sulfonated fluorinated-aromatic polymers as proton exchange membranes*, E-Polymers 14 (2014), pp. 227–257.
[101] S. Banerjee, *Handbook of Specialty Fluorinated Polymers: Preparation, Properties, and Applications*, William Andrew, 2015.
[102] A.J. Lovinger, *Ferroelectric polymers*, Science. 220 (1983), pp. 1115–1121.

[103] X. Ren, N. Meng, H. Zhang, J. Wu, I. Abrahams, H. Yan et al., *Giant energy storage density in PVDF with internal stress engineered polar nanostructures*, Nano Energy 72 (2020), pp. 104662.
[104] F. Chen, Y. Zhao, Y. Zhou, J. Xu, Y. Yang and W. Yang, *Effect of solvent on the energy storage property of poly(vinylidene fluoride-hexafluoropropylene)*, Mater. Res. Express 7 (2019).
[105] P. Martins, A.C. Lopes and S. Lanceros-Mendez, *Electroactive phases of poly(vinylidene fluoride): Determination, processing and applications*, Prog. Polym. Sci. 39 (2014), pp. 683–706.
[106] W. Xia, Z. Xu, F. Wen, W. Li and Z. Zhang, *Crystalline properties dependence of dielectric and energy storage properties of poly(vinylidene fluoride-chlorotrifluoroethylene)*, Appl. Phys. Lett. 97 (2010), pp. 5–8.
[107] X. Hu, K. Yi, J. Liu and B. Chu, *High Energy Density Dielectrics Based on PVDF-Based Polymers*, Energy Technol. 6 (2018), pp. 849–864.
[108] B. Chu, X. Zhou, B. Neese, Q.M. Zhang and F. Bauer, *Relaxor ferroelectric poly(vinylidene fluoride-trifluoroethylene- chlorofluoroethylene) terpolymer for high energy density storage capacitors*, IEEE Trans. Dielectr. Electr. Insul. 13 (2006), pp. 1162–1168.
[109] Z. Zhang and T.C.M. Chung, *The structure-property relationship of poly(vinylidene difluoride)-based polymers with energy storage and loss under applied electric fields*, Macromolecules 40 (2007), pp. 9391–9397.
[110] B. Chu, X. Zhou, K. Ren, B. Neese, M. Lin, Q. Wang et al., *A dielectric polymer with high electric energy density and fast discharge speed*, Science. 313 (2006), pp. 334–336.
[111] G.N. Tomara, A.P. Kerasidou, A.C. Patsidis, P.K. Karahaliou, G.C. Psarras, S.N. Georga et al., *Dielectric response and energy storage efficiency of low content TiO_2-polymer matrix nanocomposites*, Compos. Part A Appl. Sci. Manuf. 71 (2015), pp. 204–211.
[112] Y. Zhou, Q. Li, B. Dang, Y. Yang, T. Shao, H. Li et al., *A Scalable, High-Throughput, and Environmentally Benign Approach to Polymer Dielectrics Exhibiting Significantly Improved Capacitive Performance at High Temperatures*, Adv. Mater. 30 (2018), pp. 1–7.
[113] Y. Wang, J. Cui, Q. Yuan, Y. Niu, Y. Bai and H. Wang, *Significantly Enhanced Breakdown Strength and Energy Density in Sandwich-Structured Barium Titanate/Poly(vinylidene fluoride) Nanocomposites*, Adv. Mater. 27 (2015), pp. 6658–6663.
[114] L. Wang, F. Xiao and J. Wang, *Study on the mechanical properties of ultra-low dielectric film by tensile test*, 2021 22nd Int. Conf. Electron. Packag. Technol. ICEPT 2021 10072020028 (2021), pp. 2020–2023.
[115] R.S. Raja Rafidah, W. Rashmi, M. Khalid, W.Y. Wong and J. Priyanka, *Recent progress in the development of aromatic polymer-based proton exchange membranes for fuel cell applications*, Polymers (Basel). 12 (2020), 1061.
[116] S.K.M. Haque, J.A. Ardila-Rey, Y. Umar, A.A. Mas'ud, F. Muhammad-Sukki, B.H. Jume et al., *Application and suitability of polymeric materials as insulators in electrical equipment*, Energies 14 (2021), 2758.
[117] M.J. Parnian, S. Rowshanzamir and F. Gashoul, *Comprehensive investigation of physicochemical and electrochemical properties of sulfonated poly (ether ether ketone) membranes with different degrees of sulfonation for proton exchange membrane fuel cell applications*, Energy 125 (2017), pp. 614–628.
[118] E. Ogungbemi, O. Ijaodola, F.N. Khatib, T. Wilberforce, Z. El Hassan, J. Thompson et al., *Fuel cell membranes – Pros and cons*, Energy 172 (2019), pp. 155–172.
[119] X. Yan, G. He, X. Wu and J. Benziger, *Ion and water transport in functionalized PEEK membranes*, J. Memb. Sci. 429 (2013), pp. 13–22.
[120] C. Koning, L. Teuwen, E.W. Meijer and J. Moonen, *Synthesis and properties of α,θ-diaminoalkane based polyimides*, Polymer (Guildf). 35 (1994), pp. 4889–4895.
[121] P. Zhang, K. Zhang, X. Chen, S. Dou, J. Zhao and Y. Li, *Mechanical, dielectric and thermal properties of polyimide films with sandwich structure*, Compos. Struct. 261 (2021).
[122] W. Xie, X. Tang, Y. Yan, Q. Zhang and T.M. Tritt, *Unique nanostructures and enhanced thermoelectric performance of melt-spun BiSbTe alloys*, Appl. Phys. Lett. 94 (2009), pp. 102111.
[123] Y. Wang, X. Zhou, Q. Chen, B. Chu and Q. Zhang, *Recent development of high energy density polymers for dielectric capacitors*, IEEE Trans. Dielectr. Electr. Insul. 17 (2010), pp. 1036–1042.
[124] G. Pilania, C. Wang, X. Jiang, S. Rajasekaran and R. Ramprasad, *Accelerating materials property predictions using machine learning*, Sci. Rep. 3 (2013), pp. 1–6.
[125] G. Pilania, C.C. Wang, K. Wu, N. Sukumar, C. Breneman, G. Sotzing et al., *New group IV chemical motifs for improved dielectric permittivity of polyethylene*, J. Chem. Inf. Model. 53 (2013), pp. 879–886.

[126] P. Simon, Y. Gogotsi, P. Simon, Y. Gogotsi and N. Materials, *Materials for electrochemical capacitors To cite this version: HAL Id: hal-02417326*, 7 (2019), pp. 845–854.

[127] V.S. Puli, M. Ejaz, R. Elupula, M. Kothakonda, S. Adireddy, R.S. Katiyar et al., *Core-shell like structured barium zirconium titanate-barium calcium titanate–poly(methyl methacrylate) nanocomposites for dielectric energy storage capacitors*, Polymer (Guildf). 105 (2016), pp. 35–42.

[128] Q. Huang, H. Luo, C. Chen, X. Zhou, K. Zhou and D. Zhang, *Enhanced energy density in P(VDF-HFP) nanocomposites with gradient dielectric fillers and interfacial polarization*, J. Alloys Compd. 696 (2017), pp. 1220–1227.

[129] P. Maji and R.B. Choudhary, *Facile synthesis, dielectric properties and electrocatalytic activities of PMMA-NiFe2O4 nanocomposite*, Mater. Chem. Phys. 193 (2017), pp. 391–400.

[130] T. Nardi, N. Mora, F. Rachidi and Y. Leterrier, *Graded-permittivity polymer nanocomposites as superior dielectrics*, Compos. Sci. Technol. 129 (2016), pp. 1–9.

[131] D. Ponnamma, H. Parangusan, K. Deshmukh, P. Kar, A. Muzaffar, S.K.K. Pasha et al., *Green synthesized materials for sensor, actuator, energy storage and energy generation: a review*, Polym. Technol. Mater. 59 (2020), pp. 1–62.

[132] Y. Hao, X. Wang, K. Bi, J. Zhang, Y. Huang, L. Wu et al., *Significantly enhanced energy storage performance promoted by ultimate sized ferroelectric BaTiO3 fillers in nanocomposite films*, Nano Energy 31 (2017), pp. 49–56.

[133] Z. Pan, B. Liu, J. Zhai, L. Yao, K. Yang and B. Shen, *$NaNbO_3$ two-dimensional platelets induced highly energy storage density in trilayered architecture composites*, Nano Energy 40 (2017), pp. 587–595.

[134] Y.C. Lu, S. Yu, X. Zeng, R. Sun and C.P. Wong, *High energy density polymer nanocomposites with Y-doped barium strontium titanate nanoparticles as fillers*, IET Nanodielectrics 1 (2018), pp. 137–142.

[135] K. Zou, C. He, Y. Yu, J. Huang, Z. Fan, Y. Lu et al., *Ultrahigh Energy Efficiency and Large Discharge Energy Density in Flexible Dielectric Nanocomposites with $Pb_{0.97}La_{0.02}(Zr_{0.5}Sn_xTi_{0.5-x})O_3$ Antiferroelectric Nanofillers*, ACS Appl. Mater. Interfaces 12 (2020), pp. 12847–12856.

[136] T. Zhang, X. Chen, Q. Zhang and Q.M. Zhang, *Dielectric enhancement over a broad temperature by nanofiller at ultra-low volume content in poly(ether methyl ether urea)*, Appl. Phys. Lett. 117 (2020).

[137] Y. Lin, C. Sun, S. Zhan, Y. Zhang and Q. Yuan, *Ultrahigh Discharge Efficiency and High Energy Density in Sandwich Structure $K_{0.5}Na_{0.5}NbO_3$ Nanofibers/Poly(vinylidene fluoride) Composites*, Adv. Mater. Interfaces 7 (2020), pp. 1–7.

[138] C. Li, L. Shi, W. Yang, Y. Zhou, X. Li, C. Zhang et al., *All Polymer Dielectric Films for Achieving High Energy Density Film Capacitors by Blending Poly(Vinylidene Fluoride-Trifluoroethylene-Chlorofluoroethylene) with Aromatic Polythiourea*, Nanoscale Res. Lett. 15 (2020), 1–9.

[139] D. Xu, W. Xu, T. Seery, H. Zhang, C. Zhou, J. Pang et al., *Rational Design of Soluble Polyaramid for High-Efficiency Energy Storage Dielectric Materials at Elevated Temperatures*, Macromol. Mater. Eng. 305 (2020), pp. 1–5.

[140] S. Luo, J. Yu, S. Yu, R. Sun, L. Cao, W.H. Liao et al., *Significantly Enhanced Electrostatic Energy Storage Performance of Flexible Polymer Composites by Introducing Highly Insulating-Ferroelectric Microhybrids as Fillers*, Adv. Energy Mater. 9 (2019), pp. 1–7.

[141] M. Serhan, M. Sprowls, D. Jackemeyer, M. Long, I.D. Perez, W. Maret et al., *Total iron measurement in human serum with a smartphone, AIChE Annu. Meet. Conf. Proc. 2019–November* (2019).

[142] Q. Li, K. Han, M.R. Gadinski, G. Zhang and Q. Wang, *High Energy and Power Density Capacitors from Solution-Processed Ternary Ferroelectric Polymer Nanocomposites*, Adv. Mater. 26 (2014), pp. 6244–6249.

[143] C. Muralidhar and P.K.C. Pillai, *Pyroelectric behavior in barium titanate/polyvinylidene fluoride composites*, IEEE Trans. Electr. Insul. EI-21 (1986), pp. 501–504.

[144] Y.U. Wang, D.Q. Tan and J. Krahn, *Computational study of dielectric composites with core-shell filler particles*, J. Appl. Phys. 110 (2011).

[145] L. Xie, X. Huang, B.W. Li, C. Zhi, T. Tanaka and P. Jiang, *Core-satellite Ag@$BaTiO_3$ nanoassemblies for fabrication of polymer nanocomposites with high discharged energy density, high breakdown strength and low dielectric loss*, Phys. Chem. Chem. Phys. 15 (2013), pp. 17560–17569.

[146] K. Yang, X. Huang, J. He and P. Jiang, *Strawberry-like Core-Shell Ag@Polydopamine@$BaTiO_3$ Hybrid Nanoparticles for High-K Polymer Nanocomposites with High Energy Density and Low Dielectric Loss*, Adv. Mater. Interfaces 2 (2015), pp. 1–10.

[147] L. Wang, X. Huang, Y. Zhu and P. Jiang, *Enhancing electrical energy storage capability of dielectric polymer nanocomposites via the room temperature Coulomb blockade effect of ultra-small platinum nanoparticles*, Phys. Chem. Chem. Phys. 20 (2018), pp. 5001–5011.

[148] K. Bi, M. Bi, Y. Hao, W. Luo, Z. Cai, X. Wang et al., *Ultrafine core-shell $BaTiO_3$@SiO_2 structures for nanocomposite capacitors with high energy density*, Nano Energy 51 (2018), pp. 513–523.

[149] S. Luo, S. Yu, R. Sun and C.P. Wong, *Nano Ag-deposited $BaTiO_3$ hybrid particles as fillers for polymeric dielectric composites: Toward high dielectric constant and suppressed Loss*, ACS Appl. Mater. Interfaces 6 (2014), pp. 176–182.

[150] C.W. Nan, Y. Shen and J. Ma, *Physical properties of composites near percolation*, Annu. Rev. Mater. Res. 40 (2010), pp. 131–151.

[151] L.A. Fredin, Z. Li, M.T. Lanagan, M.A. Ratner and T.J. Marks, *Sustainable high capacitance at high frequencies: Metallic aluminum-polypropylene nanocomposites*, ACS Nano 7 (2013), pp. 396–407.

[152] M. Arbatti, X. Shan and Z. Cheng, *Ceramic-polymer composites with high dielectric constant*, Adv. Mater. 19 (2007), pp. 1369–1372.

[153] F.E. Bouharras, M. Raihane and B. Ameduri, *Recent progress on core-shell structured $BaTiO_3$@polymer/fluorinated polymers nanocomposites for high energy storage: Synthesis, dielectric properties and applications*, Prog. Mater. Sci. 113 (2020), pp. 100670.

[154] N. Jayasundere and B. V. Smith, *Dielectric constant for binary piezoelectric 0-3 composites*, J. Appl. Phys. 73 (1993), pp. 2462–2466.

[155] T. Yamada, T. Ueda and T. Kitayama, *Piezoelectricity of a high-content lead zirconate titanate/polymer composite*, J. Appl. Phys. 53 (1982), pp. 4328–4332.

[156] A.H. Sihvola, *Self-Consistency Aspects of Dielectric Mixing Theories*, IEEE Trans. Geosci. Remote Sens. 27 (1989), pp. 403–415.

[157] J. Chen, X. Gao and W. Song, *Effect of various carbon nanofillers and different filler aspect ratios on the thermal conductivity of epoxy matrix nanocomposites*, Results Phys. 15 (2019), pp. 102771.

[158] T. Zhou, J.W. Zha, R.Y. Cui, B.H. Fan, J.K. Yuan and Z.M. Dang, *Improving dielectric properties of $BaTiO_3$/ferroelectric polymer composites by employing surface hydroxylated $BaTiO_3$ nanoparticles*, ACS Appl. Mater. Interfaces 3 (2011), pp. 2184–2188.

[159] C. Ehrhardt, C. Fettkenhauer, J. Glenneberg, W. Münchgesang, C. Pientschke, T. Großmann et al., *$BaTiO_3$-P(VDF-HFP) nanocomposite dielectrics - Influence of surface modification and dispersion additives*, Mater. Sci. Eng. B Solid-State Mater. Adv. Technol. 178 (2013), pp. 881–888.

[160] H. Tang, Y. Lin and H.A. Sodano, *Synthesis of high aspect ratio $BaTiO_3$ nanowires for high energy density nanocomposite capacitors*, Adv. Energy Mater. 3 (2013), pp. 451–456.

[161] X. Huang, L. Xie, K. Yang, C. Wu, P. Jiang, S. Li et al., *Role of interface in highly filled epoxy/$BaTiO_3$ nanocomposites. Part II-effect of nanoparticle surface chemistry on processing, thermal expansion, energy storage and breakdown strength of the nanocomposites*, IEEE Trans. Dielectr. Electr. Insul. 21 (2014), pp. 480–487.

[162] Y. Shen, Y. Hu, W. Chen, J. Wang, Y. Guan, J. Du et al., *Modulation of topological structure induces ultrahigh energy density of graphene/$Ba_{0.6}Sr_{0.4}TiO_3$ nanofiber/polymer nanocomposites*, Nano Energy 18 (2015), pp. 176–186.

[163] Y. Zhu, Y. Zhu, X. Huang, J. Chen, Q. Li, J. He et al., *High Energy Density Polymer Dielectrics Interlayered by Assembled Boron Nitride Nanosheets*, Adv. Energy Mater. 9 (2019), pp. 1–10.

[164] X. Huang and P. Jiang, *Core-shell structured high- k polymer nanocomposites for energy storage and dielectric applications*, Adv. Mater. 27 (2015), pp. 546–554.

[165] S.A. Paniagua, Y. Kim, K. Henry, R. Kumar, J.W. Perry and S.R. Marder, *Surface-initiated polymerization from barium titanate nanoparticles for hybrid dielectric capacitors*, ACS Appl. Mater. Interfaces 6 (2014), pp. 3477–3482.

[166] M.N. Tchoul, S.P. Fillery, H. Koerner, L.F. Drummy, F.T. Oyerokun, P.A. Mirau et al., *Assemblies of titanium dioxide-polystyrene hybrid nanoparticles for dielectric applications*, Chem. Mater. 22 (2010), pp. 1749–1759.

[167] Z. Li, L.A. Fredin, P. Tewari, S.A. Dibenedetto, M.T. Lanagan, M.A. Ratner et al., *In situ catalytic encapsulation of core-shell nanoparticles having variable shell thickness: Dielectric and energy storage properties of high-permittivity metal oxide nanocomposites*, Chem. Mater. 22 (2010), pp. 5154–5164.

[168] X. Pang, Y. He, B. Jiang, J. Iocozzia, L. Zhao, H. Guo et al., *Block copolymer/ferroelectric nanoparticle nanocomposites*, Nanoscale 5 (2013), pp. 8695–8702.

[169] Z.M. Dang, S.S. You, J.W. Zha, H.T. Song and S.T. Li, *Effect of shell-layer thickness on dielectric properties in Ag@TiO_2 core@shell nanoparticles filled ferroelectric poly(vinylidene fluoride) composites*, Phys. Status Solidi Appl. Mater. Sci. 207 (2010), pp. 739–742.

[170] S. Liu, S. Xue, B. Shen and J. Zhai, *Reduced energy loss in poly(vinylidene fluoride) nanocomposites by filling with a small loading of core-shell structured* $BaTiO_3/SiO_2$ *nanofibers*, Appl. Phys. Lett. 107 (2015).

[171] J. Jiang, Z. Shen, J. Qian, Z. Dan, M. Guo, Y. He et al., *Synergy of micro-/mesoscopic interfaces in multilayered polymer nanocomposites induces ultrahigh energy density for capacitive energy storage*, Nano Energy 62 (2019), pp. 220–229.

[172] X. Wu, X. Chen, Q.M. Zhang and D.Q. Tan, *Advanced dielectric polymers for energy storage*, Energy Storage Mater. 44 (2022), pp. 29–47.

[173] H. Li, F. Liu, B. Fan, D. Ai, Z. Peng and Q. Wang, *Nanostructured Ferroelectric-Polymer Composites for Capacitive Energy Storage*, Small Methods 2 (2018), pp. 1–18.

[174] J. Hu, S. Zhang and B. Tang, *2D filler-reinforced polymer nanocomposite dielectrics for high-K dielectric and energy storage applications*, Energy Storage Mater. 34 (2021), pp. 260–281.

[175] Q. Li and S. Cheng, *Polymer nanocomposites for high-energy-density capacitor dielectrics: Fundamentals and recent progress*, IEEE Electr. Insul. Mag. 36 (2020), pp. 7–28.

[176] Z. Han and Q. Wang, *Recent progress on dielectric polymers and composites for capacitive energy storage*. iEnergy 1 (1) (2022), pp. 50–71.

[177] J. Li, J. Claude, L.E. Norena-Franco, S. Il Seok and Q. Wang, *Electrical energy storage in ferroelectric polymer nanocomposites containing surface-functionalized* $BaTiO_3$ *nanoparticles*, Chem. Mater. 20 (2008), pp. 6304–6306.

[178] K. Yu, Y. Niu, Y. Zhou, Y. Bai and H. Wang, *Nanocomposites of surface-modified* $BaTiO_3$ *nanoparticles filled ferroelectric polymer with enhanced energy density*, J. Am. Ceram. Soc. 96 (2013), pp. 2519–2524.

[179] K. Yu, H. Wang, Y. Zhou, Y. Bai and Y. Niu, *Enhanced dielectric properties of* $BaTiO_3$*/poly(vinylidene fluoride) nanocomposites for energy storage applications*, J. Appl. Phys. 113 (2013).

[180] H. Luo, D. Zhang, C. Jiang, X. Yuan, C. Chen and K. Zhou, *Improved dielectric properties and energy storage density of poly(vinylidene fluoride-co-hexafluoropropylene) nanocomposite with hydantoin epoxy resin coated* $BaTiO_3$, ACS Appl. Mater. Interfaces 7 (2015), pp. 8061–8069.

[181] Z. Pan, L. Yao, J. Zhai, B. Shen and H. Wang, *Significantly improved dielectric properties and energy density of polymer nanocomposites via small loaded of* $BaTiO_3$ *nanotubes*, Compos. Sci. Technol. 147 (2017), pp. 30–38.

[182] Y. Song, Y. Shen, H. Liu, Y. Lin, M. Li and C.W. Nan, *Enhanced dielectric and ferroelectric properties induced by dopamine-modified* $BaTiO_3$ *nanofibers in flexible poly(vinylidene fluoride-trifluoroethylene) nanocomposites*, J. Mater. Chem. 22 (2012), pp. 8063–8068.

[183] X. Zhang, Y. Shen, Q. Zhang, L. Gu, Y. Hu, J. Du et al., *Ultrahigh energy density of polymer nanocomposites containing* $BaTiO_3@TiO_2$ *nanofibers by atomic-scale interface engineering*, Adv. Mater. 27 (2015), pp. 819–824.

[184] Z.H. Shen, J.J. Wang, X. Zhang, Y. Lin, C.W. Nan, L.Q. Chen et al., *Space charge effects on the dielectric response of polymer nanocomposites*, Appl. Phys. Lett. 111 (2017).

[185] S. Chen, Y. Cheng, Q. Xie, B. Xiao, Z. Wang, J. Liu et al., *Enhanced breakdown strength of aligned-sodium-titanate- nanowire/epoxy nanocomposites and their anisotropic dielectric properties*, Compos. Part A Appl. Sci. Manuf. 120 (2019), pp. 84–94.

[186] Y. Feng, Q. Wu, Q. Deng, C. Peng, J. Hu and Z. Xu, *High dielectric and breakdown properties obtained in a PVDF based nanocomposite with sandwich structure at high temperature: Via all-2D design*, J. Mater. Chem. C 7 (2019), pp. 6744–6751.

[187] S. Ding and Y. Kong, *Significantly Improved Breakdown Strength and Energy Density in Polymer Composites with High Filler Loading by Constructing Isoelectronic Traps*, ACS Appl. Electron. Mater. 1 (2019), pp. 1752–1755.

[188] Prateek, V.K. Thakur and R.K. Gupta, *Recent Progress on Ferroelectric Polymer-Based Nanocomposites for High Energy Density Capacitors: Synthesis, Dielectric Properties, and Future Aspects*, Chem. Rev. 116 (2016), pp. 4260–4317.

[189] H. Zhang, M.A. Marwat, B. Xie, M. Ashtar, K. Liu, Y. Zhu et al., *Polymer Matrix Nanocomposites with 1D Ceramic Nanofillers for Energy Storage Capacitor Applications*, ACS Appl. Mater. Interfaces 12 (2020), pp. 1–37.

[190] L. Sun, Z. Shi, H. Wang, K. Zhang, D. Dastan, K. Sun et al., *Ultrahigh discharge efficiency and improved energy density in rationally designed bilayer polyetherimide-*$BaTiO_3$*/P(VDF-HFP) composites*, J. Mater. Chem. A 8 (2020), pp. 5750–5757.

[191] L.S. Schadler and J.K. Nelson, *Polymer nanodielectrics - Short history and future perspective*, J. Appl. Phys. 128 (2020).

[192] F. Paquin, J. Rivnay, A. Salleo, N. Stingelin and C. Silva, *Multi-phase semicrystalline microstructures drive exciton dissociation in neat plastic semiconductors*, J. Mater. Chem. C 3 (2015), pp. 10715–10722.

[193] Z. Bao, C. Hou, Z. Shen, H. Sun, G. Zhang, Z. Luo et al., *Negatively Charged Nanosheets Significantly Enhance the Energy-Storage Capability of Polymer-Based Nanocomposites*, Adv. Mater. 32 (2020), pp. 1–10.

[194] S.L. Zhong, Z.M. Dang, W.Y. Zhou and H.W. Cai, *Past and future on nanodielectrics*, IET Nanodielectrics 1 (2018), pp. 41–47.

[195] T.J. Lewis, *Interfaces are the dominant feature of dielectrics at the nanometric level*, IEEE Trans. Dielectr. Electr. Insul. 11 (2004), pp. 739–753.

[196] S. Park, C.H. Kim, W.J. Lee, S. Sung and M.H. Yoon, *Sol-gel metal oxide dielectrics for all-solution-processed electronics*, Mater. Sci. Eng. R Reports 114 (2017), pp. 1–22.

[197] X. Huang, B. Sun, Y. Zhu, S. Li and P. Jiang, *High-K polymer nanocomposites with 1D filler for dielectric and energy storage applications*, Prog. Mater. Sci. 100 (2019), pp. 187–225.

[198] S. Li, L. Yang, W. Liu and W. Wang, *Chapter 9 Dielectric Breakdown of Polymer Nanocomposites*, 2016.

[199] Prateek, R. Bhunia, S. Siddiqui, A. Garg and R.K. Gupta, *Significantly Enhanced Energy Density by Tailoring the Interface in Hierarchically Structured TiO_2 -$BaTiO_3$ -TiO_2 Nanofillers in PVDF-Based Thin-Film Polymer Nanocomposites*, ACS Appl. Mater. Interfaces 11 (2019), pp. 14329–14339.

[200] N. Kamezawa, D. Nagao, H. Ishii and M. Konno, *Transparent, highly dielectric poly(vinylidene fluoride) nanocomposite film homogeneously incorporating $BaTiO_3$ nanoparticles with fluoroalkylsilane surface modifier*, Eur. Polym. J. 66 (2015), pp. 528–532.

[201] J.C. Pandey and M. Singh, *Dielectric polymer nanocomposites: Past advances and future prospects in electrical insulation perspective*, SPE Polym. 2 (2021), pp. 236–256.

[202] G. Polizos, E. Tuncer, V. Tomer, I. Sauers, C.A. Randall and E. Manias, *Dielectric Spectroscopy of Polymer-Based Nanocomposite Dielectrics with Tailored Interfaces and Structured Spatial Distribution of Fillers*, Vol. 2013, 2013.

[203] J. Chen, Y. Wang, Q. Yuan, X. Xu, Y. Niu, Q. Wang et al., *Multilayered ferroelectric polymer films incorporating low-dielectric-constant components for concurrent enhancement of energy density and charge–discharge efficiency*, Nano Energy 54 (2018), pp. 288–296.

[204] A. Sharma, S. Basu and N. Gupta, *Investigation of the interface in nanodielectrics using electrostatic force microscopy*, IEEE Trans. Dielectr. Electr. Insul. 27 (2020), pp. 433–441.

[205] A. Sharma, S. Basu and N. Gupta, *Detection of charge around a nanoparticle in a nanocomposite using electrostatic force microscopy*, IEEE Trans. Dielectr. Electr. Insul. 27 (2020), pp. 866–872.

[206] H.S. Khare and D.L. Burris, *A quantitative method for measuring nanocomposite dispersion*, Polymer (Guildf). 51 (2010), pp. 719–729.

[207] W. Zhou, T. Li, M. Yuan, B. Li, S. Zhong, Z. Li et al., *Decoupling of inter-particle polarization and intra-particle polarization in core-shell structured nanocomposites towards improved dielectric performance*, Energy Storage Mater. 42 (2021), pp. 1–11.

[208] S. Chen, S. Chen, R. Qiao, H. Xu, Z. Liu, H. Luo et al., *Enhanced dielectric constant of PVDF-based nanocomposites with one-dimensional core-shell polypyrrole/sepiolite nanofibers*, Compos. Part A Appl. Sci. Manuf. 145 (2021), pp. 106384.

[209] D. Shen, M. Wang, Y. Wu, Z. Liu, Y. Cao, T. Wang et al., *Enhanced thermal conductivity of epoxy composites with core-shell SiC@SiO_2 nanowires*, High Volt. 2 (2017), pp. 154–160.

[210] Z. Wang, Y. Cheng, M. Yang, J. Huang, D. Cao, S. Chen et al., *Dielectric properties and thermal conductivity of epoxy composites using core/shell structured Si/SiO2/Polydopamine*, Compos. Part B Eng. 140 (2018), pp. 83–90.

[211] D. Mao, J. Chen, L. Ren, K. Zhang, M.M.F. Yuen, X. Zeng et al., *Spherical core-shell Al@Al_2O_3 filled epoxy resin composites as high-performance thermal interface materials*, Compos. Part A Appl. Sci. Manuf. 123 (2019), pp. 260–269.

[212] W. Zhou, Y. Kou, M. Yuan, B. Li, H. Cai, Z. Li et al., *Polymer composites filled with core@double-shell structured fillers: Effects of multiple shells on dielectric and thermal properties*, Compos. Sci. Technol. 181 (2019), 107686.

[213] Y. Wang, L. Zhu, J. Zhou, B. Jia, Y. Jiang, J. Wang et al., *Dielectric properties and thermal conductivity of epoxy resin composite modified by Zn/ZnO/Al_2O_3 core–shell particles*, Polym. Bull. 76 (2019), pp. 3957–3970.

[214] A. Rybak, *Functional polymer composite with core-shell ceramic filler: Ii. rheology, thermal, mechanical, and dielectric properties*, Polymers (Basel). 13 (2021), 2161.

[215] W. Xu, G. Yang, L. Jin, J. Liu, Y. Zhang, Z. Zhang et al., *High- k Polymer Nanocomposites Filled with Hyperbranched Phthalocyanine-Coated $BaTiO_3$ for High-Temperature and Elevated Field Applications*, ACS Appl. Mater. Interfaces 10 (2018), pp. 11233–11241.
[216] Z. Li, F. Liu, G. Yang, H. Li, L. Dong, C. Xiong et al., *Enhanced energy storage performance of ferroelectric polymer nanocomposites at relatively low electric fields induced by surface modified $BaTiO_3$ nanofibers*, Compos. Sci. Technol. 164 (2018), pp. 214–221.
[217] M.A. Marwat, B. Xie, Y. Zhu, P. Fan, W. Ma, H. Liu et al., *Largely enhanced discharge energy density in linear polymer nanocomposites by designing a sandwich structure*, Compos. Part A Appl. Sci. Manuf. 121 (2019), pp. 115–122.
[218] L. Wu, K. Wu, D. Liu, R. Huang, J. Huo, F. Chen et al., *Largely enhanced energy storage density of poly(vinylidene fluoride) nanocomposites based on surface hydroxylation of boron nitride nanosheets*, J. Mater. Chem. A 6 (2018), pp. 7573–7584.
[219] K. Yang, X. Huang, Y. Huang, L. Xie and P. Jiang, *Polymerization: Toward Ferroelectric Polymer Nanocomposites with*, Chem. Mater. 25 (2013), pp. 2327–2338.
[220] B. Chu, M. Lin, B. Neese, X. Zhou, Q. Chen and Q.M. Zhang, *Large enhancement in polarization response and energy density of poly(vinylidene fluoride-trifluoroethylene-chlorofluoroethylene) by interface effect in nanocomposites*, Appl. Phys. Lett. 91 (2007).
[221] X. Hao, Y. Wang, J. Yang, S. An and J. Xu, *High energy-storage performance in $Pb_{0.91}La_{0.09}(Ti_{0.65}Zr_{0.35})O_3$ relaxor ferroelectric thin films*, J. Appl. Phys. 112 (2012), pp. 6.
[222] J. Chen, X. Wang, X. Yu, L. Yao, Z. Duan, Y. Fan et al., *High dielectric constant and low dielectric loss poly(vinylidene fluoride) nanocomposites: Via a small loading of two-dimensional Bi_2Te_3@Al_2O_3 hexagonal nanoplates*, J. Mater. Chem. C 6 (2018), pp. 271–279.
[223] Y. Poplavko, *Broadband Dielectric Spectroscopy*, 2021.
[224] P. Hu, S. Gao, Y. Zhang, L. Zhang and C. Wang, *Surface modified $BaTiO_3$ nanoparticles by titanate coupling agent induce significantly enhanced breakdown strength and larger energy density in PVDF nanocomposite*, Compos. Sci. Technol. 156 (2018), pp. 109–116.
[225] J. Wang, J. Hu, L. Yang, K. Zhu, B.W. Li, Q. Sun et al., *High discharged energy density of polymer nanocomposites induced by Nd-doped $BaTiO_3$ nanoparticles*, J. Mater. 4 (2018), pp. 44–50.
[226] P.J. Wang, D. Zhou, H.H. Guo, W.F. Liu, J.Z. Su, M. Sen Fu et al., *Ultrahigh enhancement rate of the energy density of flexible polymer nanocomposites using core-shell $BaTiO_3$@MgO structures as the filler*, J. Mater. Chem. A 8 (2020), pp. 11124–11132.
[227] Q. Sun, P. Mao, L. Zhang, J. Wang, Y. Zhao and F. Kang, *Significantly enhanced dielectric and energy storage performance of AlN/$KNbO_3$/PVDF sandwich-structured composites via introducing the AlN/PVDF insulating layers*, Ceram. Int. 46 (2020), pp. 9990–9996.
[228] B. Xie, Q. Wang, Q. Zhang, Z. Liu, J. Lu, H. Zhang et al., *High Energy Storage Performance of PMMA Nanocomposites Utilizing Hierarchically Structured Nanowires Based on Interface Engineering*, ACS Appl. Mater. Interfaces 13 (2021), pp. 27382–27391.
[229] M. Tanaka, K. Kobayashi, K. Mizumoto and K. Yamaguchi, *Clinical aspects of intraductal papillary mucinous neoplasm of the pancreas*, J. Gastroenterol. 40 (2005), pp. 669–675.
[230] Y. Lin, Y. Zhang, C. Sun, S. Zhan, Q. Yuan and H. Yang, *Energy storage performance in polymer dielectrics by introducing 2D $SrBi_4Ti_4O_{15}$ nanosheets*, Ceram. Int. 46 (2020), pp. 15270–15275.
[231] L. Qi, L. Petersson and T. Liu, *Review of Recent Activities on Dielectric Films for Capacitor Applications*, J. Int. Counc. Electr. Eng. 4 (2014), pp. 1–6.
[232] M. Feng, Y. Feng, T. Zhang, J. Li, Q. Chen, Q. Chi et al., *Recent Advances in Multilayer-Structure Dielectrics for Energy Storage Application*, Adv. Sci. 8 (2021).
[233] Y. Cui, X. Wang, Q. Chi, J. Dong, T. Ma, C. Zhang et al., *Sandwich structured BT-Fe_3O_4/PVDF composites with excellent dielectric properties and energy density*, J. Mater. Sci. Mater. Electron. 28 (2017), pp. 11900–11906.
[234] Y. Ma, H. Luo, R. Guo, K. Zhou and D. Zhang, *Enhanced performance in multilayer-structured nanocomposites using $BaTiO_3$ and $Ba_{0.8}Sr_{0.2}TiO$ decorated graphene hybrids*, Ceram. Int. 44 (2018), pp. 20871–20876.
[235] J. Fu, Y. Hou, M. Zheng and M. Zhu, *Dielectric and energy harvesting properties of $FeTiNbO_6$ /PVDF composites with reinforced sandwich structure*, Polym. Compos. 40 (2019), pp. E570–E578.
[236] Q. Chi, Z. Gao, T. Zhang, C. Zhang, Y. Zhang, Q. Chen et al., *Excellent Energy Storage Properties with High-Temperature Stability in Sandwich-Structured Polyimide-Based Composite Films*, ACS Sustain. Chem. Eng. 7 (2019), pp. 748–757.

[237] R. Guo, H. Luo, M. Yan, X. Zhou, K. Zhou and D. Zhang, *Significantly enhanced breakdown strength and energy density in sandwich-structured nanocomposites with low-level $BaTiO_3$ nanowires*, Nano Energy 79 (2021), pp. 105412.

[238] J.K. Nelson and J.C. Fothergill, *Internal charge behavior of nanocomposites*, Nanotechnology 15 (2004), pp. 586–595.

[239] J.Y. Pei, J.W. Zha, W.Y. Zhou, S.J. Wang, S.L. Zhong, L.J. Yin et al., *Enhancement of breakdown strength of multilayer polymer film through electric field redistribution and defect modification*, Appl. Phys. Lett. 114 (2019).

[240] J. Li, Z. Shen, X. Chen, S. Yang, W. Zhou, M. Wang et al., *Grain-orientation-engineered multilayer ceramic capacitors for energy storage applications*, Nat. Mater. 19 (2020), pp. 999–1005.

[241] B. Wang, L. Liu, L. Huang, L. Chi, G. Liang, L. Yuan et al., *Fabrication and origin of high-K carbon nanotube/epoxy composites with low dielectric loss through layer-by-layer casting technique*, Carbon N. Y. 85 (2015), pp. 28–37.

[242] Y. Feng, M.-L. Li, W.-L. Li, T.-D. Zhang, Y. Zhao and W.-D. Fei, *Polymer/metal multi-layers structured composites: A route to high dielectric constant and suppressed dielectric loss*, Appl. Phys. Lett. 112 (2018), pp. 22901.

[243] Y. Feng, M.L. Li, W.L. Li, T.D. Zhang, Y. Zhao and W.D. Fei, *Polymer/metal multi-layers structured composites: A route to high dielectric constant and suppressed dielectric loss*, Appl. Phys. Lett. 112 (2018).

[244] Y. Zhang, W. Li, S. Xu, Z. Wang, Y. Zhao, J. Li et al., *Interlayer coupling to enhance the energy storage performance of $Na_{0.5}Bi_{0.5}TiO_3$-$SrTiO_3$ multilayer films with the electric field amplifying effect*, J. Mater. Chem. A 6 (2018), pp. 24550–24559.

[245] H. Ohshima, *Electrical Phenomena at Interfaces and Biointerfaces: Fundamentals and Applications in Nano-, Bio-, and Environmental Sciences*, John Wiley & Sons, 2012.

17 Metal Oxide-based High-K Dielectrics for Flexible and Stretchable Electronics

Satyanarayan Dhal and Arpita Patro

17.1 INTRODUCTION

17.1.1 WHAT ARE STRETCHABLE ELECTRONICS?

Stretchable electronics is a rapidly growing field that combines the properties of traditional electronics with the flexibility and stretchability of polymers and other soft materials. The goal of this field is to develop electronic devices that can bend, stretch, and deform without losing their functionality. This technology has the potential to revolutionize a wide range of industries, from wearable devices to healthcare and robotics.

One of the key challenges in developing stretchable electronics is finding ways to create electronic components that can withstand mechanical deformation without breaking or losing their functionality. Researchers have developed a variety of different techniques to overcome this challenge, including using flexible and stretchable substrates, creating stretchable interconnects, and integrating stretchable components with soft materials. One of the most promising approaches is the use of "soft lithography" techniques to create stretchable electronic devices. This method uses a combination of soft materials, such as polymers and elastomers, and micro-fabrication techniques to create electronic components that can stretch and deform while maintaining their functionality. The use of soft materials allows for a greater degree of flexibility, while microfabrication techniques allow for precise control over the size and shape of the electronic components. Another approach is the use of stretchable interconnects, which are designed to allow electronic components to stretch and deform without breaking the electrical connections between them. Researchers have developed a variety of different stretchable interconnects, including those made from metal, carbon nanotubes, and conductive polymers. Stretchable electronics have a large range of potential applications, including wearable devices, such as fitness trackers and smart watches, which can conform to the shape of the wearer's body. Additionally, stretchable electronics can be used in the field of healthcare, such as in the development of implantable medical devices that can stretch and deform with the movement of the body. It also has applications in robotics, where stretchable electronics can be used to create robots that are more compliant and better able to adapt to their environment.

17.1.2 WHAT ARE FLEXIBLE ELECTRONICS?

Flexible electronics is a field of technology that involves the development of electronic devices and systems that can bend and flex without losing their functionality. This skill is based on the utilization of flexible and bendable materials, such as polymers, and the incorporation of electronic components on these materials. The goal is to create electronic devices that can conform to the shape of their environment and adapt to different conditions. One of the key challenges in developing flexible electronics is finding ways to create electronic components that can withstand mechanical deformation without breaking or losing their functionality. Researchers have developed a variety of different techniques to overcome this challenge, including using flexible substrates, creating flexible

DOI: 10.1201/9781003352365-17

interconnects, and integrating flexible components with soft materials. One of the utmost capable approaches is the utilization of "flexible substrates," which are thin, bendable materials that can support electronic components. Researchers have developed a variety of flexible substrates, including those made from polymers, metal foils, and glass. These substrates can be used to create flexible electronic devices, such as flexible displays, sensors, and batteries. Another approach is the use of "flexible interconnects," which are designed to allow electronic components to bend and flex without breaking the electrical connections between them. Researchers have developed a variety of flexible interconnects, including those made from metal, carbon nanotubes, and conductive polymers. These interconnects can be used to create flexible electronic devices. Flexible electronics have a wide range of potential applications, including wearable devices, such as fitness trackers, smart watches, and smart clothing, which can conform to the shape of the wearer's body. Additionally, flexible electronics can be used in the field of healthcare, such as in the development of implantable medical devices that can bend and flex with the movement of the body. It also has applications in the field of robotics, where flexible electronics can be used to create robots that are more compliant and better able to adapt to their environment. High dielectric materials are a type of material that have a high dielectric constant, which is a degree of a material's capability to store electrical energy in an electric field. These materials are of great interest in the field of flexible electronics because they can be used to create highly compact, energy-efficient electronic devices.

17.1.3 High-K Dielectric for Flexible and Stretchable Electronics for the Betterment of the Readers

High-K dielectric is a material used in the development of flexible[1] and stretchable[2] electronics.[3] It has a high dielectric constant, which means it can store more electrical charge than traditional dielectrics. This property makes it ideal for use in flexible and stretchable devices as it allows for more compact and efficient designs. The use of high-K dielectric in flexible and stretchable electronics has the potential to significantly improve the performance and reliability of these devices, making them more accessible and usable for a wide range of applications. As a result, the adoption of High-K dielectric materials is likely to play a key role in the ongoing development and improvement of flexible and stretchable electronics technology.

17.1.4 Significance of High-K Dielectric Materials in Stretchable and Flexible Electronics

One of the key advantages of high dielectric materials is their ability to reduce the size and weight of electronic devices. Traditional electronic devices use metal-oxide-semiconductor (MOS) capacitors to store electrical charge. However, MOS capacitors have a relatively low dielectric constant, which limits their ability to store electrical energy. High dielectric materials, on the other hand, have a much higher dielectric constant, which allows them to store more electrical energy in a smaller volume. This means that electronic devices can be made smaller and lighter, while still maintaining the same level of performance. Another advantage of high dielectric materials is their ability to improve the performance of electronic devices. High dielectric materials can be used to create high-frequency electronic devices, such as RFID tags, which can operate at much higher frequencies than traditional electronic devices. Additionally, high dielectric materials can be used to create high-efficiency power converters, which can convert electrical energy more efficiently than traditional power converters. There are several types of high dielectric materials that are currently being researched for use in flexible electronics. One of the most promising is a class of materials known as ferroelectric materials. These materials include barium titanate and lead zirconate titanate, which have been shown to have dielectric constants that are several orders of magnitude higher than traditional dielectrics. Other high dielectric materials that are being researched include metal-insulator-metal (MIM) capacitors, which are made from thin layers of metal separated by a dielectric material. These capacitors have been shown to have dielectric constants that are several times higher than traditional dielectrics. Additionally, high dielectric materials, such as Barium Strontium Titanate (BST) and lead magnesium niobate-lead titanate (PMN-PT), are also being researched for use in flexible electronics.

17.1.5 Significance of Metal Oxide High-K Dielectric Materials in Stretchable and Flexible Electronics

Metal oxide based high-K dielectrics are a class of materials that have been researched for use in flexible and stretchable electronics. They have a high dielectric constant, which allows for a smaller physical size of the electronic components, making them appropriate for usage in flexible and stretchable devices.

High-K dielectrics are metal oxide materials that have a dielectric constant (K) that is larger than that of silicon dioxide (SiO_2). This allows for a reduction in the thickness of the dielectric layer, which leads to a decrease in the capacitance and an increase in the breakdown voltage. Examples of metal oxide-based high-K dielectrics include hafnium oxide (HfO_2), aluminum oxide (Al_2O_3), and zinc oxide (ZnO). These materials have a high dielectric constant and a high bandgap, which make them suitable for use in flexible and stretchable electronic devices. The metal oxide-based high-K dielectrics can be deposited using different techniques, such as atomic layer deposition (ALD), chemical vapor deposition (CVD), and spin-coating. One of the key challenges in using metal oxide-based high-K dielectrics in flexible and stretchable electronics is their mechanical properties. They are brittle and rigid, which makes them difficult to use in flexible and stretchable devices. To overcome this, researchers are developing techniques to deposit metal oxide-based high-K dielectrics on flexible and stretchable substrates, such as plastics and rubbers. Research in this area is ongoing, and new developments are expected in the near future, including the development of new metal oxide-based high-K dielectrics with improved mechanical properties, as well as new fabrication techniques that will enable the mass production of flexible and stretchable electronic devices.

In summary, metal oxide-based high-K dielectrics are a promising class of materials for use in flexible and stretchable electronics. They have a high dielectric constant and a high bandgap, which make them suitable for use in these types of devices. However, their mechanical properties remain a challenge, thus, research is ongoing to overcome this.

Hafnium oxide (HfO_2)[4, 5, 6] is a metal oxide-based high-K dielectric that has been researched for use in flexible and stretchable electronics. It has a high dielectric constant of around 25, which is much higher than that of silicon dioxide (SiO_2), which is around 3.9. This high dielectric constant allows for a smaller physical size of the electronic components, making it suitable for use in flexible and stretchable devices (as shown in Figure 17.1).

Hafnium oxide has a high bandgap of around 6 eV, which makes it suitable for use in electronic devices that operate at high voltage and high temperature. Additionally, it has a high thermal stability, making it suitable for use in electronic devices that are exposed to high temperatures. One of the key challenges in using hafnium oxide in flexible and stretchable electronics is its mechanical properties. It is a brittle and rigid material, which makes it difficult to use in flexible and stretchable devices. To overcome this, researchers are developing techniques to deposit hafnium oxide on flexible and stretchable substrates, such as plastics and rubbers. One of the techniques is Atomic Layer Deposition (ALD) which is a chemical vapor deposition technique that enables precise control over the thickness and composition of the deposited film, which is crucial for the application of hafnium oxide in flexible and stretchable electronics. Research in this area is ongoing, and new developments are expected in the near future. For example, researchers are exploring the use of hafnium oxide in flexible and stretchable transistors, which are a key component of flexible and stretchable electronic devices.

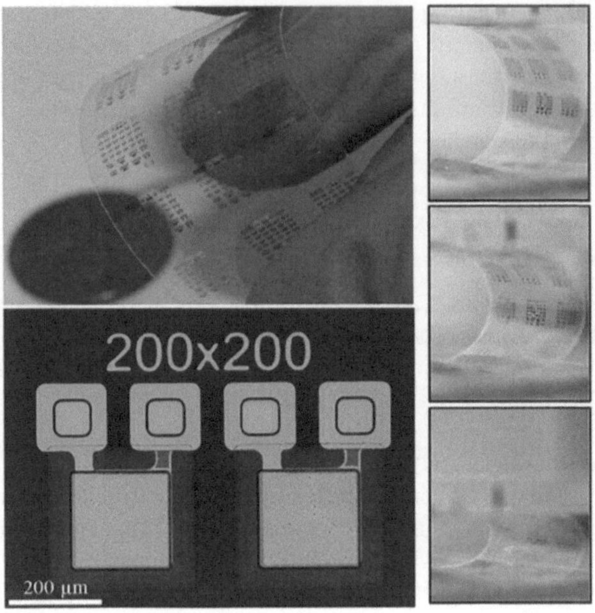

FIGURE 17.1 Hafnium oxide Thin-film capacitors on an innovative unstiffening polymer substrate in which gold was used for the top and bottom contacts. "Reprinted from Microelectronic Engineering, 249, Ovidio Rodriguez-Lopez, Edgar Guerrero Ruiz, Alexander J. Polednik, Adriana C. Duran-Martinez, Aldo Garcia-Sandoval, WalterVoit, Gerardo Gutierrez-Heredia, Electrical characterization of flexible hafnium oxide capacitors on deformable softening polymer substrate, 1., Copyright (2021), with permission from Elsevier."

For example, Lopez et al. looked into low-temperature hafnium oxide thin-film capacitors on an innovative unstiffening polymer substrate, studying their electrical performance and compatibility. For the fabrication of metal-insulator-metal HfO_2 capacitors, atomic layer deposition (ALD) was used to deposit HfO_2 as the dielectric material at 100°C, and gold was used for the top and bottom contacts. In order to create the HfO_2 capacitors, various dielectric thicknesses (50, 40, 30, and 20 nm) were used for the silicon and softening polymer substrates, respectively. With analteration in HfO_2 thickness from 20 to 50 nm, the dielectric constant went from 12 to 17. In addition, the polymer substrate had a 3.9% 2% difference in capacitance density and a 3.4% 2% difference in dielectric constant from the silicon substrate, respectively. Leakage current is also greater in polymer substrate devices than in Si substrate devices, indicating more defects in the dielectric film. Lastly, the HfO_2 capacitors' resistance to mechanical stress was measured by bending the devices on the polymer substrate with a 5-mm bending radius (up to 105 cycles). The findings demonstrate that highly stable and mechanically resilient HfO_2 thin-film capacitors can be fabricated on a softening polymer substrate. The findings of this study may ultimately aid in the manufacture of bendable biomedical devices on plastic substrates.

In summary, hafnium oxide (HfO_2) is a metal oxide-based high-K dielectric that has been researched for use in flexible and stretchable electronics. Its high dielectric constant, high bandgap, and high thermal stability make it suitable for use in these types of devices. However, its mechanical properties remain a challenge, and research is ongoing to overcome this issue, for example, by depositing hafnium oxide on flexible and stretchable substrates, using techniques such as Atomic Layer Deposition (ALD). Aluminum oxide (Al_2O_3) is a metal oxide-based high-K dielectric that has been researched for use in flexible and stretchable electronics.[7] It has a high dielectric constant of around 9.6, which is higher than that of silicon dioxide (SiO_2), which is around 3.9. This high dielectric constant allows for a smaller physical size of the electronic components, making it suitable for use in flexible and stretchable devices (as shown in Figure 17.2).

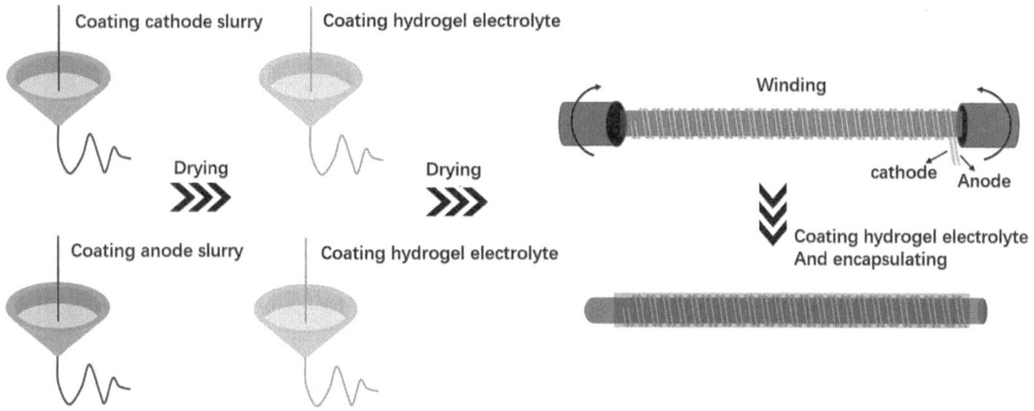

FIGURE 17.2 Stretchable fiber-shaped aqueous aluminum ion batteries. (Reprinted from Xiong, T., He, B., Zhou, T., Wang, Z., Wang, Z., Xin, J., Zhang, H., Zhou, X., Liu, Y. and Wei, L., 2022. Stretchable fiber-shaped aqueous aluminum ion batteries. EcoMat, 4(5), p.e12218., Copyright (2022), with permission from Elsevier.)

Research has advanced in the application of aluminum oxide (Al_2O_3), a metal oxide-based high-k dielectric, for use in flexible and stretchable electronics. Its dielectric constant of 9.6 is significantly greater than that of silicon dioxide (SiO_2), which is only about 3.9. With such a high dielectric constant, electronic components can be made to a smaller size, opening the door to their use in stretchable and flexible devices. Due to its high bandgap of 8.8 eV, aluminum oxide can be used in high-voltage, high-temperature electronic devices. It is also highly resistant to heat, making it ideal for use in electronics. Aluminum oxide's mechanical properties, particularly its brittleness and rigidity, present a significant challenge to its application in stretchable and flexible electronics. To address this, researchers are exploring deposition techniques such as **atomic layer deposition (ALD)**, a chemical vapor deposition method that enables precise control over the thickness and composition of the deposited film. This approach facilitates the integration of aluminum oxide with flexible substrates like plastics and rubbers.

As a **metal oxide-based high-k dielectric**, aluminum oxide (Al_2O_3) is valued for its high dielectric constant, wide bandgap, and excellent thermal stability, making it a promising candidate for flexible and stretchable devices, such as stretchable transistors. However, its mechanical limitations remain a significant obstacle, and ongoing research focuses on overcoming these challenges through innovative substrate engineering and advanced deposition techniques.

In summary, aluminum oxide shows great potential in the field of flexible and stretchable electronics, with ALD playing a pivotal role in its adaptation. Despite mechanical challenges, active research is expected to yield breakthroughs that enhance its practical applicability.

The goal of flexible and stretchable electronics (FSE) is to create flexible and stretchable optoelectronic devices by constructing electronic circuits on "soft" substrates. Flexible electronics require a systematic study of device substrates.[8] Due to its lightweight, compatibility with textiles, tissues, and organs, and greater comfort to users, FSE devices[9] are predicted to enable the introduction of value-added goods. Foldable displays, photovoltaic cells, robots, and wearable medical gadgets are just a few examples of the next generation of consumer electronics[10] that are predicted to experience explosive growth in the next few years. Market research indicates that the global market for flexible and stretchy electronics would expand at a CAGR about 22 percent and 121 percent, respectively, throughout the coming years. The trend in electronics today is toward smaller, more flexible devices that are easier to carry around. This breakthrough not only paves the way for more cost-effective fabrication methods, but also opens up new applications across the board for wearable electronics, from smart textiles to skin mount devices. Flexible electronics[11] based on amorphous silicon, organic, and oxide semiconductors have been realized by employing micrometer-thin substrates. However, stretchable electronics are needed for applications like epidermal electronics, smart implants, and artificial electronic skins for robots.

The state-of-the-art in elastic electronics can be broken down into three categories: first, conductive interconnection lines can be realized by employing pure elastic conductors, air-bridge structures, metal lines on pre-stretched substrates, or patterned into different geometries. Examples include metal films on porous polydimethylsiloxane and carbon nanotubes entrenched in a rubber matrix. To add, oxide transistors rolled to elastic PDMS substrates that can be overextended by 5% without compromising their performance. Fabrication on elastomeric substrates with stiff islands and the use of "wavy" layouts using pre-stretched substrates are two other methods that allow for more stretchable active devices made from brittle materials. In order to reduce the stress on the gadgets, stiff islands are used. Realizing conformal electronics that is stretchable in multiple dimensions is even more difficult than fabricating stretchable electronics devices.

Fiber-based electronics are ideal for many e-textile and e-skin optoelectronic applications due to their lightweight, small size, and mechanical flexibility/stretchability. These wearable devices must be manufactured cost-effectively while ensuring high reliability, multifunctionality, and durability.

Wang et al.[13] demonstrated the use of the blow-spinning technique to fabricate **3D-inorganic nanofiber network (FN) films** from materials like semiconducting indium-gallium-zinc oxide (IGZO), copper oxide, and conducting indium-tin oxide (ITO) and copper metal. These FN-based thin-film transistors, particularly those made with IGZO, exhibit exceptional gas sensing performance at room temperature and maintain their functionality with minimal degradation after 1,000 bending cycles.

Metal oxide FNs possess excellent elasticity, enabling their integration with elastomers to produce multifunctional single-sensing resistors and fully monolithic e-skin devices. These devices can detect and differentiate various stimuli, including analytes, light, strain, pressure, temperature, humidity, body movement, and respiratory functions. Their high sensitivity, rapid response time, and precise detection limits position FN-based devices as a promising solution for flexible and wearable electronics.[12]

It would be ideal for many e-textile/e-skin optoelectronic applications if electronics were built on fibers since this would make them lighter and mechanically flexible/stretchable. These wearable devices have specific manufacturing requirements, including low cost, high dependability, multifunctionality, and durability.

Due to their great transparency, huge area, and mass manufacture of metal oxide semiconductors, thin-film transistors made from them have been the subject of much study. Large-scale integrated circuit development is made possible by the material's compatibility with standard semiconductor procedures, including photolithography of the metal oxide. Additionally, in recent years, sensor applications and neuromorphic devices have been developed, thanks to pairings with various materials. Similarly, Jeon et al.[14] have described in much detail the emerging applications for thin-film transistors, such as flexible/stretchable devices, integrated circuits, biosensors, and neuromorphic devices, etc. Recently developed thin-film transistors based on metal oxides that are very compatible for integration into newly reported applications are also reviewed.

Scaling beyond the Si CMOS constraint has generated persistent research interest in Ge-based CMOS, however Ge-based flexible electronics are still infrequently considered. The primary difficulty in producing Ge-based flexible electronics is the lack of suitable dielectrics, which must possess both high dielectric/Ge interface quality and particular temperature compatibility with the polymer supports. Concerns have also been raised about the trend away from the vacuum manufacturing technique and toward the use of solution-processed high-K dielectrics. Lou et al.[15] have described a methodical approach to optimizing the device performance of polyimide-based, solution-processed high-K dielectric films for intrinsic Ge nanowire transistors. Initially, after comparing the electrical characteristics, surface roughness, and process temperature of a number of solution-processed high-K dielectrics (ZrOx, HfOx, TiOx, AlOx, YOx, and LaOx), YOx thin film was found to be the most suited choice. The precursor solution concentration of 0.2 M, the annealing temperature of 500°C, and the annealing environment of O_2 all contributed to the successful achievement of the exceptional YOx/Ge interface quality. Last but not least, single intrinsic flexible Ge nanowire transistors were shown, with good operating properties similar to devices on rigid substrates, thanks to careful dielectric screening and interface engineering. Further, the devices exhibit high stability, highlighting their promising future in flexible electronics.

In order to create high-performance, flexible electronics, a higher dielectric constant and increased mechanical stability in the dielectrics themselves are necessary. Using n-octadecylphosphonic acid (ODPA) functionalized aluminium titanate (AT) nanoparticles as dopants in poly(4-vinylphenol), Zhou et al.[16] reported on new polymer nanocomposite dielectrics (PVP). By providing increased interfacial adhesion, the nanoparticle surface layer results in composite films with enhanced dielectric characteristics. Nanocomposite layers may be deposited using a simple solution technique, and their combination of variable dielectric permittivity and low leakage current makes them particularly well suited for use on flexible substrates. The electrical performance of air-stable p-type and n-type transistors manufactured on the composite dielectric layer was much better than that of transistors fabricated on pure polymer dielectric. Complementary inverters built using these transistors have high signal gain and fast switching. We analyzed the capacitors', transistors', and inverters' electrical performance systematically on stresses of varying magnitudes on flexible substrates. All of these devices passed the flexibility test with flying colors, showing that the composite dielectric layer is a promising material for use in future conformable sensors, portable displays, and other flexible electronic applications that require mechanical stability and environmental robustness.

The domains of soft robotics, flexible electronics, and artificial intelligence are all showing significant interest in ionomers. However, because of a lack of knowledge of the interactions between polymerics and ionic species, modulating the inherent dynamics of ionomer molecules to produce desired characteristics is highly essential but problematic. Using a unique ionic confinement engineering strategy, Wang et al. were able to achieve great stretchability (>10,000% elongation) by adjusting the composition of a copolymer (PTAE-Fe) to promote dynamic ionic crosslinking. To achieve an efficiency of 95% in self-healing, PTAE-ionic Fe's functional group content was greatly increased. A good flame retardant, the ionomer's heat release rate is decreased by 69.1% thanks to the ionic confinement technique, and the material has a low dielectric-k property (Dk 2.5). According to the ionic confinement technique, a stretchy device infused with eutectic gallium-indium (EGaln) and low Dk flexible ink were also developed. Multiple energy and storage systems might gain from using the ionomer in its current form of fabrication.

Similarly, for portable thin film transistor (TFT) devices and circuits requiring low voltage and low power consumption, electric double layer (EDL) electrolyte dielectrics have garnered a lot of attention because of their high gate modulation efficiency. Using proton conducting inorganic perhydropolysilazane (PHPS) derivatives as the EDL dielectrics at low frequencies (1 kHz) that are still processable in humidity even at quite low temperature (100 °C) and relatively short reaction time (10 min), fully printed single-walled carbon nanotube (SWCNT) TFTs are fabricated by Li et al.[17] Full-ink-printing techniques allow for SWCNT TFTs with superior electrical features, including low working voltage (1 V), high on/off ratio (106), modest subthreshold swing (70–110 mV/dec), strong stability, and average carrier mobility of 6.3 $cm^2 V^{-1} s^{-1}$. In addition, by modifying the concentrations of the printed sc-SWCNT ink and the electron doping technique, it is shown to be possible to precisely regulate the threshold voltage and polarity of the SWCNT TFTs (from depleted p-type to boosted p-type and n-type, and well-balanced ambipolar) (using ethanolamine as the electron doping agent). As-prepared SWCNT TFTs are used in completely printed flexible complementary metal oxide semiconductor (CMOS) and CMOS-like inverters (Figure 17.3), and are documented that have high voltage gains (up to 45 at VDD = 0.7 V) and reasonably substantial noise margins, including depleted and enhanced p-type, enhanced n-type, ambipolar devices, ultralow power consumption, and high voltage gains of flexible printed CMOS-like and CMOS inverters.

Furthermore, Laila M. Al-Harbi et al. employed cast synthesis to rapidly produce PEO (Polyethylene oxide), CMC (Carboxymethyl cellulose), and ZnO/GO NPs as nanofiller for

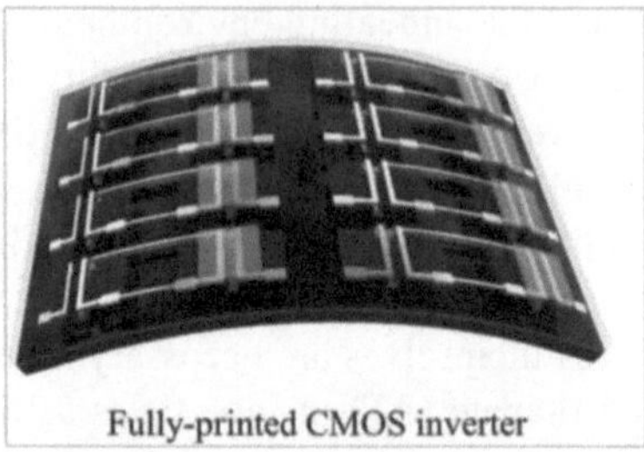

FIGURE 17.3 Image depicting low-temperature and efficiently processed PHPS-derivative dielectrics, good-performance, fully printed flexible SWCNT TFTs. (Reprinted from Li, X., Wang, X., Deng, J., Li, M., Shao, S. and Zhao, J., Printed carbon nanotube thin film transistors based on perhydropolysilazane-derived dielectrics for low power flexible electronics. Carbon, 191, pp. 267–276. Copyright (2022), with permission from Elsevier.)

flexible composite sheets. They proved that the PEO/CMC mixture worked well with the ZnO/GO NPs by interacting with them. For the purpose of estimating the optical characteristics of pure PEO/CMC mix and nanocomposites films, such as the energy band gap and the Urbach energy, the UV-Vis optical absorbance was utilized. Optical characteristics, such as the Urbach energy and energy band gap, were calculated. The energy band gap narrowed and the Urbach energy grew as the fraction of ZnO/GO NPs in the sample rose. Polymer nanocomposites' ionic conductivity has been measured by AC-impedance spectroscopy. PEO/CMC film has a higher ionic conductivity of 109 S/cm, which may be improved upon by adding 8% ZnO/GO nanofiller to get 107 S/cm at room temperature. In contrast to the trend seen for AC electrical conductivity, the values of ϵ of PEO/CMC- ZnO/GO NPs nanocomposites decrease with increasing frequency of the electric field. The values of the nanocomposites were found to rise with the elevation of ZnO/GO NPs concentration. Industrial applications include electric and optoelectronic devices that may benefit from the enhanced optical, dielectric, and electrical characteristics of a PEO/CMC-8wt%ZnO/GO nanocomposite.

17.1.6 Difficulties

The following is a summary of the difficulties encountered when creating stretchable active electronic devices. Flexible substrates must have the required thermal, mechanical, and chemical stability, as well as a surface roughness that is suitable for the device fabrication process. And while wrinkles force electronics to bend at radii on the micrometer scale, elastomers with stiff islands must contend with the delamination issue brought on by stress localization at the interfaces. Here, two methods are explored: one using wave electronics, the other using locally reinforced composite substrates. Both methods yield stretchable electronic devices made of inorganic materials that can withstand strains of over 200% without breaking. Moreover, the developed technology is used to showcase the fundamental building blocks of a stretchable electronic system. This demonstrates that the proposed technology has the potential to be used in electronic skins and smart implants.

In addition to products that can be bent in only one direction, FSE can also be used to make smart implants and epidermal electronics that can bend and deform.[18] The best way to enable FSE devices is to divide them into small functional groups that are separated in space by conductors that can be stretched elastomers, ribbons that pop up, or wires that twist and turn.[19] By Using such a design, silicon complementary metal oxide semiconductor (CMOS) circuits can be combined with a high level of stretchability (>100%). Although several of these technologies have been shown to stretch very well. as well as being able to bend in two directions, advanced integrated circuits have to deal with a lot of problems. Because of this, the use of electronic materials that don't react to mechanical stress makes it possible to make better devices. Also, prices go down because of low temperatures, since a lot of polymers could be used as bases.

17.1.6.1 Why Metal Oxide TFT Gate Dielectrics (MOTFT) Are Promising

High charge-carrier mobility, high optical transmittance, and excellent chemical/mechanical stability make metal oxide semiconductors like ZnO, In_2O_3, and SnO_2 very appealing. Since the first demonstration of an amorphous TFT semiconductor in 2004, there has been significant advancement in the field of metal oxide semiconductors,[20] particularly the amorphous ones, which are a very promising class of electronic material. Amorphous metal oxide (a-MO) semiconductors[21] compete with amorphous and polycrystalline silicon, not only for traditional display and energy applications,[22] but also for new fields like paper electronics[23] owing to their excellent electrical and optical properties, remarkable uniformity, and mechanical flexibility. It's significant that a-MO films can be applied using both expensive techniques like sputtering and MOCVD and inexpensive techniques like spin-coating and printing.[24]

17.1.6.2 Developments on MOTFTs Based on Inorganic Dielectrics

In 2004, transparent electronic devices made on flexible substrates are expected to meet new technological needs that can't be met with silicon-based electronics. Organic semiconductors and hydrogenated amorphous silicon (a-Si:H) have been widely studied for flexible electronics, and they have shown the ability to be manufactured into flexible solar cells and TFTs. Device performance is hampered, however, by the channel materials' poor mobilities. Furthermore, there is a lack of enthusiasm for Si-based devices in the field of transparent circuits. Due to their narrow bandgap, they cannot be seen through. From the perspectives of processing temperature and uniformity of device properties, amorphous semiconductors are favored over polycrystalline ones for active layers. However, the carrier mobility of amorphous semiconductors is lower than that of single-crystalline Si. A new semiconducting material from the In-Ga-Zn-O system that is a transparent amorphous oxide semiconductor (a-IGZO) for the active channel in thin-film transistors with clear films (TTFTs) have shown a remarkable presence in these active research areas. A pulsed laser deposition laser was used to create a-IGZO films from a polycrystalline InGaZn04 target at room temperature in an oxygen atmosphere. The a-IGZO is put on polyethylene terephthalate at room temperature and shows Hall effect showing mobility greater than 10 $cm^2\ V^{-1}\ s^{-1}$, which is a thousand times more than the case for hydrogenated amorphous silicon. TTFTs made on polyethylene terephthalate sheets have saturation mobilities of 6–9 $cm^2\ V^{-1}\ s^{-1}$, and the features of the device are stable when the TTFT sheet is bent over and over. Bending has essentially no effect on TTFT performance, with the saturation current decreasing slightly. Flexible TTFTs made from an amorphous oxide semiconductor called a-IGZO can be made and work at room temperature. TTFTs were made from cheap polymer films and had good performance, like saturation mobility between 6 and 9 $cm^2\ V^{-1}\ s^{-1}$, a low leak current 10^{-10} A, and a rate of on-to-off 10^3 even during and after bending. The active a-IGZO layer in this was made with pulsed laser deposition. Large-area uniform coatings can be made with sputtering or metal-organic chemical vapour deposition (MOCVD).

As a first step in fabricating a flexible AMOLED (Active-matrix organic light-emitting diodes) backplane, two important requirements need to be met. Making a uniform and stable TFT backplane that has good electrical performance and stability is the first necessity. The second is to use flexible substrate materials and procedures that are compatible with current TFT production facilities, yield high devices and substrates, and mimic the properties of conventional "glasslike" masks such as metal foils and plastics. Poor surface smoothness and particle-free handling and a large coefficient of thermal expansion continue to be significant challenges in case plastics. Taking care of this, Park et al. described the electrical and physical properties of a-IGZO (amorphous indium gallium zinc oxide) thin film transistors manufactured on a polyimide substrate PI, which has a high glass transition temperature of T_g = 360 °C and a low coefficient of thermal expansion of 3.4 ppm/K. In the experiments, PI substrates with thicknesses ranging from 10 m to several tens of micrometers were used. During the whole process of making a TFT or AMOLED device, the substrates were attached to a 370 × 470 mm^2 conventional alkali-free glass carrier that was 0.7 mm thick. The substrates were only separated from the carrier when the device was completely made.

By using plasma-enhanced chemical vapor deposition (PECVD), an inorganic buffer layer made of SiN_x and SiO_2 was put on the whole surface of the substrate. The Mo gate layer was put down with sputtering, and the pattern was made with photolithography. Using a polycrystalline $In_2Ga_2ZnO_7$ target and dc sputtering at room temperature, an a-IGZO film 50 nm thick was grown on the gate dielectric. The Mo source and drain electrodes with a thickness were then made by sputtering, and their shapes were made with photolithography and wet etching. In this way, the 6.5 in. flexible full color AMOLED displays on a polymer substrate, have been made. The good thermal properties of the PI substrate material made it possible for a-IGZO TFTs on flexible polymer substrate to have very good, uniform electrical performance. Transistor field-effect transistors (TFTs) on a flexible PI substrate display 15.1 cm^2/V s field-effect mobility is quite high, with low off-state current. Although both PI and glass substrates demonstrate the same TFT performance, the bias-stress stability on PI must be investigated before it can be used for an AMOLED display. TFT backplane mechanical deformation resistance must also be evaluated with electrical performances. Critical failure strain in TFT layers, which corresponds to the commencement of mechanical damage in the device's most brittle layer structure, is the primary metric for evaluating device quality. It has also been demonstrated that the flexible a-IGZO TFTs and AMOLED displays can withstand a significant bending deformation without electrical properties being compromised. A -IGZO TFTs on suitable polymer substrates, such as a PI substrate with low CTE, provide an excellent technological solution for mechanically flexible electronics, and in particular for high-performance flexible AMOLED displays.

Figure 17.4 Images are shown on a 6.5-inch, full-color, flexible, top-emitting AMOLED panel when the panel is bent about 2 cm. The AMOLED display has a thickness of less than 0.1 mm. It has an aperture ratio of 53%, a top-emissive OLED construction, and a resolution of 160XRGBX272 pixels at 85 ppi. Each pixel is driven by a circuit consisting of just two TFTs and one storage capacitor. The AMOLED display can be flat or curved and still show still images and video. A 2-cm bend in the panel does not affect the clarity of the displayed image or the reliability of the electrical components.

Ultra-flexible electronics proposed by Giovanni Salvatore et al., would be made possible by using an engineered substrate whose thickness and properties are tailored to the specific

FIGURE 17.4 A 6.5-inch, full-color, flexible, top-emitting AMOLED panel when the panel is bent about 2 cm. Reprinted from Park, J.S., Kim, T.W., Stryakhilev, D., Lee, J.S., An, S.G., Pyo, Y.S., Lee, D.B., Mo, Y.G., Jin, D.U. and Chung, H.K., 2009. Flexible full color organic light-emitting diode display on polyimide plastic substrate driven by amorphous indium gallium zinc oxide thin-film transistors. Applied Physics Letters, 95(1), p.013503. Copyright (2009), with permission from AIP Publishing.

application. To facilitate the release, transfer, and adherence of electronics onto biological tissues, plastic foils, and any shaped rigid or flexible objects, they suggested a wafer-scale technology for fabricating electronics on an engineered substrate. Micromachined sensors, transistors, and circuits are then placed on top of a parylene film 1 m thick, which is separated from a carrier chip when a PVA layer is dissolved in water. They realized electronics on top of a bilayer polymer substrate, which can be selectively dissolved to enable the release from hosting support, the transfer and the adhesion onto almost any surface, and the removal after use. Transparent and non-transparent transistors based on amorphous-indium gallium zinc oxide (a-IGZO) have mobility as high as $16\,cm^2\,V^{-1}\,s^{-1}$ and $24\,cm^2V^{-1}\,s^{-1}$. Contact lenses made of plastic can now house transparent and lightweight devices that can be used to track intraocular pressure in patients with glaucoma. Since this scheme may be implemented at the wafer scale using a-IGZO, it may provide a valid alternative to previously reported methods by striking a balance between large-area integration and performance.[25] These electronics are ultra-flexible, transparent, conformable, lightweight, and biocompatible all at once. In order to monitor intraocular pressure and identify glaucoma in its early stages, transparent transistors and strain gauge sensors were put into plastic contact lenses. Future uses are also envisioned for ultralight solar cells, implantable devices, smart skin, and electronic textiles. Flexible devices to a pre-stretched PDMS elastomeric substrate can also be created. The released IGZO devices in a wavelike arrangement show a mobility of more than $10\,cm^2\,V^1\,s^1$ and a strain of more than 200%. Experiments were performed with a 70% elongation for up to 2,000 stretching and relaxation cycles as a form of stress. Elastic TFTs were repeatedly stretched and relaxed and showed only minor changes in threshold voltage (50 mV) and mobility (4%). Some disadvantages are also there. For instance, the long production times, and high manufacturing costs, are offset by their ability to produce highly robust a-metal Oxide films at low temperatures and endow the a-metal Oxide TFTs with high performance. Some of these restrictions may be surmounted by fabrication techniques based on solution processing. One that shows promise for continuous, low-cost production is spray pyrolysis in its various forms.

Spray-combustion synthesis (SCS) is a method that was reported that enables low-temperature MO film growth, including the gate dielectric, the semiconductor, and the electrical contacts by combining efficient spray-pyrolysis with a highly exothermic combustion synthesis. This method allows for mobilities and electrical uniformities comparable to magnetron-sputtered films while producing high-density, macroscopically continuous films for a wide range of TFT materials. Electron mobilities of up to $42.5\,cm^2\,V^1\,s^1$ and $32.5\,cm^2\,V^1\,s^1$ have been achieved in SCS-derived IZO and IGZO TFTs with SCS ZrO_2 gate dielectrics and Al contacts, respectively, operating at voltages of 3 V. In addition, the operational stability of the flexible TFTs built on Ary-Lite substrates using ZITO as the gate dielectric is excellent, and the maximum mobilities reach $9\,cm^2\,V^1\,s^1$. Despite being bent multiple times at a radius of 10 mm, these devices have not stopped working.

17.1.6.3 Developments on MOTFTs Based on Electrolyte Dielectrics

Printing technology has many benefits for material patterning, such as high throughput, low cost, and low material waste. The flexible substrate was a 50-m-thick Kapton film with photolithographically etched Au source-drain electrodes (W/L = 500 m/25 m). The ZnO active layer, ion-gel gate insulator, and poly(3,4-ethylenedioxythiophene):poly(styrenesulfonate) (PEDOT: PSS) gate electrode were all printed using a commercial AJP. The $Zn(OH)_2$ (0.5 M) was dissolved in aqueous ammonium hydroxide to make the ZnO precursor ink. High-performance n-channel electrolyte-gated transistors on Kapton were made achievable by annealing the devices at 250°C after ZnO printing. Changing the number of printing passes allowed one to adjust the ZnO film's thickness. A single printing pass produced a ZnO film with a minimum thickness of about 23 nm. It was found that the number of printing passes increased with increasing film thickness for up to 10 passes. As a function of annealing temperature (as-printed, 100°C, 250°C, and 500°C), the transfer properties of printed ZnO electrolyte-gated transistors on SiO_2/Si substrates were investigated. The voltages used by every device were all less than 2 V. Devices with as-printed and 100°C annealed ZnO EGTs displayed poor electrical characteristics, including low ON-current levels (10^6A) and significant hysteresis. However, the device performance was significantly enhanced when the annealing temperature was raised to 250°C.

The carrier mobility was estimated to be 2.4 $cm^2 V^{-1} s^{-1}$ and the ON-current level increased up to 10^4 A with slight hysteresis. The device demonstrated somewhat improved carrier mobility (= 2.6 $cm^2 V^{-1} s^{-1}$) and low hysteresis at the highest annealing temperature (500°C). The operation of EGTs is based on the development of high-capacitance electrical double layers (EDLs) at the interfaces between the ion-gel and gate electrodes, while the gate is biased. In the case of ZnO EGTs, positive surface charges on the gate electrode attract negative ions. Meanwhile, positive ions at the ion-gel/ZnO interface cause the accumulation of mobile electrons. The total EDL capacitance is very large, of order 3.8 μ F/cm^2 (obtained from displacement current Adv.Mater.2013, 25, 3413–3418).

Hence, on flexible Kapton substrates, high-performance ZnO EGTs produced using an aerosol jet were demonstrated. The printable ZnO ink Zn (OH)2aqueous ammonia solution was used, and the printed film could be converted into a reliable ZnO active layer by annealing at 250°C. Low-voltage electrolyte-gated transistors could be achieved by using high-capacitance ion-gel gate insulators. With field-effect electron mobility of 1.6 $cm^2 V^{-1} s^{-1}$, threshold voltages of 1 V, and subthreshold swings of 81 mV/dec, the ZnO electrolyte-gated transistors demonstrated excellent n-type device performance. Furthermore, under static and dynamic tensile loads of up to 0.2%, the electrolyte-gated transistors showed good continuous operation stability.

17.1.6.4 MOTFTs Based on Hybrid Dielectrics

Lai et al. demonstrated that thin film transistors made on amorphous indium-gallium-zinc oxide (a-IGZO TFTs) have received a lot of attention lately. a-IGZO TFTs provide high mobility in comparison to other materials, have consistent structures, and are compatible with modern wide-area manufacturing techniques; as a result, these TFTs are appropriate for flexible displays on lightweight plastic substrates. For use as a nanocomposite gate dielectric for an a-IGZO TFT, AlOx NPs were integrated into a polymer. By using the C-V characteristics, bending test, and positive-bias stress test, the functionality of this aIGZO TFT was assessed. A 400-nm-thick cross-linked polymer layer was coated on the PEN as a smoothing layer before fabricating TFT. Silver (Ag) gate material was deposited on the PEN through a pre-patterned metal foil (shadow mask) by thermal evaporation in 50-nm-thick. The polymer solution consisting of poly (4-vinylphenol) (PVP) and poly (melamine-co-formaldehyde) (PMF) at a ratio of 2:1 dissolved in n-butanol was used for gate dielectric. This polymer was also used as a smoothing layer. Regarding the nanocomposite polymer, aluminum oxide (Al_2O_3) NPs were added to the PVP/PMF precursor at a concentration from 0.25 to 1.0 wt. % in solution.

We have shown an aIGZO TFT with a polymer-nanoparticle composite gate dielectric working well on PEN. Mobility of up to 5.13 cm^2/V/s was attained. The electrical characteristics of the a-IGZO TFT after 100 rope turns (4 mm radius, strain 1/4 1.56%) are almost unchanged. Under positive gate voltage stress at 10 V after 1500 s, the threshold voltage and subthreshold swing are less affected by nanocomposite gate dielectric. The potential of such an a-IGZO TFT as the addressing mechanism in a flexible display is very high.

17.1.7 Disadvantages

Some potential disadvantages of stretchable electronics include:

- Limited durability: Stretchable electronics are often made from flexible materials that can degrade over time, reducing their overall lifespan.
- Manufacturing challenges: Fabricating stretchable electronic devices can be challenging, as the materials and processes used are often different from those used for traditional electronics.

- Limited performance: Stretchable electronics may not perform as well as traditional electronics, as the flexible materials used in their construction can limit their ability to conduct electricity or transmit data.
- High cost: The materials and processes used to create stretchable electronics can be costly, making them more expensive than traditional electronics.
- Complex testing: Testing and evaluating the performance of stretchable electronics can be more complex than testing traditional electronics, due to the unique properties of the materials used in their construction.
- One potential disadvantage of using high dielectric materials in stretchable electronics is that they may have poor electrical conductivity. High dielectric materials, such as polymers, are often used as insulators in stretchable electronic devices, but their electrical conductivity is often much lower than that of traditional electronic materials such as metals. This can result in reduced performance and efficiency of devices that use these materials, making them less suitable for certain high-performance applications. Additionally, high dielectric materials may also have lower thermal conductivity, which may lead to high temperature in the device and affect the performance and lifetime of the device.
- One potential disadvantage of using metal oxides in stretchable electronics is that they may have limited mechanical flexibility. Metal oxides, such as indium tin oxide (ITO), are commonly used as transparent conductive materials in stretchable electronics, but they can be brittle and prone to cracking or breaking when stretched or bent. This can limit the overall flexibility and stretchability of devices that use these materials, making them less suitable for certain applications that require high degrees of stretchability or flexibility. Additionally, some metal oxides are also not biocompatible, so they cannot be used in biomedical devices.

17.1.8 Future Prospects and Challenges[26]

In summary, stretchable electronics is an exciting and rapidly growing field that combines the properties of traditional electronics with the flexibility and stretchability of soft materials. Researchers are developing a variety of different techniques to create stretchable electronic devices that can withstand mechanical deformation without losing their functionality. With a wide range of potential applications, stretchable electronics have the potential to revolutionize a wide range of industries. Similarly, flexible electronics is a rapidly growing field that involves the development of electronic devices and systems that can bend and flex without losing their functionality. Researchers are developing a variety of different techniques to create flexible electronic devices that can withstand mechanical deformation without breaking or losing their functionality. With a wide range of potential applications, flexible electronics have the potential to revolutionize the whole world. In conclusion, high-dielectric materials are a type of material that have a high dielectric constant, which is a measure of a material's ability to store electrical energy in an electric field. These materials are of great interest in the field of flexible electronics because they can be used to create highly compact, energy-efficient electronic devices. Ferroelectric materials, metal-insulator-metal (MIM) capacitors, Barium Strontium Titanate (BST), and lead magnesium niobate-lead titanate (PMN-PT) are some of the most promising high dielectric materials that are currently being researched for use in flexible electronics.

Flexible and stretchable electronics is an emerging field that has the potential to revolutionize the way we interact with technology. With its ability to conform to various shapes and sizes, it has many potential applications in a wide range of industries, including healthcare, consumer electronics, and the Internet of Things (IoT). One of the major future prospects for flexible and stretchable electronics is in the field of wearable technology. Smartwatches, fitness trackers, and other wearable devices have already become popular, but their form factors are limited by the rigid electronic components used in their construction. Flexible and stretchable electronics would enable the development of more comfortable and ergonomic wearable devices that can conform to the wearer's body.

Another area where flexible and stretchable electronics is expected to have a significant impact is in the field of biomedical devices. These devices, such as sensors and implants, could be made more comfortable and effective by using flexible and stretchable electronics. For example, stretchable sensors could be used to monitor vital signs such as heart rate and blood pressure, while flexible implantable devices could be used to deliver drugs to specific parts of the body. The Internet of Things (IoT) is another area where flexible and stretchable electronics is expected to have a significant impact. IoT devices are becoming increasingly ubiquitous, and they are being used in a wide range of applications, such as home automation, smart cities, and industrial automation. Flexible and stretchable electronics would enable the development of more compact and robust IoT devices that can conform to various shapes and sizes. The field of energy storage is also an area where flexible and stretchable electronics are expected to play a significant role in the future. The development of flexible and stretchable batteries and supercapacitors could lead to the creation of wearable and portable electronic devices that can be powered for extended periods of time.

In summary, the future prospects for flexible and stretchable electronics are very promising. Its ability to conform to various shapes and sizes makes it suitable for a wide range of applications in various industries, such as wearable technology, biomedical devices, the Internet of Things (IoT) and energy storage. The field is still in the early stages of development and much research is ongoing to overcome the challenges and unlock its full potential. Metal oxide high dielectric materials, such as hafnium oxide (HfO_2), aluminum oxide (Al_2O_3), and zinc oxide (ZnO), are a promising class of materials for use in flexible and stretchable electronics. They have a high dielectric constant and a high bandgap, which make them suitable for use in these types of devices. However, their mechanical properties remain a challenge and research is ongoing to overcome this.

One of the future prospects for these materials is their use in flexible and stretchable transistors, which are a key component of flexible and stretchable electronic devices. Researchers are exploring the use of these materials as gate dielectrics in thin-film transistors (TFTs) to improve the electrical performance and stability of the devices.

Another area of potential future application of metal oxide high dielectric materials is in the field of energy storage devices, such as flexible and stretchable batteries and supercapacitors. These materials could be used as electrolytes or dielectrics in these devices to improve their performance and stability. In addition, these materials are also being studied for use in flexible and stretchable sensors, such as pressure sensors and strain sensors, which can be integrated into wearable devices and other applications. Researchers are also investigating new techniques to deposit these materials on flexible and stretchable substrates, such as Atomic Layer Deposition (ALD), which enables precise control over the thickness and composition of the deposited film. Overall, the future prospects for metal oxide high dielectric materials in flexible and stretchable electronics are quite promising. However, there are still challenges to be overcome, such as improving the mechanical properties of these materials to make them more suitable for use in flexible and stretchable devices. Ongoing research in this field is expected to lead to new developments and improvements in the near future.

REFERENCES

1. Zhao, Z., Liu, K., Liu, Y., Guo, Y. and Liu, Y., 2022. Intrinsically flexible displays: key materials and devices. *National Science Review*, *9*(6), p.nwac090.
2. Htwe, Y.Z.N. and Mariatti, M., 2022. Printed graphene and hybrid conductive inks for flexible, stretchable, and wearable electronics: Progress, opportunities, and challenges. *Journal of Science: Advanced Materials and Devices*, p.100435.

3. Shi, G., Lowe, S.E., Zhong, Y.L. and Zhao, H., 2020. Flexible and stretchable inorganic electronics: Conductive materials, fabrication strategy, and applicable devices. In *2D Materials for Photonic and Optoelectronic Applications* (pp. 199–252). Wood Head Publishing.
4. Rodriguez-Lopez, O., Ruiz, E.G., Polednik, A.J., Duran-Martinez, A.C., Garcia-Sandoval, A., Voit, W. and Gutierrez-Heredia, G., 2021. Electrical characterization of flexible hafnium oxide capacitors on deformable softening polymer substrate. *Microelectronic Engineering*, *249*, p.111618.
5. Kim, Y.S., Chung, H., Kwon, S., Kim, J. and Jo, W., 2022. Grain boundary passivation via balancing feedback of hole barrier modulation in HfO2-x for nanoscale flexible electronics. *Nano Convergence*, *9*(1), pp.1–13.
6. Jahn, Y.M. and Ya'Akobovitz, A., 2021. Outstanding stretchability and thickness-dependent mechanical properties of 2D HfS2, HfSe2, and hafnium oxide. *Nanoscale*, *13*(44), pp.18458–18466.
7. Xiong, T., He, B., Zhou, T., Wang, Z., Wang, Z., Xin, J., Zhang, H., Zhou, X., Liu, Y. and Wei, L., 2022. Stretchable fiber-shaped aqueous aluminum ion batteries. *EcoMat*, p.e12218.
8. Zardetto, V., Brown, T.M., Reale, A. and Di Carlo, A., 2011. Substrates for flexible electronics: A practical investigation on the electrical, film flexibility, optical, temperature, and solvent resistance properties. *Journal of Polymer Science Part B: Polymer Physics*, *49*(9), pp.638–648.
9. McCoul, D., Hu, W., Gao, M., Mehta, V. and Pei, Q., 2016. Recent advances in stretchable and transparent electronic materials. *Advanced Electronic Materials*, *2*(5), p.1500407.
10. Nathan, A., Ahnood, A., Cole, M.T., Lee, S., Suzuki, Y., Hiralal, P., Bonaccorso, F., Hasan, T., Garcia-Gancedo, L., Dyadyusha, A. and Haque, S., 2012. Flexible electronics: the next ubiquitous platform. *Proceedings of the IEEE*, *100*(Special Centennial Issue), pp.1486–1517.
11. Service, R. F., *Science* 1997, 278, 383
12. Wang, B., Thukral, A., Xie, Z., Liu, L., Zhang, X., Huang, W., Yu, X., Yu, C., Marks, T.J. and Facchetti, A., 2020. Flexible and stretchable metal oxide nanofiber networks for multimodal and monolithically integrated wearable electronics. *Nature Communications*, *11*(1), pp.1–11.
13. Wang, B., Thukral, A., Xie, Z., Liu, L., Zhang, X., Huang, W., Yu, X., Yu, C., Marks, T.J. and Facchetti, A., 2020. Flexible and stretchable metal oxide nanofiber networks for multimodal and monolithically integrated wearable electronics. *Nature Communications*, *11*(1), p.2405.
14. Jeon, Y., Lee, D. and Yoo, H., 2022. Recent advances in metal-oxide thin-film transistors: Flexible/stretchable devices, integrated circuits, biosensors, and neuromorphic applications. *Coatings*, *12*(2), p.204.
15. Lou, X., Zhang, W., Xie, Z., Yang, L., Yu, X., Liu, Y. and Chang, H., 2019. Solution-processed High-K dielectrics for improving the performance of flexible intrinsic Ge nanowire transistors: dielectrics screening, interface engineering and electrical properties. *Journal of Physics D: Applied Physics*, *52*(50), p.505103.
16. Zhou, Y., Han, S.T., Xu, Z.X., Yang, X.B., Ng, H.P., Huang, L.B. and Roy, V.A.L., 2012. Functional High-K nanocomposite dielectrics for flexible transistors and inverters with excellent mechanical properties. *Journal of Materials Chemistry*, *22*(28), pp.14246–14253.
17. Li, X., Wang, X., Deng, J., Li, M., Shao, S. and Zhao, J., 2022. Printed carbon nanotube thin film transistors based on perhydropolysilazane-derived dielectrics for low power flexible electronics. *Carbon*, *191*, pp.267–276.
18. Rogers, J.A., Someya, T. and Huang, Y., 2010. Materials and mechanics for stretchable electronics. *Science*, *327*(5973), pp.1603–1607.
19. Heremans, P., Tripathi, A.K., de Jamblinne de Meux, A., Smits, E.C., Hou, B., Pourtois, G. and Gelinck, G.H., 2016. Mechanical and electronic properties of thin-film transistors on plastic, and their integration in flexible electronic applications. *Advanced Materials*, *28*(22), pp.4266–4282.
20. Nomura, K., Ohta, H., Takagi, A., Kamiya, T., Hirano, M. and Hosono, H., 2004. Room-temperature fabrication of transparent flexible thin-film transistors using amorphous oxide semiconductors. *Nature*, *432*(7016), pp.488–492.
21. Zhou, N., Kim, M.G., Loser, S., Smith, J., Yoshida, H., Guo, X., Song, C., Jin, H., Chen, Z., Yoon, S.M. and Freeman, A.J., 2015. Amorphous oxide alloys as interfacial layers with broadly tunable electronic structures for organic photovoltaic cells. *Proceedings of the National Academy of Sciences*, *112*(26), pp.7897–7902.
22. Sun, D., Chen, C., Zhang, J., Wu, X., Chen, H. and Guo, T., 2018. High performance inkjet-printed metal oxide thin film transistors via addition of insulating polymer with proper molecular weight. *Applied Physics Letters*, *112*(1), p.012102.

23. Tobjörk, D. and Österbacka, R., 2011. Paper electronics. *Advanced Materials*, *23*(17), pp.1935–1961.
24. Yabuta, H., Sano, M., Abe, K., Aiba, T., Den, T., Kumomi, H., Nomura, K., Kamiya, T. and Hosono, H., 2006. High-mobility thin-film transistor with amorphous In Ga Zn O 4 channel fabricated by room temperature rf-magnetron sputtering. *Applied Physics Letters*, *89*(11), p.112123.
25. Salvatore, G., Münzenrieder, N., Kinkeldei, T. et al. Wafer-scale design of lightweight and transparent electronics that wraps around hairs. *Nature Communications* **5**, 2982 (2014). https://doi.org/10.1038/ncomms3982
26. Park, J.W., Kang, B.H. and Kim, H.J., 2020. A review of low-temperature solution-processed metal oxide thin-film transistors for flexible electronics. *Advanced Functional Materials*, *30*(20), p.1904632.

Index

S

T

X

Z